Molecular Genetics of Sex Determination

Molecular Genetics of Sex Determination

Edited by

Stephen S. Wachtel
Division of Reproductive Genetics
Department of Obstetrics and Gynecology
University of Tennessee, Memphis
Memphis, Tennessee

Academic Press, Inc.
A Division of Harcourt Brace & Company
San Diego ■ New York ■ Boston
London ■ Sydney ■ Tokyo ■ Toronto

This book is printed on acid-free paper. ∞

Academic Press, Inc.
525 B Street, Suite 1900, San Diego, California 92101-4495

United Kingdom Edition published by
Academic Press Limited
24–28 Oval Road, London NW1 7DX

Library of Congress Cataloging-in-Publication Data

Molecular genetics of sex determination / edited by Stephen S. Wachtel,
p. cm.
Includes bibliographical references and index.
ISBN 0-12-728960-7
1. Sex determination, Genetic. 2. Sex chromosomes. I. Wachtel, Stephen S.
[DNLM: I. Sexdetermination. 2. Chromosomes--physiology. 3. Sex Reversal, Gonadal. QS 638 M718 1993]
QH600.6.M65 1993
599' .016--dc20
DNLM/DLC
for Library of Congress 93-8302
CIP

PRINTED IN THE UNITED STATES OF AMERICA
93 94 95 96 97 98 EB 9 8 7 6 5 4 3 2 1

Contents

Contributors

Numbers in parentheses indicate the pages on which the authors' contributions begin.

Nabeel A. Affara (225), Department of Pathology, University of Cambridge, Cambridge CB2 1QP, United Kingdom

Lenore K. Beitel (341), Lady Davis Institute for Medical Research, Sir Mortimer B. Davis - Jewish General Hospital, McGill University, Montreal, Quebec, Canada H3T 1E2

Richard L. Cate (439), Biogen Inc., Cambridge, Massachusetts 02142

Desmond W. Cooper (143), School of Biological Sciences, Macquarie University, Sydney, New South Wales 2109, Australia

Christopher Crawford (399), Department of Pediatrics, Cornell University Medical College, New York, New York 10021

Claire Cronmiller (171), Department of Biology, University of Virginia, Charlottesville, Virginia 22901

Patricia K. Donahoe (457), Pediatric Surgical Research Laboratories and the Department of Surgery, Massachusetts General Hospital and Harvard Medical School, Boston, Massachusetts, 02114

Malcolm A. Ferguson-Smith (225), Department of Pathology, University of Cambridge, Cambridge CB2 1QP, United Kingdom

Jennifer A. Marshall Graves (143), Department of Genetics and Human Variation, La Trobe University, Melbourne, Victoria 3083, Australia

John Gubbay (43), Howard Hughes Research Laboratory, The Rockefeller University, New York, New York 10021

Michael L. Gustafson (457), Pediatric Surgical Research Laboratories and the Department of Surgery, Massachusetts General Hospital and Harvard Medical School, Boston, Massachusetts, 02114

Gregory K. Hampikian[1] (143), Worcester Foundation for Experimental Biology, Shrewsbury, Massachusetts 01545

Sandrine Imbeaud (439), Unité de Recherches sur l'Endocrinologie du Développement, INSERM, École Normale Supérieure, 92120 Paris, France

Nathalie Josso (439), Unité de Recherches sur l'Endocrinologie du Développement, INSERM, École Normale Supérieure, 92120 Paris, France

Morris Kaufman (341), Lady Davis Institute for Medical Research, Sir Mortimer B. Davis - Jewish General Hospital, McGill University, Montreal, Quebec, Canada H3T 1E2

Robin Lovell-Badge (43), Laboratory of Developmental Genetics, National Institute for Medical Research, London NW7 1AA, United Kingdom

Mary F. Lyon (123), Genetics Division, MRC Radiobiology Unit, Chilton, Didcot, Oxon OX11 ORD, United Kingdom

Anne McLaren (69), Wellcome/CRC Institute, Cambridge CB2 1QR, United Kindgom

Ulrich Müller (205), Institut für Humangenetik, Justus-Liebig-Universität Giessen, D-35392 Giessen, Germany

Claude M. Nagamine (83), Department of Cell Biology, Vanderbilt University School of Medicine, Nashville, Tennessee 37232

E. Kirk Neely (311), Department of Pediatrics, Stanford University School of Medicine, Stanford, California 94305

Maria I. New (399), Department of Pediatrics, Cornell University Medical College, New York, New York 10021

Susumu Ohno (107), Theoretical Biology Department, Beckman Research Institute of The City of Hope, Duarte, California 91010

Jean-Yves Picard (439), Unité de Recherches sur l'Endocrinologie du Développement, INSERM, École Normale Supérieure, 92120 Paris, France

Leonard Pinsky (341), Lady Davis Institute for Medical Research, Sir Mortimer B. Davis - Jewish General Hospital, McGill University, Montreal, Quebec, Canada H3T 1E2

Ron G. Rosenfeld (311), Department of Pediatrics, Stanford University School of Medicine, Stanford, California 94305

Helen K. Salz (171), Department of Genetics, Case Western Reserve School of Medicine, Cleveland, Ohio 44106

[1] Present address: Biological Sciences, Clayton State College, Morrow, Georgia 30260.

Joe Leigh Simpson (287), Division of Reproductive Genetics, Department of Obstetrics and Gynecology, University of Tennessee, Memphis, Memphis, Tennessee 38163

Andrew H. Sinclair (23), Centre for Child Growth and Hormone Research, Royal Children's Hospital, Parkville, Victoria 3052, Australia

Terrence R. Tiersch (1), Division of Reproductive Genetics, Department of Obstetrics and Gynecology, University of Tennessee, Memphis, Memphis, Tennessee 38163

Mark Trifiro (341), Lady Davis Institute for Medical Research, Sir Mortimer B. Davis - Jewish General Hospital, McGill University, Montreal, Quebec, Canada H3T 1E2

Stephen S. Wachtel (1, 267, 287), Division of Reproductive Genetics, Department of Obstetrics and Gynecology, University of Tennessee, Memphis, Memphis, Tennessee 38163

Milo Zachmann (367), Department of Pediatrics, University of Zurich, Kinderspital, CH-8032 Zurich, Switzerland

Preface

The Wright brothers made their first flight in 1903. In 1947, just 44 years later, Chuck Yeager broke the sound barrier in the rocket-powered Bell X1, and 22 years after *that*, Neil Armstrong took his first steps on the moon. The rate of technological advance seems dizzying, but it is not limited to flying machines: among the biologic disciplines, genetics in particular has kept pace with aeronautics.

When the Wright brothers made their first flight, William Sutton, a graduate student, proposed that chromosomes carry genes. When Yeager was piloting the Bell X1, Alfred Hershey, Max Delbrück, and Salvador Luria were laying the groundwork for modern molecular biology. They received the Nobel Prize for their phage work in 1969, the year in which Neil Armstrong walked on the moon. The function of the human Y chromosome had been inferred only 10 years earlier, and at that time no Y-linked gene had been identified. Yet by mid-1990 the *SRY* testis-determining gene had been cloned, and in 1992 a detailed map of the human Y chromosome was published.

This volume is intended for those wanting to keep abreast of the fast-moving biology of sex determination in an era of accelerated discovery and prolific output. The chapters were completed within the months immediately preceding publication, and as a result, the references are up-to-date and comprehensive, including the literature of 1992 and most of 1993. We invited contributions from the preeminent authorities in each field—those closest to the significant and groundbreaking achievements that form the subject matter of this book. Ohno's Law is discussed by Ohno, and the Lyon hypothesis, by Lyon. Andrew Sinclair tells how he cloned the *SRY* sex-determining gene, and John Gubbay and Robin Lovell-Badge describe the isolation of the corresponding *Sry* gene and production of the first transgenic XX male mice. And so on.

The volume is organized into two parts. The first part includes chapters on the basic biology of sex determination and summarizes the pioneering work in mice, marsupials, and *Drosophila.* The second part includes chapters on human genetics and clinical aspects, and covers the various syndromes of abnormal sex differentiation. As the single current and authoritative work in its field, *Molecular Genetics of Sex Determination* is capital reading for reproductive biologists and graduate students, medical students and physicians, all of whom will find valuable insights in a dynamic and provocative biologic arena.

▮ Stephen S. Wachtel ▮

1

The Search for the Male-Determining Gene

Stephen S. Wachtel and Terrence R. Tiersch
Division of Reproductive Genetics
Department of Obstetrics and Gynecology
University of Tennessee, Memphis
Memphis, Tennessee 38163

Molecular Genetics of Sex Determination

I. Introduction

Several traits, including hairy ears, have been attributed to genes on the human Y chromosome, but sex determination was not included among them until recently. Less than 36 years ago, a prominent geneticist said: "That the (human) Y chromosome has a function of its own, is attested by its very existence. What it is still must be discovered" (Stern, 1957). In fact, the major function of the Y remained uncertain until 1959, when four papers appeared showing that the mammalian Y chromosome determines male sex: the report by Welshons and Russell (1959) showing that XO mice become females; the report by Jacobs and Strong (1959) describing the 47,XXY karyotype in a man with Klinefelter syndrome; the report by Ford *et al.* (1959b) describing the 45,X karyotype in a woman with Turner syndrome; and another by Ford *et al.* (1959a) describing two supernumerary chromosomes—an X and a 21—in a 48,XXY+21 man with Klinefelter and Down syndromes. Those papers provided strong evidence for male-determining genes on the Y chromosome. They indicated, moveover, that a single Y chromosome is sufficient to induce differentiation of the embryonic testis—even in the presence of two or more X chromosomes (so sex determination in the human is different from that in *Drosophila*). Indeed, 48,XXXY and 49,XXXXY males also have been described, severely retarded, but unmistakably male.

Today we know that the sex-determining action of the Y chromosome is limited in particular to the development of the testis (reviewed in Wolf *et al.,* 1992). Further male differentiation is due to the action of testicular hormones and their metabolites (Fig. 1). Thus in the presence of the Y, the gonad becomes a testis: the testis secretes testosterone, which governs timely differentiation of the epididymides, vasa deferentia, and seminal vesicles, and anti-Müllerian hormone, which blocks development of the uterus and fallopian tubes. Dihydrotestosterone, the 5α-reductase metabolite of testosterone, and a more potent androgen, induces orderly differentiation of the external genitalia. In the mutational absence of androgen or of the androgen receptor, the embryo develops the female phenotype, despite presence of the Y chromosome and of the testes (see Chapters 15 and 16, this volume).

It is a dictum of biology that genes of fundamental significance tend to be conserved. On that basis a number of genes and sequences have been proposed as candidates for the elusive Y-situated testis-determining gene, now called *TDF* for testis-determining factor. Among them are *HYS,* the determinant of serological histocompatibility-Y (H-Y) antigen; Bkm (banded krait minor) DNA, a satellite cloned origi-

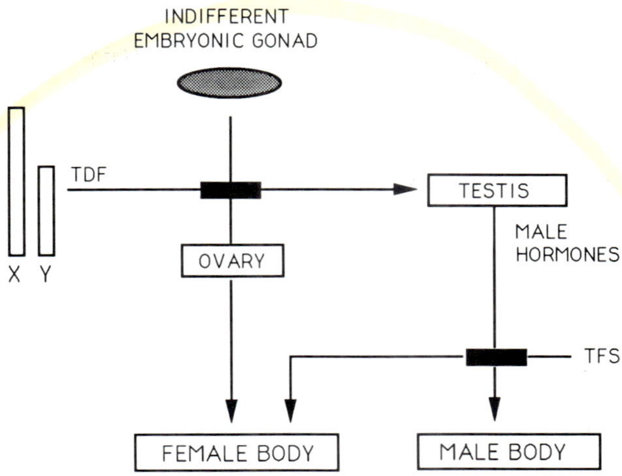

Figure 1 Sex determination in the human. In normal XX embryos, the gonad becomes an ovary and the body develops along female lines. In the presence of the Y-situated testis-determining gene, *TDF,* the gonad becomes a testis. The testis secretes testosterone, which masculinizes the Wolffian ducts, and anti-Müllerian hormone, which blocks formation of the female (Müllerian) ducts; the end result is the male phenotype. In cases of mutation of the X-linked gene encoding the androgen receptor, the response to testosterone (and its metabolite, dihydrotestosterone) is inadequate, and the embryo develops as a female despite presence of XY chromosomes and testes (TFS, testicular feminization syndrome).

nally from the sex-determining W chromosome of a snake; *ZFY* (zinc finger Y); and *SRY* (sex-determining region Y). In this chapter we shall discuss each of these candidates, with emphasis on *ZFY* and *SRY*. We shall consider the role of *TDF* in normal sex determination and the effect of perturbations of *TDF* in XX male syndrome, XX true hermaphroditism, and XY gonadal dysgenesis.

II. H-Y Antigen

Male-specific antigens have been identified by the methods of transplantation, cell-mediated cytotoxicity, and serology. The H-Y antigen that causes rejection of male skin grafts in female mice is probably the same as the male antigen identified in cell-mediated cytotoxicity tests, but there is evidence that it is different from the male antigen recognized by antibody from male-grafted females; the latter is sometimes referred to as serological H-Y antigen and sometimes as SDM, for serologically detected male antigen (Simpson *et al.*, 1987).

When antibody from male-grafted female mice was found to react with cells from the heterogametic (XY) sex in mammals, birds, and amphibians, it was proposed that the H-Y serological antigen was the product of the mammalian testis-determining gene (Wachtel *et al.*, 1975). In fact, since the initial report of phylogenetic conservation of serological H-Y antigen, the molecule has been found in more than 80 species representing each of the major vertebrate classes (Table I).

We have already noted that the sex-determining function of the mammalian Y chromosome is limited to gonadal differentiation; hence embryonic castration leads to development of the female phenotype, even in the presence of a Y chromosome (Jost, 1947). It follows that the function of the *TDF* gene is also limited to gonadal differentiation. But that allows the prediction that the product of *TDF* should always be present in cases of testicular development, or, in terms of the serological H-Y antigen, that testicular differentiation in mammals should always occur in association with the H-Y^+ cellular phenotype, regardless of body sex or chromosomal sex.

To evaluate the association of serological H-Y antigen with testicular development, the molecule was therefore surveyed in cases of aberrant sex differentiation—especially cases of sex reversal such as XX male syndrome, XX true hermaphroditism, and XY gonadal dysgenesis. This is a standard practice in evaluation of candidates for *TDF*. In addition to providing information about the gene or sequence in question, this ap-

Table I
Phylogenetic Conservation of Serological H-Y Antigen

Class	Species	Sex chromosomes	H-Ys in
Mammalia	Human	XX/XY	Male
	Mouse	XX/XY	Male
Aves	Chicken	ZW/ZZ	Female
	Quail	ZW/ZZ	Female
Reptilia	Copperhead	ZW/ZZ	Female
	Alligator	TSD[a]	Male, female
Amphibia	Leopard frog	XX/XY	Male
	Clawed frog	ZW/ZZ	Female
Osteichthyes	Guppy	XX/XY	Male
Chondrichthyes	Horned shark	?	Male

[a] TSD, temperature-dependent sex determination.

proach can often provide insight concerning the etiology of the sex-reversed condition itself.

The initial experimental evidence seemed to support the idea that sex-specific antigens were responsible for male differentiation (Ohno *et al.*, 1979; Wachtel, 1983), but 9 years after the original proposal, McLaren *et al.* (1984) described XX male mice lacking the H-Y antigen in cell-mediated cytotoxicity tests, and more recently, Goldberg *et al.* (1991) reported that these mice also fail to express the male antigen identified in serological tests. Hence neither male antigen could be responsible for testicular development, and the search for *TDF* has been directed elsewhere.

Yet there may be more to the H-Y story. Serological H-Y induces seminiferous tubules in bovine fetal ovaries in culture (Ohno *et al.*, 1979). Thus when Behringer *et al.* 1990) found seminiferous tubules in the ovaries of transgenic female mice chronically expressing the human anti-Müllerian hormone (AMH), the question arose whether H-Y (SDM) might be related to AMH. H-Y antibodies *did* react with purified AMH in serological tests, and the reactions were inhibited by absorption of the antibodies with male cells (Müller *et al.*, 1993). H-Y (SDM) and AMH both are secreted by Sertoli cells and both may be encoded by autosomal genes under control of the Y chromosome. Since both can sex reverse ovarian cells *in vitro,* and both are implicated in sex reversal of ovarian cells *in vivo,* it was inferred that H-Y (SDM) and AMH are in fact similar molecules with a common epitope.

III. Bkm Satellite DNA

The human Y chromosome contains long repetitive DNA sequences known as satellites. These occur typically in the heterochromatic regions of the chromosome, which are in general genetically inert. Within these regions, the G : C and A : T nucleotide pairs are distributed unequally; for that reason, satellite DNA has a density that is different from the density of DNA in which the base pairs are equally represented. Satellite DNA thus can be identified and isolated from the main body of DNA by analytical centrifugation along a density gradient (Fig. 2). One of the more common patterns occurring in satellite DNA is the repetitive tetranucleotide GATA, which is highly conserved in evolution.

In snakes and birds, the female is the heterogametic sex, and carries the ZW chromosome pair. This is in contrast to the situation in mammals, in which the male is the heterogametic sex (XY). Like the Y chromosome, the W is largely heterochromatic, and contains many se-

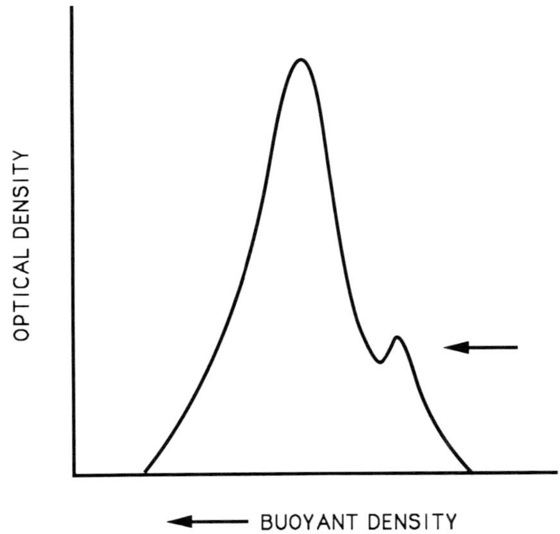

Figure 2 Satellite DNA. The buoyant density of double-stranded DNA—determined by centrifugation through a cesium chloride density gradient—is a function of G-C content. The larger peak above includes about 90% of the genome, centered about a value of 1.70 g cm^{-3} corresponding to an average G-C content of about 42% (which is typical of mammals). The smaller peak at the right (arrow), representing about 10% of the genome, is a *satellite* DNA with a G-C content of about 30% and a buoyant density of 1.69 g cm^{-3}. Satellites may have densities larger or smaller than the density of the main band.

quences of repetitive DNA. In fact several satellites have been isolated from the W chromosome of the banded krait, a snake. One of the satellites, consisting mostly of GATA repeats, was called banded krait minor satellite DNA (Singh *et al.*, 1981).

The Bkm satellite was found to be poorly represented among the autosomes of the banded krait and related ophidians: it was concentrated on the W chromosome instead. A comparative study of phylogenetically primitive, intermediate, and advanced snakes revealed that the association of Bkm with the W chromosome was correlated with the amount of heterochromatin on the W; this was related in turn to the evolution of sex chromosome heteromorphism within the snakes (Singh *et al.*, 1976).

The discovery of Bkm-related sequences in taxa as diverse as reptiles, birds, mammals, and even insects signaled widespread phylogenetic distribution of Bkm, and thus its potential involvement in sex determination (Table II). In the mouse Bkm was found on the autosomes in males and females. But rich clusters of Bkm were identified in the sex-

Table II
Conservation of Bkm

Class	Species	Mechanism of sex determination	Sex-specific fragments in
Mammalia	Human	XX/XY	—[a]
	Mouse	XX/XY	Male
Aves	Chicken	ZW/ZZ	Female
	Japanese quail	ZW/ZZ	Female
Reptilia	Green turtle[b]	TSD[c]	Male, female
	Kemp's ridley[b]	TSD	Male
	Banded krait	ZW/ZZ	Female
	Cobra[b]	ZW/ZZ	Female
Osteichthyes	Rainbow trout	XX/XY	—
	Channel catfish[b]	XX/XY	—
Insecta	Mealworm moth	ZW/ZZ	—
	Drosophila[d]	XX/XY[e]	—

[a] Although Bkm sequences are found throughout the genome, sex-specific bands are not evident in Southern blots.
[b] Heteromorphic sex chromosomes are not apparent in these species, although the W of the cobra can be identified by differential staining for heterochromatin.
[c] TSD, temperature-dependent sex determination.
[d] *Drosophila melanogaster*, the fruit fly.
[e] The sex of *Drosophila* is determined by the ratio of autosomes to X chromosomes.

determining region of the Y chromosome (Jones and Singh, 1981). This prompted the hypothesis that Bkm was involved in mammalian sex determination. The hypothesis was tested by study of the *Sxr* mutation (sex reversed), which causes XX mice to become phenotypic males with small, sterile, testes (Cattanach *et al.*, 1971; and see Chapter 4, this volume).

The condition is transmitted by carrier XY*Sxr* males, which produce four kinds of sperm, and thus four kinds of offspring: normal females, fertile XY noncarrier males, fertile XY*Sxr* carrier males, and sterile XX*Sxr* sex-reversed males. Of these, only the XX females lack the clusters of Bkm found in males (Table III). By using Bkm to probe preparations of fixed chromosomes (*in situ* hybridization), Singh and Jones (1982) showed that one of the X chromosomes in sex-reversed XX*Sxr* males contained a cluster of Bkm such as that normally found in the sex-determining region of the Y. It was inferred that *Sxr* sex reversal was due to a nonreciprocal recurrent Y–X translocation in the germline of the father. Because the translocation was small enough to have escaped detection by standard cytogenetic methods (Y–X crossover has

Table III
Four Kinds of Offspring Produced by XY*Sxr* Carrier Males

Sperm	Progeny	Sex phenotype	Bkm foci[a]
X	XX	Fertile female	None
Y	XY	Noncarrier male	Y
X*Sxr*	XX*Sxr*	Sex-reversed male[b]	X
Y*Sxr*	XY*Sxr*	Carrier male	Y

[a] Rich clusters of Bkm normally asociated with the sex-determining region of the mouse Y chromosome (Singh and Jones, 1982).
[b] XX males are sterile.

since been confirmed cytogentically), this indicated that the Bkm-rich area of the Y and the testis-determining gene were very closely linked or identical.

The work of Singh and Jones (1982) resolved the mechanism of XX sex reversal in the mouse and opened the door for similar studies in XX males of the human. As in the mouse, there was strong evidence for transfer of Y material to one of the X chromosomes in most human XX males (see Chapters 10 and 12). Yet later studies indicated that Bkm sequences were not likely to play a significant role in the sex determination of man, because they were found scattered throughout the genome and were not concentrated in the sex-determining region of the Y chromosome (Kiel-Metzger *et al.*, 1985). With the work on Bkm, however, the study of mammalian sex determination had now passed beyond the bounds of immunology and into the realm of molecular biology.

IV. The Zinc Finger Y Gene (*ZFY*)

Aberrant inheritance of the Xg(a) phenotype in human XX males (see de la Chapelle *et al.*, 1964, and Chapter 11 this volume) could be due to X–Y crossover involving the male-determining gene of the Y and the *Xg* gene of the X. On that basis, Ferguson-Smith (1966) proposed that XX male syndrome and XX true hermaphroditism both are the result of Y to X translocation (Fig. 3) (and we have seen that this is in fact the case in XX*Sxr* males of the mouse). The difference between XX male syndrome and XX true hermaphroditism in the human could accordingly be due to variable X-chromosome inactivation. If the X;Y translocation chromosome (the one carrying *TDF*) were activated in most cells in one gonad, that gonad would become a testis, and if the normal X

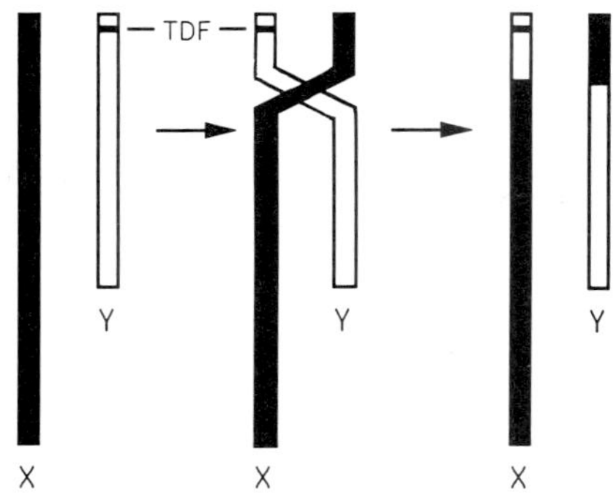

Figure 3 Translocation of the male-determining gene. Crossover of a portion of the Y chromosome to the X during meiosis. If the translocated portion contains the testis-determining gene, *TDF*, and if an egg is fertilized by the sperm carrying the X;Y translocation chromosome, the resulting embryo becomes an XX male. X–Y crossovers have been confirmed in most, but not all, 46,XX males.

were activated in most cells in the other gonad, that one would become an ovary.

Y–X translocation has now been confirmed—at least in the majority of XX males that have been studied (Petit *et al.*, 1987)—and this has enabled geneticists to construct deletion maps of the Y chromosome, for if Y–X crossover involved different amounts of DNA in different XX males, sequences closer to *TDF* should be found in a greater proportion of XX males than should those that are farther away.

By flow cytometry, Müller *et al.* (1986) obtained populations of chromosomes enriched for the Y. The Y-chromosome DNA was digested with restriction enzymes to obtain a series of Y fragments that could be used as probes to detect corresponding Y-specific sequences in XX males. As expected, the different fragments were found in different frequencies among the several XX males surveyed. For example, sequences Y-156 and Y-190 were found in about half of the XX males, whereas Y-280 and Y-286 were found in about three-fourths of the XX males (Table IV). Müller (1987) thus inferred that Y-280 and Y-286 are closer to the *TDF* gene than are Y-156 and Y-190; several other Y sequences were mapped in this way (see Fig. 1 in Chapter 10, this volume). A similar map was constructed by Vergnaud *et al.* (1986) using an alternative series of Y-specific probes in samples of DNA from 27 patients, including XX males and XX true hermaphrodites.

By this method, Page *et al.* (1987) constructed a map of the short arm of the Y chromosome, consisting of 13 deletion intervals. One of the

Table IV

Examples of Hybridization of Y-Specific Sequences in DNA from 12 XX Males

Probe	460	462	548	756	775	693	GM2626	GM2670	102	510	GM1189	481
Y-156	+	+	+	+	+	+	−	−	−	−	−	−
Y-190	+	+	+	+	+	+	−	−	−	−	−	−
Y-223a	+	+	+	+	+	+	+	+	−	−	−	−
Y-228	+	+	+	+	+	+	+	+	−	−	−	−
Y-280	+	+	+	+	+	+	+	+	+	+	−	−
Y-286	+	+	+	+	+	+	+	+	+	+	−	−
Y-198	−	−	−	−	−	−	−	−	+	−	−	−
Y-253	−	−	−	−	−	−	−	−	+	−	−	−
Y-202	−	−	−	−	−	−	−	−	−	−	−	−
Y-221	−	−	−	−	−	−	−	−	−	−	−	−
Y-294	−	−	−	−	−	−	−	−	−	−	−	−

Note: Modified after the table in Müller (1987). Reprinted with permission of the author and the Company of Biologists, Ltd. The data in this table can be used to generate maps such as the one in Fig. 1 in Chapter 10, this volume.

intervals—called "1A2" and comprising 140 kb (140,000 DNA base pairs, or about 0.2% of the Y chromosome)—was found to be present in an XX male who carried only 300 kb or about 0.5% of the Y, and absent in an X,t(Y;22) female (a woman with a Y;22 translocation) who lacked only 160 kb or about 0.3% of the Y. The authors concluded that interval 1A2 contained "an essential portion if not all" of the testis determining gene *TDF*. They reckoned that the gene might be pinpointed by a search for highly conserved DNA sequences within 1A2.

They accordingly prepared a series of "Noah's ark blots." Probes representing collectively the entire 1A2 deletion interval were developed and tested in Southern blots with samples of DNA from males and females of a variety of species. Many of the probes identified corresponding sequences in DNA from humans and great apes, but under stringent conditions, did not identify sequences from more distant mammals. But one of the probes—pDP1007, consisting of a 1.3-kb *Hin*dIII fragment—identified a single- or low-copy DNA sequence in males from every mammal tested, including gorilla, rhesus monkey, owl monkey, rabbit, dog, goat, horse, and bovine. This signaled occurrence of a sequence corresponding to pDP1007 on the Y chromosome in each of these species.

Besides hybridizing with the male-specific fragment, pDP1007 hybridized with a second fragment present in mammalian males and females. The second fragment was presumed to be X-chromosomal because more (twice as much?) was present in females than in males. (The pDP1007 probe also hybridized with DNA from male and female of the chicken.) The amino acid sequence identified by pDP1007 was compared with previously determined amino acid sequences and was found to resemble closely the sequences of proteins with multiple tandemly repeated "finger" domains. Finger proteins are regulatory; some finger proteins bind DNA and regulate transcription (conversion of the DNA sequence into messenger RNA).

On the basis of these observations, Page *et al.* (1987) speculated that the finger protein encoded by pDP1007 was the product of the mammalian Y-linked testis-determining gene. They suggested that the pDP1007 protein might bind to DNA in a sequence-specific manner, thereby regulating transcription of another gene in the pathway leading ultimately to differentiation of the male gonad. The gene was named *ZFY*, for zinc finger Y, because zinc cations form complexes with the finger domains reminiscent of the manner in which iron porphyrin forms complexes with the subunits of hemoglobin. The corresponding X-linked homolog was called *ZFX*, and it was duly noted that any *ZFY*-based model for sex determination in mammals would have to accommodate the presence of an X-linked homolog in males *and* females.

The discovery of *ZFY* was followed by a flurry of experiments aimed at ascertaining the function of the putative testis determinant. In one study, Bull *et al.* (1988) used the pDP1007 probe to look for related sequences in reptiles with genetic sex determination and in reptiles in which the sex of the embryo is determined by the temperature of the incubating egg. Positive hybridization, indicating presence of *ZFY*-like sequences, was obtained in Southern blots, but the results were the same for male and female. Was the *ZFY*-like sequence *expressed* in both sexes? To answer that question, messenger RNA was extracted from embryos of the snapping turtle maintained at female-producing and at male-producing temperatures (the sex of the snapping turtle is determined by temperature). As in the other tests, the pDP1007 probe hybridized with mRNA from both sexes, indicating transcription of the *ZFY*-like sequence in male and female (Table V).

Sinclair *et al.* (1988) studied *ZFY* in several marsupials including kangaroo, wallaby, and others. In these animals, as in the eutherian mammals, sex is determined by a gene on the Y chromosome, but when marsupial DNA was tested with the pDP1007 probe, sequences corresponding to *ZFY* were found in males and females. *In situ* hybridization

Table V
Conservation of *ZFY* and *SRY*

Class	Species	Mechanism of sex determination	*ZFY* hybridization in	*SRY* hybridization in
Mammalia	Human	XX/XY	Male (female)[a]	Male
	Mouse	XX/XY	Male (female)[a]	Male
	Kangaroo	XX/XY	Male, female	Male
	Wallaby	XX/XY	Male, female	Male
Aves	Chicken	ZZ/ZW	Male, female	Male, female
	Cockatiel	ZZ/ZW	Male, female	NT[b]
Reptilia	Leopard gecko	TSD[c]	Male, female	NT
	Alligator	TSD	Male (female?)[d]	Male, female
	Cottonmouth	ZW/ZZ	Male, female	NT
	Crotalus atrox[e]	ZW/ZZ	NT	Male, female
Osteichthyes	Catfish	XX/XY	Male, female	Male, female
	Wreckfish[f]	Sex changing	Male, female	Male, female
Agnatha	Lamprey	?	NT	Male, female

[a] The related sequence, *ZFX* (*Zfx*), is found on the X chromosome in these species.
[b] NT, not tested.
[c] TSD, temperature-dependent sex determination.
[d] The female was not tested.
[e] Western diamondback rattlesnake.
[f] *Anthias squamipinnis*, a coral reef species in which all fish begin life as females.

confirmed that *ZFY*-homologous sequences are autosomal in marsupials. Given the phylogenetic relationship of marsupials and eutherian mammals, this was an unexpected finding. It was inferred that *ZFY* is not the primary sex-determining gene in marsupials, and it was concluded that the genetics of sex determination are different in marsupials and eutherians, or, if they were the same, that *ZFY* could not be *TDF* in man (Table V).

In the mouse, two *ZFY* homologs were found on the Y chromosome (called *Zfy-1* and *Zfy-2*) and one on the X (*Zfx*) (Page, 1988; also see Nagamine *et al.*, 1989, and Chapter 5 this volume). The *Zfy-1* and *Zfy-2* genes are both present in XX*Sxr* males. However, only *Zfy-1* is present in XX*Sxr'* (or XXSxr^b) males, which carry a deleted, albeit male-determining, *Sxr* region. It could be inferred that *Zfy-2* is not necessary for testicular development in the mouse (Page, 1988). More recently, Koopman *et al.* (1989) found that *Zfy-1* but not *Zfy-2* is transcribed during organogenesis of the mouse testis, and that *neither* gene was expressed in mutant embryonic testes lacking germ cells. These findings seemed to imply that neither gene functioned in development of the murine testis per se, but they did not exclude a role for *Zfy-1* (and *ZFY*) in development of the male germ cell.

There were other reasons for believing that *ZFY* is not the same as *TDF*. In their original paper, Page *et al.* (1987) noted that "all XX true hermaphrodites and a few XX males" that they had studied lacked the 1A2 interval, including the sequences identified by pDP1007. They suggested that testicular differentiation in such cases was secondary to mutation in autosomal or X-linked testis-determining genes. Later, Verga and Erickson (1989) described another XX male lacking *ZFY*. In view of the homology of *ZFY* and *ZFX*, those authors suggested that maleness in this case was the result of expression of both copies of *ZFX* (because of partial failure of X-inactivation). They mentioned the alternative possibility that *ZFY* was not the critical sex-determining gene, which took on added significance with the discovery that *ZFX* escapes X-inactivation (Schneider-Gädicke *et al.*, 1989).

There still remained the possibility of mutation of downstream autosomal or X-linked genes. But XX males with downstream mutations would be expected to lack evidence of Y–X crossover, so the discovery of four *ZFY*-negative XX males who carried Y-specific sequences next to the pseudoautosomal boundary now seemed to eliminate *ZFY* as a candidate for *TDF*, and redefined the region of the Y chromosome in which *TDF* must lie (Palmer *et al.*, 1989). A search of that region, comprising some 35 kb adjacent to the pseudoautosomal boundary, yielded a new gene, found to be Y-specific and, like *ZFY*, conservative across a broad spectrum of mammalian species. The new gene was called *SRY* (sex-determining region Y) in the human and *Sry* in the mouse.

V. The Sex-Determining Region Y Gene (*SRY*)

The *SRY* gene was discovered in the laboratory of Peter Goodfellow in London (Sinclair *et al.,* 1990). Initially DNA from the 35-kb region near the pseudoautosomal boundary was digested and subcloned into fragments of about 4 kb. Then each of the 4-kb fragments was cut into smaller pieces in the range of about 500–1000 bp (0.5–1 kb). Of the fragments generated in this manner, 50 were used as probes in samples of DNA from males and females of the human, mouse, and bovine, and from human–hamster somatic cell hybrids containing the human X or Y chromosome. Seven of the probes identified single-copy male-specific sequences in *Eco*R1-digested human DNA, but only one could also identify male-specific sequences in the mouse and bovine. This probe was designated pY53.3.

A 0.9-kb subclone of pY53.3—which hybridized most conspicuously with Y-specific DNA—was used in the ensuing search for *TDF.* First, Sinclair *et al.* (1990) probed “Noah’s Ark” blots containing male and female DNA from human, chimpanzee, rabbit, pig, horse, bovine, and tiger. Unique male-specific sequences were detected in each species (although at reduced stringency, other fragments were detected in males and females, signaling occurrence of pY53.3-related autosomal or X-linked fragments). Next the sequence of the 2.1-kb fragment was determined, whereby two open reading frames were identified.

Analysis of the protein encoded by the nucleotide sequence of one of the open reading frames (223 amino acids) revealed a “striking” homology to part of the Mc mating-type protein of the fission yeast *Schizosaccharomyces pombe,* and a conservative DNA binding motif found in the nuclear high-mobility-group proteins HMG1 and HMG2. The conservative motif—spanning 80 amino acids—was compared to the protein that would be predicted by the nucleotides of the corresponding rabbit sequence. Of the 80 conservative amino acids in the human, 64 (80%) were also present in the rabbit, but only 54% homology was found outside of the conservative motif. Finally, the 0.9-kb *Hin*cII fragment of probe pY53.3 was used to evaluate presence of RNA transcripts in samples of human ovary, testis, and male lung and kidney. In a Northern blot, pY53.3 identified a 1.1-kb transcript in testis but not in the other tissues (see also Su and Lau, 1993).

Taken together, these observations confirmed occurrence of the new gene (*SRY*) in the testis-determining region of the human Y chromosome. Given its location, the conserved DNA binding motif, and its testis-specific expression, the authors proposed that *SRY* was in fact *TDF* (Sinclair *et al.*, 1990). They cautioned that there was room for other genes in the 35-kb region containing *SRY*, and that final proof of identity of *SRY* and *TDF* would rest on the analysis of *SRY* in women with XY gonadal dysgenesis, or on the production of an *SRY*-transgenic XX male (see Chapter 2, this volume).

In an accompanying paper, Gubbay *et al.* (1990) described the murine homolog of *SRY*. When the human sequence, pY53.3, was used to probe Southern blots containing DNA from male and female mice, a strongly hybridizing 3.5-kb male-specific fragment was identified. This fragment was also present in XX*Sxr* and XXSxr^b male mice. Sxr^b, a deletion variant of *Sxr*, is the smallest part of the mouse Y chromosome known to contain the male-determining gene. Since in addition to *Zfy-1* and repetitive sequences such as Bkm only *Sry* was found within the Sxr^b region, pY53.3 had now defined a new gene in the minimal region of the Y chromosome known to determine sex in mouse and human. Moreover, the pY53.3 probe failed to hybridize with DNA from a line of XY female mice carrying a mutation within *Tdy* (the murine homolog of *TDF*): the 3.5-kb band was absent in Southern blots, indicating deletion of the sequence with which pY53.3 had hybridized in males.

When the mouse homolog of pY53.3 was cloned, it was found to contain a 237-bp sequence characterized by 80% sequence homology with pY53.3. As with the corresponding human sequence, the mouse clone exhibited a remarkable similarity to the highly conserved 80 amino acid motif in the Mc protein of the yeast *S. pombe*. RNA transcripts of *Sry* were found in adult testis and in urogenital ridge from 11.5-day-old male embryos, but not in liver or in urogenital ridge from female embryos. Gubbay *et al.* (1990) thus inferred that *Sry* is "a good candidate" for *Tdy* (see Chapter 3, this volume).

The sex-determining role of *SRY* (*Sry*) was borne out in subsequent studies. For example, several laboratories set out to find mutants within *SRY* in women with XY gonadal dysgenesis (in whom the testes fail to develop and the gonads are represented by inert streaks of fibrous tissue). After studying 11 cases of XY gonadal dysgenesis, Berta *et al.* (1990) described such two mutations: both were single base changes within the conservative domain. One was a *de novo* mutation and one was shared by the father of the patient. It was concluded that the latter was "fortuitously associated with, or predisposing towards sex reversal." A similar finding was reported by Jäger *et al.* (1990): four nucleotides were deleted from within the conservative motif in 1 of 12 women with

XY gonadal dysgenesis (see Pivnick *et al.*, 1992, and Chapter 13, this volume).

As for timely expression of the *Sry* gene, Koopman *et al.* (1990) found that transcription of *Sry* was limited to the genital ridge of the mouse fetus, first occurring with the onset of testicular organogenesis (10.5 days), and becoming less pronounced at about 12.5 days. *Sry* transcripts were not detected thereafter. Transcription did not require the presence of the fetal germ cells. Moreover, expression of *Sry* in mutant W^e/W^e (germ cell negative) mouse embryos was no different from that in wild-type embryos, unlike the situation described for *Zfy-1*.

In our own laboratory, Tiersch *et al.* (1991) reacted a probe from the conserved motif of the *SRY* gene with DNA from 23 different species from five different classes representing 400 million years of vertebrate evolution. Cross-hybridization occurred in each case—in species with genetic modes of sex determination and in species with temperature-regulated sex determination (Table V; Fig. 4). The data were compatible with persistence of a critical function, but there was no evidence of sex-specificity outside of the mammals (see Coriat *et al.*, 1993), so it would be informative to determine whether *SRY* homologs are *expressed* in both sexes in these animals.

VI. *SRY* and the Search for *TDF*

It would seem that the search for the testis-determining gene has come to an end, for *SRY* evidently satisfies the requirements of that gene. It is widely conserved in evolution; a homolog is present in the marsupial Y (Foster *et al.*, 1992); it is located in the smallest region of the Y chromosome that is sex-determining; it is present in most XX males; mutations within *SRY* have been identified in sex-reversed XY women, and the *Sry* homolog is deleted in a line of XY females of the mouse; it exhibits the characteristics of a regulatory gene; it is expressed in the gonadal ridge in the mouse embryo and precisely at the time of testicular organogenesis; its transcription is curtailed after formation of the seminiferous tubules; its expression is correlated with a somatic cell lineage in the

---

Figure 4 Conservation of sequences corresponding to *SRY*. Bottom: Southern blot of *Hin*dIII-digested DNA from males and females of nine different vertebrate species, showing hybridization with a 320-bp fragment amplified from *SRY*. In each pair of animals, the male DNA was on the left. Lanes 1,2: human; lanes 3,4: moluccan cockatoo; lanes 5,6: chicken; lanes 7,8: Japanese quail; lanes 9,10: American alligator; lanes 11,12: green sea turtle (*Chelonia mydas*); lanes 13,14: channel catfish; lanes 15,16: gizzard shad; lanes 17,18:

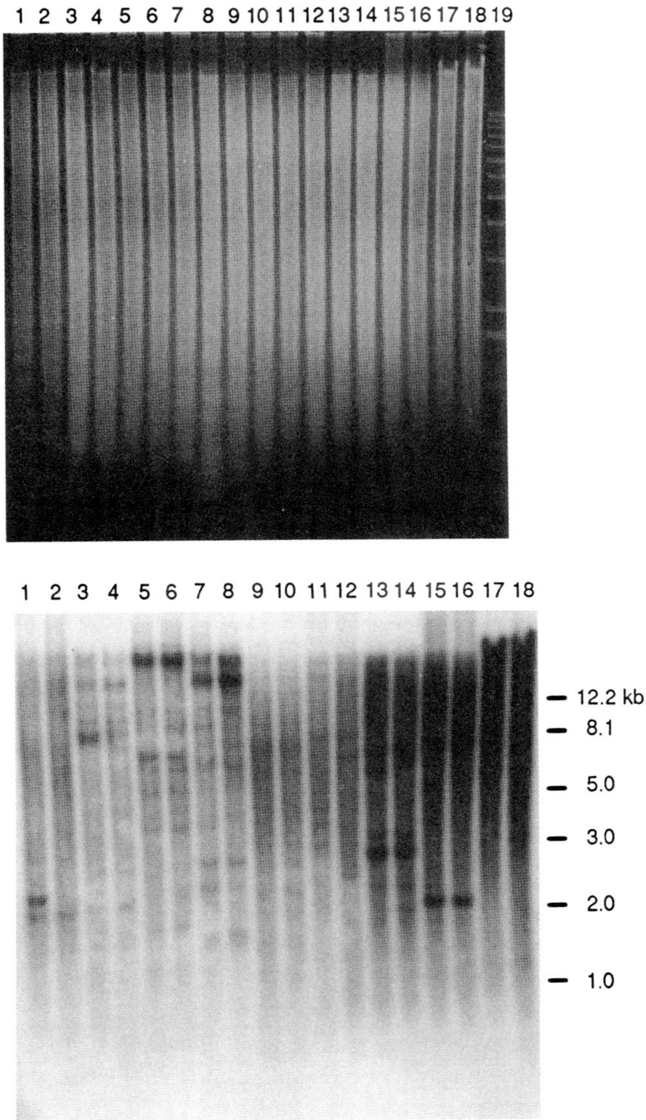

sea lamprey. A male-specific fragment of 2.1 kb was identified in the human, but sex-specific fragments were not observed in any of the nonmammals. Differences in hybridization patterns in male and female catfish were caused by polymorphisms and were not associated with gender. Differences in intensity of bands in males and females of the birds (lanes 3,4,7,8) were not found consistently. Top: agarose gel used to prepare Southern blot, after staining with ethidium bromide. Lane 19 contains size markers. See Table V. From Tiersch *et al.* (1991), by permission.

developing gonad, not with the germ cells; and finally, although sex reversal did not occur in every case, microinsertion of a 14-kb segment containing *Sry* was sufficient to induce development of testes in at least some transgenic XX embryos (Koopman *et al.*, 1991).

VII. *TDF* in Normal and Abnormal Sex Determination

The precise role of the Y-linked testis-determining gene remains to be elucidated. It would seem likely that the protein product of *TDF* is a regulatory factor, probably a DNA binding element that regulates expression of a secondary or "downstream" testis determinant. The downstream locus could be X-linked or autosomal. Assuming identity of *SRY* and *TDF*, this would account for the apparent absence of mutations within *SRY* in most cases of XY gonadal dysgenesis and the apparent absence of *SRY* in certain XX males and XX true hermaphrodites (Ferguson-Smith *et al.*, 1990). Existence of downstream sex-determining genes would also account for autosomal and X-linked forms of XY gonadal dysgenesis (Simpson, 1989; Tommerup *et al.*, 1993; and see Chapter 13, this volume).

Perhaps the earliest recognizable event in the embryology of the male gonad is the appearance of the Sertoli cells. The Sertoli cells make physical contact with one another. They aggregate and surround the germ cells, thereby forming the seminiferous cords that characterize the newly differentiated testis. Further differentiation involves the appearance of Leydig cells from the interstitial cells of the gonad, and the development of the tunica albuginea, the fibrous outer covering of the testis. We have said that the Sertoli cells secrete the anti-Müllerian hormone and the Leydig cells secrete testosterone. It follows that the product of the *TDF* gene initiates a cascade of events, the first of which is differentiation of the Sertoli cells. But how that first differentiative event is governed is still a mystery.

Yet we have gained certain insights into the etiology of abnormal sex differentiation. We can say, for example, that many cases of XY gonadal dysgenesis are not due to mutations within the conservative motif of *SRY*, but are due to mutations outside of the conservative motif, or alternatively to mutations in other X-linked or autosomal testis-determining genes (Pivnick *et al.*, 1992; but see Hawkins *et al.*, 1992). Constitutive activation of these "secondary" genes would explain sex reversal in the few XX males in whom *SRY* sequences have not been identified (five such males have been described by Ferguson-Smith *et al.*, 1990).

As a concluding point, the etiology of XX true hermaphroditism is enigmatic, for *SRY* sequences have not been detected in the great majority of XX true hermaphrodites that have been tested, and the presumptive X-linked and autosomal genes of the testis-determining pathway have yet to be identified. Although absence of *SRY* in blood cells need not rule out differential expression of *SRY* in *gonadal* tissue, the same would be true of a downstream product induced by *SRY*. A mutant X-linked gene, subject to inactivation, would readily explain male development in at least some Y(−) true hermaphrodites and Y(−) XX males. A mutant autosomal gene with variable penetrance could explain male development in some others.

VIII. Summary

The *SRY* gene seems to satisfy the requirements for *TDF*. The various abnormalities of sex differentiation manifest in conditions such as XX male syndrome, and XY gonadal dysgenesis can accordingly be explained as transcriptional or translational errors within *SRY*. For example, testicular differentiation in XX males is due to Y–X crossover and transfer of *SRY* to the X or, more rarely, to constitutive mutation (activation in the absence of a regulatory signal) of a secondary, downstream, testis determinant. And failure of male differentiation in women with XY gonadal dysgenesis is due to mutation or deletion in *SRY* itself or in downstream autosomal or X-linked genes required for testicular development. But the etiology of true hermaphroditism remains to be clarified.

References

Behringer, R. R., Cate, R. L., Froelick, G. R., Palmiter, R. D., and Brinster, R. L. (1990). Abnormal sexual development in transgenic mice chronically expressing Müllerian inhibiting substance. *Nature (London)* **345,** 167–170.

Berta, P., Hawkins, J. R., Sinclair, A. H., Taylor, A., Griffiths, B. L., Goodfellow, P. N., and Fellous, M. (1990). Genetic evidence equating *SRY* and the testis-determining factor. *Nature (London)* **348,** 448–450.

Bull, J. J., Hillis, D. M., and O'Steen, S. (1988). Mammalian ZFY sequences exist in reptiles regardless of sex-determining mechanism. *Science* **242,** 567–569.

Cattanach, B. M., Pollard, C. E., and Hawkes, S. G. (1971). Sex-reversed mice: XX and XO males. *Cytogenetics* **10,** 318–337.

Coriat, A.-M., Müller, U., Harry, J. L., Uwanogho, D., and Sharpe, P. Y. (1993). PCR amplification of SRY-related gene sequences reveals evolutionary conservation of SRY-box motif. *PCR Methods and Applications* **2,** 218–222.

de la Chapelle, A., Hortling, H., Niemi, M., and Wennström, J. (1964). XX sex chromosomes in a human male. First case. *Acta Med. Scand., Suppl.* **412,** 25–38.

Ferguson-Smith, M. A. (1966). X-Y chromosomal interchange in the aetiology of true hermaphroditism and of the XX Klinefelter's syndrome. *Lancet* **2,** 475–476.

Ferguson-Smith, M. A., North, M. A., Affara, N. A., and Briggs, H. (1990). The secret of sex. *Lancet* **336,** 809–810.

Ford, C. E., Jones, K. W., Miller, O. J., Mittwoch, U., Penrose, L. S., Ridler, M., and Shapiro, A. (1959a). The chromosomes in a patient showing both mongolism and the Klinefelter syndrome. *Lancet* **1,** 709–710.

Ford, C. E., Jones, K. W., Polani, P. E., de Almeida, J. C., and Briggs, J. H. (1959b). A sex-chromosome anomaly in a case of gonadal dysgenesis (Turner's syndrome). *Lancet* **1,** 711–713.

Foster, J. W., Brennan, F. E., Hampikian, G. K., Goodfellow, P. N., Sinclair, A. H., Lovell-Badge, R., Selwood, L., Renfree, M. B., Cooper, D. W., and Graves, J. A. M. (1992). Evolution of sex determination and the Y chromosome: SRY-related sequences in marsupials. *Nature (London)* **359,** 531–533.

Goldberg, E. H., McLaren, A., and Reilly, B. (1991). Male antigen identified serologically does not identify a factor responsible for testicular development. *J. Reprod. Immunol.* **20,** 305–309.

Gubbay, J., Collignon, J., Koopman, P., Capel, B., Economou, A., Münsterberg, A., Vivian, N., Goodfellow, P., and Lovell-Badge, R. (1990). A gene mapping to the sex-determining region of the mouse Y chromosome is a member of a novel family of embryonically expressed genes. *Nature (London)* **346,** 245–250.

Hawkins, J. R., Taylor, A., Goodfellow, P. N., Migeon, C. J., Smith, K. D., and Berkovitz, G. D. (1992). Evidence for increased prevalence of SRY mutations in XY females with complete rather than partial gonadal dysgenesis. *Am. J. Hum. Genet.* **51,** 979–984.

Jacobs, P. A., and Strong, J. A. (1959). A case of human intersexuality having a possible XXY sex-determining mechanism. *Nature (London)* **183,** 302–303.

Jäger, R. J., Anvret, M., Hall, K., and Scherer, G. (1990). A human XY female with a frame shift mutation in the candidate testis-determining gene SRY. *Nature (London)* **348,** 452–454.

Jones, K. W., and Singh, L. (1981). Conserved repeated DNA sequences in vertebrate sex chromosomes. *Hum. Genet.* **58,** 46–53.

Jost, A. (1947). Recherches sur la différenciation sexuelle de l'embryon de Lapin. III. Role des gonades foetales dans la différenciation somatique. *Arch. Anat. Microsc. Morphol. Exp.* **36,** 271–316.

Kiel-Metzger, K., Warren, G., Wilson, G. N., and Erickson, R. P. (1985). Evidence that the human Y chromosome does not contain clustered DNA sequences (Bkm) associated with heterogametic sex determination in other vertebrates. *N. Engl. J. Med.* **313,** 242–245.

Koopman, P., Gubbay, J., Collignon, J., and Lovell-Badge, R. (1989). *Zfy* gene expression patterns are not compatible with a primary role in mouse sex determination. *Nature (London)* **342,** 940–942.

Koopman, P., Münsterberg, A., Blanche, C., Vivian, N., and Lovell-Badge, R. (1990). Expression of a candidate sex-determining gene during mouse testis differentiation. *Nature (London)* **348,** 450–452.

Koopman, P., Gubbay, J., Vivian, N., Goodfellow, P., and Lovell-Badge, R. (1991). Male development of chromosomally female mice transgenic for Sry. *Nature (London)* **351,** 117–121.

McLaren, A., Simpson, E., Tomonari, K., Chandler, P., and Hogg, H. (1984). Male sexual differentiation in mice lacking H-Y antigen. *Nature (London)* **312,** 552–555.

Müller, U. (1987). Mapping of testis-determining locus on Yp by the molecular genetic analysis of XX males and XY females. *Development (Cambridge, UK), Suppl.* **101,** 51–58.

Müller, U., Donlon, T., Schmid, T., Fitch, M., Richter, N., Lalande, C., and Latt, S. M. (1986). Deletion mapping of the testis determining locus with DNA probes in 46,XX males and 46,XY and 46,X,dic(Y) females. *Nucleic Acids Res.* **14,** 6489–6505.

Müller, U., Wachtel, S. S., Jaswaney, V., and Goldberg, E. H. (1993). H-Y (SDM) antibody specifically binds Müllerian inhibiting substance. *Hum. Genet.* **91,** 515–518.

Nagamine, C. M., Chan, K., Kozak, C. A., and Lau, Y.-F. (1989). Chromosome mapping and expression of a putative testis-determining gene in mouse. *Science* **243,** 80–83.

Ohno, S., Nagai, Y., Ciccarese, S., and Iwata, H. (1979). Testis-organizing H-Y antigen and the primary sex-determining mechanism of mammals. *Recent Prog. Horm. Res.* **35,** 449–476.

Page, D. C. (1988). Is ZFY the sex-determining gene on the human Y chromosome? *Philos. Trans. R. Soc. London, Ser. B* **322,** 155–157.

Page, D. C., Mosher, R., Simpson, E. M., Fisher, E. M. C., Mardon, G., Pollack, J., McGillivray, B., de la Chapelle, A., and Brown, L. G. (1987). The sex-determining region of the human Y chromosome encodes a finger protein. *Cell (Cambridge, Mass.)* **51,** 1091–1104.

Palmer, M. S., Sinclair, A. H., Berta, P., Ellis, N. A., Goodfellow, P. N., Abbas, N. E., and Fellous, M. (1989). Genetic evidence that ZFY is not the testis-determining factor. *Nature (London)* **342,** 937–939.

Petit, C., de la Chapelle, A., Levilliers, J., Castillo, S., Noel, B., and Weissenbach, J. (1987). An abnormal X-Y interchange accounts for most but not all cases of human XX maleness. *Cell (Cambridge, Mass.)* **49,** 595–602.

Pivnick, E. K., Wachtel, S., Woods, D., Simpson, J. L., and Bishop, C. E. (1992). Mutations in the conserved domain of SRY are uncommon in XY gonadal dysgenesis. *Hum. Genet.* **90,** 308–310.

Schneider-Gädicke, A., Beer-Romero, P., Brown, L. G., Nussbaum, R., and Page D. C. (1989). ZFX has a gene structure similar to ZFY, the putative human sex determinant, and escapes X inactivation. *Cell (Cambridge, Mass.)* **57,** 1247–1258.

Simpson, E., Chandler, P., Goulmy, E., Disteche, C. M., Ferguson-Smith, M. A., and Page, D. C. (1987). Separation of the genetic loci for the H-Y antigen and for testis determination on human Y chromosome. *Nature (London)* **326,** 876–878.

Simpson, J. L. (1989). Genetic heterogeneity in XY sex reversal: Potential pitfalls in isolating the testis-determining factor (TDF). *In* S. S. Wachtel, ed.), "Evolutionary Mechanisms in Sex Determination" pp. 264–277. CRC Press, Boca Raton, FL.

Sinclair, A. H., Poster, J. W., Spencer, J. A., Page, D. C., Palmer, M., Goodfellow, P. N., and Graves, J. A. M. (1988). Sequences homologous to ZFY, a candidate human sex-determining gene, are autosomal in marsupials. *Nature (London)* **336,** 780–783.

Sinclair, A. H., Berta, P., Palmer, M. S., Hawkins, J. R., Griffiths, B. L., Smith, M. J., Foster, J. W., Firschauf, A.-M., Lovell-Badge, R., and Goodfellow, P. (1990). A gene from the human sex-determining region encodes a protein with homology to a conserved DNA-binding motif. *Nature (London)* **346,** 240–244.

Singh, L., and Jones, K. W. (1982). Sex reversal in the mouse (*Mus musculus*) is caused by a recurrent nonreciprocal crossover involving the X and an aberrant Y chromosome. *Cell (Cambridge, Mass.)* **28,** 205–216.

Singh, L., Purdom, I. F., and Jones, K. W. (1976). Satellite DNA and evolution of sex chromosomes. *Chromosoma* **59,** 43–62.

Singh, L., Purdom, I. F., and Jones, K. W. (1981). Conserved sex-chromosome-associated nucleotide sequences in eukaryotes. *Cold Spring Harbor Symp. Quant. Biol.* **45,** 805–814.

Stern, C. (1957). The problem of complete Y-linkage in man. *Am. J. Hum. Genet.* **9,** 147–166.

Su, H., and Lau, C. Y.-F. (1993). Identification of the transcriptional unit, structural organization, and promoter sequence of the human sex-determining region Y (*SRY*) gene, using a reverse genetic approach. *Am. J. Hum. Genet.* **52,** 24–38.

Tiersch, T. R., Mitchell, M. J., and Wachtel, S. S. (1991). Studies on the phylogenetic conservation of the *SRY* gene. *Hum. Genet.* **87,** 571–573.

Tommerup, N., Schempp, W., Meinecke, P., Pederson, S., Bolund, L., Brandt, C., Goodpasture, C., Guldberg, P., Held, K. R., Reinwein, H., Saugstad, O. D., Scherer, G., Skjeldal, O., Toder, R., Westvik, J., van der Hagen, C. B., and Wolf, V. (1993). Assignment of an autosomal sex reversal locus (SRA1) and campomelic dysplasia (CMPD1) to 17q24.3–q25.1. *Nat. Genet.* **4,** 170–174.

Verga, V., and Erickson, R. P. (1989). Minireview: Is Zinc-Finger Y the sex-determining gene? *Am. J. Hum. Genet.* **45,** 671–674.

Vergnaud, G., Page, D. C., Simmler, M.-C., Brown, L., Rouyer, F., Noel, B., Botstein, D., de la Chapelle, A., and Weissenbach, J. (1986). A deletion map of the human Y chromosome based on DNA hybridization. *Am. J. Hum. Genet.* **38,** 109–124.

Wachtel, S. S. (1983). "H-Y Antigen and the Biology of Sex Determination." Grune & Stratton, New York.

Wachtel, S. S., Ohno, S., Koo, G. C., and Boyse, E. A. (1975). Possible role for H-Y antigen in the primary determination of sex. *Nature* (*London*) **257,** 235–236.

Welshons, W. J., and Russell, L. B. (1959). The Y-chromosome as the bearer of male determining factors in the mouse. *Proc. Natl. Acad. Sci. U.S.A.* **45,** 560–566.

Wolf, U., Schempp, W., and Scherer, G. (1992). Molecular biology of the human Y chromosome. *Rev. Physiol. Biochem. Pharmacol.* **121,** 147–213.

2

The Cloning of *SRY*

Andrew H. Sinclair
Centre for Child Growth and Hormone Research
Royal Children's Hospital
Parkville, Victoria 3052
Australia

Molecular Genetics of Sex Determination

I. Mammalian Sex Determination

The critical factor in mammalian sex determination and differentiation is the development of the gonad. In a series of classic experiments, Alfred Jost demonstrated the central role of the testis in male development. When male rabbit embryos were castrated *in utero,* subsequent development was female; when female rabbit embryos were castrated at the same stage, subsequent development was also female. These results (reviewed by Jost *et al.*, 1973) suggested that the basic mammalian body plan is female but presence of testes determines whether the embryo will develop as a male. Further work showed that the testes produce testosterone and anti-Müllerian hormone (AMH), which are responsible for inducing the male secondary sex characteristics. Consequently, sex determination can be reduced to its essence: the choice between forming testis or ovary. Once gonadal sex is determined, the subsequent processes known as sex differentiation commence.

In mammals, sex determination is chromosomally based: females have two X chromosomes and males have an X and a Y. Ford *et al.* (1959) showed that individuals with Turner syndrome had a single X chromosome and were female. In the same year, Jacobs and Strong (1959) showed that patients with Klinefelter syndrome had two X chromosomes, and a Y, and were male. These results implied that the Y chromosome carried a gene required for testis development. In humans, this gene was later called the testis determining factor (*TDF*) because of the uncertain nature of the genetic material involved. Mouse geneticists were less coy, however, and called it the testis determining Y gene (*Tdy*).

The role of *TDF* as a "switch" in sex determination is often misunderstood. Evidently, many genes are required in the pathway leading to testis formation. Some of these genes are autosomal and some, X-linked, but in the absence of *TDF,* the pathway will not produce testes. In the presence of *TDF* the pathway is completed and testes are produced, ultimately leading to male sex differentiation.

In summary, the main event in mammalian sex differentiation is the formation of testes, which is controlled by the Y-linked *TDF* gene. In this chapter we will briefly review the search for *TDF,* the identification of *SRY,* and show that *SRY* can be equated to *TDF.*

II. The Human Y Chromosome

The human Y chromosome is a bipartite structure with two different functions. At male meiosis the Y chromosome must pair with the X chromosome and it must encode *TDF* (Ellis, 1991). These functions are mediated by distinct regions on the Y chromosome. Pairing with the X chromosome occurs within a region shared between the short arms of the X and Y chromosomes known as the pseudoautosomal region (PAR) (Burgoyne, 1982). An obligate recombination event within the PAR is necessary to ensure correct meiotic segregation (Ellis, 1991). There is a distinct delineation between the two regions marked by the pseudoautosomal boundary (PAB) (Ellis *et al.*, 1989). The boundary separates the pseudoautosomal region from the Y-specific region which, by definition, does not participate in homologous recombination with the X chromosome and must carry *TDF*.

III. Early Candidates for *TDF*

For more than 10 years, many believed that the male-specific H-Y antigen was the product of *TDF* (Wachtel *et al.*, 1975), but the discovery of male mice failing to express H-Y excluded the *Hya* gene as a possible candidate (McLaren *et al.*, 1984). Another candidate was the *Bkm* repeat sequence first isolated from a snake called the banded krait (Jones and Singh, 1981). Although *Bkm* sequences were found on the sex chromosomes of many mammals, their absence from the human Y excluded them as a candidate for *TDF* (Kiel-Metzger *et al.*, 1985; reviewed in Chapter 1, this volume).

IV. Searching for *TDF* on the Human Y Chromosome

In order to find *TDF* on the Y chromosome, three types of detailed maps were constructed: a meiotic map of the pseudoautosomal region (Goodfellow *et al.*, 1986; Weissenbach *et al.*, 1987); a deletion map based on the analysis of sex-reversed XX males and XY females (Vergnaud *et al.*, 1986; Guellaen *et al.*, 1984), and a long-range restriction map linking the first two maps (Pritchard *et al.*, 1987). By studying individuals with Y-chromosome deletions it was possible to localize *TDF* to the short arm (Yp) (Goodfellow *et al.*, 1985). However, a more precise localization was made possible by examining the chromosomes of XX males. These sex-reversed individuals appear to lack the Y chromosome but have a male phenotype with infertile testes. XX males occur mainly because of an abnormal recombination event between the X and Y chromosomes at male meiosis. This aberrant exchange extends beyond the pseudoautosomal region into the Y-specific region and results in transfer of *TDF* to the X chromosome (Ferguson-Smith, 1966; Guellaen *et al.*, 1984; Petit *et al.*, 1987). Transfer of varying portions of the Y chromosome to the X in different XX males allowed construction of a deletion map. This allowed *TDF* to be mapped to the most distal part of the Y-specific region, adjacent to the pseudoautosomal boundary. The meiotic map of the pseudoautosomal region indicated that *MIC2* was the closest pseudoautosomal gene to *TDF* (Goodfellow *et al.*, 1986). In order to link the deletion map and the meiotic map (which are based on different principles) a long-range restriction map was constructed (Pritchard *et al.*, 1987). This map began at the 5′ CpG-rich island of the *MIC2* gene in the pseudoautosomal region, spanned the boundary, and identified a CpG island in the Y-specific region. As CpG-rich islands are often associated with genes, it was predicted that this Y-specific island might define a new candidate for *TDF* (Pritchard *et al.*, 1987).

V. Walking on the Y Chromosome: Identification of *ZFY*

On the basis of this information, Goodfellow and colleagues initiated a Y-chromosome walk from *MIC2* across the pseudoautosomal boundary into the Y-specific region toward the CpG island identified by Pritchard *et al.* (1987). This walk finished 5 kb distal (telomeric) to the CpG island that marked a possible candidate for *TDF*. At the same time Page and colleagues (1987) also concluded that *TDF* must lie within this region. They began walking along the Y chromosome from a point proximal (centromeric) to the CpG island identified by Pritchard *et al.* (1987). In addition they used two patients to create a deletion map of the Y-specific region around *TDF* (Fig. 1). One patient, an XX male (LGL203) contained only 280 kb of the Y-specific region. A portion of this Y-specific region was deleted in an XY female (WHT 1013) with a Y ; 22 translocation (Fig. 1). These patients and the position of the CpG island enabled Page *et al.* (1987) to narrow their search to a 150-kb region in which *TDF* must reside. After a 230-kb chromosome walk, Page *et al.* (1987) cloned this region and identified a gene that met many of the criteria expected for *TDF*. It was conserved on the Y chromosome of all eutherian mammals tested and it encoded a protein with 13 "zinc fingers," a nuclear localization signal and an activation domain—features consistent with its being a transcription factor. The gene was called *ZFY* (zinc finger Y); because of its putative function as a transcription factor and location, it was a good candidate for *TDF*.

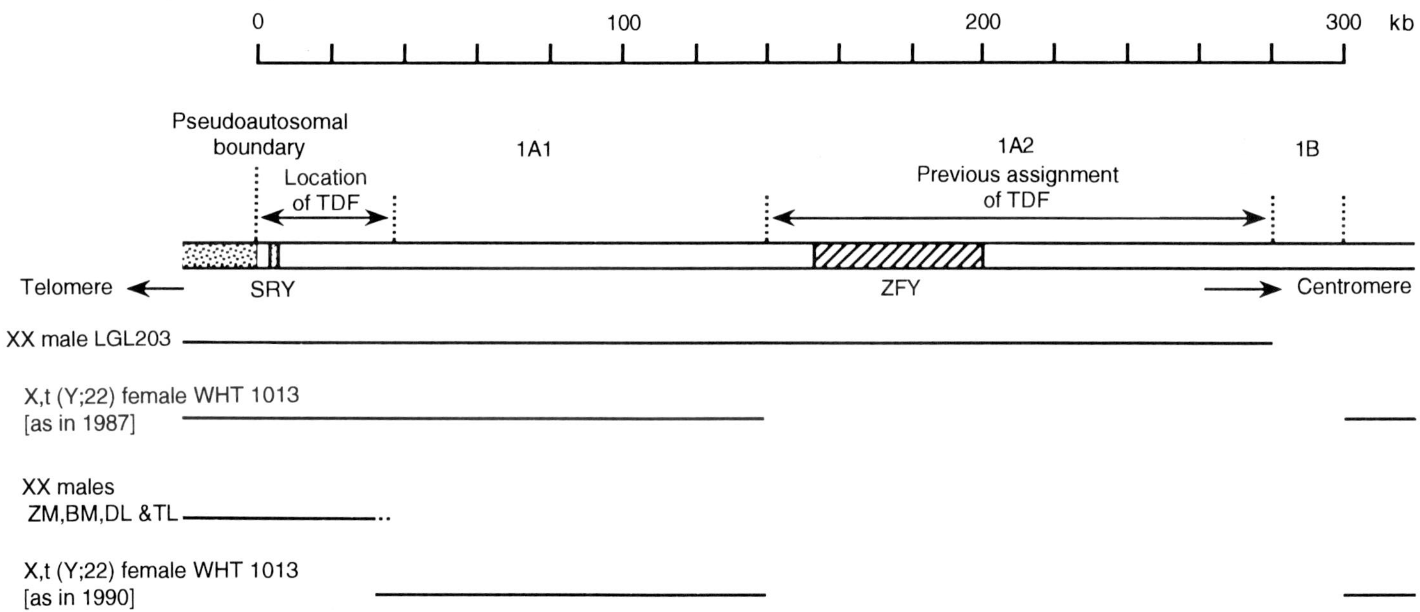

Figure 1 Schematic diagram of the sex-determining region of the Y chromosome. The extent of DNA present in XX males and XY females is indicated as a line. Note the difference in the same XY female between the years 1987 and 1990 (see Page *et al.*, 1990). Reprinted by permission. From Hawkins and Sinclair (1991).

VI. Difficulties in Equating *ZFY* and *TDF*: The Role of Marsupials

Other features of *ZFY* began to emerge that were less consistent with a role in sex determination. A related gene, *ZFX,* was discovered on the X chromosome of all eutherian mammals. The sequence similarities between *ZFY* and *ZFX* (Palmer *et al.,* 1990) led to the proposition that sex determination might be controlled by gene dosage (Schneider-Gadicke *et al.,* 1989). If *ZFX* were X-inactivated, two copies would be expressed in the male and only one in the female. This hypothesis became untenable, however, when it was found that *ZFX* escaped X-inactivation (Schneider-Gadicke *et al.,* 1989). Another puzzling feature of *ZFY* was its expression profile: it was expressed in all human male tissues tested—both fetal and adult (Schneider-Gadicke *et al.,* 1989). As a further test of its candidacy *ZFY* was used for *in situ* hybridization experiments on the chromosomes of metatherian (marsupial) mammals (Sinclair *et al.,* 1988). The Y chromosome is known to be male-determining in marsupials, and these animals provide an ideal mammalian "outgroup" with which to test the conservation of *ZFY.* It came as a surprise to find that *ZFY* was not on either sex chromosome in the marsupial species examined but was autosomal instead. This implied either that marsupials have a sex-determining mechanism very different from that of eutherian mammals or that ZFY is not the testis-determining gene (Sinclair *et al.,* 1988). The results of the marsupial experiments called for reevaluation of *ZFY* as a candidate for *TDF* and, in particular, reevaluation of the data that mapped the sex-determining region of the Y chromosome.

VII. Reevaluation of the Y-Chromosome Deletion Map

If many of the features of *ZFY* were incompatible with its being *TDF,* there still was the mapping data that appeared incontrovertible. The deletion map of the Y chromosome defined the sex-determining region as the 150-kb interval 1A2 (Page *et al.,* 1987) (Fig. 1). By definition, *TDF* should reside in this region, and the only gene found in that interval was *ZFY.* The data on XX male LGL203 that marked the proximal end of the sex-determining region could not be questioned. The only concern was the XY female patient (WHT1013) with a Y ; 22 translocation who was used by Page *et al.* (1987) in constructing the map (Fig. 1). Goodfellow and colleagues began to propose alternative explanations for sex reversal in this patient. It was possible that *TDF* was present but inactivated in this patient due to a position effect caused by the Y; 22 translocation. If this were true, *TDF* should lie somewhere between *ZFY* and the pseudoautosomal boundary.

VIII. Mapping the Sex-Determining Region

It was known that some XX males could be *ZFY* negative. If *ZFY* were *TDF,* the reason these individuals were male was the presence of "downsteam" gain-of-function mutations. If, however, *TDF* lay between *ZFY* and the pseudoautosomal boundary, some XX males should lack *ZFY* and still retain a small portion of the Y chromosome close to the boundary. Consequently, a panel of 14 *ZFY*-negative XX males and intersexes (all of whom had testes) were surveyed for presence of the Y-pseudoautosomal boundary. Among these *ZFY*-negative subjects, four (ZM,BM,DL,TL) had retained the Y-specific region adjacent to the boundary (Fig. 1) (Palmer *et al.,* 1989). These results excluded *ZFY* as a candidate for *TDF* and, importantly, redefined the sex-determining region of the Y chromosome to a region of 60 kb, later limited to 35 kb, adjacent to the pseudoautosomal boundary (Fig. 1) (Sinclair *et al.,* 1990). The discrepancy between this newly defined sex-determining region and the previously defined region was resolved when a second deletion extending 45 kb into the Y-specific region was found in XY female, WHT1013 (Page *et al.,* 1990) (Fig. 1).

The range of phenotypes observed in these four sex-reversed individuals varied from normal male with undescended testes, lacking germ cells, to hermaphroditic with bilateral ovotestes and uterus. An explanation for this phenotypic variation is that another gene on the Y chromosome contributes to, but is not absolutely necessary for, testicular development. Alternatively, the proximity of the breakpoints on the Y chromosome partially disrupts *TDF* expression, or the translocated Y fragment on the paternal X chromosome is affected by spreading X-inactivation.

IX. *SRY,* A New Candidate for *TDF*

The following strategy was employed to search the target 35 kb of Y-chromosomal DNA for the testis-determining gene. This region was encompassed within the chromosome walk initiated by Goodfellow and colleagues but initial analysis had not detected any genes. On this second occasion the region was subcloned into fragments of 4 kb and digested with a frequent cutter enzyme such as *Rsa*I to produce fragments in the range 0.5 to 1 kb (Sinclair *et al.*, 1990). In this manner, 50 probes were generated. Each of these probes was hybridized to Southern blots of DNA from humans (males and females), bovines (males and females), mice (males and females), and human–hamster somatic cell hybrids containing the human X or Y chromosome. All of the probes were tested with and without prehybridization to total human DNA in order to suppress the effect of repetitive sequences. Despite this precaution, most probes did not detect unique sequences but hybridized to repetitive elements distributed throughout the human and bovine genome (Sinclair *et al.*, 1990).

Seven probes detected single-copy Y-specific bands in human DNA (pYNB, pY53.3, pY53.1α, pY53.1β, pY53.1γ, pY53.2, and pY4.1β; see Fig. 2). However, only one of these probes, pY53.3, also reacted with Y-specific bands in the murine and bovine genomes. We named the putative gene defined by this probe *SRY* (sex-determining region Y) (Sinclair *et al.*, 1990). At this stage we had no idea of the gene's function

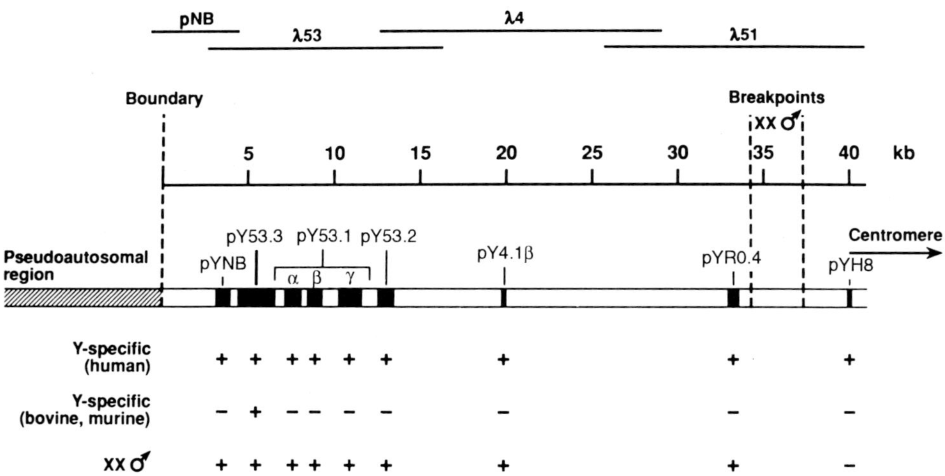

Figure 2 Map of distal short arm of human Y chromosome: shaded region (left), pseudoautosomal region; broken line, boundary between pseudoautosomal and Y-specific region. Top, the three overlapping λ clones λ51, λ4, and λ53 and the plasmid pNB obtained from walking along the Y chromosome. Breakpoints of the XX males, broken lines at 35 kb. Black boxes, probes detecting single-copy Y-specific human DNA fragments (+). When these probes were hybridized to male and female DNA from bovine and murine genomes, only pY53.3 detected Y-specific fragments (+). Reprinted by permission. From Sinclair *et al.* (1990).

so the name was based solely on the location of the gene. The mouse equivalent, *Sry*, was present in *Sxr′* mice, which have the smallest part of the Y chromosome known to be sex-determining and it was deleted from a mutant Y chromosome that is no longer sex-determining (Gubbay *et al.*, 1990).

Sequences homologous to *SRY* were present on the Y chromosome of all eutherian mammals tested (Fig. 3) (Sinclair *et al.*, 1990). Crucially, Foster *et al.* (1992) were able to show that *SRY* is also present on the Y chromosome of marsupial species (Foster *et al.*, 1992). The conservation of Y-specific sequences among different mammalian species is very rare. This degree of conservation among eutherian and metatherian mammals suggests that *SRY* has a functional role on the mammalian Y chromosome. At lower stringency, other *SRY* fragments were apparent and these were shared between males and females. These fragments were not present in an X-only human hybrid, suggesting that the other sequences detected by *SRY* are autosomal (Sinclair *et al.*, 1990).

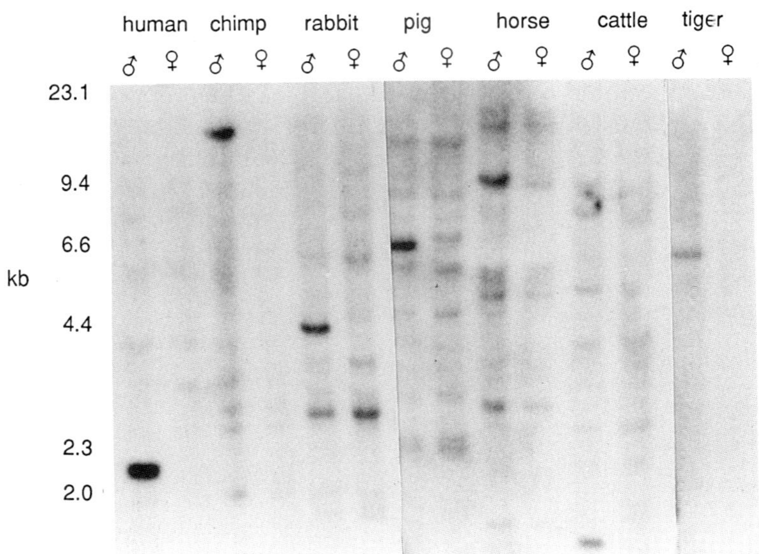

Figure 3 Noah's Ark blot of *Hin*dIII-digested DNA from male and female pairs of eutherian mammals, hybridized with *SRY*. The probe detected male-specific fragments in human (2.1 kb), chimpanzee (~18 kb), rabbit (4.2 kb), pig (6.6 kb), horse (~10 kb), cattle (1.6 kb), and tiger (6.3 kb). Rabbit female lane is slightly overloaded. Reprinted by permission. From Sinclair *et al.* (1990).

X. Sequence Analysis of *SRY*

The nucleotide sequence of *SRY* comprised two open reading frames of 99 and 223 amino acids, which overlapped in different frames, reading 5′→3′ from the centromere toward the pseudoautosomal boundary (Sinclair *et al.,* 1990). A search of the EMBL DNA sequence database failed to disclose any related sequences. We screened the PIR protein database, using predicted proteins encoded by these open reading frames. The smaller open reading frame did not contain any homologous sequences but the longer open reading frame showed a high degree of similarity to several proteins. Homology was found with the Mc protein encoded by the *mat* 3-M gene of the fission yeast *Schizosaccharomyces pombe* (Kelly *et al.,* 1988) (Fig. 4). The putative *SRY* protein also showed homology with a protein encoded by a gene at the *Neurospora crassa a* mating-type region (Staben and Yanofsky, 1990), a conserved motif found in the nonhistone proteins, high mobility groups 1 and 2 (HMG1 and HMG2; see Kolodrubetz, 1990), the human nucleolar transcription factor *hUBF* (human upstream binding factor; Jantzen *et al.,* 1990), and the transcription factors *LEF-1* and *TCF-1* (Travis *et al.,* 1991; van de Wetering *et al.,* 1991). In all these proteins, the common motif is a 79 amino acid sequence, known as the HMG box, which is associated with DNA binding activity. A DNA binding capacity for the *SRY* protein would be consistent with a role in gene regulation.

The previously mentioned autosomal fragments detected by *SRY* on Noah's Ark blots were later found to be *SRY*-related autosomal genes (now called *SOX* genes). These *SOX* genes also share the HMG box motif and form an *SRY*-related subgroup. Recent studies have shown

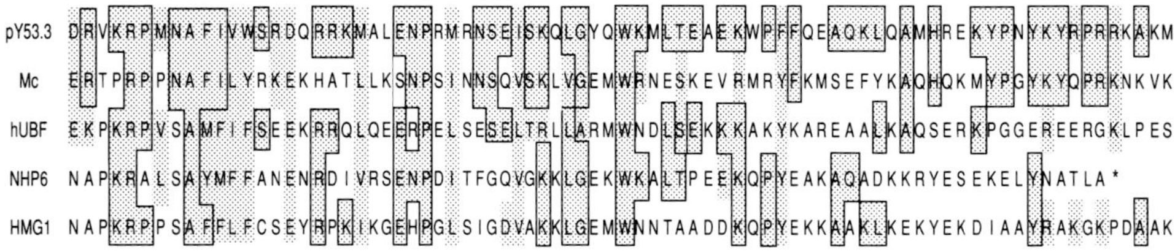

Figure 4 Conserved 79 amino acids (single-letter code) of pY53.3 (*SRY*) human compared with Mc protein from *S. pombe* (Mc), human upstream binding factor (hUBF), nonhistone chromosomal protein (NHP6) from *S. cerevisiae,* and high mobility group protein 1 (HMG1). Boxed regions represent identical amino acids, and shaded regions, conservative amino acid changes with respect to human pY53.3 (*SRY*) sequence. Reprinted by permission. From Sinclair *et al.* (1990).

that these *Sox* genes are not expressed in the genital ridge in the mouse but have a role in neuronal differentiation (J. Collignon, personal communication) and in spermatogenesis (Denny *et al.*, 1992a,b).

To examine the functional importance of the conserved motif in *SRY* we also cloned and sequenced the rabbit Y-specific homolog (Sinclair *et al.*, 1990) and the mouse homolog, *Sry* (Gubbay *et al.*, 1990). Within the conserved protein motif, human, rabbit, and mouse show ~90% overall similarity, which drops to ~50% outside this region. Recent studies on the *SRY* homolog isolated from marsupial species show ~85% similarity to the human *SRY* HMG box motif (Foster *et al.*, 1992, and see Chapter 8 this volume). This degree of homology within the HMG box motif suggests that it is the coding sequence of a gene.

The SRY protein interacts with a nucleotide sequence similar to the *LEF-1* and *TCF-1* binding site, but the binding affinity appears rather low (Harley *et al.*, 1992; Nasrin *et al.*, 1991). Subsequent analysis by Giese *et al.* (1992) showed that SRY protein binds to the minor groove of the double helix and recognizes a higher affinity binding site defined by the nucleotide sequence 5′-CCATTGTTCT. In addition, they found that SRY protein is capable of bending DNA 85°, thereby bringing otherwise distinct regions of the genome into juxtaposition and possibly augmenting protein–protein contacts (Giese *et al.*, 1992). It is not known whether bending of the DNA helix plays a major role in the mechanism by which SRY protein regulates transcription.

XI. Tissue Distribution and Expression of *SRY*

In order to determine the tissue distribution and expression of *SRY*, a Northern blot of poly(A) RNA from human ovary, testis, lung (male), and kidney (male) was constructed. The Northern blot was hybridized with *SRY* and detected a ~1.1-kb fragment in adult testis only (Fig. 5) (Sinclair *et al.*, 1990). This indicates that *SRY* encodes a testis-specific transcript. Using 3′RACE (rapid amplification of complementary DNA ends) PCR (polymerase chain reaction) from adult testis poly(A) RNA, we found a poly(A) tract 15 bases downstream from the potential polyadenylation signal confirming this as the 3′ end of a Y-specific transcript. As yet, the 5′ end of the *SRY* transcript is unresolved and will require the isolation and sequencing of transcripts. Despite enormous efforts to isolate a cDNA transcript and the screening of many testis cDNA libraries, only one human *SRY* cDNA clone has been found. Sequence analysis of the cDNA transcript did not shed light on the 5′ end of *SRY* but did show that the genomic and cDNA sequences were identi-

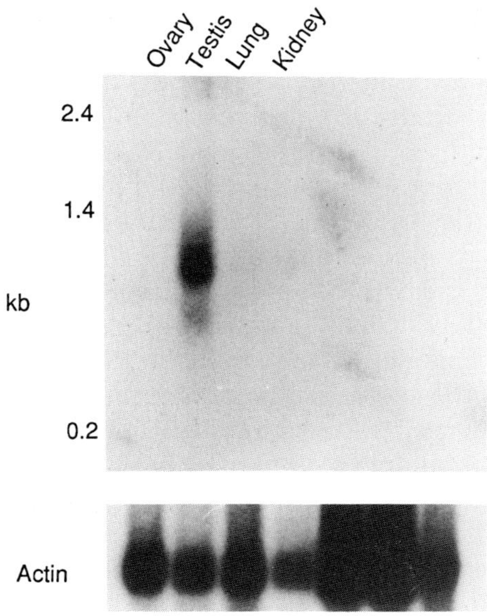

Figure 5 Northern blot of poly(A)$^+$ RNA from ovary, adult testis, male lung, and male kidney, hybridized with *SRY*. The probe detected a transcript of ~1.1 kb in adult testis but no other tissue tested. The lower portion is the same filter stripped and reprobed with β-actin, confirming presence of poly(A)$^+$ RNA in all tissue samples. Reprinted by permission. From Sinclair *et al.* (1990).

cal, suggesting that SRY comprises one exon (A. H. Sinclair *et al.*, unpublished data; also see Su and Lau, 1993). In the fetal mouse, Koopman *et al.* (1990) showed that *Sry* is confined to the genital ridge, is germ cell independent, and is expressed from 10.5 to 11.5 days postcoitum, which coincides with the start of testicular development (Koopman *et al.*, 1990).

The 35 kb of Y-specific sequences adjacent to the boundary contains many repetitive sequences that could have obscured small exons, particularly if they are poorly conserved between species. However, a comparison of the sequence from a mouse 14.5-kb *Sry* genomic clone (Gubbay *et al.*, 1992) with 25 kb of genomic sequence from the human *SRY* locus failed to reveal any conserved genes (A. H. Sinclair and D. I. Jackson, unpublished data).

In summary, *SRY* is a testis-specific transcribed gene present in the sex-determining region of human and mouse. Sequences homologous to *SRY* are Y-specific in all eutherian and metatherian mammals tested. *SRY* encodes a protein that binds a specific nucleotide sequence in the minor groove of the DNA helix and causes bending of the DNA. All of the above-mentioned features are consistent with a role in developmental regulation of the testis and imply that *SRY* is an excellent candidate for the testis-determining gene, *TDF*.

XII. Genetic Evidence Equating *SRY* and *TDF*: Analysis of XY Females

If *SRY* is the testis-determining gene, mutations in *SRY* should prevent testis formation and result in a female rather than a male phenotype. To test this hypothesis, we analyzed XY females, some of whom would be expected to carry mutations in the *SRY* gene. Initially, *SRY* sequences were amplified by PCR from 11 XY females and 50 normal male controls (Berta *et al.*, 1990). The amplified PCR products were codigested with two restriction enzymes to give fragments in the optimal (200 bp) size range, and analyzed by use of the single-strand conformation polymorphism (SSCP) assay, which identifies sequence differences as indicated by band (mobility) shifts.

The normal male sequences gave the same pattern as those from nine of the XY females. Two of the XY females (AA and JN) showed band-shifted SSCP patterns for *SRY* (Berta *et al.*, 1990) (Fig. 6). The father of AA gave a normal pattern but the father of JN gave an abnormal pattern, similar to that in his daughter, JN (Fig. 6). DNA from these two individuals and their fathers was amplified, cloned, and sequenced. The sequencing data supported the SSCP result showing that individual AA had a G-to-A base change. At the amino acid level, this indicated a conservative change from methionine to isoleucine within the conserved binding motif of *SRY* (Berta *et al.*, 1990) (Fig. 7). The father of AA had a normal SRY sequence, indicating that AA had suffered a *de novo* mutation in her *SRY* gene. This *de novo* mutation demonstrates that *SRY* is required for testis development and normal sex determination of the male. A similar conclusion was reached by Jäger *et al.* (1990), who described an XY female with a *de novo* frameshift mutation in the conserved HMG box domain of *SRY*.

In the above-mentioned studies, 23 XY females were analyzed. Of those, 3 had mutations in *SRY*, and 2 of the mutations were *de novo* (Berta *et al.*, 1990; Jager *et al.*, 1990). These studies were extended by an analysis of an additional 17 XY females (Hawkins *et al.*, 1992). At the time of writing, 7 *de novo* mutations in *SRY* have been identified among some 60 XY females (see Chapter 13, this volume). All of these mutations lie within the HMG box domain of the *SRY* gene and presumably affect

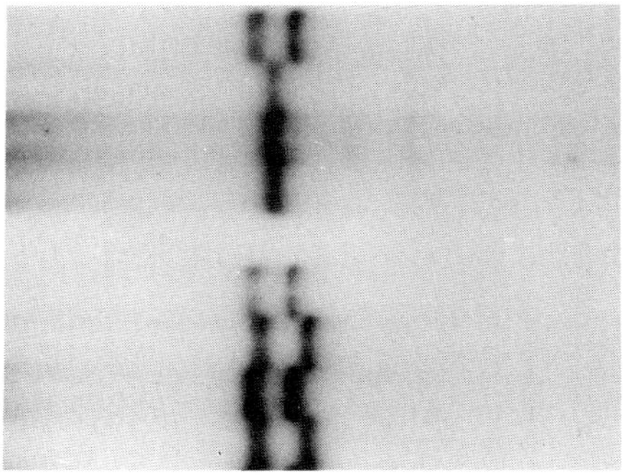

Figure 6 SSCP analysis of *SRY* in XY females. In the upper portion is family N: DNA from a normal male cell line PGF; DNA from the father is flanked on each side by DNA of his XY daughter, JN. In the lower portion is family A: DNA from the father is flanked by that of his XY daughter, AA (both AA and JN have XY gonadal dysgenes). Fifty normal males were tested as controls and no variation was detected. Reprinted by permission from Berta *et al.* (1990).

DNA binding. Absence of testicular development in XY females who do not have *SRY* mutations can be attributed to mutations in other genes in the testis-determining pathway. However, the promoter of *SRY* has not yet been identified, and it is likely that some mutations that affect this region will also be found.

Less expected was the finding that the XY female, JN, shares a mutation in *SRY* with her father: A G-to-C base change causing a valine to leucine amino acid substitution (Berta *et al.*, 1990) (Fig. 7). We proposed three explanations for the sequence variant found in JN. First, the variant could be conditionally sex reversing and dependent on other genetic or environmental factors. Second, the variant could be sex reversing and the father, mosaic for wild-type or variant sequences. Third, the variant could be a low-frequency polymorphism and sex reversal in JN, due to an autosomal or X-linked sex-reversing gene. Further analysis of this family revealed that the father had an XY female sister (McElreavery *et al.*, 1991), which precludes the possibility that he is mosaic. Subsequent biochemical evidence, presented below, has shown that the variant is conditionally sex reversing.

Figure 7 (A) The open reading frame of the genomic clone of SRY is shown extending from 354-1022 bp of pY53.3. The conserved motif, which encodes a potential DNA binding protein, extends from 582 to 821 bp. The primers XES-7 and XES-2 were used to amplify a 609-bp region. The entire amplification product was sequenced, but only the

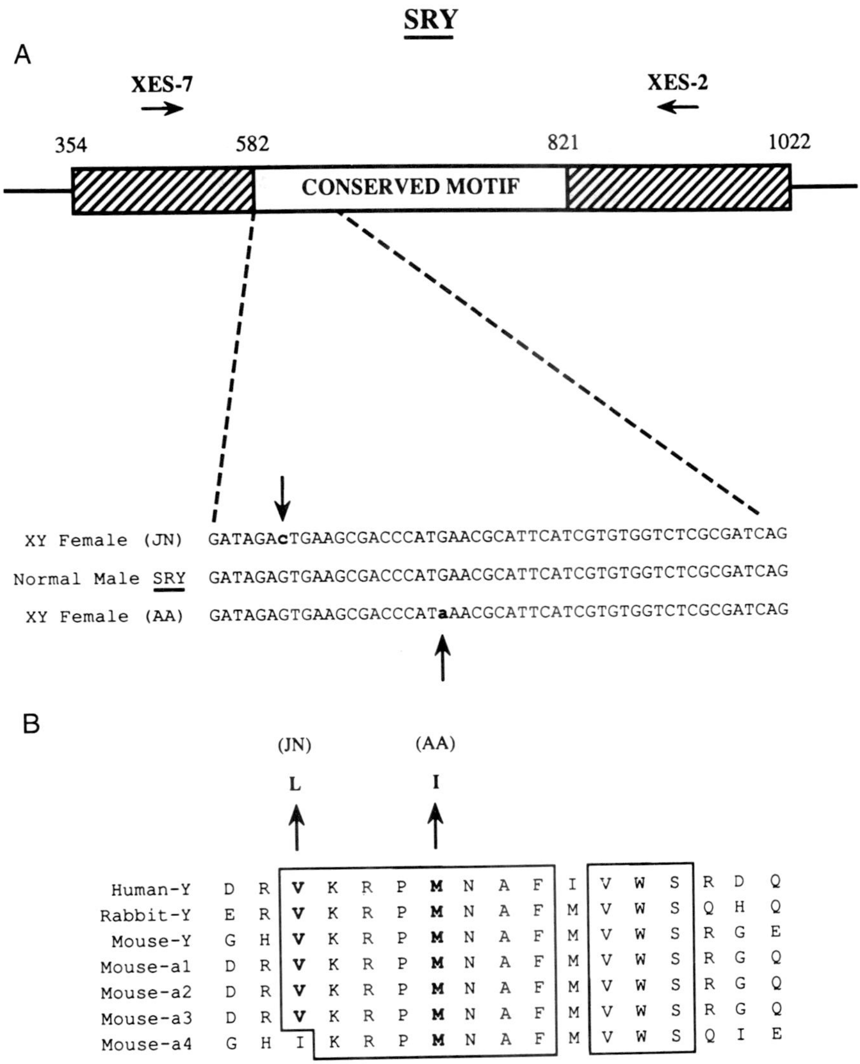

small region containing the changes in the XY females is displayed. The dotted line indicates the location within the open reading frame of the nucleotide sequence shown. Top line, nucleotide sequence for the XY female. JN with base changes from G to C indicated by arrow. Middle line, normal male *SRY* nucleotide sequence. Bottom line, XY female, AA, with a base change from G to A indicated by arrow. (B) The amino acid sequence of *SRY* for the human-Y, rabbit-Y, mouse-Y, and mouse autosomal *SRY*-like genes a_1, a_2, a_3, a_4. Boxed shaded regions show identical amino acids conserved across these species. In the XY female, JN, the variant causes a change from valine to leucine, whereas in the XY female, AA, the mutation causes a change from methionine to isoleucine. Reprinted by permission. From Berta *et al.* (1990).

As previously discussed, Harley *et al.* (1992) used gel retardation experiments to analyze the DNA binding capacity of SRY protein. Recombinant SRY protein was produced from five XY females with mutations in the *SRY* gene. Three of these five mutations were *de novo* and two were inherited, one in patient JN. Normal SRY protein binds to a specific heptameric oligonucleotide (known to be the target sequence for *TCF-1*); this produces a band shift. The DNA binding capacity of SRY protein was lost in four of the five mutants and dramatically reduced in the fifth (Harley *et al.*, 1992). Patient JN was one of the mutants in which DNA binding capacity had been abolished. This indicates that the variant originally observed in this individual is not a polymorphism but is an inherited mutation that is conditionally sex reversing. The mutant that showed reduced DNA binding capacity was also an inherited mutation. In this case, the reduction, rather than abolition, of binding explains the reduced penetrance of this inherited sex-reversing mutation. Together, these results demonstrate that the DNA binding activity of *SRY* is required for its function in testis formation, and male sex determination.

XIII. *SRY* is the Testis-Determining Gene

The analysis of mutations in XY females does not tell us whether *SRY* is the only gene on the Y chromosome required for male sex determination. This was addressed by injecting *Sry* sequences into chromosomally female mouse embryos. Some of the resulting mice, transgenic for *Sry*, developed as XX males (Koopman *et al.*, 1991; see Chapter 3, this volume). This shows that *Sry* is the only gene on the Y chromosome required for male sex determination. Hence, it can be stated that *SRY* is *TDF*. The cloning of *SRY* is the starting point for unraveling of the testis-determining pathway and, consequently, elucidation of the molecular events involved in mammalian sex determination.

Acknowledgments

The success of the recent work on sex determination is due to the collaboration between the laboratories of Peter Goodfellow and Robin Lovell-Badge, which brought together expertise in human genetics and mouse developmental biology.

References

Berta, P., Hawkins, J. R., Sinclair, A. H., Taylor, A., Griffiths, B. L., Goodfellow, P N., and Fellous, M. (1990). Genetic evidence equating SRY and the testis determining factor. *Nature (London)* **348,** 448–450.

Burgoyne, P. S. (1982). Genetic homology and crossing over in the X and Y chromosomes of mammals. *Hum. Genet.* **61,** 85–90.

Denny, P., Swift, S., Brand, N., Dabhade, N., Barton, P., and Ashworth, A. (1992a). A conserved family of genes related to the testis determining gene, *SRY*. *Nucleic Acids Res.* **20,** 2887.

Denny, P., Swift, S., Connor, F., and Ashworth, A. (1992b). An SRY-related gene expressed during spermatogenesis in the mouse encodes a sequence-specific DNA-binding protein. *EMBO J.* **11,** 3705–3712.

Ellis, N. A. (1991). The human Y chromosome. *Semin. Dev. Biol.* **2**(4), 231–241.

Ellis, N. A., Goodfellow, P. J., Pym, B., Smith, M., Palmer, M., Frischauf, A.-M., and Goodfellow, P. N. (1989). The pseudoautosomal boundary in man is defined by an *Alu* repeat sequence inserted on the Y chromosome. *Nature (London)* **337,** 81–84.

Ferguson-Smith, M. A. (1966). X-Y chromosomal interchange in the aetiology of true hermaphroditism and XX Klinefelter's syndrome. *Lancet* **2,** 475–476.

Ford, C. E., Jones, K. W., Polani, P., De Almedia, J. C., and Brigg, J. H. (1959). A sex chromosome anomaly in a case of gonadal sex dysgenesis (Turner's syndrome). *Lancet* **1,** 711.

Foster, J. W., Brennan, F. E., Hampikian, G. K., Goodfellow, P. N., Sinclair, A. H., Lovell-Badge, R., Selwood, L., Renfree, M. B., Cooper, D. W., and Graves, J. A. M. (1992). Evolution of sex determination and the Y chromosome: *SRY*-related sequences in marsupials. *Nature (London)* **359,** 531–533.

Giese, K., Cox, J., and Grosschedl, R. (1992). The HMG domain of Lymphoid Enhancer Factor 1 bends DNA and facilitates assembly of functional nucleoprotein structures. *Cell (Cambridge, Mass.)* **69,** 185–195.

Goodfellow, P. J., Darling, S. M., Thomas, N. S., and Goodfellow, P. N. (1986). A pseudoautosomal gene in man. *Science* **234,** 740–743.

Goodfellow, P. N., Darling, S. M., and Wolfe, J. (1985). The human Y chromosome. *J. Med. Genet.* **22,** 332–344.

Gubbay, J., Collignon, J., Koopman, P., Capel, B., Economou, A., Munsterberg, A., Vivian, N., Goodfellow, P. N., and Lovell-Badge, R. (1990). A gene mapping to the sex determining region of the mouse Y chromosome is a member of a novel family of embryonically expressed genes. *Nature (London)* **346,** 245–250.

Gubbay, J., Vivian, N., Economou, A., Jackson, D., Goodfellow, P. N., and Lovell-Badge, R. (1992). Inverted repeat structure of the *Sry* locus in mice. *Proc. Natl. Acad. Sci. U.S.A.* **89,** 7953–7957.

Guellaen, G., Casanova, M., Bishop, C., Geldwerth, D., Audre, G., Fellous, M., and Weissenbach, J. (1984). Human XX males with Y single copy DNA fragments. *Nature (London)* **307,** 172–173.

Harley, V. R., Jackson, D. I., Hextall, P. J., Hawkins, J. R., Berkovitz, G. D., Sockanathan, S., Lovell-Badge, R., and Goodfellow, P. N., (1992). DNA binding activity of recombinant SRY from normal males and XY females. *Science* **255,** 453–456.

Hawkins, J. R., and Sinclair, A. H. (1991). Identification and isolation of *SRY*, a gene from the sex determining region of the Y chromosome. *Semin. Dev. Biol.* **2,** 251–257.

Hawkins, J. R., Taylor, A., Berta, P., Levilliers, J., Van der Auwera, B., and Goodfellow,

P. N. (1992). Mutational analysis of *SRY:* nonsense and missense mutations in XY sex reversal. *Hum. Genet.* **88,** 471–474.

Jacobs, P. A., and Strong, J. A. (1959). A case of human intersexuality having a possible XXY sex determining mechanism. *Nature (London)* **183,** 302–303.

Jäger, R. J., Anvret, M., Hall, K., and Scherer, G. (1990). A human XY female with a frame shift mutation in the candidate testis determining gene *SRY. Nature (London)* **348,** 452–454.

Jantzen, H. M., Admon, A., Bell, S. P., and Tjian, R. (1990). Nucleolar transcription factor hUBF contains a DNA-binding motif with homology to HMG proteins. *Nature (London)* **344,** 830–836.

Jones, K. W., and Singh, L. (1981). Conserved repeated DNA sequences in vertebrate sex chromosomes. *Hum. Genet.* **58,** 46–53.

Jost, A., Vigier, B., Prepin, J., and Perchellet, J. P. (1973). Studies on sex differentiation in mammals. *Recent Prog. Horm. Res.* **29,** 1–41.

Kelly, M., Burke, J., Smith, M., Klar, A., and Beach, D. (1988). Four mating-type genes control sexual differentiation in the fission yeast. *EMBO J.* **7,** 1537–1547.

Kiel-Metzger, K., Warren, G., Wilson, G. N., and Erickson, R. P. (1985). Evidence that the human Y chromosome does not contain clustered DNA sequences (Bkm) associated with heterogametic sex determination in other vertebrates. *N. Engl. J. Med.* **313,** 242–245.

Kolodrubetz, D. (1990). Consensus sequence for HMG-1 like DNA binding domains. *Nucleic Acids Res.* **18,** 5565.

Koopman, P., Munsterberg, A., Capel, B., Vivian, N., and Lovell-Badge, R. (1990). Expression of a candidate sex determining gene during mouse testis differentiation. *Nature (London)* **248,** 450–452.

Koopman, P., Gubbay, J., Vivian, N., Goodfellow, P. N., and Lovell-Badge, R. (1991). Male development of chromosomally female mice transgenic for *Sry. Nature (London)* **351,** 117–121.

McElreavy, K., Vilain, E., Abbas, N., Brauner, R., Nichoul-Fekete, C., Richaud, F., Rappaport, R., Raymond, J. P., Toublanc, J. E., and Fellous, M. (1991). Conditional mutation in the testis determining region suggests a second sex determining gene in humans. *Proc. Int. Congr. Hum. Genet., 8th,* Abstr. p. 413.

McLaren, A., Simpson, E., Tomornari, K., Chandler, P., and Hogg, H. (1984). Male sexual differentiation in mice lacking H-Y antigen. *Nature (London)* **312,** 552–555.

Nasrin, N., Buggs, C., Kong, X. F., Carnazza, J., Goebel, M., and Alexander-Bridges, M. (1991). DNA-binding properties of the product of the testis-determining gene and a related protein. *Nature (London)* **354,** 317–320.

Page, D. C., Mosher, R., Simpson, E. M., Fisher, E. M. C., Mardon, G., Pollack, J., McGillivray, B., de la Chapelle, A., and Brown, L. G. (1987). The sex-determining region of the human Y chromosome encodes a finger protein. *Cell (Cambridge, Mass.)* **51,** 1091–1104.

Page, D. C., Fisher, E. M. C., McGillivray, B., and Brown, L. G. (1990). Additional deletion in sex-determining region of human Y chromosome resolves paradox of X,t(Y:22) females. *Nature (London)* **346,** 279–281.

Palmer, M. S., Sinclair, A. H., Berta, P., Ellis, N. A., Goodfellow, P. N., Abbas, N. E., and Fellous, M. (1989). Genetic evidence that ZFY is not the testis determining factor. *Nature (London)* **342,** 937–939.

Palmer, M. S., Berta, P., Sinclair, A. H., Pym, B., and Goodfellow, P. N. (1990). Characterization and comparison of human ZFY and ZFX transcripts. *Proc. Natl. Acad. Sci. U.S.A.* **87,** 1681–1685.

Petit, C., de la Chapelle, A., Leveilliers, J., Castello, S., Noel, B., and Weissenbach, J. (1987). An abnormal terminal exchange accounts for most but not all cases of human XX maleness. *Cell (Cambridge, Mass.)* **49,** 595–602.

Pritchard, C. A., Goodfellow, P. J., and Goodfellow, P. N. (1987). Mapping the limits of the human pseudoautosomal region and a candidate sequence for the male-determining gene. *Nature (London)* **328,** 273–275.

Schneider-Gadicke, A., Beer-Romero, P., Brown, L. G., Nussbaum, R., and Page, D. C. (1989). ZFX has a gene structure similar to *ZFY,* the putative human sex determinant, and escapes X-inactivation. *Cell (Cambridge, Mass.)* **57,** 1247–1258.

Sinclair, A. H., Foster, J. W., Spencer, J. A., Page, D. C., Palmer, M., Goodfellow, P. N., and Graves, J. A. M. (1988). Sequences homologous to *ZFY,* a candidate for the sex determining gene, are autosomal in marsupials. *Nature (London)* **336,** 780–783.

Sinclair, A. H., Berta, P., Palmer, M. S., Hawkins, J. R., Griffiths, B. L., Smith, M. J., Foster, J. W., Frischauf, A.-M., Lovell-Badge, R., and Goodfellow, P. N. (1990). A gene from the human sex determining region encodes a protein with homology to a conserved DNA binding motif. *Nature (London)* **346,** 240–244.

Staben, C., and Yanofsky, C. (1990). *Neurospora crassa* a mating-type region. *Proc. Natl. Acad. Sci. U.S.A.* **87,** 4917–4921.

Su, H., and Lau, C. Y.-F. (1993). Identification of the transcriptional unit, structural organization, and promoter sequence of the human sex-determining region Y (SRY) gene, using a reverse genetic approach. *Am. J. Hum. Genet.* **52,** 24–38.

Travis, A., Amsterdam, A., Bélanger, C., and Grosschedl, R. (1991). LEF-1, a gene encoding a lymphoid specific protein with an HMG domain, regulates T-cell receptor and enhancer function. *Genes Dev.* **5,** 880–894.

van de Wetering, M., Oosterwegel, M., Dooijes, D., and Clevers, H. (1991). Identification and cloning of TCF-1, a T lymphocyte-specific transcription factor containing a sequence-specific HMG box. *EMBO J.* **10,** 123–132.

Vergnaud, G., Page, D. C., Simmler, M.-C., Brown, L., Rouyer, F., Noel, B., Botstein, D., de la Chapelle, A., and Weissenbach, J. (1986). A deletion map of the human Y chromosome based on DNA hybridization. *Am. J. Hum. Genet.* **38,** 109–124.

Wachtel, S. S., Ohno, S., Koo, G. C., and Boyse, E. A. (1975). Possible role for H-Y antigen in the primary determination of sex. *Nature (London)* **257,** 235–236.

Weissenbach, J., Levilliers, J., Petit, C., Rouyer, F., and Simmler, M.-C. (1987). Normal and abnormal interchanges between the human X and Y chromosomes. *Development (Cambridge, UK)* **101S,** 67–74.

3

The Mouse Y Chromosome

John Gubbay
Howard Hughes Research Laboratory
The Rockefeller University
New York, New York 10021

Robin Lovell-Badge
Laboratory of Developmental Genetics
National Institute for Medical Research
London NW7 1AA
United Kingdom

Molecular Genetics of Sex Determination

I. Introduction

The mammalian Y chromosome is somewhat unique in both its structure and its function. Absent in 50% of the population, it clearly carries no information required for the development of viable females. In fact, genes on the unique portion of the Y chromosome seem to be involved almost exclusively in the processes of primary sex determination and spermatogenesis.

The Y chromosome acts as a dominant genetic switch to determine male development (Ford *et al.*, 1959; Jacobs and Strong, 1959; Welshons and Russell, 1959). It does this by initiating a pattern of gene expression that results in the differentiation of the gonadal primordia into testes rather than ovaries. The hormonal output of the testes then gives rise to the plethora of male-specific characteristics that subsequently develop (Capel and Lovell-Badge, 1993).

The small mouse Y chromosome is composed of a minute short arm (Yp) and a long arm (Yq). The telomeric end of Yq is known as the pseudoautosomal region and shares homology with a region of the X chromosome. This region is responsible for the obligatory pairing and exchange of the X and Y chromosomes during male meiosis. The pairing event is critical to ensure proper segregation of the sex chromosomes during spermatogenesis and is also responsible for the 1:1 sex ratio found in mammals. It is also essential for the progression of male meiosis (Burgoyne *et al.*, 1992). As it is very unusual for recombination to occur outside the pseudoautosomal region, the vast majority of the Y chromosome is not amenable to normal meiotic mapping. Thus the generation of mice harboring Y-chromosomal rearrangements has been invaluable to the molecular and genetic analysis of this chromosome. In particular, the sex-reversed mouse (*Sxr*) model has proved to be highly informative.

II. Sex Reversed (*Sxr*)

Sxr was first described by Cattanach *et al.* (1971) as an autosomally inherited form of sex reversal in which sterile XX males and XY carrier males and normal males and females are formed. Evidence that *Sxr* was not inherited autosomally came from the work of Singh and Jones (1982). From the banded krait snake these workers had isolated a female-specific satellite DNA sequence known as *Bkm* (banded krait minor satellite) (Singh *et al.*, 1980). Unlike mammals these snakes show female heterogamety. Southern analysis indicated that *Bkm*-homologous sequences in the mouse were Y linked. In addition, DNA from an XX sex-reversed male

(XX*Sxr*) showed the same pattern of *Bkm* hybridization as a normal XY male did. This strongly suggested that rather than being due to an autosomal mutation, sex reversal was in fact due to the inheritance of a fragment of the Y chromosome that must include its sex-determining function. This fragment is referred to as *Sxr*.

An explanation for the origin of *Sxr* became apparent when the region of the Y chromosome giving rise to *Sxr* was accurately defined (McLaren *et al.*, 1988; Roberts *et al.*, 1988). Roberts *et al.* (1988) made use of a single-copy sequence that recognizes a Y-specific band on Southern blots also present in DNA from XX*Sxr* male mice. *In situ* hybridization with this sequence as a probe has shown that the peak of hybridization on the normal Y chromosome corresponds to the very small short arm of the Y. An XY carrier male (XY*Sxr*) showed two peaks of signal, one defining Yp, and the other fragment at the telomere of Yq. As the mouse pseudoautosomal region is at the end of the long arm, this pattern is consistent with the presence of *Sxr* distal to the Y-chromosome pseudoautosomal region in XY*Sxr* carrier males. This suggests a simple explanation for the origin of *Sxr* involving the duplication and translocation of a large part (or all) of Yp to the distal tip of the Y pseudoautosomal region. During meiosis in a mouse carrying such a Y*Sxr* chromosome, the obligatory crossover within the X and Y pseudoautosomal regions will transfer *Sxr* to the distal tip of the X chromosome. As *Sxr,* and consequently the Y-chromosome sex-determining region, is now X-linked, mice inheriting such an X*Sxr* chromosome along with a normal X chromosome will be sex-reversed males.

Sxr includes a number of Y-linked functions including the structural gene for H-Y antigen or a gene controlling its expression (*Hya*) (Simpson *et al.*, 1981). H-Y antigen is a minor histocompatibility antigen (Eichwald and Silmser, 1955; Billingham and Silvers, 1960) considered a candidate for the testis-determining gene for a long time because its expression seemed always to be associated with maleness and the presence of the Y chromosome (Wachtel *et al.*, 1975). Its association with *Sxr* was considered support for this hypothesis.

Can functions mapping to the remainder of the Y chromosome be identified by studying the phenotype of XX*Sxr* males? XX*Sxr* males differ from normal males not only in lacking most of the Y chromosome but also in having two X chromosomes. In order to assess fully the phenotypic effects of the Y-chromosome deficiencies in sex-reversed males, it is necessary to eliminate this second variable. This is particularly important when analyzing spermatogenesis because the presence of two X chromosomes is incompatible with male germ cell survival beyond the perinatal period (Lyon, 1970; McLaren, 1983). *Sxr* males with single X chromosomes (XO*Sxr*) were first described by Cattanach *et al.*, 1971. Viable

XO*Sxr* mice can be produced by mating XO females with XY*Sxr* carrier males. Any XO*Sxr* germ cells produced by these mice can then be compared to XO germ cells. The fate of XO germ cells developing in a testis has been studied in mosaic male mice where an XO component has arisen by mitotic nondisjunction. XO germ cells present in XO/XY/XYY mosaic males show a severe spermatogenic impairment consistent with there being a block to spermatogonial proliferation such that only a few germ cells reach early meiotic prophase (Levy and Burgoyne, 1986). In contrast, all stages of spermatogenesis were represented in the testes of XO*Sxr* mice, although the later stages are severely depleted and the few sperm produced are often diploid and/or abnormal, such that the mice are sterile. These findings led Burgoyne *et al.* (1986) to suggest that *Sxr* carries a gene involved in spermatogenesis, which they have called *Spy*. The failure of XO*Sxr* germ cells to progress fully to functional sperm indicates that other Y-encoded sequences besides *Spy* are required for spermatogenesis (see later and Burgoyne, 1991).

In 1984, McLaren *et al.* discovered a variant of *Sxr* termed *Sxr′* that had retained sex-determining function. XX*Sxr′* mice were male but had lost the *Hya* gene; they did not express H-Y antigen. This finding eliminated *Hya* as a candidate for the testis determining gene. When XO*Sxr′* germ cells were analyzed they were found to have the same severe spermatogenic block that XO germ cells have. This indicated that *Sxr′* lacked *Spy* as well (Burgoyne *et al.*, 1986). Southern analysis with a molecular probe detecting several loci within *Sxr* suggested that *Sxr′* was a deleted version of *Sxr* (Bishop and Mitchell, 1991). Sex determination is therefore the only gene function known to be encoded by *Sxr′*. (*Sxr* and *Sxr′* are now formally referred to as Sxr^a and Sxr^b, respectively.)

A. Molecular Characterization of *Sxr*

Besides the functions described above that are associated with the *Sxr* fragment, a number of cloned genes have also been mapped to this region. One of these, known as *Sry* (sex-determining region Y chromosome), is the only Y-linked sequence that is both necessary and sufficient to cause male development and thus accounts for the sex-determining function of the Y chromosome and the *Sxr* fragment (Gubbay *et al.*, 1990; Koopman *et al.*, 1991b). The *Sry* gene is discussed in greater detail in Section IV.

Zfy-1 and *Zfy*-2 are two genes, mapping to *Sxr,* that are highly related at the sequence level. They code for proteins containing zinc finger DNA-binding domains (Page *et al.*, 1987; Mardon *et al.*, 1989). In the mouse there is a closely related X-linked homolog (*Zfx*) and a related pseudogene on chromosome 10 (*Zfa*) (Nagamine *et al.*, 1989; Ashworth *et al.*, 1990).

Another pair of related sequences mapping to *Sxr* have been called *A1S9Y-1* and *A1S9Y-2*. *A1S9* (Zacksenhaus and Sheinin, 1988, 1990a; Brown *et al.*, 1989) is a human X-linked gene that complements a temperature-sensitive cell cycle mutation in mouse L cells. It encodes the ubiquitin-activating enzyme, E1 (Pelliniemi, 1975; McGrath *et al.*, 1990; Handley *et al.*, 1991), which catalyzes the first step in ubiquitin-mediated protein degradation in eukaryotic cells. Kay *et al.* (1991) and Mitchell *et al.* (1991) have shown that two Y-linked mouse homologs of this gene map to the *Sxr* portion of the mouse Y chromosome. These genes have been renamed *Ube1-y1* and *Ube1-y2*.

The presence of two *Zfy* genes and two sequences related to *A1S9* is due to a duplication event that occurred during rodent evolution (Simpson and Page, 1991) and encompassed the single ancestral copies of both genes. Simpson and Page (1991) have also shown that *Sxr*^{b} is indeed a modified version of *Sxr*^{a} carrying a deletion and that this deletion was generated by unequal crossing-over within homologous sequences in each copy of the duplication. Interestingly the breakpoints of this deletion map within the *Zfy* genes themselves such that *Sxr*^{b} contains a fusion gene consisting of the 5′ promoter region of *Zfy*-2 fused to the coding region of *Zfy*-1. Mitchell and Bishop (1992) further analyzed the region deleted in *Sxr*^{b} by initiating a chromosome walk from three loci known to be deleted in *Sxr*^{b}. They obtained three nonoverlapping contigs covering, in total, a region of 270 kb.

Two of these contigs (defined by probes Zf2 and T5) map to each end of the deleted region as each contig includes part of a *Zfy* gene, and the remaining contig (defined by probe *Sx1*) maps to the central portion of the deletion. Based on the length of these contigs and an estimate of the length of a *Zfy* gene, the minimum amount of DNA deleted in the *Sxr*^{b} fragment is thought to be 320 kb. The nature of the three contigs has given some indication of the complexity of the mouse Y chromosome in this region. Whereas the *Sx1* contig defines mostly single-copy loci, the other contigs reveal the presence of multiple homologous loci throughout the *Sxr*^{a} region that are not deleted in *Sxr*^{b}. Indeed some probes from these contigs define regions that have homologs not just in *Sxr* but on other parts of the Y chromosome and on the X or autosomes. The authors speculate that these may be insertion elements that could have mediated some of the multiple rearrangements that have given rise to the present day mouse Y chromosome (Fig. 1).

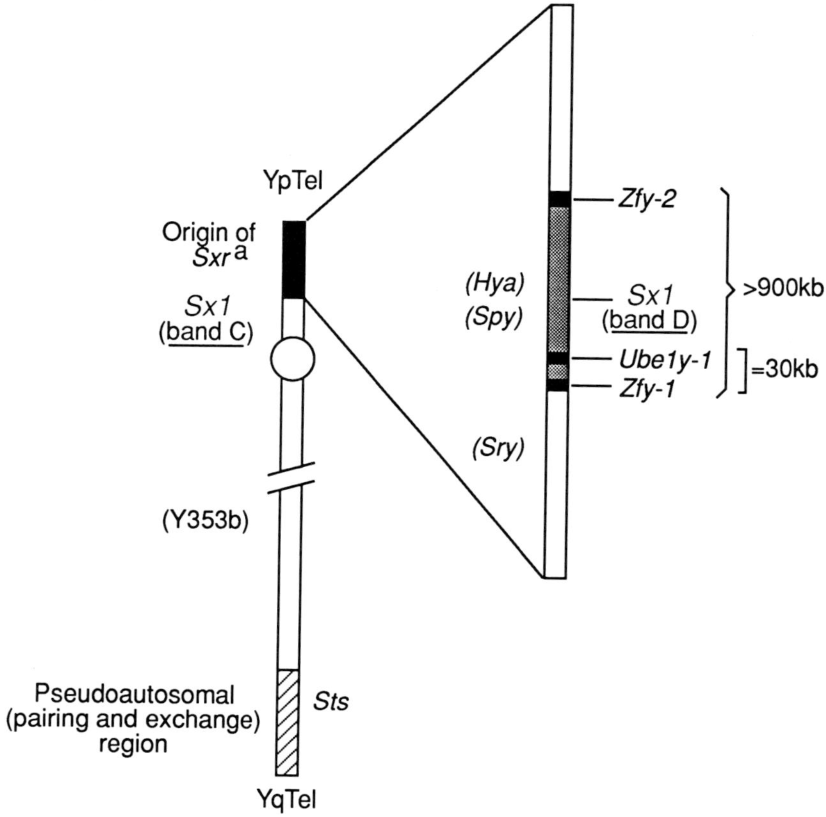

Figure 1 The mouse Y chromosome. Genes and sequences mentioned in the text are indicated on a cartoon of the mouse Y chromosome. The position of those in parentheses is not precisely mapped. For example, *Sry* has not been linked to *Zfy*-1, but it is thought to lie between *Zfy*-1 and *Sx1* repeat sequences close to the centromere.

Sry does not appear to have been duplicated on the mouse Y chromosome. In addition, the fact that it is retained in *Sxr*^{b} indicates that it does not map to the region between *Zfy*-1 and *Zfy*-2. However, the *Sry* open reading frame has been shown to reside within a long (>15.5 kb) inverted repeat sequence (Gubbay *et al.*, 1992). This unusual structure is found in other *Mus* species such as *M. spretus* (N. Vivian, unpublished), but is probably not a general feature of *Sry* genes, being absent from the human and rabbit loci.

B. Speculation on the Functions of Genes Mapping to *Sxr*

1. The *Zfy* Gene Family

Although the function of the *Zfy* gene family still remains elusive, some clues may be obtained by considering the expression patterns of the genes, especially in terms of differences between humans and mice. *ZFY* and *ZFX*, the two members of this family in humans, and the mouse *Zfx* gene, are expressed in essentially all cells. This suggests a general cellular function rather than a role in a specific developmental event (Lau and Chan, 1989; Schneider-Gädicke *et al.*, 1989; Ashworth *et al.*, 1990). In contrast, mouse *Zfy*-1 and *Zfy*-2 have a restricted pattern of expression. *Zfy*-1 expression is dependent on the presence of germ cells in the adult and the embryo, whereas *Zfy*-2, which also appears to be expressed in male germ cells, only becomes abundant in the testis between 7 and 14 days postnatally (Koopman *et al.*, 1989; Nagamine *et al.*, 1989, 1990). In addition, Koopman *et al.* (1989) have shown that *Zfy*-1 but not *Zfy*-2 is expressed in embryonic stem (ES) cells. The differences in the expression patterns of the two genes can be interpreted in several ways. The two genes may have diverged sufficiently from a single ancestor, so that they have distinct functions associated with their respective sites of expression. Alternatively, they may play identical roles but their regulatory sequences have diverged, resulting in overlapping but slightly different patterns of expression. Mardon and Page (1989) have proposed that *Zfy*-1 and *Zfy*-2 may be functionally interchangeable on the basis of their similar sequences.

Although the pattern of expression of the *Zfy* gene family is different in man and mouse, if all these genes are considered to be interchangeable, then equivalent dosage of the family as a whole is maintained between the sexes in both cases, albeit in different ways. In humans, where *ZFX* has been found not to undergo X inactivation, *ZFY* is ubiquitously expressed (Schneider-Gädicke *et al.*, 1989), thus ensuring that two *ZFY*-like genes are functional in every cell in both sexes, whereas in mouse, where *Zfx* does undergo X inactivation, the *Zfy* genes (and *Zfa*) are not expressed in the vast majority of cells (Koopman *et al.*, 1989; Ashworth *et al.*, 1991). In fact, *Zfy*-1 expression seems to correlate with cell types that in the female would not show X inactivation, such as ES cells and germ cells within the fetal gonad. It is possible that expression occurs in these sites simply to compensate dosage between males and females.

Alternatively, selective pressure to maintain these genes on the Y chromosome during mammalian evolution suggests a distinct male-specific function. The germ cell-dependent expression of the *Zfy* genes may reflect a role in spermatogenesis. The spermatogenesis gene *Spy* is present in *Sxr* but deleted in *Sxr*^{b}. Given that *Sxr*^{b} carries a *Zfy*-1/2 fusion gene in which the normal function of either or both of these genes may have been disrupted, both genes can be considered candidates for *Spy*.

2. *Ube1y-1*

Ube1y-1 represents yet another candidate gene waiting to occupy the mantle of *Spy*. It maps to *Sxr*^{a} and is absent from the *Sxr*^{b} fragment (in contrast, *Ube1y-2* is retained within *Sxr*^{b} but has been shown to be a nonfunctional pseudogene (Tucker *et al.*, 1992). The expression of *Ube1y-1* is testis-specific, but it is absent from the testes of XX*Sxr* mice which lack germ cells (Burgoyne *et al.*, 1986), implying that its expression is germ cell dependent. Given its good sequence homology to human *A1S9*, which has been shown to be essential for normal cell cycle progression (Mayer *et al.*, 1990; Zacksenhaus and Sheinin, 1990b), it has been suggested that *Ube1y-1* may play a similar role in male germ cells in the mouse. If *Ube1y-1* is *Spy*, its postulated role in the cell cycle could explain the block in spermatogonial proliferation that occurs in its absence. Categorical proof of the nature of *Spy* will emerge only when (and if) one or more of the candidate genes are shown to be able to complement the *Sxr*^{b} defect in transgenic mouse experiments. A homolog of *Ube1y-1* is also present on the marsupial Y chromosome (Mitchell *et al.*, 1992) in contrast to the marsupial homologs of *Zfy*, which are autosomal (Sinclair *et al.*, 1988). The Metatherian (marsupial) and Eutherian (placental) Y chromosomes must have diverged from each other some 120–150 million years ago. This extreme conservation of *Ube1y-1* indicates that it may play a critical role in males. The only other gene for which such conservation has been observed is *Sry* (Foster *et al.*, 1992). However, no homolog of *Ube1y-1* can be detected on the human Y chromosome (Mitchell *et al.*, 1991). Mitchell *et al.* (1992) therefore proposed that in this case the activity of the X-linked allele may compensate for the absence of a Y-linked gene.

3. Other Potential Functions

The remaining function known to map to *Sxr* and be absent in *Sxr*^{b} is *Hya*. Simpson *et al.* (1987) have shown that the human *HYA* and *ZFY* loci are genetically separable, and it seems reasonable to presume that the same must be true in the mouse. In addition, the restricted expression pattern of the *Zfy* genes is inconsistent with the widespread detection of H-Y antigen, which presumably reflects a widespread expression of *Hya*. The latter argument can also be used to preclude *Ube1y-1* as being a candidate for *Hya*.

Tsunoda *et al.* (1985) have reported a growth rate difference between XX and XY preimplantation embryos. Burgoyne (1991, 1993) has shown that this effect depends on the strain of origin of the Y chromosome and has therefore postulated the existence of a Y-linked gene responsible for accelerating the early development of XY blastocysts. The location of this gene on the Y chromosome is unknown. However, the finding that *Zfy*-1 is expressed in ES cells, which are derived from and approximate to cells of the inner cell mass of the blastocyst, would be consistent with it being responsible for this effect. This hypothesis could be pursued by correlating the growth rate variations associated with Y chromosomes exhibiting differences in *Zfy*-1.

III. The Role of Yq in Spermatogenesis

The long arm of the mouse Y chromosome has not been subjected to the same level of molecular analysis as the short arm in the form of the *Sxr* rearrangement. However, this section provides evidence that the long arm is critical for spermatogenesis in the mouse. XO Sxr^{a} mice that do not lack *Spy* are nevertheless sterile. The defect in spermatogenesis manifests itself between diplotene and the second meiotic metaphase such that almost no round spermatids are produced. This observation suggests that Yq may be involved in the latter stages of spermatogenesis. Miklos (1974) and subsequently Burgoyne (e.g., see Sutcliffe *et al.*, 1991) proposed that the presence of unpaired sex chromosomes may be deleterious to spermatogenesis. It has been possible to assess the role of Yq in providing a pairing partner during spermatogenesis due to the availability of a rearranged Y chromosome consisting essentially of the centromere attached to the pairing region (known as Y^{*x}) (Eicher and Washburn, 1983). Burgoyne *et al.* (1992) generated X*Sxr* Y^{*x} mice and found that the sperm counts and testis weights of these animals are near normal, confirming the importance of X–Y pairing in spermatogenesis. However, the sperm produced by these mice all had abnormal heads, indicating that Yq must encode another function related to sperm head morphology. This property of Yq has been investigated by an analysis of mice with partial deletions of Yq. Moriwaki *et al.* (Styrna *et al.*, 1991) and subsequently Conway *et al.* (1993) have shown that partial long arm deletion increases the frequency of abnormal sperm but does not have an all or nothing effect on spermatogenesis. This indicates that the information required for sperm head morphology may be present in multiple copies along the length of Yq. Bishop and Hatat (1987) identified a multiple-copy long arm sequence that is transcribed in adult testis—known as Y353B. Conway *et al.* (1993) went on to show that this transcript is found in round spermatids and that the level of transcription is sensitive to the number of copies of Y353B present. Thus Y353B is a candidate for the gene involved in sperm head morphology.

IV. What Can We Learn from the Human Y Chromosome?

The most striking feature of a comparison between the mouse and the human Y chromosomes is the catalog of differences to be found. These differences may provide clues to the evolution of the sex chromosomes and explain certain biological features that differ between man and mouse.

The human pseudoautosomal region is at the tip of the short arm (Pearson and Bobrow, 1970; Chen and Falek, 1971; Ellis and Goodfellow, 1989) in contrast to its location in mouse at the tip of the long arm. If the function of this region is simply to provide a site for X–Y pairing during male meiosis, it may not depend on the specific sequences. Indeed this may explain the differences in pseudoautosomal gene content between man and mouse. The only gene known to map to the mouse pseudoautosomal region is steroid sulfatase (*STS*). As expected for a gene from this region, a second copy has been found on the mouse X chromosome, and both genes are active (Keitges *et al.*, 1985). The long arm of the human Y chromosome encodes an *STS* pseudogene, the functional homolog being X-linked (Yen *et al.*, 1988). It is possible that *STS* was once pseudoautosomal in human ancestors as well as in the mouse, but has become relocated to the long arm by an inversion event on the human Y chromosome. Two genes have been definitively shown to be pseudoautosomal in humans. These are the receptors for the granulocyte–macrophage colony-stimulating factor (*GM-R*) (Gough *et al.*, 1990) and the structural gene for the cell surface antigen (*MIC2*) (Darling *et al.*, 1986), which is involved in T-cell adhesion (Gelin *et al.*, 1989).

The organization of the human and mouse Y chromosomes also differs with respect to *Sry*. In the case of humans, *SRY* has been shown to lie a mere 5 kb away from the pseudoautosomal boundary. Given that aberrant crossover events can sometimes transfer Y-specific sequences that are near the boundary to the X chromosome, giving rise to *SRY*-positive XX males, the location of *SRY* seemed ideal for those attempting to isolate the gene by positional cloning, but somewhat precarious in terms of its normal function. Although *Sry* is essentially as far away from the mouse pseudoautosomal region as possible, the existence of the *Sxr* mutation illustrates that even in this case *Sry* can become X-linked!

Many of the other Y-linked genes described in this chapter differ in their organization between man and mouse. The mapping data are summarized in Table I. Table I also includes a gene not previously mentioned in this text, known as *RPS4Y*, that encodes a ribosomal protein and has functional copies on the human X and Y chromosomes (Fisher *et al.*, 1990). Although it does have a homolog on the mouse X chromosome, *Rps4X*, there is no Y-linked copy in the mouse. This situation seems to mirror that of *Ube1y-1* which has an X-linked homolog in humans but lacks an equivalent on the human Y chromosome.

As discussed in a previous section, the differences in the expression pattern of the mouse *Zfy* genes and human *ZFY* (the former being expressed only in testis and the latter in all cells), and the fact that human *ZFX* escapes X-inactivation whereas mouse *Zfx* does not, can be rationalized by noting that this allows man and mouse to achieve an equal "*Zf*" gene dose between males and females. The same pattern of X-inactivation characterizes the human and mouse *Ube-X* genes [the latter being subject to inactivation whereas the former escapes it (Kay *et al.*, 1991)]. Given that *Ube1y-1* is transcriptionally inactive in all tissues except testis, this would lead to a balanced dose of the genes between both sexes in the mouse. Paradoxically, the absence of a Y-linked homolog in man coupled with a failure of *UBE-X* to respond to X-inactivation can potentially lead to females receiving two *UBE-X* doses versus only a single dose in males. However, it remains possible that a highly diverged homolog of *UBE-X* or even a different gene altogether that can functionally complement *UBE-X* may reside on the human Y chromosome.

Ashworth *et al.* (1991) have shown that mouse *Rps4X* is subject to X-inactivation in contrast to its human homolog (Fisher *et al.*, 1990). In this case equivalent gene dose between males and females in the mouse is ensured by the absence of a Y-linked gene.

In two of the cases described above (*Rps4Y* and *Zfy*) a common theme is that the Y-linked member of each of these gene families either is absent in

Table I

	Marsupial	Mouse	Human
SRY	+	+	+
ZFY	−	++	+
RPS4Y	?	−	+
Ube1Y-1	+	+(+)	−
HYA	?	+(Y^p)	+(Y^q)

the mouse or fails to be expressed in the vast majority of cells of the organism. One consequence of this is that an XO mouse will express equivalent amounts of these genes in both normal males and females. In contrast, an XO human will suffer haploid insufficiency for these gene products, as in XX females neither gene undergoes inactivation and in XY males the Y-linked copy of each gene is ubiquitously expressed. XO mice are perfectly viable, whereas 99% of human XO fetuses do not survive to birth and those that do suffer from a range of stigmata known as Turner syndrome (Turner, 1938; see Chapter *14*). The difference in the way dosage compensation is achieved for these genes between man and mouse may provide a molecular explanation for the different viability of XO mice compared to humans. Indeed, Fisher *et al.* (1990) have suggested that *RPS4X* may be the gene responsible for Turner syndrome.

If we compare genes known to be present on the Y chromosomes of marsupials, mice, and man, *Sry* and perhaps *Hya* are the only genes that appear to have an obligatory location on the Y chromosome. *Hya* remains to be characterized molecularly, and its function, except as a transplantation antigen, has yet to be defined. We are however beginning to understand how *Sry* acts in sex determination.

V. The *Sry* Gene

A. *Sry* is Sex Determining

There is now a large body of evidence proving that *Sry* is equivalent to the mouse Y-linked testis-determining gene (formerly known as *Tdy*). In particular, two lines of work have provided direct proof of this equivalence: Harley *et al.* (1992) demonstrated an association of mutations within the coding region of *SRY* (the human homolog of *Sry*) with male-to-female sex reversal in humans, whereas Koopman *et al.* (1991b) demonstrated female to male sex reversal in XX transgenic mice carrying the *Sry* gene in the form of a 14.6-kb genomic fragment.

B. *Sry* Expression

In order to understand the role played by *Sry* in testis determination it is necessary to consider its expression pattern in the context of how the gonads develop. The genital ridges arise as a thickening on the ventral–medial surface of each mesonephros (primitive kidney), from about 10 days postcoitum (dpc). They increase in volume but remain indifferent, i.e., morphologically identical between the two sexes, until 11.5 dpc. Between 11.5 and 12.5 dpc, the developing testis takes on a characteristic stripey appearance, due to the differentiation and alignment of Sertoli cells into epithelial testis cords. No overt sign of differentiation occurs in the developing ovary until several days later when follicle cells organize around oocytes. Koopman *et al.* (1990) have shown that *Sry* is expressed for a limited period between 10.5 and 12.0 dpc. This expression is in cells of the genital ridge as it develops, with no sign of expression beforehand at 9.5 dpc and only traces sometimes present at 12.5 dpc, when testis cord formation is apparent. The transient nature of the *Sry* expression suggests that *Sry* initiates testis development but is not required for long-term maintenance of testis-specific gene expression.

The mammalian gonad is thought to be derived from four cell lineages, each of which is effectively bipotential (for example, see Lovell-Badge, 1993). These comprise the germ cells, which migrate into the genital ridge between the 10th and 11th day of development, and three somatic cell types: (*a*) the supporting cell precursors, which give rise to Sertoli cells in the male and follicle cells in the female; (*b*) steroid-producing cell precursors, which give rise to Leydig cells in the male, theca or interstitial cells in the female; and (*c*) connective tissue cells, which include those that give vascular tissue (particularly prominent in testis), tunica, and peritubular myoid cells in the testis, and tunica and stromal cells in the ovary.

Testes develop normally in embryos homozygous for mutations at the *W* (or white spotting) locus, which compromise the ability of germ cells to proliferate and colonize the genital ridges (Mintz and Russell, 1957). The germ cells are therefore irrelevant to testis determination. Expression of *Sry* is unaffected in *W/W* mutant embryos, confirming that *Sry* is expressed by one or more of the somatic cell types (Koopman *et al.*, 1990).

Of the three somatic cell lineages, the supporting cells appear to be the most critical for testis determination. Burgoyne *et al.* (1988) and Palmer and Burgoyne (1991b) looked at the sex chromosome constitution of cells within the testes of XX ↔ XY chimeras and found that Sertoli cells were almost exclusively XY, whereas the proportion of XX and XY cells in the other gonadal lineages was the same as that found elsewhere in the embryo. This suggested that the testis-determining gene acts cell-autonomously within the supporting cell precursors. Presumably these are the cells that express *Sry*, and this expression is responsible for their differentiation along the Sertoli cell pathway.

However, some additional data suggest that *Sry* expression alone is not sufficient for Sertoli cells to complete their differentiation. Buehr *et al.* (1993) have shown that 11.5 dpc male genital ridges isolated from their mesonephros fail to organize to give testis cords. It appears that the organization of pre-Sertoli cells into cords is dependent on a population of cells that migrate into the gonad from the mesonephros.

Equally, the finding that XY follicle cells can be present in female XX ↔ XY chimeras (these embryos develop as females if the proportion of XY cells is low) also suggests that *Sry* cannot exert its effect solely by cell autonomous action. Thus perhaps *Sry* acts to initiate Sertoli cell differentiation, the success of which would then depend on receiving appropriate and sufficient signals from either the pre-Sertoli cells themselves or another population, such as that derived from the mesonephros.

Sry is also expressed in the adult mouse testis, where a transcript of about 1.3 kb is seen on Northern blots. This expression is dependent on the presence of germ cells and probably occurs in round spermatids (Koopman *et al.*, 1990; Capel *et al.*, 1993). Recent data show that this transcript is a circular RNA molecule that has probably arisen from an unusual splicing event. This transcript is clearly different from that in the genital ridge and is probably not translated (Capel *et al.*, 1993).

1. The Interaction of SRY protein with DNA

SRY was one of the first members of what is now a very large and still expanding group of DNA-binding proteins that possess a domain referred to as an HMG box. This protein motif was first described as a region of homology between the high mobility group nonhistone protein HMG1, and hUBF, a cofactor of RNA polymerase I required for ribosomal RNA transcription (Jantzen *et al.*, 1990). Members of this protein family fall into two groups: (*a*) those that are expressed ubiquitously, bind DNA with little or no sequence specificity, and tend to have more than one HMG box, and (*b*) those that show a restricted tissue distribution, bind to specific sequences at high affinity, and have only a single HMG box (Ner, 1992). HMG1 and hUBF are characteristic of the first group. HMG1 is a highly abundant and ubiquitous chromatin-associated protein that binds linear DNA with low affinity and without apparent sequence specificity (Bianchi *et al.*, 1992). Despite being associated with its target genes, hUBF has no clear consensus DNA sequence motif to which it binds, and its linear interaction with DNA is fairly weak (Jantzen *et al.*, 1990). SRY is characteristic of the second class, showing both tissue and sequence specificity. Two other well-characterized members of this class are TCF1 and LEF1 (or TCF1α), which are expressed in lymphoid cells and are involved in transcriptional regulation of, for example, T-cell-specific genes such as *CD3* and *TCR*α (Travis *et al.*, 1991; van de Wetering *et al.*, 1991). It is therefore also likely that SRY functions to regulate the expression of specific target genes involved in gonadal differentiation (Fig. 2).

In vitro studies have shown that the HMG box, which is approximately 79 amino acids in length, is the region of interaction with DNA for all of these factors and is sufficient to confer the sequence-specific binding of hUBF, LEF1 and Sry (Jantzen *et al.*, 1990; Giese *et al.*, 1991; Nasrin *et al.*, 1991). Both SRY and TCF1 have been shown to bind the same target sequence, AACAAAG. However, more recent studies indicate that the sequence AACAAT is the highest affinity site for human and mouse SRY (Denny *et al.*, 1992; Giese *et al.*, 1992; Harley *et al.*, 1992; V. R. Harley *et al.*, unpublished results).

More detailed studies of the mechanism of HMG box DNA-binding have been undertaken by two groups. Grosschedl and colleagues (Giese *et al.*, 1992) and subsequently Bianchi and co-workers (Ferrari *et al.*, 1992) have found that LEF1 and SRY, and probably other members of this class, induce a dramatic bend in target DNA on binding. This bend is on the order of 130°, which is more than any other known transcription factor. Only IHF, the *E. coli* integration host factor is known to bend DNA by a similar degree (140°). Also, like IHF, Sry and LEF1 contact DNA in the minor groove, which is unusual compared to known transcription factors (Giese *et al.*, 1991; van de Wetering *et al.*, 1991).

HMG1 binds to four-way DNA junctions and cruciform DNA struc-

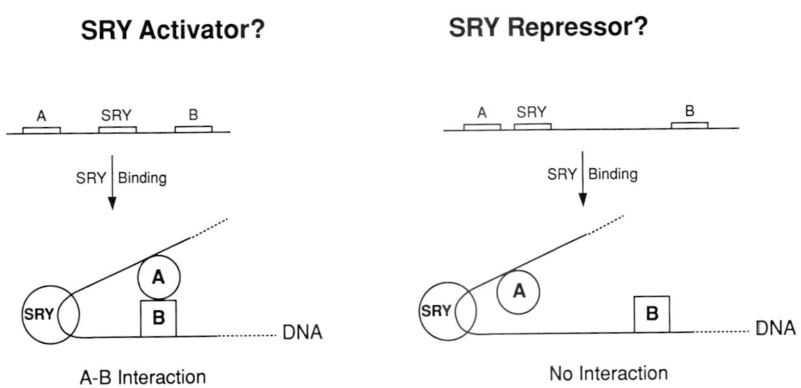

Figure 2 SRY as a modulator of gene expression. Two theoretical models of how the action of SRY on DNA could influence the interaction of other transcription factors. The bending induced by SRY could allow two other proteins to interact, and this in turn could initiate transcription of the target gene. Alternatively the bending could stop the interaction of the two factors, which in turn would prevent transcription. There are many other possible ways this type of interaction could occur, including those where the non-HMG box domains of SRY participate in the interaction with the other factors.

tures at much higher affinities than it does to linear DNA. These cruciform structures, which may be made artificially by using four synthetic oligomers, also form at inverted repeats and at Holliday junctions during recombination. They may also resemble DNA structures present in higher order chromatin. The cruciforms produced *in vitro* have two angles of 60° and two of 120°, and HMG1 is thought to bind to the former, which would be to the bend induced by SRY and LEF1 in linear DNA (Bianchi *et al.*, 1992). It was also found in these studies that SRY will bind the same cruciform structures as HMG1, without sequence specificity and with affinities as high as its binding with AACAAT motifs in linear DNA (Ferrari *et al.*, 1992). It may be that all HMG boxes recognize and bind structured DNA, but the binding of linear DNA in a sequence-specific manner, and its subsequent bending, is an extra property of the SRY/LEF1 class.

Although SRY appears to bind the same structures as HMG1, it seems unlikely that it would so within the context of the nucleus. HMG1 is an abundant protein and can be assumed to occupy all such sites. SRY is therefore likely to bind only its free specific target sequences in linear DNA. Until target genes are identified it is only possible to speculate how the dramatic bend, induced in DNA by SRY may affect transcription, but a comparison with LEF1 may be pertinent. Cotransformation studies with *LEF1* and a target gene construct show that it has weak activation properties by itself (Giese *et al.*, 1992). However, the LEF1 binding site is flanked by sites for two other transcription factors. The bend in DNA induced by LEF1 binding appears to bring these two sites together, allowing the factors bound to them to interact. This interaction then activates transcription. The distance between the sites for all three factors is critical, as is the degree of bending induced by LEF1. One could equally imagine schemes by which the bending prevented the binding or interaction of other factors, leading to target gene repression.

HMG box-containing factors, such as *Sry*, could therefore facilitate or prevent the action of other specific transcription factors on target gene expression, while having no direct influence by themselves on RNA polymerase activity. This could provide a molecular explanation for why *Sry* is not sufficient to induce Sertoli cell differentiation during normal testis determination.

C. Transgenic Mouse Studies

The ability of an *Sry* transgene to cause male development in chromosomally female transgenic embryos provided the formal proof that *Sry* is the Y-encoded testis-determining gene (Koopman *et al.*, 1991b). It also showed that no other gene specific to the Y chromosome is required for normal male development. The XX *Sry* transgenics are full phenotypic males, although they are sterile as expected, due to the presence of two X chromosomes, and the absence of Y-encoded sequences required for spermatogenesis, such as *Spy*. In these transgenic experiments the 14.6-kb *Sry* genomic fragment failed to cause sex reversal in all cases. In fact only about 25% of XX transgenic mice develop as males. One explanation for this failure could be mosaicism for the presence of the transgene due to integration after the first cleavage division, a common finding in transgenics. This would lead to *Sry* being expressed in an insufficient number of supporting cell precursors. Alternatively, it is known that transgenes often show very different levels of expression from one line to another, probably due to the effect of genomic sequences at the site of integration. Since the purpose of these experiments was to assay function and not simply tissue-specific expression, the timing or level of expression of the *Sry* transgene may be ineffective in some cases. The study of one particular transgenic line, termed 32.10 (N. Vivian, P. Koopman, and R. Lovell-Badge, unpublished), is consistent with this explanation and may reveal important details about the regulation of *Sry*. The founder animal was a fertile female that transmitted the transgene to normal daughters and sons, ruling out mosaicism as a possible reason for the failure to sex reverse. Subsequent breeding of these animals with wild-type or with other transgenic siblings has led to instances of sex reversal, as well as several cases of intersex offspring. Preliminary results showed that *Sry* is being expressed in the genital ridge of transgenic XX embryos from this line at 11.5 dpc. The level of expression appeared lower than normal and careful measurement of the number of transcripts over a time course of development is required. But it seems likely that *Sry* expression is at a threshold level, and homozygosity for the transgene increases the incidence of sex reversal.

The high incidence of XX female transgenic mice, whose developmental fate was not perturbed by the presence of *Sry*, and the data concerning transgenic line 32.10 complement a growing body of evidence suggesting that testis determination is exquisitely sensitive to the level of *Sry* transcripts present in the embryo, and that this level is itself sensitive to a variety of effects. McLaren and Monk (1982) demonstrated that in T16H/X*Sxr*^{a} mice in which the *Sxr*^{a} fragment is carried on an X chromosome that is subject to X-inactivation in all cells, female development can occur. This indicates that *Sry* expression may be sensitive to the effects of X-inactivation spreading into the *Sxr*^{a} fragment. This is in contrast to *Hya*, which continued to be expressed (McLaren *et al.*, 1984; Goldberg, 1988). Female development did not result in all cases, with males and hermaphrodites also being observed. If the effects of X-inactivation on *Sry* are somewhat variable, male development would occur in cases where the gene was not sufficiently shut down in all or some cells. In the case of hermaphrodites, the level of *Sry* is presumably very close to its required threshold resulting in the intermediate phenotype.

In humans, hermaphroditic or intersex development has been observed in some XX individuals carrying very short lengths (~35 kb) of Y-derived sequences, including *SRY*, attached to an X chromosome (Palmer *et al.*, 1989). In these cases the extreme proximity of *SRY* to the breakpoint between Y-derived and X-derived sequences may make its activity highly sensitive to position effects at the new locus and to any spreading of X-inactivation. In contrast, complete male development tends to occur in XX individuals that have inherited much larger amounts of Y-derived DNA containing *SRY*, suggesting that these additional Y-linked sequences can act as a buffer, protecting *SRY* expression from the effect of its new chromosomal position. In the *Sry* transgenic mice this buffer is clearly lacking, with the gene being associated with very little of its endogenous flanking sequences.

Another instance in which female development can occur in the presence of an intact *Sry* gene has emerged from the recent studies of a line of XY female mice (Capel *et al.*, 1993). The molecular defect in these mice is the loss of a number of copies of the Yp-specific repeat sequence known as *Sx1*. This deletion seems to have abolished *Sry* expression despite its considerable distance from the gene.

There is also evidence to suggest that the timing of *Sry* expression is critical. Eicher *et al.* (1982) showed that when the Y chromosome from *Mus musculus poschiavinus* (referred to here as Y^{POS}) was bred into the C57BL/6 background, sex-reversed females, hermaphrodites, and some XY males were produced. Eicher and Washburn (1986) demonstrated that sex reversal not only depended on the presence of Y^{POS} but also required C57BL/6 alleles to be present at a number of other loci (as yet undefined), which must be autosomal. The failure of Y^{POS} to induce testis differentiation on the C57BL/6 background can be explained by a timing mismatch hypothesis according to which the Y^{POS} *Sry* gene acts too late, such that ovary determination preempts testis determination. In this scheme the autosomal loci described by Eicher and Washburn (1986) may be involved in ovary determination with the C57BL/6 allele being early acting with respect to Y^{POS}. Palmer and Burgoyne (1991a) tested this hypothesis by showing that in the presence of the Y^{POS} chromosome, testis development is delayed by 14 h, suggesting that timing mismatch is indeed involved in this form of sex reversal. The simplest molecular interpretation of these data is that Y^{POS} encodes an *Sry* gene that is expressed later than that normally associated with C57BL/6, or that the level of expression is low, such that the critical threshold is reached later. This brings us back to the discussion of transgenic line 32.10, in which homozygosity for the *Sry* transgene appears to increase the incidence of sex reversal, implying that two copies of the transgene are required to exceed the required threshold for *Sry*. However, not all the homozygotes are male and some heterozygotes are, which suggests that genetic background effects may be segregating in these mice which have a mixed C57BL/10 and CBA origin. Could it be that, due to position effects, *Sry* expression from this particular transgene is delayed, such that sex reversal is dependent on the presence of the correct combination of sex determining alleles at other loci in the genome?

Transgenic mice have also been generated using genomic fragments carrying the human *SRY* gene. Koopman *et al.* (1991b) and R. D. Palmiter (personal communication) have found no evidence of sex reversal despite apparently normal expression of the transgene. The mouse and human SRY proteins both bind the same target DNA sequences *in vitro* and appear to induce similar degrees of bending. This suggests that further specificity must be achieved *in vivo*.

Many of the HMG box-containing genes have other domains that are well conserved among different species. Such sequences could be involved in transcriptional activation and interaction with other proteins. However, when mouse and human SRY sequences are compared there is no homology outside the HMG box, a feature evident when SRYs from any two species are compared. In addition, although the HMG box clearly represents a conserved domain between SRY proteins from different species, the degree of homology within the box is much less than that found when other HMG box-containing proteins are compared. A comparison of all SRY-related HMG box-containing genes (these so-called *Sox* genes are located on chromosomes other than the Y) in fact reveals no particular amino acid residues that are characteristic of the SRY proteins (Foster *et al.*, 1992). With respect to failure of the human *SRY* gene to function in transgenic mice, three possible explanations are worth exploring: (*a*) the variation within the HMG box is significant and the HMG boxes of human and mouse *Sry* interact with target sequences in subtly different ways not detectable *in vitro;* (*b*) the variation between the HMG boxes affects interaction with other proteins and not DNA; (*c*) this variation is not significant, and the failure of the human gene to work in mice is due to the variation in the non-HMG box parts of the protein. In any case, the data suggest that there is very rapid evolution of the SRY sequence, which could affect protein–DNA or protein–protein interactions. Indeed, comparisons among closely related species suggest that there may be selection for change in the SRY amino acid sequence (S. Whitfield *et al.*,

1993). Carrying out domain swap experiments between human and mouse genes in transgenic mice should help to resolve this problem and determine which parts of the protein are important.

The cloning of *Sry* has revealed the enormous influence a single gene can have in controlling the developmental fate of an organism. To understand this influence the challenge is now to identify the genes with which *Sry* interacts (positively or negatively) and thus to begin to unravel the pathway of Sertoli cell differentiation.

Acknowledgments

We are grateful to Blanche Capel for helpful comments and to Marilyn Brennan for invaluable help during the production of the manuscript.

References

Ashworth, A., Skene, B., Swift, S., and Lovell-Badge, R. (1990). Zfa is an expressed retroposon derived from an alternative transcript of the Zfx gene. *EMBO J.* **9,** 1529–1534.

Ashworth, A., Rastan, S., Lovell-Badge, R., and Kay, G. (1991). X-chromosome inactivation may explain the difference in viability of XO humans and mice. *Nature (London)* **351,** 406–408.

Bianchi, M. E., Falciola, L., Ferrari, S., and Lilley, D. M. J. (1992). The DNA binding site of HMG1 protein is composed of two similar segments (HMG boxes), both of which have counterparts in other eukaryotic regulatory proteins. *EMBO J.* **11,** 1055–1063.

Billingham, R. E., and Silvers, W. K. (1960). Studies on tolerance of the Y chromosome antigen in mice. *J. Immunol.* **85,** 14–26.

Bishop, C., and Hatat, D. (1987). Molecular cloning and sequence analysis of a mouse Y chromosome RNA transcript expressed in the testis. *Nucleic Acids Res.* **15,** 2959–2969.

Bishop, C. E., and Mitchell, M. J. (1991). The molecular biology of the mouse Y chromosome. *Semin. Dev. Biol.* **2,** 241–249.

Brown, C. J., Powers, V. E., Munroe, D. L., Sheinin, R., and Willard, H. F. (1989). A gene on the short arm of the human X chromosome complements murine tsA1S9 DNA synthesis mutation. *Somatic Cell Mol. Genet.* **15,** 173–178.

Buehr, M., Gu, S., and McLaren, A. (1992). Mesonephric contribution to testis differentiation in the fetal mouse. *Development (Cambridge, UK)* **117,** 273–281.

Burgoyne, P. S. (1991). Y-chromosome function in mammalian development. *Adv. Dev. Biol.* **1,** 1–29.

Burgoyne, P. S. (1993). A Y-chromosomal effect on blastocyst cell number in mice. *Development* **117,** 341–345.

Burgoyne, P. S., Levy, E. R., and McLaren, A. (1986). Spermatogenic failure in male mice lacking H-Y antigen. *Nature (London)* **320,** 170–172.

Burgoyne, P. S., Buehr, M., Koopman, P., Rossant, J., and McLaren, A. (1988). Cell autonomous action of the testis-determining gene: Sertoli cells are exclusively XY in XX XY chimaeric mouse testes. *Development (Cambridge, UK)* **102,** 443–450.

Burgoyne, P. S., Mahadevaiah, S. K., Sutcliffe, M. J., and Palmer, S. J. (1992). Fertility in mice requires X-Y pairing and a Y-chromosomal "spermiogenesis" gene mapping to the long arm. *Cell (Cambridge, Mass.)* **71,** 391–398.

Capel, B., and Lovell-Badge, R. (1993). The Sry gene and sex determination in mammals. *Adv. Dev. Biol.* **2,** 1–35.

Capel, B., Swain, A., Nicolis, S., Hacker, A., Walters, M., Koopman, P., Goodfellow, P and Lovell-Badge, R. (1993). Circular transcripts of the testis determining gene *Sry* in adult mouse testis. *Cell* (*Cambridge, Mass.*) **73,** 1019–1030.

Capel, B., Rasberry, C., Dyson, J., Bishop, C. E., Simpson, E., Vivian, N., Lovell-Badge, R., Rastan, S. and Cattanach, B. M. (1993). Sex reversal due to the deletion of Y chromosomal sequences located outside the testis determining region. *Nat. Genet.* (Submitted).

Cattanach, B. M., Pollard, C. E., and Hawkes, S. G. (1971). Sex reversed mice: XX and XO males. *Cytogenetics* **10,** 318–337.

Cattanach, B. M. *et al.* (1993). In preparation.

Chen, A., and Falek, A. (1971). Cytological evidence for the association of the short arms of the X and the Y in the human male. *Nature* (*London*) **232,** 555–556.

Conway, S. J., Mahadevaiah, S. K., Darling, S. M., Capel, B., and Burgoyne, P. S. (1993). Spermatid specific expression of a multiple copy Y chromosomal DNA sequence which may be required for development of the sperm head. *Genes Dev.* (submitted).

Darling, S. M., Banting, G. S., Pym, B., Wolfe, J., and Goodfellow, P. N. (1986). Cloning an expressed gene shared by the human sex chromosomes. *Proc. Natl. Acad. Sci. U.S.A.* **83,** 135–139.

Denny, P., Swift, S., Connor, F., and Ashworth, A. (1992). A testis specific gene related to SRY encodes a sequence- specific DNA binding protein. *EMBO J.* **10,** 3705–3712.

Eicher, E. M., and Washburn, L. L. (1983). Inherited sex reversal in mice: Identification of a new primary sex-determining gene. *J. Exp. Zool.* **228,** 297–304.

Eicher, E. M., and Washburn, L. L. (1986). Genetic control of primary sex determination in mice. *Annu. Rev. Genet.* **20,** 327–360.

Eicher, E. M., Washburn, L. L., Whitney, J. B. III, and Morrow, K. E. (1982). *Mus poschiavinus* Y chromosome in the C57BL/6J murine genome causes sex reversal. *Science* **217,** 535–537.

Eichwald, E. J., and Silmser, C. R. (1955). Untitled communication. *Transplant. Bull.* **2,** 146–149.

Ellis, N., and Goodfellow, P. N. (1989). The mammalian pseudoautosomal region. *Trends Genet.* **5,** 406–410.

Ferrari, S., Harley, V. R., Pontiggia, A., Goodfellow, P. N., Lovell-Badge, R., and Bianchi, M. E. (1992). SRY, like HMG1, recognizes sharp angles in DNA. *EMBO J.* **11,** 4497–4506.

Fisher, E. M. C., Beer-Romero, P., Brown, L. G., Ridley, A., McNeil, J. A., Lawrence, J. B., Willard, H. F., Bieber, F. R., and Page, D. C. (1990). Homologous ribosomal protein genes on the human X and Y chromosomes: Escape from X inactivation and possible implications for Turner syndrome. *Cell* (*Cambridge, Mass.*) **63,** 1205–1218.

Ford, C. E., Jones, K. W., Polani, P. E., de Almeida, J. C., and Briggs, J. H. (1959). A sex-chromosome anomaly in a case of gonadal dysgenesis (Turner's syndrome). *Lancet* **I,** 711–713.

Foster, J. W., Brennan, F. E., Hampikian, G. K., Goodfellow, P. N., Sinclair, A. H., Lovell-Badge, R., Selwood, L., Renfree, M. B., Cooper, D. W., and Graves, J. A. M. (1992). Evolution of sex determination and the Y chromosome: SRY-related sequences in marsupials. *Nature* (*London*) **359,** 531–533.

Gelin, C., Aubrit, F., Phalipon, A., Raynal, B., Cole, S., Kaczorek, M., and Bernard, A. (1989). The E2 antigen, a 32 kd glycoprotein involved in T cell adhesion processes, is the MIC2 gene product. *EMBO J.* **8,** 3253–3259.

Giese, K., Amsterdam, A., and Grosschedl, R. (1991). DNA-binding properties of the HMG domain of the lymphoid-specific transcriptional regulator LEF-1. *Genes Dev.* **5,** 2567–2578.

Giese, K., Cox, J., and Grosschedl, R. (1992). The HMG domain of lymphoid enhancer factor 1 bends DNA and facilitates assembly of functional nucleoprotein structures. *Cell (Cambridge, Mass.)* **69,** 1–20.

Goldberg, E. H. (1988). H-Y antigen and sex determination. *Philos. Trans. R. Soc. London, Ser. B* **322,** 73–81.

Gough, N. M., Gearing, D. P., Nicola, N. A., Baker, E., Pritchard, M., Callen, D. F., and Sutherland, G. R. (1990). Localization of the human GM-CSF receptor gene to the X-Y pseudoautosomal region. *Nature (London)* **345,** 734–736.

Gubbay, J., Collignon, J., Koopman, P., Capel, B., Economou, A., Mnsterberg, A., Vivian, N., Goodfellow, P., and Lovell-Badge, R. (1990). A gene mapping to the sex-determining region of the mouse Y chromosome is a member of a novel family of embryonically expressed genes. *Nature (London)* **346,** 245–250.

Gubbay, J., Vivian, N., Economou, A., Jackson, D., Goodfellow, P., and Lovell-Badge, R. (1992). Inverted repeat structure of the Sry locus in mice. *Proc. Natl. Acad. Sci. U.S.A.* **89,** 7953–7957.

Handley, P. M., Mueckler, M., Siegel, N. R., Ciechanover, A., and Schwartz, A. L. (1991). Molecular cloning, sequence, and tissue distribution of the human ubiquitin-activating enzyme E1. *Proc. Natl. Acad. Sci. U.S.A.* **88,** 273–279.

Harley, V. R., Jackson, D., Hextall, P., Hawkins, J. R., Berkovitz, G. D., Sockanathan, S., Lovell-Badge, R., and Goodfellow, P. N. (1992). DNA binding activity of recombinant SRY from normal males and XY females. *Science* **255,** 453–456.

Jacobs, P. A., and Strong, J. A. (1959). A case of human intersexuality having a possible XXY sex-determining mechanism. *Nature (London)* **183,** 302–303.

Jantzen, H.-M., Admon, A., Bell, S. P., and Tjian, R. (1990). Nucleolar transcription factor hUBF contains a DNA-binding motif with homology to HMG proteins. *Nature (London)* **344,** 830–836.

Kay, G. F., Ashworth, A., Penny, G. D., Dunlop, M., Swift, S., Brockdorff, N., and Rastan, S. (1991). A candidate spermatogenesis gene on the mouse Y chromosome is homologous to ubiquitin-activating enzyme E1. *Nature (London)* **354,** 486–489.

Keitges, E., Rivest, M., Sinisalco, M., and Gartler, S. M. (1985). X-linkage of the steroid sulfatase gene to the sex reversed mutation in the mouse. *Genetics* **116,** 464–468.

Koopman, P., Gubbay, J., Collignon, J., and Lovell-Badge, R. (1989). Zfy gene expression patterns are not compatible with a primary role in mouse sex determination. *Nature (London)* **342,** 940–942.

Koopman, P., Münsterberg, A., Capel, B., Vivian, N., and Lovell-Badge, R. (1990). Expression of a candidate sex-determining gene during mouse testis differentiation. *Nature (London)* **348,** 450–452.

Koopman, P., Ashworth, A., and Lovell-Badge, R. (1991a). The ZFY gene family in humans and mice. *Trends Genet.* **7,** 132–136.

Koopman, P., Gubbay, J., Vivian, N., Goodfellow, P., and Lovell-Badge, R. (1991b). Male development of chromosomally female mice transgenic for Sry. *Nature (London)* **351,** 117–121.

Lau, Y.-F. C., and Chan, K. (1989). The putative testis-determining factor and related genes are expressed as discrete-sized transcripts in adult gonadal and somatic tissues. *Am. J. Hum. Genet.* **45,** 942–952.

Levy, E., and Burgoyne, P. S. (1986). The fate of XO germ cells in the testes of XO/XY and XO/XY/XYY mouse mosaics: Evidence for a spermatogenesis gene on the mouse Y chromosome. *Cytogenet. Cell Genet.* **42,** 208–213.

Lovell-Badge, R. (1993). Sex determining gene expression during embryogenesis. *Philos. Trans. R. Soc. London, Ser. B* **339,** 159–164.

Lyon, M. F. (1970). The activity of the sex chromosomes in mammals. *Sci. Prog. (Oxford)* **58,** 117–130.

Mardon, G., and Page, D. C. (1989). The sex-determining region of the mouse Y chromosome encodes a protein with a highly acidic domain and 13 zinc fingers. *Cell (Cambridge, Mass.)* **56,** 765–770.

Mardon, G., Mosher, R., Disteche, C. M., Nishioka, Y., McLaren, A., and Page, D. C. (1989). Duplication, deletion, and polymorphism in the sex-determining region of the mouse Y chromosome. *Science* **243,** 78–80.

Mayer, A., Gropper, R., Schwartz, A. L., and Ciechanover, A. (1990). Purification, characterization and rapid inactivation of thermolabile ubiquitin-activating enzyme from the mammalian cell cycle mutant ts85. *J. Biol. Chem.* **264,** 2060–2068.

McGrath, J. P., Jentsch, S., and Varshavsky, A. (1990). UBA1: An essential yeast gene encoding ubiquitin-activating enzyme. *EMBO J.* **10,** 227–236.

McLaren, A. (1983). Does the chromosomal sex of a mouse germ cell effect its development. *In* "Current Problems in Germ Cell Differentiation" (A. McLaren and C. C. Wylie, eds.), pp. 225–240. Cambridge Univ. Press, Cambridge.

McLaren, A., and Monk, M. (1982). Fertile females produced by inactivation from X chromosome of 'sex-reversed' mice. *Nature (London)* **300,** 446–448.

McLaren, A., Simpson, E., Tomonari, K., Chandler, P., and Hogg, H. (1984). Male sexual differentiation in mice lacking H-Y antigen. *Nature (London)* **312,** 552–555.

McLaren, A., Simpson, E., Epplen, J. T., Studer, R., Koopman, P., Evans, E. P., and Burgoyne, P. S. (1988). Location of the genes controlling H-Y antigen expression and testis determination on the mouse Y chromosome. *Proc. Natl. Acad. Sci. U.S.A.* **85,** 6442–6445.

Miklos, G. L. G. (1974). Sex chromosome pairing and male fertility. *Cytogenet. Cell Genet.* **13,** 558–577.

Mintz, B., and Russell, E. S. (1957). Gene-Induced embryological modifications of primordial germ cells in the mouse. *J. Exp. Zool* **134,** 207–237.

Mitchell, M. J., and Bishop, C. E. (1992). A structural analysis of the Sxr region of the mouse Y chromosome. *Genomics* **12,** 26–34.

Mitchell, M. J., Woods, D. R., Tucker, P. K., Opp, J. S., and Bishop, C. E. (1991). Homology of a candidate spermatogenic gene from the mouse Y chromosome to the ubiquitin-activating enzyme E1. *Nature (London)* **354,** 483–486.

Mitchell, M. J., Woods, D. R., Wilcox, S. A., Graves, J. A. M., and Bishop, C. E. (1992). Marsupial Y chromosome encodes a homologue of the mouse Y-linked candidate spermatogenesis gene Ube1y. *Nature (London)* **359,** 528–531.

Nagamine, C. M., Chan, K., Kozak, C. A., and Lau, Y.-F. (1989). Chromosome mapping and expression of a putative testis-determining gene in mouse. *Science* **243,** 80–83.

Nagamine, C. M., Chan, K., Hake, L. E., and Lau, Y.-F. C. (1990). The two candidate testis-determining Y genes (Zfy-1 and Zfy-2) are differentially expressed in fetal and adult mouse tissues. *Genes Dev.* **4,** 63–74.

Nasrin, N., Buggs, C., Kong, X. F., Carnazza, J., Goebl, M., and Alexander-Bridges, M. (1991). DNA-binding properties of the product of the testis-determining gene and a related protein. *Nature (London)* **354,** 317–320.

Ner, S. S. (1992). HMGs everywhere. *Curr. Biol.* **2,** 208–210.

Page, D. C., Mosher, R., Simpson, E. M., Fisher, E. M. C., Mardon, G., Pollack, J., McGillivray, B., de la Chapelle, A., and Brown, L. G. (1987). The sex-determining region of the human Y chromosome encodes a finger protein. *Cell (Cambridge, Mass.)* **51,** 1091–1104.

Palmer, S. J., and Burgoyne, P. S. (1991a). The *Mus musculus* domesticus Tdy allele acts later than the *Mus musculus musculus* Tdy allele: A basis for XY sex reversal in C57Bl/6-Ypos mice. *Development (Cambridge, UK)* **113,** 709–714.

Palmer, S. J., and Burgoyne, P. S. (1991b). XY follicle cells in the ovaries of XO/XY and XO/XY/XYY mosaic mice. *Development (Cambridge, UK)* **111,** 1017–1020.

Palmer, M. S., Sinclair, A. H., Berta, P., Ellis, N. A., Goodfellow, P. N., Abbas, N. E., and Fellous, M. (1989). Genetic evidence that ZFY is not the testis-determining factor. *Nature (London)* **342,** 937–939.

Pearson, P. L., and Bobrow, M. (1970). Definitive evidence for the short arm of the Y chromosome associating with the X during meiosis in the human male. *Nature (London)* **226,** 959–961.

Pelliniemi, L. J. (1975). Ultrastructure of the early ovary and testis in pig embryos. *Am. J. Anat.* **144,** 89–112.

Roberts, C., Weith, A., Passage, E., Michot, J. L., Mattei, M. G., and Bishop, C. E. (1988). Molecular and cytogenetic evidence for the location of Tdy and Hya on the mouse Y chromosome short arm. *Proc. Natl. Acad. Sci. U.S.A.* **85,** 6446–6449.

Schneider-Gädicke, A., Beer-Romero, P., Brown, L. G., Nussbaum, R., and Page, D. C. (1989). The ZFX gene on the human X chromosome escapes X-inactivation and is closely related to ZFY, the putative sex determinant on the Y chromosome. *Cell (Cambridge, Mass.)* **57,** 1247–1258.

Simpson, E. M., and Page, D. C. (1991). An interstitial deletion in mouse Y chromosomal DNA created a transcribed *Zfy* fusion gene. *Genomics* **11,** 609–620.

Simpson, E. M., Edwards, P., Wachtel, S., McLaren, A., and Chandler, P. (1981). H-Y antigen in Sxr mice detected by H-2 restricted cytotoxic T cells. *Immunogenetics* **13,** 355–358.

Simpson, E. M., Chandler, P., Goulmy, E., Disteche, C. M., Ferguson-Smith, M. A., and Page, D. C. (1987). Separation of the genetic loci for the H-Y antigen and for testis determination on the human Y chromosome. *Nature (London)* **326,** 876–878.

Sinclair, A. H., Foster, J. W., Spencer, J. A., Page, D. C., Palmer, M., Goodfellow, P. N., and Graves, J. A. M. (1988). Sequences homologous to ZFY, a candidate human sex-determining gene, are autosomal in marsupials. *Nature (London)* **336,** 780–783.

Singh, L., and Jones, K. W. (1982). Sex reversal in the mouse (*Mus musculus*) is caused by a recurrent nonreciprocal crossover involving the X and an aberrant Y chromosome. *Cell (Cambridge, Mass.)* **28,** 205–216.

Singh, L., Purdom, I. F., and Jones, K. W. (1980). Sex chromosome associated satellite DNA: Evolution and conservation. *Chromosoma* **79,** 137–157.

Styrna, J., Imai, H. T., and Moriwaki, K. (1991). An increased level of sperm abnormalities in mice with a partial deletion of the Y chromosome. *Genet. Res.* **57,** 195–199.

Sutcliffe, M. J., Darling, M. S., and Burgoyne, P. S. (1991). Spermatogenesis in XY, XYSxr^a and XOSxr^a mice: A quantitative analysis of spermatogenesis throughout puberty. *Mol. Reprod. Dev.* **30,** 81–89.

Travis, A., Amsterdam, A., Bélanger, C., and Grosschedl, R. (1991). LEF-1, a gene encoding a lymphoid-specific protein with an HMG domain, regulates T-cell receptor alpha enhancer function. *Genes Dev.* **5,** 880–894.

Tsunoda, Y., Tokunaga, T., and Sugie, T. (1985). Altered sex-ratio of live young after transfer of fast-developing and slow-developing mouse embryos. *Gamete Res.* **12,** 301–304.

Tucker, P. K., Phillips, K. S., and Lundrigan, B. (1992). A mouse Y chromosome pseudogene is related to human ubiquitin activiating enzyme E1. Mamm. *Genome* **3,** 28–35.

Turner, H. H. (1938). A syndrome of infantilism, congenital webbed neck and cutibus vulgus. *Endocrinology (Baltimore)* **23,** 566–574.

van de Wetering, M., Oosterwegel, M., Dooijes, D., and Clevers, H. (1991). Identification and cloning of TCF-1, a T lymphocyte-specific transcription factor containing a sequence-specific HMG box. *EMBO J.* **10,** 123–132.

Wachtel, S. S., Ohno, S., Koo, G. C., and Boyse, E. A. (1975). Possible role for H-Y antigen in the primary determination of sex. *Nature (London)* **257,** 235–236.

Welshons, W. J., and Russell, L. B. (1959). The Y chromosome as the bearer of male determining factors in the mouse. *Proc. Natl. Acad. Sci. U.S.A.* **45,** 560–566.

Whitfield, S., Lovell-Badge, R., and Goodfellow, P. N. (1993). Rapid sequence evolution of the sex determining gene SRY. *Nature (London)* (In Press).

Yen, P. H., Marsh, B., Allen, E., Tsai, S. P., Ellison, J., Connolly, L., Neiswanger, K., and Shapiro, L. J. (1988). The human X-linked steroids sulfatase gene and a Y-encoded pseudogene: Evidence for an inversion of the Y chromosome during primate evolution. *Cell (Cambridge, Mass.)* **55,** 1123–1135.

Zacksenhaus, E., and Sheinin, R. (1988). Identification of a human gene complementing tsA1S9 mouse L-cell defect in DNA replication following DNA-mediated gene transfer. *Somatic Cell Mol. Genet.* **14,** 371–379.

Zacksenhaus, E., and Sheinin, R. (1990a). Molecular cloning of the human A1S9 locus: An X-linked gene essential for progression through S phase of the cell cycle. *Somatic Cell Mol. Genet.* **15,** 545–553.

Zacksenhaus, E., and Sheinin, R. (1990b). Molecular cloning, primary structure and expression of the human X linked A1S9 cDNA which complements the ts A1S9 mouse L cell defect in DNA replication. *EMBO J.* **9,** 2923–2929.

4

XX Sex Reversal in the Mouse

Anne McLaren
Wellcome/CRC Institute
Cambridge CB2 1QR
United Kingdom

Molecular Genetics of Sex Determination

I. Introduction

XX (sex-reversed) male mammals are always sterile, with testes devoid of germ cells in the adult, since spermatogonia carrying two X chromosomes (i.e., XX, XXY, XXYY) degenerate before puberty (in mice, soon after birth). The cause of this loss is not yet known. One of the two X chromosomes is inactivated in all somatic XX cells in mammals, and in the early development of the germ line. The inactive X has been shown in mice to be reactivated before birth in male as well as in female germ cells (McLaren and Monk, 1981), so it may be that some X-linked gene acts as a cell-lethal in spermatogonia when expressed in two doses.

Because of their desire to have children, many infertile men attend infertility clinics, and hence are subjected to chromosomal investigation. Numerous XX males have been identified in this way. The reasons for the particularly high frequency of XX men are discussed in Chapter 11 in this volume. Sterile male mice, on the other hand, are seldom if ever subjected to routine karyotyping: there are many possible causes of infertility, and in most mouse colonies the offending male would simply be discarded. An XX male mouse would pass unnoticed unless an X-linked coat color marker was segregating in the stock, giving a variegated phenotype in XX individuals.

II. Discovery of Sxr

When two such variegated male mice (brothers) were noticed, it was indeed fortunate that the causative factor could be transmitted by a fertile carrier; the condition was termed Sxr (sex-reversed) and half of all XY brothers were found to transmit it (Cattanach *et al.*, 1971). Since the mode of inheritance was autosomal rather than sex-linked, the most likely explanation of the anomaly was thought to be that the male-determining region of the Y chromosome had been translocated onto an autosome, but extensive investigations failed to yield any evidence of autosomal linkage.

More than 10 years elapsed before Sxr's true location was discovered. Using the snake-derived highly repetitive Bkm DNA sequences to carry out *in situ* hybridization on mouse metaphase chromosomes, Singh and Jones (1982) found that normal male mice showed a single focus of radioactively labeled probe at one end of the Y chromosome. In X/Y Sxr males, a second focus was seen, at the opposite end of the Y chromosome, whereas in X/X Sxr sex-reversed males a focus of hybridization was found on a medium-sized chromosome, thought to be the X.

Clearly Sxr was associated with a rearranged Y chromosome. Burgoyne (1982) had postulated that, in mammals, a single obligatory crossover event takes place between chromatids of the mammalian X and Y chromosomes when they pair during male meiosis, so that any gene located distal to the pairing region on either sex chromosome would have a 50% chance of being transferred to the other (50% because the crossover involves only one of the two chromatids making up the chromosome). Burgoyne termed such a location "pseudoautosomal," since the mode of inheritance would not indicate sex-linkage, but would mimic that of an autosomal gene. Sxr, he suggested, would be a likely candidate. A similar explanation of the transmission of Sxr was put forward by Eicher (1982) and by Hansmann (1982). Later in the same year Evans *et al.* (1982) confirmed cytologically that the Y chromosome of X/Y Sxr males had an extra piece attached to the distal tip, and that a similar piece was attached to one of the two X chromosomes in X/X Sxr males. This piece must have arisen by the duplication and transposition of part of the Y chromosome that included the testis-determining region.

We have found that Sxr recombines freely between the X and Y chromosomes during male meiosis, with a recombination frequency not significantly different from 50% in either direction (McLaren *et al.* 1992). Cattanach *et al.* (1990), on the other hand, reported that the frequency of Sxr recombination from the X onto the Y chromosome was much below 50%. The explanation of the discrepancy may lie in the different genetic origin of the Y chromosomes in the two studies.

III. Male-Specific Antigen

Male mice express certain histocompatibility antigens absent in females, whether recognized by T lymphocytes (transplantation antigen: Eichwald and Silmser, 1955) or by B lymphocytes (serological response: Goldberg *et al.*, 1971). Initially all such male-specific antigens were included under the generic term H-Y antigen, since breeding tests established that both were controlled by a gene or genes on the Y chromosome. No convincing evidence has been brought forward to establish that T and B lymphocytes are recognizing the same antigen; the term H-Y antigen is therefore now usually restricted to the antigen or antigens recognized by T lymphocytes, to distinguish it from serologically detected male (SDM) antigen.

XX Sxr male mice were shown to express both H-Y (transplantation) antigen and SDM antigen (Bennett *et al.*, 1977; Simpson *et al.*, 1981).

This finding established that the gene or genes controlling the expression of male-specific antigens is located in Sxr, i.e., in that region of the Y that has been duplicated and transposed and that is responsible also for testis determination. The association between the testis-determining factor and male-specific antigens lent added weight to the attractive hypothesis put forward earlier by Wachtel *et al.* (1975), that H-Y antigen (in the generic sense) was itself the testis-determining substance. The so-called "H-Y hypothesis" gained widespread acceptance, but as we shall see below has not stood the test of time.

IV. Sex Reversal Sex-Reversed

If the Sxr region is attached to an X chromosome, what happens to it when the X chromosome is inactivated? To answer this question we made use of an X-autosome translocation, T(X;16) 16H (hereinafter termed T16H), which is preferentially expressed in the adult female (Lyon *et al.*, 1964). If the inactivation spread into the attached fragment to inactivate the postulated Y-borne testis-determining gene *Tdy* in every cell, the embryo would be expected to develop as a female. Normally, X inactivation is a random process, so that in an XX Sxr embryo, no more than 50% of cells on average would have Sxr attached to an inactive X. In T16H/X females, the normal X is inactive in 100% of cells. It turned out that T16H/XSxr individuals were neither all females, nor all males. Although all were of the same sex chromosome constitution, some developed as females, some as males, and some as intersexes (McLaren and Monk, 1982; Cattanach *et al.*, 1982). The proportion of females to males in T16H/X Sxr individuals varied according to the strain of the male transmitting Sxr (McLaren, 1986).

The sex-reversed sex-reversed females from this cross were of particular interest. They proved to express H-Y antigen (Simpson *et al.*, 1984), but were fully fertile and transmitted Sxr to half their progeny: evidently neither the expression of "male-specific" antigen nor the presence of *Tdy* interfered with normal female reproduction. Sxr recombined freely during female meiosis between the X chromosome to which it was attached and the T16H X-autosome translocation (McLaren *et al.*, 1992). The existence of fertile Sxr-carrying females allowed the production of X Sxr/Y Sxr males whose progeny, whether XX or XY, were all male (McLaren and Burgoyne, 1983).

To explain the variation in phenotypic sex among individuals of uniform sex chromosome constitution, we postulated that T16H/X Sxr mice are mosaics. In one cell population the inactivated state has spread to include *Tdy*, so the cells are female-determining; in another population, *Tdy* has escaped inactivation. Phenotypic sex will then depend on the relative proportion of male-determining and female-determining cells in the gonadal primordium.

The hypothesis that T16H/X Sxr individuals are mosaics has been questioned (see, for example, Ohno, 1985). We recently tested it by setting up a clonal analysis of one T16H/X Sxr female (Scott *et al.*, 1990). Single hemopoietic stem cells from bone marrow were rendered immortal by exposure to Abelson virus *in vitro*, and grown up until the clone was sufficiently large to be assayed for H-Y expression by an H-Y specific T-cell clone proliferation assay (Tomonari, 1983). Since the female was heterozygous at the H-2 locus, two T-cell clones were used, one specific for H-2^k and the other for H-2^b. Because the immunological response of T cells to H-Y antigen is H-2-restricted, each T-cell clone recognizes H-Y antigen only in the context of the *H-2* haplotype that it carries. The H-2^k-specific T-cell clone gave uniformly positive results: all the cells grown from the T16H/X Sxr female were positive for H-Y antigen as presented by H-2^k. On the other hand, nearly half the T16H/X Sxr female's cells proved to be negative for H-Y antigen presented by H-2^b. This result not only supports the mosaic hypothesis, it also suggests that there are not one but two loci in the Sxr region controlling H-Y expression, each specific for a different H-2 haplotype. The *Hya*(k) locus may be further from the point of attachment to the X chromosome, since it appeared to escape inactivation. The relative position of *Tdy* and the *Hya*(b) locus cannot be ascertained from this experiment.

V. A Variant Sxr Fragment

When tested either by H-Y-specific T-cell clone proliferation assays or by H-Y-specific T-cell-mediated cytotoxic tests, T16H/X Sxr females proved in general to express H-Y antigen as effectively as do T16H/X Sxr males or X/X Sxr males (Simpson *et al.*, 1984). One exceptional T16H/X Sxr female, however, failed to show any evidence of H-Y expression on either test. This female was the daughter of a T16H/X mother and an XSxr/Y Sxr homozygous father. Breeding experiments established that the failure to express H-Y was due to a heritable mutation in the Sxr region (McLaren *et al.*, 1984). We initially termed the variant region Sxr′; when further variants were identified, it seemed more appropriate to use the designations Sxr^a and Sxr^b for Sxr and Sxr′, respectively (McLaren *et al.*, 1988b).

X/X Sxr^b males and T16H/X Sxr^b males or females are uniformly negative for H-Y antigen by *in vitro* testing. X/X Sxr^b males rejected skin grafts from normal H-Y-positive XY males (Simpson *et al.*, 1986); XX females rejected skin grafts from XY or X/X Sxr^a males of the same inbred strain, but accepted skin grafts from X/X Sxr^b males (McLaren *et al.*, 1988a). This showed that X/X Sxr^b males did not merely express an altered form of male-specific antigen; they expressed no male-specific antigen that could be recognized by T lymphocytes. X/X Sxr^b males also differed from XY and X/X Sxr^a males and resembled XX females, in that their tissues did not provoke any serological response in XX females, as judged by an *in vitro* sperm cytotoxicity assay for SDM (Goldberg *et al.*, 1991). Thus they appear to lack not only H-Y but also SDM antigen. We conclude that the presence of male-specific antigens is not required for primary testis determination, and hence that the "H-Y hypothesis" of Wachtel *et al.* (1975) can no longer be upheld.

The conclusion that H-Y antigen is not the primary testis determinant is underlined by the finding that the human Y chromosome carries the gene controlling H-Y antigen expression on its long arm (Simpson *et al.*, 1987), whereas the testis-determining region is located on the distal part of the short arm, adjacent to the X–Y pairing region (Page, 1986).

By using DNA markers that map to the Sxr region, Bishop *et al.* (1988) established that the generation of Sxr^b from Sxr^a involved a deletion event within Sxr. Pulsed-field gel electrophoresis studies suggest that more than 330 kb of DNA was deleted, of which 270 kb has now been isolated (Mitchell and Bishop, 1992). However, enough of the Sxr region remains in Sxr^b to be detectable cytologically (I. Hansmann, personal communication).

VI. DNA Fingerprinting of Sxr

The DNA of the Sxr region and its normal homolog on the Y chromosome is rich in repeated nucleotide sequences of composition GATA and GACA. These sequences constitute the conserved portion of the snake satellite DNA probe Bkm that was found by Singh and Jones (1982) to hybridize preferentially with the testis-determining region of the mouse Y chromosome and with Sxr. Schafer *et al.* (1986) have used hybridization with labeled synthetic oligonucleotide GATA/GACA probes as a method of "fingerprinting" DNA after cutting with restriction enzymes and separation of fragments by gel electrophoresis. The part of the gel specific for the testis-determining region (native, or transposed as in Sxr) shows restriction fragment length polymorphism

(RFLP). In this way we analyzed the DNA from many mice belonging to the two inbred stocks in which X/Y Sxr^a and X/Y Sxr^b males have been respectively back-crossed onto C57BL females. The band patterns (DNA fingerprints) for Sxr^a and Sxr^b proved to be different from one another, and the normal Y chromosomes from the two stocks were also different from each other, as well as from Sxr^a and Sxr^b (Epplen *et al.*, 1988). DNA's from X/Y Sxr^a and X/Y Sxr^b males showed the patterns from Sxr^a or Sxr^b superimposed on the normal Y-chromosome pattern for that stock. In each of the two stocks we also detected one newly arisen variant Sxr pattern (Sxr^d, Sxr^e), suggesting that new variation is generated at a rather high frequency.

VII. Location of Tdy on the Mouse Y Chromosome

In our back-crossed Sxr^b stock, an XX male was detected that unexpectedly expressed H-Y antigen (McLaren *et al.*, 1988b). His XX brothers were H-Y negative, as have been all other X/X Sxr^b males tested, and his karyotype was clearly XX not XY. DNA fingerprinting with GATA/GACA probes revealed that his RFLP pattern was identical to that of the normal Y chromosome from the Sxr^b stock, and distinct from Sxr^b, Sxr^a, and the Sxr^a stock Y chromosome. It appeared that in this individual the Sxr^b fragment that should have been transferred to the X chromosome in the preceding male meiosis had been replaced by the native Y-chromosome homolog. Since the native homolog includes the *Hya* locus, it is then not surprising that the mouse expressed H-Y antigen.

Chandley and Speed (1987) showed by electron microscopy that Y Sxr chromosomes may sometimes loop around so that the Sxr region pairs with its native homolog, in addition to the pairing of the Y chromosome with the X. If a double crossover occurred, between the paired regions of the Y chromosome as well as between the X and the Y, the native testis-determining region of the Y could be transferred to the X. However, analysis showed that if the native testis-determining region were located in the proximal portion of the long arm of the Y chromosome, as had hitherto been assumed, such a double crossover would give rise to a dicentric chromosome.

We therefore postulated (McLaren *et al.*, 1988b) that *Tdy* and *Hya* in the mouse are both located on the very tiny short arm of the Y chromosome (Ford, 1966). Direct evidence that this postulate was correct came from *in situ* hybridization studies by Roberts *et al.* (1988), using a radioactively labeled DNA probe that hybridizes specifically to the Sxr region and its native homolog.

VIII. Recombination with Sxr

The high proportion of the repeated DNA sequences in the Sxr region would be expected to result in a high incidence of mismatched pairing and hence unequal recombination events. In our earlier study (Epplen *et al.*, 1988), the rate of unequal recombination at one specific locus was about 3%. Harbers *et al.* (1987) reported an even higher rate of unequal recombination (7%) in the pseudoautosomal region of the mouse Y chromosome. An unequal crossover within the Sxr region is likely to have led to the generation of Sxr^b from Sxr^a, since the original H-Y-negative T16H/X Sxr female was sired by a homozygous X Sxr/Y Sxr male. Bishop *et al.* (1988) have pointed out that Sxr itself could have arisen by nonhomologous exchange between the short arm of the Y chromosome and the distal pseudoautosomal region of the same chromosome.

Since we now have several markers (*Hya, Spy, Sby, Zfy-2,* and a DNA marker Sx1/D) located within that region of Sxr that is deleted in Sxr^b, we examined the descendants of an $XSxr^b/YSxr^a$ male to see whether we could detect recombination between markers, due to mismatched pairing and unequal crossing-over (McLaren *et al.*, 1992). In 159 progeny we detected three variant Sxr regions (1.9%) presumed to be due to unequal crossing-over between the two original Sxr regions. Among the descendants we detected a further four variants probably derived from illegitimate pairing and crossing-over between Sxr^b and the homologous region on the short arm of the Y chromosome.

IX. Spermatogenesis Gene in Sxr

For decades, testis determination was the only genetic function known to be associated with the Y chromosome. A Y-borne testis-determining gene was postulated, termed *Tdy* in the mouse. Then H-Y antigen was discovered: The Y-borne gene controlling it was termed *Hya. Spy,* a gene required for normal spermatogenesis, is also located on the Y chromosome.

Spermatogenesis cannot be studied in XX males: Since all XX spermatogonia degenerate within a few days of birth, the testis of an adult X/X Sxr male mouse is usually devoid of germ cells. Occasionally, growing oocytes may be observed within the testis tubules of prepubertal X/X Sxr males (McLaren, 1980); these are derived from germ cells that entered meiosis before birth, rather than developing as T prospermatogonia (McLaren, 1981). In T16H/X Sxr embryos, X-inactivation mosaicism at the *Tdy* locus may achieve a sufficiently high level of female-producing cells to give rise to overt intersexual or female development. Perhaps in X/X Sxr embryos mosaicism exists, but with a much lower level of female-producing cells, such that the development of some testis cords may be marginally disrupted. The inhibitory effect on entry of germ cells into meiosis that testis cords normally exert may then be less than complete.

Occasionally also, the testis of an adult X/X Sxr^a male includes one or more tubules in which active spermatogenesis is taking place. The germ cells in such tubules have been shown to be XO rather than XX in chromosome constitution (Lyon *et al.*, 1981): presumably one of the two X chromosomes has been lost from a germ cell before birth, and the resulting XO spermatogonia have survived to populate the tubule.

In X Sxr^a/0 males, active spermatogenesis takes place in all tubules and all stages are represented, but few spermatozoa are produced and the mice are sterile (Cattanach *et al.*, 1971). Most of the spermatids are diploid, and the few spermatozoa that are produced, whether diploid or haploid, are abnormal (Levy and Burgoyne, 1986a). The defect seems to involve the lack of a meiotic pairing partner for the $XSxr^a$ chromosome and the lack of genetic information required for spermiogenesis mapping to the long arm of the Y (Burgoyne *et al.*, 1992).

In X Sxr^b/0 males, the spermatogenic impairment is much more severe: Few spermatogonia can be found in the postpubertal testis, and none develop beyond prophase of the first meiotic division (Burgoyne *et al.*, 1986). XO germ cells in XO/XY/XYY mosaic males suffered a very similar impairment (Levy and Burgoyne, 1986b); since the somatic component of the testis was largely XY in these mosaics, the developmental failure of both XO and X Sxr^b/0 germ cells was presumed to be due to the absence of a Y-borne gene that acts cell-autonomously in the germline during normal spermatogenesis. The gene, termed *Spy,* must be located, like *Hya,* in the region of Sxr that is deleted from Sxr^b. Sutcliffe and Burgoyne (1989) have established, by quantitative analysis of spermatogenic stages, that *Spy* is required for normal survival and proliferation of differentiating type A spermatogonia.

A gene mapping to the region deleted in Sxr^b has recently been isolated and shown to have strong homology to the X-linked human ubiquitin-activating enzyme E1 (Mitchell *et al.*, 1991; Kay *et al.*, 1991). The new gene, *Sby,* is a strong candidate for the Y-linked spermatogenesis gene *Spy.*

X. Zinc Finger Genes in Sxr

Page *et al.* (1987) reported the cloning of a gene (*ZFY*) from the presumed testis-determining region of the human Y chromosome. Since it was highly conserved, and coded for a zinc-finger protein thought to be involved in DNA binding and gene regulation, it seemed a good candidate for the human testis-determining gene. In all mammals examined, a similar but not identical gene was found on the X chromosome; in the mouse there was also an autosomal homolog, and two genes of very similar sequence on the Y chromosome, termed *Zfy-1* and *Zfy-2* (Mardon *et al.*, 1989; Nagamine *et al.*, 1989). Both genes mapped to the Sxr region, but *Zfy-2* appeared to be missing from Sxr^b. *Zfy-1* was therefore regarded as the candidate gene for *Tdy*.

The predicted amino acid sequences of the zinc finger regions of *Zfy-1* and *Zfy-2* are so similar that it seems unlikely that either is a pseudogene (Mardon and Page, 1989; Ashworth *et al.*, 1989). However, of the two Y chromosomes commonly found in mouse inbred strains, one (the *Mus musculus* Y) carries a *Zfy-2* gene that lacks 18 nucleotides present in *Zfy-1* and in the *Zfy-2* of the *M. domesticus* Y (Nagamine *et al.*, 1990). This deletion might well render nonfunctional finger 2 of the total of 13 zinc fingers.

Simpson and Page (1991) have since reported that the Sxr^b deletion spans a region of DNA extending from the 5′ promoter region of *Zfy-1*. The gene remaining in Sxr^b is therefore not *Zfy-1*, but a hybrid gene *Zfy-2/1* in which the 5′ promoter region of *Zfy-2* has been fused to the body of *Zfy-1*. The deletion event probably occurred as a result of unequal crossing-over in the germline of the XSxr/YSxr father of the original Sxr^b-carrying female.

Northern analysis has shown both *Zfy-1* and *Zfy-2* to be strongly expressed in the adult testis (Mardon and Page, 1989; Nagamine *et al.*, 1990). Expression was associated with the presence of germ cells. Using the more sensitive polymerase chain reaction, Koopman *et al.* (1990) were able to demonstrate expression of *Zfy-1* in the genital ridges of male mouse embryos at the time that primary testis determination would be expected to occur. No expression of *Zfy-2* was detected, but Nagamine *et al.* (1990) have reported some transcription of both *Zfy-1* and *Zfy-2* in the fetal testis, as well as in other tissues.

Although the time of expression of *Zfy-1* in the fetal gonad is consistent with a role in primary testis determination, the tissue distribution of expression is not. When genital ridges were taken from *W/W* embryos, which are almost entirely devoid of germ cells, no *Zfy-1* transcripts could be detected (Koopman *et al.*, 1989) but testis determination occurred nonetheless. *Zfy-1* expression appears to be confined to the germ cell lineage, which plays no direct role in testis determination; *Zfy-1* cannot therefore be the primary testis-determining gene. A similar conclusion was reached on genetic grounds in the human: Some XX men were reported who possess Y-chromosomal sequences but lack *ZFY* (Palmer *et al.*, 1989), so the search for the Y-borne testis-determining gene shifted to a different region of the human Y.

The discovery of the human gene *SRY* (Sinclair *et al.*, 1990) and the isolation of its mouse homolog *Sry* (Gubbay *et al.*, 1990) is in Chapters 2 and 3 in this volume. Final proof that *Sry* is indeed the only testis-determining gene on the mouse Y chromosome was provided by the finding that a 14-kb DNA fragment containing *Sry*, inserted into the mouse genome as a transgene, effectively sex-reversed XX embryos into phenotypic males (Koopman *et al.*, 1991).

References

Ashworth, A., Swift, S., and Affara, N. (1989). Sequence of cDNA for murine *Zfy-1*, a candidate for *Tdy*. *Nucleic Acids Res.* **17,** 2864.

Bennett, D., Mathieson, B. J., Scheid, M., Yanagisawa, K., Boyse, E. A., Wachtel, S., and Cattanach, B. M. (1977). Serological evidence for H-Y antigen in Sxr, XX sex-reversed phenotypic males. *Nature* (*London*) **265,** 255–257.

Bishop, C. E., Weith, A., Mattei, M. G., and Roberts, C. (1988). Molecular aspects of sex determination in mice: An alternative model for the origin of the Sxr region. *In* "Sex Determination in Mouse and Man" (A. McLaren and M. A. Ferguson-Smith, eds.), pp. 119–124. Royal Society, London.

Burgoyne, P. S. (1982). Genetic homology and crossing over in the X and Y chromosomes of mammals. *Hum. Genet.* **61,** 85–90.

Burgoyne, P. S., Levy, E. R., and McLaren, A. (1986). Spermatogenic failure in male mice lacking H-Y antigen. *Nature* (*London*) **320,** 170–172.

Burgoyne, P. S., Mahadevaiah, S. K., Sutcliffe, M. J., and Palmer, S. J. (1992). Fertility in mice requires X-Y pairing and a Y chromosomal "spermiogenesis" gene mapping to the long arm. *Cell* (*Cambridge, Mass.*) **71,** 391–398.

Cattanach, B. M., Pollard, C. E., and Hawkes, S. G. (1971). Sex-reversed mice: XX and XO males. *Cytogenetics* **10,** 318–337.

Cattanach, B. M., Evans, E. P., Burtenshaw, M. D., and Barlow, J. (1982). Male, female and intersex development in mice of identical chromosome constitution. *Nature* (*London*) **300,** 445–446.

Cattanach, B. M., Rasberry, C., Burtenshaw, M. D., and Evans, E. P. (1990). Illegitimate pairing of the X and Y chromosomes in *Sxr* mice. *Genet. Res.* **56,** 121–128.

Chandley, A. C., and Speed, R. M. (1987). Cytological evidence that the Sxr fragment of XYSxr mice pairs homologously at meiotic prophase with the proximal testis-determining region. *Chromosoma* **95,** 345–349.

Eicher, E. M. (1982). Primary sex determining genes in mice: A brief review. *In* "Prospects for Sexing Mammalian Sperm" (R. P. Amann and G. E. J. Seidal eds.), pp. 121–135. Colorado Associated University Press, Boulder.

Eichwald, E. J., and Silmser, C. R. (1955). Untitled communication. *Transplant. Bull.* **2,** 148–149.

Epplen, J. T., Studer, R., and McLaren, A. (1988). Heterogeneity in the *Sxr* (*sex-reversal*) locus of the mouse as revealed by synthetic GA(T/C)A probes. *Genet. Res.* **51,** 239–246.

Evans, E. P., Burtenshaw, M. D., and Cattanach, B. M. (1982). Meiotic crossing-over between the X and Y chromosomes of male mice carrying the sex-reversing (Sxr) factor. *Nature (London)* **300,** 443–445.

Ford, C. E. (1966). The murine Y-chromosome as a marker. *Transplantation* **4,** 333–335.

Goldberg, E. H., Boyse, E. A., Bennett, D., Scheid, M., and Carswell, E. A. (1971). Serological demonstration of H-Y (male) antigen on mouse sperm. *Nature (London)* **232,** 478–480.

Goldberg, E. H., McLaren, A., and Reilly, B. (1991). Male antigen defined serologically does not identify a factor responsible for testicular development. *J. Reprod. Immunol.* **20,** 305–309.

Gubbay, J., Collignon, J., Koopman, P., Capel, B., Economou, A., Munsterberg, A., Vivian, N., Goodfellow, P., and Lovell-Badge, R. (1990). A gene mapping to the sex-determining region of the mouse Y chromosome is a member of a novel family of embryonically expressed genes. *Nature (London)* **346,** 245–250.

Hansmann, I. (1982). Sex reversal in the mouse. *Cell (Cambridge, Mass.)* **30,** 331–332.

Harbers, K., Soriano, P., Müller, U., and Jaenisch, R. (1986). High frequency of unequal recombination in pseudoautosomal region shown by proviral insertion in transgenic mouse. *Nature (London)* **324,** 682–685.

Kay, G. F., Ashworth, A., Penny, G. D., Dunlop, M., Swift, S., Brockdorff, N., and Rastan, S. (1991). A candidate spermatogenesis gene on the mouse Y chromosome is homologous to ubiquitin-activating enzyme E1. *Nature (London)* **354,** 486–489.

Koopman, P., Gubbay, J., Collignon, J., and Lovell-Badge, R. (1989). *Zfy* gene expression patterns are not compatible with a primary role in mouse sex determination. *Nature (London)* **342,** 940–942.

Koopman, P., Münsterberg, A., Capel, B., Vivian, N., and Lovell-Badge, R. (1990). Expression of a candidate sex-determining gene during mouse testis differentiation. *Nature (London)* **348,** 450–452.

Koopman, P., Gubbay, J., Vivian, N., Goodfellow, P., and Lovell-Badge, R. (1991). Male development of chromosomally female mice transgenic for Sry. *Nature (London)* **351,** 117–121.

Levy, E. R., and Burgoyne, P. S. (1986a). Diploid spermatids: A manifestation of spermatogenic impairment in XOSxr and T31H/+ male mice. *Cytogenet. Cell Genet.* **42,** 159–163.

Levy, E. R., and Burgoyne, P. S. (1986b). The fate of XO germ cells in the testes of XO/XY and XO/XY/XYY mouse mosaics: Evidence for a spermatogenesis gene on the mouse Y chromosome. *Cytogenet. Cell Genet.* **42,** 208–213.

Lyon, M. F., Searle, A. G., Ford, C. E., and Ohno, S. (1964). A mouse translocation suppressing sex-linked variegation. *Cytogenetics* **3,** 306–323.

Lyon, M. F., Cattanach, B. M., and Charlton, H. M. (1981). Genes affecting sex differentiation in mammals. *In* "Mechanisms of Sex Differentiation in Animals and Man" (C. R. Austin and R. G. Edwards, eds.). Academic Press, London.

Mardon, G., and Page, D. C. (1989). The sex-determining region of the mouse Y chromosome encodes a protein with a highly acidic domain and 13 zinc fingers. *Cell (Cambridge, Mass.)* **56,** 765–777.

Mardon, G., Mosher, R., Disteche, C. M., Nishioka, Y., McLaren, A., and Page, D. C. (1989). Duplication, deletion and polymorphism in the sex-determining region of the mouse Y chromosome. *Science* **243,** 78–80.

McLaren, A. (1980). Oocytes in the testis. *Nature (London)* **283,** 688–689.

McLaren, A. (1981). The fate of germ cells in the testis of fetal *Sex-reversed* mice. *J. Reprod. Fertil.* **61,** 461–467.

McLaren, A. (1986). Sex ratio and testis size in mice carrying Sxr and T(X;16) 16H *Dev. Genet.* **7,** 177–185.

McLaren, A., and Burgoyne, P. S. (1983). Daughterless XSxr/YSxr mice. *Genet. Res.* **42,** 348–349.

McLaren, A., and Monk, M. (1981). X-chromosome activity in the germ cells of Sex-reversed mouse embryos. *J. Reprod. Fertil.* **63,** 533–537.

McLaren, A., and Monk, M. (1982). Fertile females produced by inactivation of an X chromosome of 'sex-reversed' mice. *Nature (London)* **300,** 446–448.

McLaren, A., Simpson, E., Tomonari, K., Chandler, P., and Hogg, H. (1984). Male sexual differentiation in mice lacking H-Y antigen. *Nature (London)* **312,** 552–555.

McLaren, A., Hunt, R., and Simpson, E. (1988a). Absence of any male-specific antigen recognized by T lymphocytes in X/XSxr′ male mice. *Immunology* **63,** 447–449.

McLaren, A., Simpson, E., Epplen, J. T., Studer, R., Koopman, P., Evans, E. P., and Burgoyne, P. S. (1988b). Location of the genes controlling H-Y antigen expression and testis determination on the mouse Y chromosome. *Proc. Natl. Acad. Sci. U.S.A.* **85,** 6642–6645.

McLaren, A., Simpson, E., Bishop, C. E., Mitchell, M. J. and Darling, S. M. (1992). Recombination between the X and Y chromosomes and the Sxr region of the mouse. *Genet. Res.* **60,** 175–184.

Mitchell, M. J., and Bishop, C. E. (1992). A structural analysis of the Sxr region of the mouse Y chromosome. *Genomics* **12,** 26–34.

Mitchell, M. J., Woods, D. R., Tucker, P. K., Opp, J. S., and Bishop, C. E. (1991). Homology of a candidate spermatogenic gene to the ubiquitin activating enzyme E1. *Nature (London)* **354,** 483–486.

Nagamine, C. M., Chan, K., Kozak, C. A., and Lau, Y.-F. C. (1989). Chromosome mapping and expression of a putative testis-determining gene in mouse. *Science* **243,** 80–83.

Nagamine, C. M., Chan, K., Hake, L. E., and Lau, Y.-F. C. (1990). The two candidate testis-determining Y genes (*Zfy-1* and *Zfy-2*) are differentially expressed in fetal and adult mouse tissues. *Genes Dev.* **4,** 63–74.

Ohno, S. (1985). The Y-linked testis determining gene and H-Y plasma membrane antigen gene: Are they one and the same? *Endocr. Rev.* **6,** 421–431.

Page, D. C. (1986). Sex reversal: Deletion mapping the male-determining function of the human Y chromosome. *Cold Spring Harbor Symp. Quant. Biol.* **51,** 229–235.

Page, D. C., Mosher, R., Simpson, E. M., Fisher, E. M. C., Mardon, G., Pollack, J., McGilivray, B., de la Chapelle, A., and Brown, L. G. (1987). The sex determining region of the human Y chromosome encodes a finger protein. *Cell (Cambridge, Mass.)* **51,** 1091–1104.

Palmer, M. S., Sinclair, A. H., Berta, P., Goodfellow, P. N., Ellis, N. A., Abbas, N. E., and Fellous, M. (1989). Genetic evidence that *ZFY* is not the testis determining factor. *Nature (London)* **342,** 937–939.

Roberts, C., Weith, A., Passage, E., Michet, J. L., Mattei, M. G., and Bishop, C. E. (1988). Molecular and cytogenetic evidence for the location of Tdy and Hya on the mouse Y chromosome short arm. *Proc. Natl. Acad. Sci. U.S.A.* **85,** 6646–6649.

Schafer, R., Ali, S., and Epplen, J. T. (1986). The organization of the evolutionarily conserved GATA/GACA repeats in the mouse genome. *Chromosoma* **93,** 502–510.

Scott, D., McLaren, A., Dyson, J., and Simpson, E. (1990). Selective X inactivation affecting the expression of different epitopes of the Hya gene in mouse B cell clones. *Immunogenetics* **33,** 54–61.

Simpson, E., Edwards, P., Wachtel, S. S., McLaren, A., and Chandler, P. (1981). H-Y antigen in Sxr mice detected by H-2-restricted cytotoxic T cells. *Immunogenetics* **13,** 355–358.

Simpson, E., McLaren, A., Chandler, P., and Tomonari, K. (1984). Expression of H-Y antigen by female mice carrying Sxr. *Transplantation* **37,** 17–21.

Simpson, E., Chandler, P., Hunt, R., Hogg, H., Tomonari, K., and McLaren, A. (1986). H-Y status of X/XSxr′ male mice: In vivo tests. *Immunology* **57,** 345–349.

Simpson, E., Chandler, P., Goulmy, E., Disteche, C. M., Ferguson-Smith, M. A., and Page, D. C. (1987). Separation of the genetic loci for the H-Y antigen and for testis determination on human Y chromosome. *Nature (London)* **326,** 876–878.

Simpson, E. M., and Page, D. C. (1991). An interstitial deletion in mouse Y chromosomal DNA created a transcribed *Zfy* fusion gene. *Genomics* **11,** 601–608.

Sinclair, A. H., Berta, P., Palmer, M. S., Hawkins, J. R., Griffiths, B. L., Smith, M. J., Foster, J. W., Frischauf, A.-M., Lovell-Badge, R., and Goodfellow, P. N. (1990). A gene from the human sex-determining region encodes a protein with homology to a conserved DNA-binding motif. *Nature (London)* **346,** 240–244.

Singh, L., and Jones, K. W. (1982). Sex reversal in the mouse (*Mus musculus*) is caused by a recurrent nonreciprocal crossover involving the X and an aberrant Y chromosome. *Cell (Cambridge, Mass.)* **28,** 205–216.

Sutcliffe, M. J., and Burgoyne, P. S. (1989). Analysis of the testes of H-Y negative XOSxr^{b} mice suggests that the spermatogenesis gene (*Spy*) acts during the differentiation of the A spermatogonia. *Development (Cambridge, UK)* **107,** 373–380.

Tomonari, K. (1983). Antigen and MHC restriction specificity of two types of cloned male-specific T cell lines. *J. Immunol.* **131,** 1641–1645.

Wachtel, S. S., Ohno, S., Koo, G. C., and Boyse, E. (1975). Possible role for H-Y antigen in the primary determination of sex. *Nature (London)* **257,** 235–236.

5

XY Sex Reversal in the Mouse

Claude M. Nagamine
Department of Cell Biology
Vanderbilt University School of Medicine
Nashville, Tennessee 37232

Molecular Genetics of Sex Determination

I. Introduction

In mammals, the Y chromosome is responsible for testis determination. Since in the absence of a Y, the fetal gonads develop into ovaries, one can view ovarian differentiation as being the "default" pathway. A gene on the Y is hypothesized to switch the developing gonads from the pathway of ovarian differentiation to that of testis differentiation by triggering a cascade of testis-determining genes (these downstream testis-determining genes do not have to be on the Y). The developing testes, in their turn, secrete hormones such as testosterone and Müllerian-inhibiting substance (anti-Müllerian hormone), which result in the male phenotype (McLaren, 1988, 1991).

The hypothetical Y-chromosome gene that initiates testis determination is designated testis determining factor (*TDF*) in humans and testis determining-Y (*Tdy*) in the mouse. Convincing evidence suggests that the sex determining region-Y gene (*SRY* and *Sry* are the human and mouse designations), a member of the high mobility group (HMG) family of transcription factors, is *TDF*/*Tdy* (Gubbay *et al.*, 1990a; Sinclair *et al.*, 1990). First, *Sry* is evolutionarily conserved on the Y of metatherian and eutherian mammals (Foster *et al.*, 1992; Sinclair *et al.*, 1990). Second, expression of *Sry* correlates with the initiation of testis determination in mice (Koopman *et al.*, 1990). Third, *Sry* deletions, or missense or nonsense mutations in the DNA-binding HMG box domain of *SRY* are found in XY sex-reversed females in humans (Berta *et al.*, 1990; Harley *et al.*, 1992; Hawkins *et al.*, 1992; Jäger *et al.*, 1990; McElreavey *et al.*, 1992; Nasrin *et al.*, 1991; Page *et al.*, 1990; Vilain *et al.*, 1992). Fourth, transgenic insertion of a 14-kb genomic sequence containing *Sry* can induce XX mouse embryos to develop as males (Koopman *et al.*, 1991). Exactly how *SRY*/*Sry* initiates testis determination remains to be understood and the genes upstream and downstream of *SRY*/*Sry* in the testis differentiation cascade have not been identified. If the complex gene interactions involved in sex determination in the fruit fly *Drosophila melanogaster* and the nematode *Caenorhabditis elegans* (reviewed in Hodgkin, 1990, 1992; Kuwabara and Kimble, 1992) are any indication of the complexity of sex determination in mammals, then this challenge may prove to be formidable.

One way to understand normal testis determination fully is to study cases of abnormal testis determination, for example, XY sex reversal. In this review, XY sex reversal refers to presence of nonmosaic XY karyotypes in hermaphrodites (one or both gonads composed of testicular and ovarian tissues) or phenotypic females (bilateral ovaries or streak gonads). Mosaic karyotypes, for example, XO/XY, are excluded since a low proportion of XY cells in an XO/XY gonad can sometimes result in hermaphroditism. This is clearly demonstrated by hermaphroditism in some XX/XY aggregation mouse chimeras (Patek *et al.*, 1991; Tarkowski, 1964) and in XO/XY mosaics of the BALB/cWt mouse strain (Eicher *et al.*, 1980; Whitten, 1975). The utility of the XY sex-reversed individual lies in its contribution to the understanding of why testis determination failed despite the presence of a Y chromosome, thereby providing insight into the regulation and function of *TDF* and genes upstream and downstream of *Tdy* in the testis differentiation cascade. The mouse models for XY sex reversal fall into two categories: XY̶ and XYd sex reversal.

II. XY̶ Sex Reversal

A. Origin of the XY̶ Mouse Strain

The XY̶ sex-reversed mouse strain originated from a chimeric mouse constructed with embryonic stem cells that had been repeatedly infected with a retrovirus. The goal was to produce insertional mutations in the genes in the testis differentiation pathway. One chimeric male sired XY females, suggesting that the testis differentiation pathway had been blocked. Because the mutation appeared to be in the testis determining gene *Tdy*, the mutation was designated Tdy^{m1}. The Y chromosome carrying the mutation was designated $Y^{Tdy\text{-}m1}$ or more simply, Y̶ (Lovell-Badge and Robertson, 1990).

B. Etiology of XY̶ Sex Reversal

The sex reversal phenomenon segregated with the Y chromosome, suggesting that the genetic lesion was on the Y and not on the X chromosome or an autosome. Karyotyping of XY̶ females and Southern blot analyses of their genomic DNA with Y-specific probes initially revealed no gross deletions or rearrangements of the Y (Gubbay *et al.*, 1990b; Lovell-Badge and Robertson, 1990). However, with the cloning of *Sry*, it was discovered that this gene was deleted in XY̶ females, providing strong evidence that *Sry* was *Tdy* (Gubbay *et al.*, 1990a, b). Mapping of the deletion suggested that it was greater than 10 kilobase pairs (kb) and that it did not include the neighboring genes *Zfy-1* and *Zfy-2*. The extent of the deletion is not known. The simplest explanation for the etiology of XY̶ sex reversal is the deletion of *Sry* and neighboring sequences. Since no proviral sequences have been found on the Y̶ chromosome it is unclear whether the deletion arose from a spontaneous mutation following, but unrelated to, the retroviral infection of the embryonic stem cells or as a direct result of a retrovirus integration and the subsequent loss of proviral sequences.

C. The X¥ Sex-Reversed Gonad

Unlike XYd sex reversal (Section III), no evidence of hermaphroditism has been found in the gonads of X¥ fetuses. The gross anatomy and histology of the ovaries of two adult X¥ females (4 and 12 weeks of age) have been reported, and no obvious abnormalities were noted. However, the number of oocytes was fewer than normal and a high percentage of atretic follicles was observed (Lovell-Badge and Robertson, 1990).

X¥ females are able to transmit the ¥ chromosome to their progeny. However, X¥ females have a limited reproductive lifespan and the litters produced are small. Furthermore, in addition to XY males, XX females, and X¥ females, many of the offspring from X¥ females are sex-chromosome aneuploids (X¥Y males, XX¥ females, XO females). This suggests that there is little or no pairing between the X and the ¥ chromosomes during meiosis. XX¥ females are also fertile. Surprisingly, in contrast to X¥ females, they have normal or nearly normal reproductive lifespan, and fertility. The reason for this discrepancy in fertility between X¥ and XX¥ females is not known (Lovell-Badge and Robertson, 1990).

D. Molecular Analysis of the ¥ Chromosome

The generation of X¥Y *males*, rather than X¥Y *females*, demonstrates that the genetic lesion on the ¥ chromosome can be complemented with a normal Y chromosome to induce maleness. How small a Y chromosome is necessary? To address this question, Lovell-Badge and Robertson (1990) used the sex-reversed mutation (Sxr). The Sxr mutation is believed to have originated from duplication of a section of the short arm of the Y chromosome. The duplicated region, designated here as the Sxr region, was subsequently translocated to the Y long arm, distal to the pseudoautosomal region, which pairs and recombines with the X during male meiosis (Cattanach, 1987; Cattanach *et al.*, 1971; McLaren *et al.*, 1988; Roberts *et al.*, 1988; Singh and Jones, 1982). The Sxr region includes, at minimum, the genes for the testis-determining gene *Tdy* (*Sry*) (Gubbay *et al.*, 1990a), *Zfy-1* and *Zfy-2* (Mardon *et al.*, 1989; Nagamine *et al.*, 1990), *Hya* (McLaren *et al.*, 1984), *Sdma* (Goldberg *et al.*, 1991), *Spy* (Burgoyne *et al.*, 1986), *Ube1y* (Sby, A1s9Y-1) (Disteche *et al.*, 1992; Kay *et al.*, 1991; Mitchell *et al.*, 1991, 1992), and *Bkm* repetitive sequences (Singh and Jones, 1982). An obligatory pairing and crossing-over between one X and one YSxr chromatid during male meiosis transfers the Sxr region to an X chromatid, thereby giving rise to sperm carrying an X with the Sxr region (XSxr). When these XSxr sperm fertilize normal oocytes, the zygotes develop into XXSxr *males* due to the presence of *Tdy*. The Sxr^b allele, previously called Sxr′ (McLaren *et al.*, 1984), harbors a deletion in the Sxr region that encompasses most of the coding region of *Zfy-2*, the noncoding region of *Zfy-1*, and the genes between these two endpoints (*Sdma*, *Spy*, *Ube1y*, and *Hya*) (Bishop, 1992; Simpson and Page, 1991). *Tdy* (*Sry*), however, is intact as evidenced by the fact that $XXSxr^b$ mice have testes and male phenotypes. The Sxr^b region can therefore be considered the smallest portion of the mouse Y that is capable of testis determination.

When X¥ females were mated to males carrying Sxr^b ($XYSxr^b$), the $X¥XSxr^b$ progeny had testes and a male phenotype, demonstrating that the ¥ mutation can be fully complemented by Sxr^b. This provided additional evidence that the genetic lesion on the ¥ chromosome was in *Tdy*. Furthermore, the ¥ deletion has no apparent effect on *Spy*, *Hya*, *Zfy-1*, and *Zfy-2* since $XSxr^b$¥ males are fertile and X¥ females express *H-Ya*, *Zfy-1*, and *Zfy-2* (Gubbay *et al.*, 1990b; Lovell-Badge and Robertson, 1990).

III. XYd Sex Reversal

XYd sex reversal occurs when certain *Mus musculus domesticus* Y chromosomes (Yd) are back-crossed into the C57BL/6 (B6) laboratory inbred mouse strain. Unlike X¥ sex reversal, the etiology of XYd sex reversal is complex and implicates X-chromosomal or autosomal genes in testis determination. Two of these putative testis-determining genes have been designated testis-determining, autosomal-1 (*tda-1*) (Eicher, 1982) and T-associated sex reversal (*Tas*) (Washburn and Eicher, 1983).

A. Testis-Determining, Autosomal-1 Sex Reversal

In 1982, Eicher and colleagues reported that XY hermaphrodites and XY females are generated if a *M. m. domesticus* Y chromosome (Yd) is back-crossed into B6. First generation offspring (N1) obtained from the founding cross (B6 × *M. m. domesticus* ♂) were normal. However, on the first and subsequent back-crosses to B6, XY hermaphrodites and XY females were sired (Fig. 1) (Eicher and Washburn, 1983; Eicher *et al.*, 1982). The sex reversal phenomenon was subsequently confirmed in other laboratories (Biddle and Nishioka, 1988; Gropp and Winking, 1983; Nagamine *et al.*, 1987a, b).

That *Tdy* on the *M. m. domesticus* Y chromosome was not mutated is suggested by its inducing normal testes in the N1 generation. Karyotypic analyses of XY females and hermaphrodites revealed that all had normal karyotypes with apparently normal Y chromosomes, as judged by G-banding (Eicher *et al.*, 1982). Furthermore, none were XO/XY mosaics, a condition known to cause hermaphroditism in the BALB/cWt strain (Whitten, 1975).

To determine whether the Y chromosomes of XY females were normal at the molecular level, Southern blots containing XY female DNAs were tested with Y-specific probes pY353/B (Bishop *et al.*, 1985), pCRY8/B (Roberts *et al.*, 1988), mYfin (Nagamine *et al.*, 1989, 1990), and *Sry* (Gubbay *et al.*, 1990a). All failed to detect any difference in hybridization patterns between XY females, XY hermaphrodites, and normal males

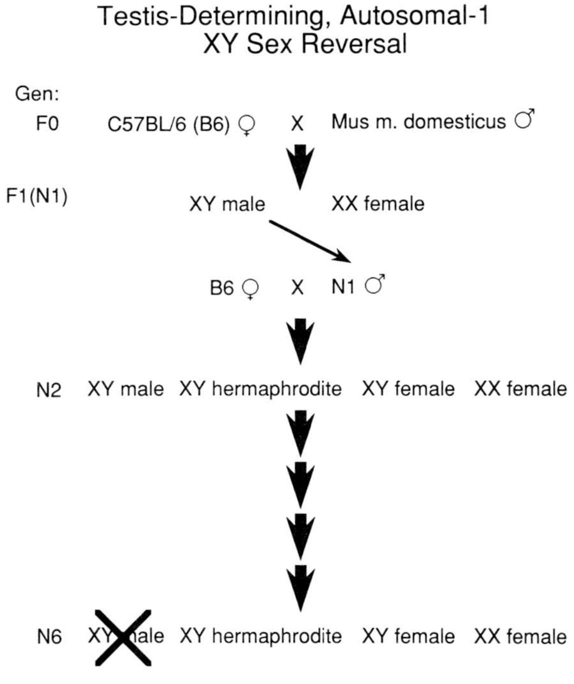

Figure 1 Back-crossing scheme illustrating XYd sex reversal. N1, N2, N6 = back-cross generations.

(C. M. Nagamine, unpublished results). This confirmed at the molecular level the cytogenetic data suggesting that the Y chromosomes of affected XY females were grossly normal.

What is the genetic cause of the sex reversal? One interpretation is that the sex reversal is the result of the testis-determining gene on the *M. m. domesticus* Y chromosome (*Tdy*^{Dom}) interacting abnormally with a recessive autosomal or pseudoautosomal testis-determining gene in the B6 strain designated *testis-determining, autosomal-1* (*tda-1*) (Eicher, 1982; Eicher and Washburn, 1983). *Tdy*^{Dom}, *tda-1*/*tda-1* fetuses develop only ovotestes or ovaries (Table I). N1 progeny are heterozygous for all autosomal loci since they obtain 50% of their autosomal genes from the B6 female and 50% from the *M. m. domesticus* male. Because all N1 males were normal, a dominant allele (*Tda-1*) was hypothesized for *M. m. domesticus*. The first back-cross results in the homozygosity of about 25% of B6 autosomal loci. XY fetuses that are homozygous for *tda-1* (*Tdy*^{Dom}, *tda-1*/*tda-1*) develop as hermaphrodites or XY females. Subsequent back-crossing results in an exponential increase of B6 genes, thereby increasing the probability that XY fetuses will be homozygous for *tda-1* and develop into XY hermaphrodites or XY females. Indeed, all XY fetuses are affected at the N6 back-cross generation (Fig. 1), when at a recombination frequency of 0.5, approximately 97% of autosomal genes should be of B6 origin (Festing, 1979).

The chromosomal location of *tda-1* is unknown (Eicher, 1988). It should be noted that during attempts to map *tda-1* a second putative autosomal gene, *tda-2*, was identified on chromosome 12 (Eicher, 1988). Nothing more will be said here about *tda-2*, since little information has been published.

Table I
Hypothesis on the Cause of Testis-Determining, Autosomal-1 Sex Reversal

Y	Genotype	XY gonad
Dom	*tda-1*/*tda-1*	Ovotestes/ovaries
Dom	*Tda-1*/*Tda-1* or *Tda-1*/*tda-1*	Testes
B6	*Tda-1*/*Tda-1* or *Tda-1*/*tda-1* or *tda-1*/*tda-1*	Testes

Note. A *M. m. domesticus* (Dom) Y chromosome and homozygosity for the C57BL/6 (B6) recessive *tda-1* allele results in the development of XY ovotestes or XY ovaries. Note that the B6 Y does not result in XYd sex reversal. This suggests that two different *Tdy* alleles exist: *Tdy*^{Dom} and *Tdy*$^{\mathrm{B6}}$

B. T-Associated Sex Reversal

1. Identification of the *Tas* Locus

T-associated sex reversal was discovered when the T hairpin tail (T^{hp}) mutation (Johnson, 1974, 1975) was being introduced into B6 (Washburn and Eicher, 1983). XY fetuses inheriting T^{hp} were found to have ovotestes or ovaries when examined at the N3 and N4 back-cross generations (Washburn *et al.,* 1990). XY siblings not inheriting T^{hp} and all XX fetuses, whether or not they inherited T^{hp}, developed normally. Like *tda-1* sex reversal, the B6 genomic background is crucial for the phenomenon. Sufficient back-crossing to B6 had to have taken place to ensure homozygosity for an unidentified B6 recessive gene (*tda-1*?, *tda-2*?). Furthermore, XY sex reversal does not occur when T^{hp} is introduced into the C3H/HeJ strain (Washburn and Eicher, 1983; Washburn *et al.*, 1990). T^{hp} is a deletion involving the proximal region of chromosome 17 (Bennett, 1975). It was proposed that the sex reversal associated with T^{hp} is due to the hemizygous expression of a gene on the normal chromosome 17 (Washburn and Eicher, 1983, 1989). This hypothetical gene, its locus demarcated by the T^{hp} deletion, was designated T-associated sex reversal (*Tas*).

2. Are *tda-1* and *Tas* XY Sex Reversal-Related Phenomena?

The T^{hp} mutation arose spontaneously in the AKR strain and is unusual in that it can be propagated only through males. Nearly all fetuses that inherit T^{hp} from a female die *in utero* or soon after birth (Cattanach, 1992; Johnson, 1974, 1975). Consequently, to back-cross T^{hp} into B6 to obtain *Tas* sex reversal, one must also back-cross the AKR Y chromosome (Table II). The Y chromosome of the AKR strain is derived from *M. m. domesticus* (Bishop *et al.*, 1985; Lamar and Palmer, 1984). As noted below (Section III,C,2), back-crossing the AKR Y chromosome into B6, giving rise to the B6.Y^{AKR} strain, results in a delay in testis differentiation such that XY fetuses have ovotestes at the fetal age of 14–16 days post coitus (dpc). (Gestation in the laboratory mouse lasts 19–20 days. Testis differentiation initiates at 12 dpc.) However, sex reversal in B6.Y^{AKR} is transient since with additional fetal development, testis differentiation catches up and no evidence of hermaphroditism is found at birth (Nagamine *et al.*, 1987b). It was previously suggested that *Tas* simply exacerbates the development of ovotestes in B6.Y^{AKR} such that the presence of ovotestes is no longer transient (Nagamine *et al.,* 1987b). Data obtained by Washburn and Eicher (1989) using the *T-Orleans* mutation, Tt^{Orl}, support this hy-

Table II
T-Associated Sex Reversal Requires Three Conditions: An AKR Y Chromosome, a C57BL/6 (B6) Genomic Background, and the Presence of Either T^{hp} or Tt^{Orl}

Y	Genomic background	T- allele	XY gonads
AKR	B6	—	Ovotestes (fetus only)
AKR	B6	T^{hp}	Ovotestes/ovaries (adult)
AKR	B6	Tt^{Orl}	Ovaries
AKR	C3H	T^{hp}	Testes
B6	B6	Tt^{Orl}	Testes

Note. As in tda-1 sex reversal, the substitution of a B6 Y results in normal testis determination. Note that an AKR Y on a B6 genomic background, in the absence of T^{hp} or Tt^{Orl}, results in transient development of ovotestes. The Tt^{Orl} allele has a more severe effect on XY sex reversal.

pothesis. Tt^{Orl} is a deletion in a partial *t* haplotype (Silver *et al.*, 1983). The deletion found in Tt^{Orl} overlaps that found in T^{hp} and, as with T^{hp}, genes on the normal chromosome 17 that are *trans* to the Tt^{Orl} deletion are expressed. However, unlike the case with T^{hp}, Tt^{Orl} can be transmitted through males or females and thus can be placed in association with the Y chromosome of any mouse strain. When Tt^{Orl} was back-crossed into B6 and placed in association with the Y chromosome of the AKR strain, all XY fetuses inheriting Tt^{Orl} developed ovaries. XY sex reversal did not occur when Tt^{Orl} was associated with the Y chromosome of the B6 strain (Table II) (Washburn and Eicher, 1989). Therefore, three conditions must be met to produce *Tas* sex reversal: (*a*) presence of an AKR Y (which is of *M. m. domesticus* origin), (*b*) a B6 genomic background with homozygosity of a recessive B6 gene (*tda-1*?, *tda-2*?), and (*c*) presence of the T^{hp} or Tt^{Orl} deletion.

It is clear that *tda-1* and *Tas* XY sex reversals both require a *M. m. domesticus* Y chromosome on the B6 genomic background. Therefore both conditions have been grouped under the general heading of XYd sex reversal.

C. Genetics of XYd Sex Reversal

1. XYd Sex Reversal and Interspecies Hybridization

One hypothesis is that XYd sex reversal is the result of the hybridization of two different species or subspecies of *Mus* (Eicher, 1988; Eicher and Washburn, 1983; Nallaseth, 1992; Washburn and Eicher, 1989). Two observations compromise this hypothesis. First, laboratory mice are not a distinct species or subspecies, but are a mixture of genomes derived from different populations of the house mouse, *M. musculus*. *Mus musculus* is currently recognized as a polytypic species composed of four subspecies: *M. m. domesticus*, *M. m. musculus*, *M. m. castaneus*, and *M. m. bactrianus* (Auffray *et al.*, 1990; Bonhomme and Guénet, 1989; She *et al.*, 1990). In Japan, hybridization between *M. m. musculus* and *M. m. castaneus* has resulted in a unique population, *M. m. molossinus* (Bonhomme *et al.*, 1989; Yonekawa *et al.*, 1988).

Although the derivation of the classical laboratory mouse strains from the wild subspecies is not known, it is clear that at least two of the progenitors were *M. m. domesticus* and *M. m. molossinus*. The evidence for this is that the vast majority of laboratory strains, including B6, have maternally inherited *M. m. domesticus* mitochondrial DNA and paternally inherited Y chromosomes from either *M. m. molossinus* (approximately 75% of laboratory strains, including B6) or *M. m. domesticus* (25%) (Bishop *et al.*, 1985; Ferris *et al.*, 1983; Nagamine *et al.*, 1992; Nishioka, 1987; Tucker *et al.*, 1992; Yonekawa *et al.*, 1980, 1982). The genetic contribution of *M. m. domesticus* to the laboratory strains is not limited to mitochondrial DNA. Indeed, numerous cytogenetic, biochemical, immunological, and molecular genetic studies suggest that many nuclear genes are derived from *M. m. domesticus* (Harada *et al.*, 1989; Moriwaki, 1987; Suzuki *et al.*, 1986; Watanabe *et al.*, 1989). Therefore, the sex reversal caused by placing the wild *M. m. domesticus* Y chromosome onto the B6 genomic background is not easily explained as the result of hybridization between two species or subspecies.

Second, not all *M. m. domesticus* Y chromosomes cause XY sex reversal when placed on the B6 genomic background (Section III,C,2.) (Biddle and Nishioka, 1988; Biddle *et al.*, 1991; Nagamine and Koo, 1989; Nagamine *et al.*, 1987b). If XY sex reversal is simply due to the incompatibility of genes between two different species or subspecies, one would expect all *M. m. domesticus* Y chromosomes to have the same effect.

It should be noted that controversy exists on whether *M. m. domesticus* is a subspecies (Auffray *et al.*, 1990; Bonhomme and Guénet, 1989; She *et al.*, 1990) or a separate species (Eicher *et al.*, 1989; Nallaseth, 1992; Tucker *et al.*, 1989, 1992). This, however, does not change the preceding arguments.

2. Yd Chromosomes and Testis Determination in B6

Recombinant DNA probes specific to the mouse Y chromosome have demonstrated that several laboratory mouse strains have *M. m. domesticus*

Ys (Bishop *et al.*, 1985; Lamar and Palmer, 1984; Nishioka and Lamothe, 1987). To date, there has been no report of XY sex reversal in these laboratory strains. To address the question whether all *M. m. domesticus* Y chromosomes are equally capable of causing sex reversal, the Yd chromosomes from the Posch-1, AKR, and SJL strains were back-crossed into B6, generating the B6.Y^{Dom}, B6.Y^{AKR}, and B6.Y^{SJL} strains, respectively. The progenitors of the Posch-1 strain were trapped in Tirano, Italy. AKR and SJL strains are laboratory mouse strains. Placing the Y chromosome of Posch-1 onto the B6 genomic background resulted in XYd sex reversal that was identical to that reported by Eicher *et al.*, (1982). Placing the Y chromosome of AKR onto the B6 genomic background resulted in XYd sex reversal but this was evident only in fetuses at ages 14–16 dpc. Morphologically and histologically the B6.Y^{AKR} fetal ovotestes were similar to those obtained from B6.Y^{Dom} except that the number of developing oocytes was considerably less in the affected regions. Surprisingly, with additional fetal development, B6.Y^{AKR} ovotestes masculinized and no evidence of hermaphroditism was seen postnatally either at the gross anatomical or histological levels (Nagamine and Koo, 1989; Nagamine *et al.*, 1987b). Thus, XYd sex reversal in B6.Y^{AKR} is transient and the ovarian regions of the gonads are lost. Back-crossing the SJL Y onto the B6 genomic background resulted in normal testis determination. Only 1/87 (1%) B6.Y^{SJL} fetuses had ovotestes. On the basis of these data one can conclude that different *M. m. domesticus* Y chromosomes have different testis-determining capabilities in B6. The Yd chromosomes of the three strains tested can be ranked in ability of inducing testes as follows: Posch-1 < AKR < SJL. These results were confirmed by other investigators (Biddle and Nishioka, 1988; Biddle *et al.*, 1991; Taketo *et al.*, 1991).

The genetic data suggested that the *M. m. domesticus* Y chromosomes of the Posch-1, AKR, and SJL strains were functionally heterogeneous with respect to testis determination in B6. Could this difference be demonstrated at the molecular level? A recombinant DNA probe specific for the murine Y chromosome (pY353/B) was used to address this question. The pY353/B probe is a 1.5-kb repetitive sequence that is spread along the entire length of the mouse Y chromosome but not the Sxr region (Bishop and Hatat, 1987; Bishop *et al.*, 1985). Southern blots containing *Eco*RI-digested genomic DNA from the B6.Y^{Dom}, AKR, and SJL/J strains were hybridized with the pY353/B probe. Three different hybridization patterns were observed, thus confirming the genetic heterogeneity of the *M. m. domesticus* Y chromosomes (Fig. 2) (Nagamine and Koo, 1989). We extended this analysis to the *M. m. domesticus* Y chromosomes of the SWR, NCS, BFM/2, WLA/76, and WMP strains. SWR and NCS are laboratory

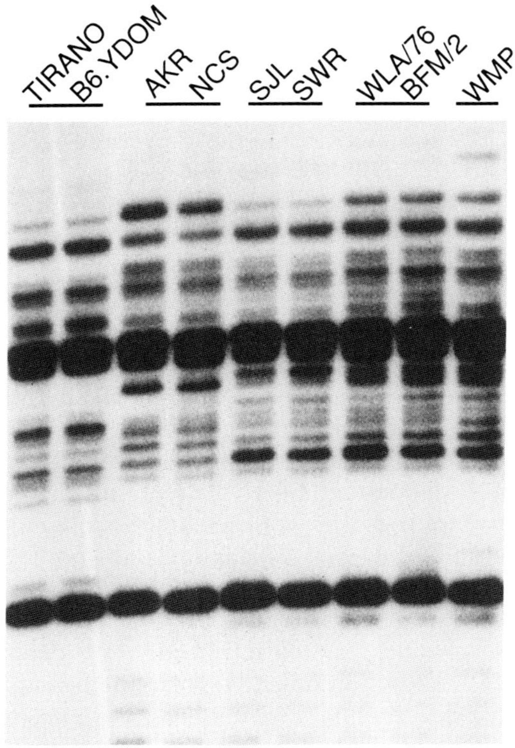

Figure 2 Southern blot hybridization patterns of different *M. m. domesticus* Y chromosomes obtained with probe pY353/B.

mouse strains. The last three strains were derived from wild populations of *M. m. domesticus* from southern France (BFM/2, WLA/76) and Tunisia (WMP) (Bonhomme and Guénet, 1989). Based on the intensity of hybridization or presence/absence of specific bands, five different hybridization patterns could be discerned (Fig. 2). Similar hybridization patterns for the SJL/J versus SWR strains and the AKR versus NCS strains predict that if the SWR and NCS Y chromosomes are back-crossed into B6, they will have testis-determining capabilities identical to those reported for the SJL/J and AKR Y chromosomes, respectively (Nagamine and Koo, 1989). Whether the WLA/76, BFM/2, and WMP Y chromosomes will result in sex reversal when placed on the B6 genomic background remains to be tested.

3. XYd Sex Reversal in Other Laboratory Mouse Strains

Is XYd sex reversal restricted to the B6 mouse strain? To address this question, B6.Y^{Dom} mice were mated to the inbred mouse strains SJL/J, AKR, BALB/c, C3H/An, C3H/HeJ, and C57BL/10J (B10), and the progeny examined as fetuses at 14 dpc and/or postnatally, for evidence of sex reversal. If we assume that XYd sex reversal is due only to the interaction of the *M. m. domesticus* Y chromosome and *Tda-1* or *tda-1* alleles, strains carrying the dominant *Tda-1* allele should produce normal F_1 males when crossed to B6.Y^{Dom} males. On the other hand, if these strains harbor the recessive *tda-1* allele, then XYd sex reversal should occur. Matings of B6.Y^{Dom} to SJL/J females resulted in normal F_1 male fetuses, suggesting that this strain has a dominant *Tda-1* allele (Nagamine *et al.*, 1987b). Since the SJL/J strain carries a *M. m. domesticus* Y chromosome, the dominant *Tda-1* allele may have come from the original male or female progenitor.

Mating AKR, BALB/c, C3H/An, and C3H/He females to B6.Y^{Dom} mice of male phenotype resulted in aberrant testicular differentiation in fetuses at 14–15 dpc, suggesting that XYd sex reversal can be transferred to these inbred strains. However, the sex reversal effect was ameliorated since fetuses with normal bilateral testes were found in all crosses except those with AKR. Furthermore, although the affected gonads were morphologically and histologically similar to B6.Y^{Dom} fetal ovotestes, they differed in that only three of nine affected gonads had developing oocytes. In addition, the number of developing oocytes was considerably lower than the number seen in B6.Y^{Dom} ovotestes. Either these strains have additional *tda-1* alleles or other genes in these strains were ameliorating the effect of the XYd sex reversal (Nagamine *et al.*, 1987b).

The B10 ♀ × B6.Y^{Dom} ♂ cross produced XY sex-reversed newborn mice, suggesting that, like B6, the B10 strain carries the recessive *tda-1* allele (C. M. Nagamine, unpublished observations). The B6 and B10 strains are sublines of the C57BL strain (Staats, 1985). Thus, it is not surprising that the B10 strain carries the *tda-1* allele. What is surprising is that a related strain, C58/J, lacks the *tda-1* allele (Eicher and Washburn, 1986). The C57BL and C58 strains were initiated in 1921, by the brother–sister matings of a single male to two different females (No. 57, No. 58). Since the C58/J strain lacks the recessive *tda-1* allele whereas both B6 and B10 have this allele, the data suggest that the *tda-1* allele may have been a mutation either in the C57BL strain or in the original female (No. 57). Alternatively, the *tda-1* allele was lost in the derivation of the C58/J strain. The NZB/BlNJ strain has also been reported to support XYd sex reversal (Eicher, 1988). However, it is unclear whether this effect is restricted to fetal stages as seen in B6.Y^{AKR} or is also seen postnatally as in B6.Y^{Dom} and (B10 × B6.Y^{Dom})F_1.

4. Stability of XYd Strains

In contrast to XY̶ females (Section II,C), adult XYd females have so far proven to be sterile (Eicher *et al.*, 1982; Taketo-Hosotani *et al.*, 1989). Propagation of the B6.Y^{Dom} strain requires, in each generation, fertile hermaphrodites (with male external genitalia) that can be back-crossed to B6 females.

Will repeated back-crossing to B6 ultimately result in only sterile XY females? Breeding records suggest that the proportion of XY females stabilized at approximately 50% of B6.Y^{Dom} XY progeny starting from about the N8 back-cross generation (Nagamine *et al.*, 1987b; Taketo-Hosotani *et al.*, 1989). A more definitive answer could be obtained by examining the progeny of overt hermaphrodites, i.e., mice with normal male external genitalia but internal organs consisting of a testicular gonad and a contralateral ovarian gonad. One can predict that if feminization increases with each back-cross generation, then overt hermaphrodites, which can be considered to be 50% feminized, should sire only other overt hermaphrodites or XY females. Five B6.Y^{Dom} overt hermaphrodites were mated to B6 females. Of their progeny, 22% (24/110) had *bilateral* testicular gonads and, therefore, were more masculinized than the sires (Nagamine and Koo, 1987). Obviously, the phenotype of the sons is not dependent on the phenotype of the fathers. In contrast to a recent report for another XYd sex-reversed strain (Nallaseth, 1992), one can conclude that the B6.Y^{Dom} strain is not in danger of extinction from the exclusive generation of XY females. Indeed, the B6.Y^{Dom} strain is now in the N24 back-cross generation.

D. The XYd Sex-Reversed Gonad

1. XYd Ovotestis

In normal mouse fetuses, testis differentiation is first detected under the dissecting microscope at 12 dpc. A tunica albuginea forms and seminiferous tubules develop in the gonad. In B6.Y^{Dom} mice, approximately 50% of the XY fetuses fail to develop any seminiferous tubules in either gonad. These gonads become XY ovaries and the fetuses become XY females (Eicher and Washburn, 1983; Nagamine *et al.*, 1987a; Taketo-Hosotani *et al.*, 1989). The remaining XY fetuses exhibit varying degrees of seminiferous tubule development.

Three points can be made about B6.Y^{Dom} XY gonads. First, seminiferous tubules initially form, if at all, in the mid-region of the fetal gonad. Tissues similar in appearance to undifferentiated gonadal tissues are present at the cranial and/or caudal poles (Eicher *et al.*, 1982). How can one determine whether the undifferentiated gonadal tissues are ovarian or simply delayed in testicular differentiation? In normal ovaries, oocytes initiate meiosis at about 15 dpc. In contrast, spermatocytes enter meiosis only postnatally. Thus the presence of meiotic germ cells can be used as a criterion for tissues following the ovarian differentiative pathway. Histological sections of B6.Y^{Dom} fetal gonads at or greater than 15 dpc revealed the presence of meiotic germ cells in the cranial and caudal poles, confirming that these tissues are not simply delayed in testicular differentiation but are actually ovarian tissues (Nagamine *et al.*, 1987a). The localization of testicular tissues to the mid-region and ovarian tissues to the cranial and caudal poles is not unique to XYd sex-reversed ovotestes. It is also a characteristic of ovotestes found in the BALB/cWt strain where hermaphroditism is due to XO/XY mosaicism (Whitten, 1975; Whitten *et al.*, 1979).

The second point is that it is not unusual for the gonads of XY fetuses in the same litter, or even the two gonads in a single XY fetus, to have different degrees of seminiferous tubule development. This suggests that the penetrance of *tda-1* XY sex reversal is incomplete. The adult XY gonad is masculinized to a greater or lesser extent, depending on the degree of seminiferous tubule development in the fetal gonad. Adult B6.Y^{Dom} gonads can be categorized into three phenotypes based on gross anatomical characteristics (Nagamine *et al.*, 1987a). "Testicular gonads" are morphologically similar to normal testes, although they may be slightly abnormal with respect to size or shape. "Ovotesticular gonads" are very small gonads with accessory ducts consisting of both Müllerian and Wolffian duct derivatives. "Ovarian gonads" appear as normal ovaries. Breeding and histological studies have demonstrated that spermatogenesis occurs in most testicular gonads and that most of the B6.Y^{Dom} mice with such gonads are fertile. However, histological sections also reveal that many testicular gonads harbor developing oocytes, usually in the rete testes. Therefore, despite their overall appearance, testicular gonads are, by definition, ovotestes (Nagamine *et al.*, 1987a).

The third point is that when the two gonads of a B6.Y^{Dom} XY fetus are compared, one finds a laterality with respect to seminiferous tubule development (Nagamine *et al.*, 1987b). In the majority of XY fetuses, the left gonad is found to contain more seminiferous tubules than the right. As a result, the left gonad is often more masculinized in adults. For example, if two testicular gonads are present, the left gonad is apt to be larger and/or more normal in appearance. Similarly, in an overt hermaphrodite, the left gonad will tend to be testicular, and the right, ovarian or ovotesticular. The laterality suggests that in the B6.Y^{Dom} fetus, the left gonad usually, but not always, develops at a faster rate than the right. A similar laterality was noted for T16H/XSxr hermaphroditic mice (Ward *et al.*, 1987, 1988). It should be noted that in human hermaphrodites the right gonad is often more masculinized than the left (Mittwoch, 1992; Mittwoch and Mahadevaiah, 1980). The cause of the laterality remains to be determined.

2. XYd Ovary

XYd ovaries are indistinguishable from XX ovaries during fetal development until about 16 dpc (Nagamine *et al.*, 1987a). Between 16 and 19 dpc, oocytes in the medullary region of XYd ovaries degenerate; as a result, this region is devoid of oocytes at birth. In normal ovaries, oocytes in the medullary region develop follicles. Upon atresia of these oocytes, the follicular cells give rise to tissues that are believed to play a key role in steroidogenesis. Thus, the premature loss of medullary oocytes in the fetal XYd ovary may contribute to an abnormal steroidogenic profile in the adult (Merchant-Larios and Taketo, 1991; Taketo-Hosotani *et al.*, 1989).

In contrast to those of the medullary region, oocytes in the cortex of the XYd ovary are not severely affected (Taketo-Hosotani *et al.*, 1989). Those oocytes develop normally and, at puberty, form Graafian follicles. However, unlike the case with normal XX ovaries where corporae luteae degenerate following ovulation, the corporae luteae of XYd ovarian gonads persist. The lutein cells in the corporae luteae undergo hyperplasia and invade surrounding ovarian stroma and follicles (Nagamine *et al.*, 1987a). After 6 weeks of age, oocytes become increasingly rare in XYd ovaries. Although XYd oocytes can be fertilized, the embryos do not survive past the blastula stage (Eicher and Washburn, 1986; Taketo-Hosotani *et al.*, 1989).

E. Theories on the Etiology of XYd Sex Reversal

1. Are *tda-1* and *Tas* Testis-Determining Genes?

The molecular basis for XYd sex reversal remains to be identified. It has been suggested that XYd sex reversal is due to a delay in initiation of testicular differentiation or the early expression of an ovary-determining gene (Burgoyne and Palmer, 1991; Eicher and Washburn, 1986; Taketo *et al.*, 1991). Eicher and colleagues have championed the view that *tda-1* and *Tas* are autosomal or pseudoautosomal testis-determining genes that are important for the normal timing of testis determination (Eicher and Washburn, 1983; Washburn *et al.*, 1990). An alternative view is that *tda-1* and *Tas* are not testis-determining genes but cause sex reversal by non-specifically retarding the growth of the XY gonad (Burgoyne and Palmer, 1991; Cattanach, 1987). This results in a delay in the initiation of testicular differentiation and enhances the possibility of ovarian differentiation. According to this hypothesis, the sex reversals seen with *tda-1* and *Tas* represent "genetic artifacts." In support of this is the observation that a deletion involving the dominant white spotting mutation (W^{19H}) can also exacerbate XYd sex reversal in B6.Y^{AKR} (Cattanach, 1987). Cloning of the *tda-1* and *Tas* genes should resolve this question.

2. XYd Sex Reversal and *Sry*

Homozygosity for a recessive B6 autosomal gene (*tda-1*) is obviously important in the initiation of XYd sex reversal. However, the fact that different *M. m. domesticus* Y chromosomes have different testis-determining capabilities on the same B6 genetic background suggests that differences must also exist on the *M. m. domesticus* Y chromosomes themselves.

Because mutations in *SRY* have been found in XY female patients, several investigators have searched for similar mutations in the *M. m. domesticus Sry* gene. The published *Sry* sequence is of *M. m. molossinus* origin (Nagamine *et al.*, 1992). A T–C transition was identified in the HMG-box domain of the *M. m. domesticus Sry* gene, resulting in the substitution of a threonine for an isoleucine (Graves and Erickson, 1992; Kunieda and Toyoda, 1992; Tucker *et al.*, 1992). However, the identification of a cytosine in the same position on the *Sry* gene of other *Mus* species makes it clear that the T–C transition is not the cause of XYd sex reversal (Graves and Erickson, 1992; Tucker *et al.*, 1992; C. M. Nagamine, unpublished data). It appears that the *Sry* gene of *M. m. molossinus* is the "mutant." The molecular basis for XYd sex reversal remains unknown.

IV. Conclusions

XY sex reversal in the mouse is represented by two mutations: XY̶ and XYd. The molecular basis for XY̶ sex reversal is the deletion of *Sry* and neighboring sequences. The molecular basis for XYd sex reversal remains unknown. The presence of certain, but not all, *M. m. domesticus* Y chromosomes on a permissive B6 genomic background is required for XYd sex reversal. Additional mutations (T^{hp}, Tt^{Orl}, W^{19H}) can exacerbate XYd sex reversal. XY and XYd sex reversals have played and will continue to play key roles in our understanding of the molecular genetics of mammalian sex determination.

Acknowledgments

This research was funded in part by NICHD Grant HD-28976.

References

Auffray, J.-C., Marshall, J. T., Thaler, L., and Bonhomme, F. (1990). Focus on the nomenclature of European species of *Mus*. *Mouse Genome* **88,** 7–8.

Bennett, D. (1975). The *T*-locus of the mouse. *Cell (Cambridge, Mass.)* **65,** 441–454.

Berta, P., Hawkins, J. R., Sinclair, A. H., Taylor, A., Griffiths, B. L., Goodfellow, P. N., and Fellous, M. (1990). Genetic evidence equating *SRY* and the testis-determining factor. *Nature (London)* **348,** 448–450.

Biddle, F. G., and Nishioka, Y. (1988). Assays of testis development in the mouse distinguish three classes of *domesticus*-type Y chromosome. *Genome* **30,** 870–878.

Biddle, F. G., Eales, B. A., and Nishioka, Y. (1991). A DNA polymorphism from five inbred strains of the mouse identifies a functional class of *domesticus*-type Y chromosome that produces the same phenotypic distribution of gonadal hermaphrodites. *Genome* **34,** 96–104.

Bishop, C. E. (1992). Mouse Y chromosome. *Mamm. Genome* **3,** S289–S293.

Bishop, C. E., and Hatat, D. (1987). Molecular cloning and sequence analysis of a mouse Y chromosome RNA transcript expressed in the testis. *Nucleic Acids Res.* **15,** 2959–2969.

Bishop, C. E., Boursot, P., Baron, B., Bonhomme, F., and Hatat, D. (1985). Most classical *Mus musculus domesticus* laboratory mouse strains carry a *Mus musculus musculus* Y chromosome. *Nature (London)* **315,** 70–72.

Bonhomme, F., and Guénet, J.-L. (1989). The wild house mouse and its relatives. *In* "Genetic Variants and Strains of the Laboratory Mouse" (M. F. Lyon and A. G. Searle, eds.), pp. 649–662. Oxford Univ. Press, Oxford.

Bonhomme, F., Miyashita, N., Boursot, P., Catalan, J., and Moriwaki, K. (1989). Genetical variation and polyphyletic origin in Japanese *Mus musculus*. *Heredity* **63,** 299–308.

Burgoyne, P. S., and Palmer, S. J. (1991). The genetics of XY sex reversal in the mouse and other mammals. *Semin. in Dev. Biol.* **2,** 277–284.

Burgoyne, P. S., Levy, E. R., and McLaren, A. (1986). Spermatogenic failure in male mice lacking H-Y antigen. *Nature (London)* **320,** 170–172.

Cattanach, B. M. (1987). Sex-reversed mice and sex determination. *Ann. N. Y. Acad. Sci.* **513,** 27–39.

Cattanach, B. M. (1992). T^{25H}: A radiation-induced deletion affecting the *T*, *qk* and *Tme* loci. *Mouse Genome* **90,** 209–210.

Cattanach, B. M., Pollard, C. E., and Hawkes, S. G. (1971). Sex-reversed mice: XX and XO males. *Cytogenetics* **10,** 318–337.

Disteche, C. M., Zacksenhaus, E., Adler, D. A., Bressler, S. L., Keitz, B. T., and Chapman, V. M. (1992). Mapping and expression of the ubiquitin-activating enzyme E1 (Ube1) gene in the mouse. *Mamm. Genome* **3,** 156–161.

Eicher, E. M. (1982). Primary sex determining genes in mice. *In* "Prospects for Sexing Mammalian Sperm" (R. P. Amann and G. E. J. Siedel, eds.), pp. 121–135. Colorado Associated University Press, Boulder.

Eicher, E. M. (1988). Autosomal genes involved in mammalian primary sex determination. *Philos. Trans. R. Soc. London Ser., B* **322,** 109–118.

Eicher, E. M., and Washburn, L. L. (1983). Inherited sex reversal in mice: Identification of a new primary sex-determining gene. *J. Exp. Zool.* **228,** 297–304.

Eicher, E. M., and Washburn, L. L. (1986). Genetic control of primary sex determination in mice. *Annu. Rev. Genet.* **20,** 327–360.

Eicher, E. M., Beamer, W. G., Washburn, L. L., and Whitten, W. K. (1980). A cytogenetic investigation of inherited true hermaphroditism in BALB/cWt mice. *Cytogenet. Cell Genet.* **28,** 104–115.

Eicher, E. M., Washburn, L. L., Whitney, J. B. I., and Morrow, K. E. (1982). *Mus poschiavinus* Y chromosome in the C57BL/6J murine genome causes sex reversal. *Science* **217,** 535–537.

Eicher, E. M., Hutchison, K. W., Phillips, S. J., Tucker, P. K., and Lee, B. K. (1989). A repeated segment on the mouse *Y* chromosome is composed of retroviral-related, Y-enriched and Y-specific sequences. *Genetics* **122,** 181–192.

Ferris, S. D., Sage, R. D., Prager, E. M., Ritte, U., and Wilson, A. C. (1983). Mitochondrial DNA evolution in mice. *Genetics* **105,** 681–721.

Festing, M. F. W. (1979). "Inbred Strains in Biomedical Research." Oxford University Press, New York.

Foster, J. W., Brennan, F. E., Hampikian, G. K., Goodfellow, P. N., Sinclair, A. H., Lovell-Badge, R., Selwood, L., Renfree, M. B., Cooper, D. W., and Graves, J. A. M. (1992). Evolution of sex determination and the Y chromosome: SRY-related sequences in marsupials. *Nature (London)* **359,** 531–533.

Goldberg, E. H., McLaren, A., and Reilly, B. (1991). Male antigen defined serologically does not identify a factor responsible for testicular development. *J. Reprod. Immunol.* **20,** 305–309.

Graves, P. E., and Erickson, R. P. (1992). An amino acid change in the DNA-binding region of Sry, found in *Mus musculus domesticus* and other species, does not explain C57BL/6J-Y^{Dom} sex reversal. *Biochem. Biophys. Res. Commun.* **185,** 310–316.

Gropp, A., and Winking, H. (1983). XY sex reversal in mice. *Genet. Res. Camb.* **41,** 314.

Gubbay, J., Collignon, J., Koopman, P., Capel, B., Economou, A., Münsterberg, A., Vivian, N., Goodfellow, P., and Lovell-Badge, R. (1990a). A gene mapping to the sex-determining region of the mouse Y chromosome is a member of a novel family of embryonically expressed genes. *Nature (London)* **346,** 245–250.

Gubbay, J., Koopman, P., Collignon, J., Burgoyne, P., and Lovell-Badge, R. (1990b). Normal structure and expression of *Zfy* genes in XY female mice mutant in *Tdy*. *Development (Cambridge, UK)* **109,** 647–653.

Harada, Y.-N., Bonhomme, F., Natsuume-Sakai, S., Tomita, T., and Moriwaki, K. (1989). Serological survey of complement factor H in common laboratory and wild mice: A new third allotype. *Immunogenetics* **29,** 148–154.

Harley, V. R., Jackson, D. I., Hextall, P. J., Hawkins, J. R., Berkovitz, G D., Sockanathan, S., Lovell-Badge, R., and Goodfellow, P. N. (1992). DNA binding activity of recombinant SRY from normal males and XY females. *Science* **255,** 453–456.

Hawkins, J. R., Taylor, A., Berta, P., Levilliers, J., Van der Auwera, B., and Goodfellow,

P. N. (1992). Mutational analysis of SRY: Nonsense and missense mutations in XY sex reversal. *Hum. Genet.* **88,** 471–474.

Hodgkin, J. (1990). Sex determination compared in Drosophila and Caenorhabditis. *Nature (London)* **344,** 721–728.

Hodgkin, J. (1992). Genetic sex determination mechanisms and evolution. *BioEssays* **14,** 253–261.

Jäger, R. J., Anvret, M., Hall, K., and Scherer, G. (1990). A human XY female with a frame shift mutation in the candidate testis-determining gene *SRY. Nature (London)* **348,** 452–454.

Johnson, D. R. (1974). Hairpin-Tail: A case of post-reductional gene action in the mouse egg? *Genetics* **76,** 795–805.

Johnson, D. R. (1975). Further observations on the hairpin-tail (T^{hp})mutation in the mouse. *Genet. Res.* **24,** 207–213.

Kay, G. F., Ashworth, A., Penny, G. D., Dunlop, M., Swift, S., Brockdorff, N., and Rastan, S. (1991). A candidate spermatogenesis gene on the mouse Y chromosome is homologous to ubiquitin-activating enzyme E1. *Nature (London)* **354,** 486–489.

Koopman, P., Münsterberg, A., Capel, B., Vivian, N., and Lovell-Badge, R. (1990). Expression of a candidate sex-determining gene during mouse testis differentiation. *Nature (London)* **348,** 450–452.

Koopman, P., Gubbay, J., Vivian, N., Goodfellow, P., and Lovell-Badge, R. (1991). Male development of chromosomally female mice transgenic for *Sry. Nature (London)* **351,** 117–121.

Kunieda, T., and Toyoda, Y. (1992). Nucleotide sequence of mouse Sry gene is different between Y chromosomes originating from *Mus musculus musculus* and *Mus musculus domesticus. Genomics* **13,** 236–237.

Kuwabara, P. E., and Kimble, J. (1992). Molecular genetics of sex determination in *C. elegans. Trends Genet.* **8,** 164–168.

Lamar, E. E., and Palmer, E. (1984). Y-encoded, species-specific DNA in mice: Evidence that the Y chromosome exists in two polymorphic forms in inbred strains. *Cell (Cambridge, Mass.)* **37,** 171–177.

Lovell-Badge, R., and Robertson, E. (1990). XY female mice resulting from a heritable mutation in the murine primary testis determining yene, *Tdy. Development (Cambridge, UK)* **109,** 635–646.

Mardon, G., Mosher, R., Disteche, C. M., Nishioka, Y., McLaren, A., and Page, D. C. (1989). Duplication, deletion, and polymorphism in the sex-determining region of the mouse Y chromosome. *Science* **243,** 78–80.

McElreavey, K. D., Vilain, E., Boucekkine, C., Vidaud, M., Jaubert, F., Richaud, F., and Fellous, M. (1992). XY sex reversal associated with a nonsense mutation in SRY. *Genomics* **13,** 838–840.

McLaren, A. (1988). Sex determination in mammals. *Trends Genet.* **4,** 153–157.

McLaren, A. (1991). Development of the mammalian gonad: The fate of the supporting cell lineage. *BioEssays* **13,** 151–156.

McLaren, A., Simpson, E., Tomonari, K., Chandler, P., and Hogg, H. (1984). Male sexual differentiation in mice lacking H-Y antigen. *Nature (London)* **312,** 552–555.

McLaren, A., Simpson, E., Epplen, J. T., Studer, R., Koopman, P., Evans, E. P., and Burgoyne, P. S. (1988). Location of the genes controlling H-Y antigen expression and testis determination on the mouse Y chromosome. *Proc. Natl. Acad. Sci. U.S.A.* **85,** 6442–6445.

Merchant-Larios, H., and Taketo, T. (1991). Testicular differentiation in mammals under normal and experimental conditions. *J. Electron Microsc. Tech.* **19,** 158–171.

Mitchell, M. J., Woods, D. R., Tucker, P. K., Opp, J. S., and Bishop, C. E. (1991). Homology of a candidate spermatogenic gene from the mouse Y chromosome to the ubiquitin-activating enzyme E1. *Nature (London)* **354,** 483–486.

Mitchell, M. J., Woods, D. R., Wilcox, S. A., Graves, J. A. M., and Bishop, C. E. (1992). Marsupial Y chromosome encodes a homologue of the mouse Y-linked candidate spermatogenesis gene *Ube1y*. *Nature (London)* **359,** 528–531.

Mittwoch, U. (1992). Sex determination and sex reversal: Genotype, phenotype, dogma and semantics. *Hum. Genet.* **89,** 467–479.

Mittwoch, U., and Mahadevaiah, S. (1980). Additional growth—a link between mammalian testes, avian ovaries, gonadal asymmetry in hermaphrodites and the expression of H-Y antigen. *Growth* **44,** 287–300.

Moriwaki, K. (1987). Genetic significance of laboratory mice in biomedical research. *In* "Animal Models: Assessing the Scope of Their Use in Biomedical Research," pp. 53-72. Alan R. Liss, Inc., New York.

Nagamine, C. M., and Koo, G. C. (1987). Evidence that the X-Y pairing/recombination region may be involved in *tda-1* inherited sex reversal. *In* "Genetic Markers of Sex Differentiation" (F. P. Haseltine, M. E. McClure, and E. H. Goldberg, eds.), pp. 127–138. Plenum, New York.

Nagamine, C. M., and Koo, G. C. (1989). *tda-1* XY sex reversal in the mouse. *In* "Evolutionary Mechanisms in Sex Determination" (S. S. Wachtel, ed.), pp. 181–187. CRC Press, Boca Raton, FL.

Nagamine, C. M., Taketo, T., and Koo, G. C. (1987a). Morphological development of the mouse gonad in *tda-1* XY sex reversal. *Differentiation (Berlin)* **33,** 214–222.

Nagamine, C. M., Taketo, T., and Koo, G. C. (1987b). Studies on the genetics of *tda-1* XY sex reversal in the mouse. *Differentiation (Berlin)* **33,** 223–231.

Nagamine, C. M., Chan, K., Kozak, C. A., and Lau, Y.-F. C. (1989). Chromosome mapping and expression of a putative testis-determining gene in mouse. *Science* **243,** 80–83.

Nagamine, C. M., Chan, K., Hake, L. E., and Lau, Y.-F. C. (1990). The two candidate testis-determining Y genes (Zfy-1 and Zfy-2) are differentially expressed in fetal and adult mouse tissues. *Genes Dev.* **4,** 63–74.

Nagamine, C. M., Nishioka, Y., Moriwaki, K., Boursot, P., Bonhomme, F., and Lau, Y.-F. C. (1992). The musculus-type Y chromosome of the laboratory mouse is of Asian origin. *Mamm. Genome* **3,** 84–91.

Nallaseth, F. S. (1992). Sequence instability and functional inactivation of murine Y chromosomes can occur on a specific genetic background. *Mol. Biol. Evol.* **9,** 331–365.

Nasrin, N., Buggs, C., Kong, X. F., Carnazza, J., Goebl, M., and Alexander-Bridges, M. (1991). DNA-binding properties of the product of the testis-determining gene and a related protein. *Nature (London)* **354,** 317–320.

Nishioka, Y. (1987). Y-chromosomal DNA polymorphism in mouse inbred strains. *Genet. Res.* **50,** 69–72.

Nishioka, Y., and Lamothe, E. (1987). The *Mus musculus musculus* type *Y* chromosome predominates in Asian house mice. *Genet. Res.* **50,** 195–198.

Page, D. C., Fisher, E. M. C., McGillivray, B., and Brown, L. G. (1990). Additional deletion in sex determining region of human Y chromosome resolves paradox of X,t(Y;22) female. *Nature (London)* **346,** 279–281.

Patek, C. E., Kerr, J. B., Gosden, R. G., Jones, K. W., Hardy, K., Muggleton-Harris, A. L., Handyside, A. H., Whittingham, D. G., and Hooper, M. L. (1991). Sex chimaerism, fertility and sex determination in the mouse. *Development (Cambridge UK)* **113,** 311–325.

Roberts, C., Weith, A., Passage, E., Michot, J. L., Mattei, M. G., and Bishop, C. E. (1988).

Molecular and cytogenetic evidence for the location of *Tdy* and *Hya* on the mouse Y chromosome short arm. *Proc. Natl. Acad. Sci. U.S.A* **85,** 6446–6449.

She, J. X., Bonhomme, F., Boursot, P., Thaler, L., and Catzeflis, F. (1990). Molecular phylogenies in the genus *Mus:* Comparative analysis of electrophoretic, scnDNA hybridization, and mtDNA RFLP data. *Biol. J. Linn. Soc.* **41,** 83–103.

Silver, L. M., Lukralle, D., and Garrels, J. I. (1983). T^{Orl} is a novel, variant form of mouse chromosome 17 with a deletion in a partial t haplotype. *Nature (London)* **301,** 422–424.

Simpson, E. M., and Page, D. C. (1991). An interstitial deletion in mouse Y chromosomal DNA created a transcribed Zfy fusion gene. *Genomics* **11,** 601–608.

Sinclair, A. H., Berta, P., Palmer, M. S., Hawkins, J. R., Griffiths, B. L., Smith, M. J., Foster, J. W., Frischauf, A.-M., Lovell-Badge, R., and Goodfellow, P. N. (1990). A gene from the human sex-determining region encodes a protein with homology to a conserved DNA-binding motif. *Nature (London)* **346,** 240–244.

Singh, L., and Jones, K. (1982). Sex reversal in the mouse (*Mus musculus*) is caused by a recurrent nonreciprocal crossover involving an X and an aberrant Y chromosome. *Cell (Cambridge, Mass.)* **28,** 205–216.

Staats, J. (1985). Standardized nomenclature for inbred strains of mice: Eighth listing. *Cancer Res.* **45,** 945–977.

Suzuki, H., Miyashita, N., Moriwaki, K., Kominami, R., Muramatsu, M., Kanehisa, T., Bonhomme, F., Petras, M. L., Yu, Z.-C., and Lu, D.-Y. (1986). Evolutionary implication of heterogeneity of the nontranscribed spacer region of ribosomal DNA repeating units in various subspecies of *Mus musculus. Mol. Biol. Evol.* **3,** 126–137.

Taketo, T., Saeed, J., Nishioka, Y., and Donahoe, P. K. (1991). Delay of testicular differentiation in the $B6.Y^{Dom}$ ovotestis demonstrated by immunocytochemical staining for Müllerian inhibiting substance. *Dev. Biol.* **146,** 386–395.

Taketo-Hosotani, T., Nishioka, Y., Nagamine, C. M., Villalpando, I., and Merchant-Larios, H. (1989). Develoment and fertility of ovaries in the $B6.Y^{Dom}$ sex-reversed female mouse. *Development (Cambridge, UK)* **107,** 95–105.

Tarkowski, A. K. (1964). True hermaphroditism in chimaeric mice. *J. Embryol. Exp. Morphol.* **12,** 735–757.

Tucker, P. K., Lee, B. K., and Eicher, E. M. (1989). Y-chromosome evolution in the subgenus *Mus* (Genus *Mus*). *Genetics* **122,** 169–179.

Tucker, P. K., Lee, B. K., Lundrigan, B. L., and Eicher, E. M. (1992). Geographic origin of the Y chromosomes in "old" inbred strains of mice. *Mamm. Genome* **3,** 254–261.

Vilain, E., McElreavey, K., Jaubert, F., Raymond, J.-P., Richaud, F., and Fellous, M. (1992). Familial case with sequence variant in the testis-determining region associated with two sex phenotypes. *Am. J. Hum. Genet.* **50,** 1008–1011.

Ward, H. B., McLaren, A., and Baker, T. G. (1987). Gonadal development in T16H/X*Sxr* hermaphrodite mice. *J. Reprod. Fertil.* **81,** 295–300.

Ward, H. B., Baker, T. G., and McLaren, A. (1988). A histological study of the gonads of T16H/X*Sxr* hermaphrodite mice. *J. Anat.* **158,** 65–76.

Washburn, L. L., and Eicher, E. M. (1983). Sex reversal in XY mice caused by dominant mutation on chromosome 17. *Nature (London)* **303,** 338–340.

Washburn, L. L., and Eicher, E. M. (1989). Normal testis determination in the mouse depends on genetic interaction of a locus on chromosome 17 and the Y chromosome. *Genetics* **123,** 173–179.

Washburn, L. L., Lee, B. K., and Eicher, E. M. (1990). Inheritance of T-associated sex reversal in mice. *Genet. Res.* **56,** 185–191.

Watanabe, T., Miyashita, N., Nishimura, M., Saitou, N., Hayashi, Y., and Moriwaki, K. (1989). Evolutionary relationships between laboratory mice and subspecies of *Mus muscu-*

lus based on the restriction fragment length variants of the chymotrypsin gene at the *Prt-2* locus. *Biochem. Genet.* **27,** 119–130.

Whitten, W. K. (1975). Chromosomal basis for hermaphrodism in mice. *In* "The Biology of Reproduction" (D. L. Market and J. Papaconstantinou, eds.), pp. 189-205. Academic Press, New York.

Whitten, W. K., Beamer, W. G., and Byskov, A. G. (1979). The morphology of fetal gonads of spontaneous mouse hermaphrodites. *J. Embryol. Exp. Morphol.* **52,** 63–78.

Yonekawa, H., Moriwaki, K., Gotoh, O., Watanabe, J., Hayashi, J.-I., Miyashita, N., Petras, M. L., and Tagashira, Y. (1980). Relationship between laboratory mice and the subspecies *Mus musculus domesticus* based on restriction endonuclease cleavage patterns of mitochondrial DNA. *Jpn. J. Genet.* **55,** 289–296.

Yonekawa, H., Moriwaki, J., Gotoh, O., Miyashita, N., Migita, S., Bonhomme, F., Hjorth, J. P., Petras, M. L., and Tagashira, Y. (1982). Origins of laboratory mice deduced from restriction patterns of mitochondrial DNA. *Differentiation (Berlin)* **22,** 222–226.

Yonekawa, H., Moriwaki, K., Gotoh, O., Miyashita, N., Matsushima, Y., Shi, L., Su Cho, W., Zhen, X.-L., and Tagashira, Y. (1988). Hybrid origin of Japanese mice "*Mus musculus molossinus*": Evidence from restriction analysis of mitochondrial DNA. *Mol. Biol. Evol.* **5,** 63–78.

6

Conservation of the X-Linkage Group *in Toto* by All Eutherian Mammals

Susumu Ohno
Theoretical Biology Department
Beckman Research Institute of The City of Hope
Duarte, California 91010

Molecular Genetics of Sex Determination

I. Introduction

The presence of males and females in the same population is usually rationalized by statements such as: "Sexual reproduction generates greater genetic diversity among the progeny by cross fertilization." Overlooked is the fact that hermaphroditic species accomplish the same end by having the same individual act as a male at one time and a female at another: Witness copulating pairs of garden snails in your flower beds. They are true hermaphrodites. Equally unjustified is the purported evolutionary trend toward bisexual reproduction. In fact, the opposite trend is obvious in the plant kingdom. Sex chromosomes as prominently dimorphic as those of mammals are commonly found in mosses and liverworts of the phylum Bryophyta, whereas the more modern flowering plants of the phylum Spermatophyta are mostly hermaphroditic, the same flower bearing pistils and stamen. Yet cross-fertilization is again the rule in modern flowering plants. In order to avoid self-fertilization, some of them, such as mustard plants of the genus *Brassica,* have even developed an elaborate self versus non-self-discrimination mechanism mediated by extremely polymorphic pollen–stigma incompatibility glycoproteins (Nasrallah *et al.,* 1987).

Over the years, we have come to believe that all attempts to rationalize what one sees in the biological world are destined ultimately to fail. For the most part, evolutionary trends appear to have been determined by the fickle finger of fate; consider the misdesign of the vertebrate eye: The retina does not face the lens but points backward. The eyes of invertebrates—those of squids and octopi—are properly designed, with the retina facing the lens. One is tempted to quote William Shakespeare here: "A thing is neither good nor bad, only thinking makes it so."

Accordingly, with an open mind as to its reason, we shall deal with the evolutionary conservation of the X-linkage group *in toto* by all eutherian mammals. Since it is a fait accompli, we might be permitted a little reminiscence.

II. Beginning of the Realization

After a partial hepatectomy, the liver regains its original size within several days. Thus, at the 24th and 48th hour after the partial hepatectomy, one can encounter waves of mitotic figures among regenerating heptocytes. In 1959, we compared male and female mitotic figures of rat hepatocytes and found the female-specific Barr sex chromatin body of the previous interphase appeared as one chromosome of substantial size condensed along its entire length in the subsequent mitotic prophase. Diploid female prophases contained one such heteropycnotic chromosome, and tetraploid prophases, two such, whereas no comparable chromosome was seen in diploid or tetraploid male prophase figures. Thus, it became evident that the Barr sex chromatin body is one of the two X-chromosomes of the female that is rendered inert by the virtue of its heteropycnotic state (Ohno *et al.*, 1959).

Ten years earlier, the Barr sex chromatin body had been found in large interphase nuclei of the neurons of female cats (Barr and Bertram, 1949). What attracted my attention was the subsequent remark, by Murray Barr, that the size of the sex chromatin body was the same in the cat and human, and all other mammalian species exhibiting sexual dimorphism of somatic interphase nuclei. Taken at face value, this suggested that the X-chromosome might be the same absolute size in diverse mammalian species. Hence, this was the first hint of conservation *in toto*. Yet, such a suspicion of conservation of one linkage group was meaningless unless the entire collection of linkage groups (the genome) has remained the same in all eutherian mammals. Indeed, the uniform genome was suggested by the observation that individual chromosomes were very small in species with high diploid chromosome numbers, such as the dog with 78 chromosomes, whereas they were very large in species with low numbers such as the Chinese hamster with 22 chromosomes.

Since the previously suggested conservation of genome size in eutherian mammals, a collaboration was established with Niels Atkin who operated a very sensitive microphotometric unit at the Mount Vernon Hospital near London. Although our results were not published until 1965, it became immediately evident that the genome size of man is indistinguishable from the genome size or more familiar mammals, such as dog, cat, horse, cattle, rat, and mouse (Atkin *et al.*, 1965). The concept of a uniform mammalian genome comprising 3×10^9 DNA base pairs (bp) was thus established. Although it was later shown that a few exceptional mammalian species loaded with constitutive heterochromatin have genomes substantially larger than the standard, whereas those few species that are almost devoid of constitutive heterochromatin have noticeably smaller genomes, these exceptions involving junk DNA validated rather than discredited the concept that all mammals are endowed with the same set of genes.

Given a mammalian genome of uniform size, if the X-chromosome has remained the same size, it should be one of the largest elements in species with high diploid chromosome numbers, whereas it should be one of the smaller elements in species with low diploid numbers. Indeed, the X chromosome was one of the largest, if not the largest, element in dog and cattle with 78 and 60 chromosomes, respectively. But we encountered a snag in species with low diploid chromosome numbers. The expectation was fulfilled only in the cat, with 38 chromosomes; the X chromosome was still very large in other species with low diploid numbers.

Together with the husband and wife team of Willy and Maria Beçak, visiting Brazilian scientists, we circumvented the above problem by realizing that the excessive baggage of the larger X chromosomes comprised constitutive heterochromatin. Hence, the concept of the original-type X and duplicate- and triplicate-type X was born (Ohno *et al.*, 1964). The Beçaks also made a historical discovery in identifying the first tetraploid species of frog. Until then, the prevailing thought was that polyploidy, which played so important a role in the evolution of plants, had no role in the evolution of animals. Subsequently, we uncovered many instances of polyploid evolution in fish and amphibians. My view is that the ancestor of mammals underwent at least one round of polyploid evolution at either the fish or the amphibian stage (Ohno, 1967). Therefore, traces of homology should remain among today's independent linkage groups. Nevertheless, polyploid evolution is irrelevant to the conservation *in toto* of the X-linkage group, since at the time of the firm establishment of the chromosomal sex determining mechanism, a mammalian ancestor forever forfeited the opportunity for further polyploid evolution.

III. Of Glucose-6-Phosphate Dehydrogenases and Mules, Hinnies, and Hares

In 1964, it was already known that the human X-linked traits, hemophilia A and B, caused by deficiency of coagulation factors VIII and IX, are also X-linked in the dog (Hutt, 1953). Hemophilia A has been identified in the Aberdeen terrier and the greyhound, whereas hemophilia B was found among the Cairn terrier. Hutt was apparently intrigued by these and other indications of homologous X-linked traits in divergent mammals, to the extent that his 1953 article was entitled "Homologous X-Linked Mutations in Man and Other Mammals." For example, the human X-linked trait known as "anhydrotic ectodermal dysplasia" is characterized by developmental absence of sweat glands, as the name implies. A similar trait, apparently X-linked, had already been described in cattle by French authors. Hutt, however, interpreted such homology as an indication that X-linkage confers selective advantage to certain individual genes. In the 1950s, the Darwinian notion of natural selection reigned supreme, for it was yet to be tempered by the realization that protein amino acid sequences change very little with time, and that the maintenance of the status quo is an all consuming concern of every multicellular organism.

At any rate, we decided to see whether X-linkage could be established in mammalian species other than the human for glucose-6-phosphase dehydrogenase (G6PD), the first enzyme of the pentose phosphate shunt. This decision was influenced in no small part by the presence at our institution of Ernest Beutler, who had just demonstrated the cohabitation of two populations of erythrocytes in women heterozygous for G6PD deficiency. Our attempts to detect electrophoretic variants of G6PD, analogous to A+ and B+ of man, in rats, mice, and other common laboratory animals, totally failed. Since it appeared that this enzyme is monomorphic in most mammalian species, we decided to utilize interspecific hybrids for this purpose.

When a jackass is crossed with a mare, a mule results. The reciprocal cross (stallion with jenny) results in a hinny. Since we already knew that G6PD of the horse could be distinguished from that of the donkey by electrophoresis, the proposition was a simple one: If G6PD were X-linked in both parental species, male mules should inherit only the horse-type G6PD from their mothers, whereas we should see only the donkey-type G6PD in male hinnies. The problem was that whereas mules were still bred in California in reasonable abundance, hinnies, which seem to combine the worst characteristics of their parental species, were not. Accordingly, we briefly engaged ourselves in purchasing and breeding of these hybrids. The net result was that the gene encoding G6PD was found to reside indeed on the equine X chromosome (Trujillo *et al.,* 1965). While the above project was being held in abeyance for the want of hinnies, we were informed by Ingmar Gustavsson that the Royal Veterinary College of Sweden, then in Stockholm, had successfully produced reciprocal hybrids between the common European hare (*Lepus europaeus*) and the northern variegating hare (*Lepus timidus*). Through use of these reciprocal hybrids, we quickly established the X-linkage of G6PD in the order Lagomorpha as well (Ohno *et al.,* 1965).

IV. Of *Tfm* Loci, Androgen Receptors, and X-Linked Genes for Primary Sex Determination

Soon, through the use of somatic hybrids, the list of proteins encoded by X-linked genes grew and the conservation *in toto* of the X-linkage group became a *fait accompli.* Yet, all of these proteins had nothing whatsoever to do with sexual development as such. Inasmuch as males and females live utilizing the same anabolic and catabolic pathways, the above was rather expected. Nevertheless, for its namesake, there has to be at least one X-linked gene that regulates sexual development. Since Jost's classical experiments on rabbit fetuses (Jost, 1947), it had become evident that the basic developmental plan of mammals is to follow the feminine pathway and that males are merely females modified by fetal exposure to testosterone.

The above tenet is highlighted by the X-linked testicular feminization syndrome long known in human genetics. In its complete form, this trait gives a beautifully feminine phenotype to XY individuals with abdominal testes. The affected individual is totally resistant to testosterone manufactured by her (his) testes. Eventually, unused testosterone is converted to estradiol by the peripheral aromatase. Thus, in this mutant, the testis is an indirect source of female hormone, estradiol, which maintains the feminine phenotype. This immediately withers upon castration. In 1970, the homologous X-linked mutation was found in the mouse (Lyon and Hawkes, 1970). After demonstrating complete resistance to administered androgen in these $X^{Tfm}Y$ mutant mice, we engaged in cooperation with Gordon Tomkins, who was doing a series of studies on how the hydrocortisone receptor mediates induction of tyrosine aminotransferase in cultured hepatoma cells. The *Tfm* defect was seen in the androgen receptor (Gehring *et al.*, 1971). Although androgen induces different sets of proteins in its divergent target organs, all inductions are mediated by only one species of receptor encoded by the X-linked *Tfm* locus. This androgen receptor shows binding affinity to both 5α-dihydrotestosterone and testosterone, albeit the affinity to the former is twofold higher than that to the latter. It follows that the X-linked locus encoding this single species of receptor is the regulatory locus controlling all aspects of masculine development that are inducible only by androgen. Furthermore, there appears to be no controlling center higher than the *Tfm* locus, for the androgen receptor is found in all cell types in males and females. The gene, whose expression is constitutive, has no need for a higher regulatory center.

In the mouse, the nature of the mutation affecting the androgen receptor locus and rendering it defunct has now been fully elucidated (Gaspar *et al.*, 1991). The mutated *Tfm* receptor is truncated, being reduced to a size of roughly 66 kDa compared to the wild-type size of 110 kDa. This reduction is due to deletion of a single base (C), which produces a reading frameshift, thereby causing a premature chain termination 46 residues downstream of the deleted site. Accordingly, the mutated *Tfm* receptor lacks DNA and androgen-binding domains.

Another lesson learned from this receptor was the importance of mesenchyme in directing testosterone-dependent organogenesis, i.e., development of Wolffian ducts into epididymides, vasa deferentia, and seminal vesicles (Cunha, 1972). When *Tfm*/+ heterozygotes were sex-reversed to males by the *Sxr* gene (Cattanach *et al.*, 1971), which is a duplicate of the Y-linked testis-determining gene that crosses over from the Y chromosome to the X during male meiosis (Singh and Jones, 1982), it was found that as long as the underlining mesenchymal cells are androgen-sensitive wild-type, Wolffian duct derivatives made of androgen-insensitive *Tfm*-epithelial cells are formed. However, being androgen-insensitive, these *Tfm* epithelial cells remained miniscule (Drews and Drews, 1975).

We now have a fairly good idea of extragonadal sex differentiation. In the absence of androgen itself or of the response to androgen, the female phenotype is manifested. All aspects of androgen-induced masculine development are thus mediated by the single locus on the X that encodes one species of androgen receptor. It is fitting that an X-linked locus regulates this manifestation of the masculine phenotype.

What about the gene that determines the fate of the indifferent embryonic gonad? It now appears that the long-sought testis-determining gene has finally been identified (Sinclair *et al.*, 1990). As it apparently encodes a DNA-binding protein, this locus too, is regulatory. Yet, it is expressed only during the critical stage of gonadal differentiation at that site (Gubbay *et al.*, 1990). Although this mode of expression is appropriate for the testis determining gene, it is clearly not at the top of the regulatory hierarchy. There ought to be another gene that restricts expression of the testis-determining gene to the fetal gonads, and it is *that* gene which should be regarded as the ultimate regulator of the primary sex-determining mechanism. Indeed, there are two known conditions under which the normal Y chromosome loses its testis determining capacity. In the wood lemming (*Myopus schisticolor*), a microscopically visible change in the short arm of the X chromosome (a small deletion, a duplication, or the combination of the two) renders the normal Y chromosome powerless, thus producing fertile XY females (Fredga *et al.*, 1976). In the other instance, it is the incompatibility between the Y-linked testis-determining gene and an autosomal gene or genes that causes XY embryos to develop as females. The cross between males of the feral tobacco mouse (*Mus musculus poschiavinus*) and females of the laboratory strain C57BL/6 (B6) produces fertile F_1 males. Therefore, the B6 X chromosome and the tobacco Y are compatible. However, when the tobacco Y is placed in the B6 background by continuous backcrossing, XY embryos begin to develop as females (Eicher *et al.*, 1982).

By encoding a DNA-binding protein, the Y-linked testis-determining gene must necessarily have a number of other genes under its regulatory command, and these regulated genes are the ones that are expected to do the actual work that culminates in testicular organogenesis. Accordingly, it is totally expected that, in addition to an X-linked gene, autosomal genes are involved in testicular organogenesis. An interesting fact is that there must be a higher regulatory gene that activates the Y-linked testis-determining gene during the crucial period of differentiation of the indifferent embryonic gonads. What if it is the gene in the short arm of the wood lemming X chromosome that activates the Y-linked testis-determining gene—at the right time and in the right place? It would follow that the X chromosome carries not only the top-ranked regulatory gene for secondary sex determination (*Tfm*) but also that for the primary sex determination. The presence of both loci on the mammalian X would indeed qualify it to be called the bona fide sex chromosome.

V. Lazarus in Evolution: Resurrection of Trichrome Vision in Higher Primates by Atavistic Mutations

Throughout my scientific career, no subject fascinated me more than the manner in which trichrome vision was regained by higher primates, from rhesus monkey to man. We are not alone in this regard, for 65 years ago, Ladd-Franklin (1929) wrote a very thought-provoking book. Only in recent years have his predictions been verified. The basic design flaw that afflicts the vertebrate eye has already been noted. Despite such misdesign, Devonian fish of 300 million years ago, the first full-fledged vertebrates, were apparently endowed with trichrome vision, for this is retained by today's teleosts as well as by birds. Thus goldfish and pigeon can be trained by color cues. It is in mammals that trichrome vision was lost, probably because insectivores, the first eutherian mammals, were nocturnal in their behavior. This is readily understandable, for at the end of the Jurassic era, prior to the Cenozoic era, those tiny insectivores had to eke out a meager living in the shadow of the dinosaurs.

At this point, let us consider the elements required for trichrome vision. Of the two types of photoreceptor cells in the retina, rods outnumber cones by more than a 10 to 1 margin. In the rods, one finds the venerable rhodopsin, which is already present in halobacterium. Although rhodopsin is purple in color, and mammalian rhodopsin has an absorption maximum at about 490 nm, it is not used for color perception, but for perception of shapes in dim light. The reason that rhodopsin cannot be used for perception of color appears to be related to the manner in which its chromophore, 13-*cis*-retinal, is held. According to the many detailed studies on the rhodopsin of *Halobacterium halobrum*, the Schiff base of 13-*cis*-retinal is bound to a lysine residue in the seventh and last transmembrane α-helix of rhodopsin. Because of this strong binding, when rhodopsin is bleached by light and bleaching has converted 13-*cis*-retinal to all-*trans*-retinal, the latter still remains bound to lysine (Keszythelyi, 1990).

In humans, the gene for rhodopsin resides in chromosome 3 (Nathans *et al.*, 1986). In order to be endowed with trichrome vision, cones should contain three different opsins; the ratio between red opsin cones, green opsin cones, and blue opsin cones is roughly 5 : 5 : 1. Of the three, blue opsin, having an absorption maximum at 420 nm, is encoded by a gene on human chromosome 7 (Nathans *et al.*, 1986). Being autosomal, this gene is likely to have remained intact in all mammals. Therefore, it is probably incorrect to say that mammals started with achromatic vision. Rather, the original vision was likely to have been monochromatic.

Ladd-Franklin (1929) had already pointed out that within the order Primates prosimians, such as the lemurs of Madagascar, are still hampered by achromatic (monochromatic) vision. The first step toward regaining trichrome vision was seen in the more primitive Platyrrhinian monkeys of the New World, e.g., marmosets and squirrel monkeys. Here, males have regained their dichromatic vision, being either red-yellow color blind (protanopia) or yellow-green color blind (deutronopia). On the other hand, many females of the Platyrrhinian monkeys were fully equipped with trichromatic vision (Ladd-Franklin, 1929). In the more evolved Catarrhinian monkeys of the Old World, including the rhesus monkey and human, even the male has regained trichromatic vision. Nevertheless, it has been an outstanding puzzle of human genetics for many years that 4% of males the world over are still hampered with dichromatic vision, being either protanopic or deutronopic.

Successful cloning of the genes for red, green, and blue opsins in recent years has completely substantiated Ladd-Franklin's foresight. When human rhodopsin was compared with human blue opsin, there was only 42% identity between these two 348-residue-long proteins. When blue opsin was then compared with either red or green opsin, both were 364 residues long; again there was only 42% identity between the two (Nathans *et al.*, 1986). This suggested that the three opsins originally derived from rhodopsin began to follow independent pathways 500 million years ago. This is, indeed, in keeping with the notion that the first vertebrates that emerged 300 million years ago in the form of Devonian fish were already equipped with trichromatic vision. When the red and green opsins of man and the Platyrrhinian monkeys were compared, however, they proved to be more than 93% identical. In brief, red and green opsins behaved like the products of polymorphic alleles of a single locus, rather than like proteins encoded by two independent loci.

On this basis, the likely evolutionary scenario that led to the regaining of trichromatic vision by Catarrhinian monkeys can be reconstructed as follows: In the ancestral synapsidian lineage of mammal-like reptiles, two independent autosomes probably carried genes for rhodopsin and blue opsin, whereas the third chromosome, the future X, carried genes for both red and green opsins. By the time the first eutherian mammals emerged in the form of nocturnal insectivores, one of the two genes on the X was lost, whereas the other degenerated to become a photoreceptor without color perception. It should be recalled that in the case of the rhodopsin, its chromophore, 3-*cis*-retinal, is very strongly held by lysine via the Schiff base, so that bleached all-*trans*-retinal remains bound to rhodopsin. This probably is the reason that rhodopsin can be used only for the perception of shapes in dim light. In the case of opsins, the Schiff base of its chromophore appears to be held rather loosely by OH groups of one or the other of the three amino acids, serine, threonine, and tyrosine (Nathans *et al.*, 1986; Neitz *et al.*, 1991). Perhaps, for this reason, all-*trans*-retinals of opsins dissociate themselves from opsins on being bleached by light; this appears critical for the perception of color.

Inasmuch as there is no shortage of lysine or three OH group-containing amino acids in the seven transmembrane α-helices of opsins, this X-linked degenerate opsin of ancestral mammals might have become rhodopsin-like in the manner of binding to its chromophore, thus losing the capacity for color perception. Because of its degenerate nature, mutations accumulated in the X-linked opsin locus; hence, it became extremely polymorphic. In Platyrrhinian monkeys, two groups of alleles of this locus regained the ability to perceive color. One, with the absorption maximum in the range 560 nm, was reborn as red opsin, whereas the other, with the absorption maximum in the range 540 nm, was reborn as green opsin. Since red opsins and green opsins are encoded by alleles of a single X-linked locus, males of the Platyrrhinian monkeys can have only dichromatic vision. Those with red opsin suffer from deutronopia (yellow-green color blindness), whereas those with green opsin suffer from protanopia (red-yellow color blindness) (Jacobs and Neitz, 1987).

In the ideal population of Platyrrhinian monkeys, the two types of X chromosome (one with the red opsin gene and the other with the green opsin gene) would have become equal in frequency, thus restoring trichromatic vision to 50% of the females, for they would have received one each of red and green opsin-carrying X chromosome from their parents. This situation in the Platyrrhinian monkeys is depicted in Fig. 1. In Catarrhinian monkeys, duplication of the opsin gene occurred on the X chromosome. Thus, one X began to carry both the red opsin gene and the green opsin gene. In the case of man, the X appears to carry even two copies of the green opsin gene (Nathans *et al.*, 1986), as depicted in Fig. 2. Nevertheless, 4% of human X chromosomes the world over have remained of the ancestral Platyrrhinian type having only a red opsin gene or a green opsin gene; hence the constant 4% incidence of protanopia and deutronopia among human males.

Aside from protanopia and deutronopia, competent ophthalmologists have long known widespread anomalies in color perception among individuals with apparent trichromatic vision; they are known as protanomaly and deutronomaly. Such individual differences in color perception

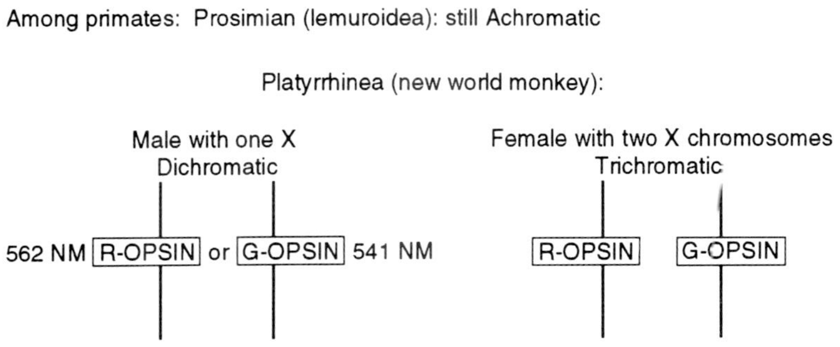

Figure 1 How the X-linked genes for red opsin and green opsin were resurrected in the course of primate evolution. Autosomally inherited blue opsin and rhodopsin are thought to have remained functional, whereas the X-linked genes for red and green opsins became defunct in nocturnal insectivores—the first eutherian mammals. Among the primates, prosimians are still endowed with this monochromatic vision. Of the two defunct X-linked genes, one became extinct, whereas the other became extremely polymorphic. Out of these extensive polymorphisms, genes that could function as red-opsins as well as those that could function as green opsins emerged in Platyrrhinian monkeys of the New World. Since these are allelic forms of the single X-linked locus, males have regained only dichromatic vision, being either protanopic or deuteroncpic. Certain females, however, have regained full trichromatic vision. Since resurrection occurred within the order Primate, one might expect similar resurrection in other orders of mammals.

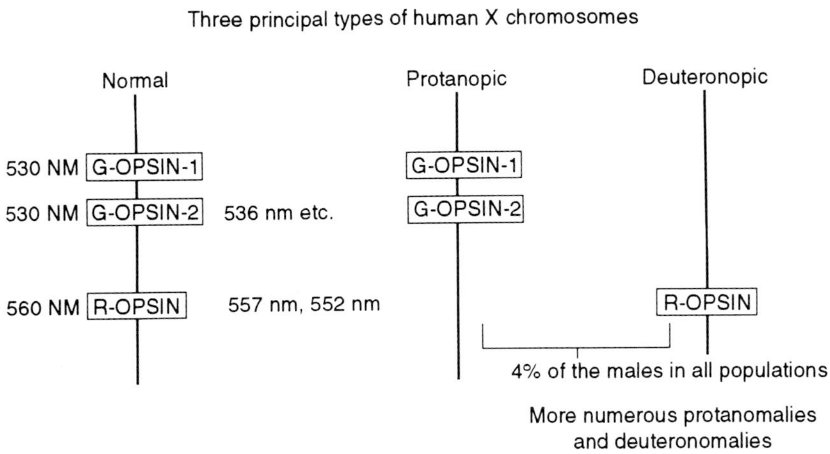

Figure 2 In Catarrhinian monkeys of the Old World, the mechanism of gene duplication placed the genes for red opsin and green opsin on the same X chromosome, thus restoring trichromatic vision even in males. In the humans, the X is said to carry two copies of the green opsin gene. However, 4% of the human X remains in the platyrrhinian state; thus males with these X chromosomes are saddled with dichromatic vision. Furthermore, due to extensive polymorphism of red as well as green opsin, individual perceptions of color in man are very divergent. This is indicated here by different absorption maxima displayed by allelic forms of red and green opsins.

can now be attributed to extensive polymorphism affecting the red opsin gene as well as the green opsin gene (Winderickx *et al.*, 1992). It is of further interest that the same extensive polymorphism was already present in Platyrrhinian monkeys. For example, red opsins from three male tamarin marmosets showed three different absorption maxima: 562, 556, and 541 nm in the study by Neitz *et al.* (1991). The last, with the absorption maximum at 541 nm is, in fact, blurring a distinction between red opsin and green opsin. Such is the consequence of resurrecting two opsin genes from a single degenerate one.

The degree of polymorphism within the genes for human red opsin and green opsin has not been fully determined, to the extent that there is no clear answer to the very pertinent question of what distinguishes red opsin from green opsin. As far as we can discern, differences are limited to 14 of the 364 positions (Fig. 3). Nevertheless, at least 8 of these 14 positions are not characterized by locus-specific residues. For example, the 65th position of red opsin is Thr, but the corresponding position of green opsin can also be occupied by Thr, although its allelic alternative is Ile (Fig. 3). At the 180th position of red opsin, allelic substitution of nonpolar Ala with OH-containing Ser apparently increases

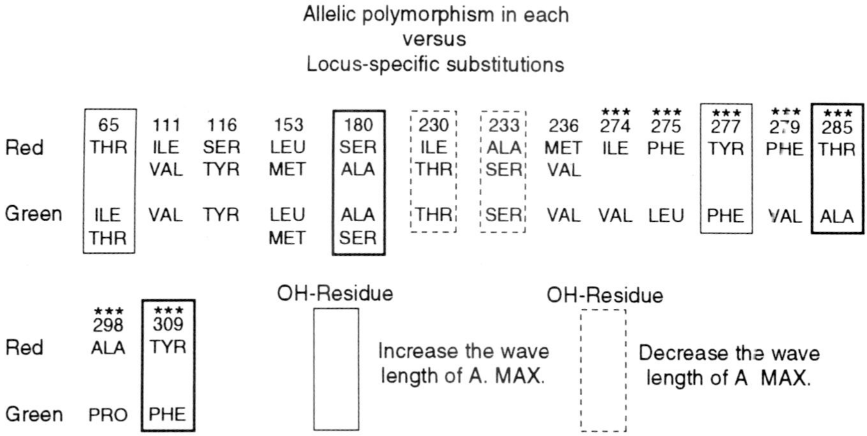

Figure 3 My best answer to the pertinent question of what relevant differences separate red opsin from green opsin in higher primates. Reports on amino acid sequences of the various allelic forms are often contradictory. Differences between red and green opsins of the same species are seen only at 14 of the 364 positions. Confusion results from the fact that 8 of these 14 positions are also involved in allelic polymorphism. The positions in which the presence of OH-group-containing residues (serine, threonine, and tyrosine) appear to increase the wavelengths of absorption maxima are inside the solid boxes, whereas the positions in which the presence of OH-group-containing residues appear to exert the converse effect are inside shaded boxes.

the wavelength of its absorption maximum. But the same allelic substitution also occurs at the corresponding position of green-opsin! At any rate, from this example of Lazarus in evolution, we have learned that (*a*) very extensive allelic polymorphism involving many positions might indicate that a locus is being resurrected; and (*b*) a 58% difference in the amino acid sequence between blue opsin on one hand and red and green opsins together on the other is not a reflection of different functional requirements, but merely that of the time (500 million years) spent as independent entities; were the need there, not only red opsin and green opsins but also blue opsin might have been resurrected from a single degenerate gene.

VI. Summary

Conservation of the X-linkage group *in toto* by all eutherian mammals is by now a *fait accompli*. This fact should be viewed merely as an extreme example of the general trend for maintenance of the status quo, which is becoming increasingly evident in vertebrate evolution (Ohno, 1973). There are two X-linked genes that play a pivotal role in sexual development. One is the *Tfm* locus, which, by encoding the one and only species of androgen receptor protein, governs all androgen-dependent sexual development, i.e., secondary sex. The other locus, yet to be defined, must necessarily control temporal and organ-specific expression of the Y-linked *SRY* gene, which is the presumed testis-determining gene.

In a pair of X-linked genes of Catarrhinian monkeys including man, encoding red and green opsins for trichrome vision, one sees extremely interesting examples of two genes resurrected from one gene that became degenerate at the beginning of mammalian evolution.

References

Atkin, N. B., Mattinson, G., Beçak, W., and Ohno, S. (1965). The comparative DNA content of 19 species of placental mammals, reptiles and birds. *Chromosoma* **17,** 1–10.

Barr, M. L., and Bertram, L. F. (1949). A morphological distinction between neurons of the male and female and the behavior of the nuclear satellite during accelerated nucleoprotein synthesis. *Nature (London)* **163,** 676–677.

Cattanach, B. M., Pollard, C. E., and Hawkes, S. G. (1971). Sex-reversed mice: XX and XO males. *Cell Genet. Cytogenet.* **10,** 318–327.

Cunha, R. (1972). Tissue interactions between epithelium and mesenchyme of urogenital and integumental origin. *Anat. Rec.* **172,** 529–542.

Drews, U., and Drews, U. (1975). Metabolic cooperation between *Tfm* and wild-type cells in mosaic mice after induction of DNA synthesis. *Cell (Cambridge, Mass.)* **6,** 475–479.

Eicher, E. M., Washburn, L. L., Whitney, J. B., III, and Morrow, K. E. (1982). *Mus poschiavinus* Y chromosome in the C57BL/6J murine genome causes sex reversal. *Science* **217,** 535–537.

Fredga, K., Gropp, A., Winking, H., and Frank, F. (1976). Fertile XX- and XY-type females in the wood lemming (*Myopus schisticolor*). *Nature (London)* **261,** 255–257.

Gaspar, M. L., Meo, T., Bourgarel, P., Guenet, J.-L., and Tosi, M. (1991). A single base deletion in the *Tfm* androgen receptor gene creates a short-lived messenger RNA that directs internal translation initiation. *Proc. Natl. Acad. Sci. U.S.A.* **88,** 8606–8610.

Gehring, U., Tomkins, G. M., and Ohno, S. (1971). Effect of the androgen insensitivity mutation on a cytoplasmic receptor for dihydrotestosterone. *Nature (London), New Biol.* **232,** 106–107.

Gubbay, J., Collignon, J., Koopman, P., Capel, B., Economou, A., Münsterberg, A., Nigel, V., Goodfellow, P., and Lovell-Badge, R. (1990). A gene mapping to the sex-determining region of the mouse Y chromosome is a member of a novel family of embryonically expressed genes. *Nature (London)* **346,** 245–250.

Hutt, F. B. (1953). Homologous sex-linked mutations in man and other mammals. *Am. Nat.* **87,** 160–162.

Jacobs, G. H., and Neitz, J. (1987). Inheritance of color vision in a new world monkey (*Saimiri sciureus*). *Proc. Natl. Acad. Sci. U.S.A. 84,* 2545–2549.

Jost, A. (1947). Sur les effects de la castration précoce de l'embryon mêle de lapin. *C. R. Soc. Biol. Ses Fil.* **141,** 126–129.

Keszthelyi, L. (1990). "Bacteriorhodopsin: Structure and Function," Bioelectrochem. III, pp. 179–208. Plenum, New York.

Ladd-Franklin, C. (1929). "Colour and Colour Theories." Harcourt Brace & Co., New York.

Lyon, M. F., and Hawkes, S. G. (1970). An X-linked gene for testicular feminization in the mouse. *Nature (London)* **227,** 1217–1219.

Nasrallah, J. B., Kao, T.-H., Dhen, C.-H., Goldberg, M. L., and Nasrallah, M. E. (1987). Amino-acid sequence of glycoproteins encoded by three alleles of the S locus of *Brassica oleracea. Nature (London)* **326,** 617–619.

Nathans, J., Thomas, D., and Hogness, D. S. (1986). Molecular genetics of human color vision: The genes encoding blue, green and red pigments. *Science* **232,** 193–202.

Neitz, M., Neitz, J., and Jacobs, G. H. (1991). Spectral tuning of pigments underlying red-green color vision. *Science* **252,** 971–974.

Ohno, S. (1967). "Sex Chromosome and Sex-Linked Genes." Springer-Verlag, Berlin and New York.

Ohno, S. (1973). Ancient linkage groups conserved in human chromosomes and the concept of frozen accidents. *Nature (London)* **244,** 259–262.

Ohno, S., Kaplan, W. D., and Kinosita, R. (1959). Formation of the sex-chromatin by a single X chromosome in live cells of *Rattus norvegicus. Exp. Cell Res.* **18,** 415–418.

Ohno, S., Beçak, W., and Beçak, M. (1964). X-autosome ration and the behavior pattern of individual X chromosomes in placental mammals. *Chromosoma* **15,** 14–30.

Ohno, S., Poole, J., and Gustavsson, I. (1965). Sex-linkage of erythrocyte glucose-6-phosphate dehydrogenase in two species of wild hares. *Science* **150,** 1737–1738.

Ohno, S., Payne, H. W., Morrison, M., and Beutler, E. (1966a). Hexose-6-phosphate dehydrogenase found in human liver. *Science* **153,** 1015–1016.

Ohno, S., Stenius, C., and Christian, L. (1966b). The XO as the normal female of the creeping vole (*Microtus oregoni*). *Chromosomes Today.* **1,** 182–187.

Sinclair, A. H., Berta, P., Palmer, M. S., Hawkins, J. R., Griffiths, B. L., Smith, M. J., Foster, J. W., Fischauf, A.-M., Lovell-Badge, R., and Goodfellow, P. N. (1990). A gene from the human sex-determining region encodes a protein with homology to a conserved DNA-binding motif. *Nature (London)* **346,** 240–244.

Singh, L., and Jones, K. W. (1982). Sex reversal in the mouse (*Mus musculus*) is caused by recurrent nonreciprocal cross-over involving the X and an aberrant Y chromosome. *Cell (Cambridge, Mass.)* **28,** 205–216.

Trujillo, J. M., Walden, B., O'Neil, P., and Anstall, H. B. (1965). Sex-linkage of glucose-6-phosphate dehydrogenase in the horse and donkey. *Science* **148,** 1603–1604.

Winderickx, J., Lindsey, D. T., Sanocki, E., Teller, D. Y., Motulsky, A. G., and Deeb, S. S. (1992). Polymorphism in red photopigment underlies variation in colour matching. *Nature (London)* **356,** 431–433.

7

X-Chromosome Inactivation

Mary F. Lyon
Genetics Division
MRC Radiobiology Unit
Chilton, Didcot
Oxon OX11 ORD
United Kingdom

Molecular Genetics of Sex Determination

It is now over 30 years since X-chromosome inactivation in mammals was first discovered, but its mechanism is still not understood. However, with the recent spectacular advances in molecular biology, there have been exciting new developments that make this a very opportune time to review the field (for other recent reviews; see Grant and Chapman, 1988; Riggs, 1990a,b; Lyon, 1992).

I. Early History of X-Chromosome Inactivation

The idea of X-chromosome inactivation first arose as a result of the coming together of various pieces of information about sex determination and sex-linked genes in mice. The sex chromatin or Barr body had been discovered in 1949 by Barr and Bertram as a dark-staining body against the nuclear membrane of female mammals. In 1959, Ohno and colleagues showed that this body was formed from a single condensed X chromosome, and simultaneously Welshons and Russell (1959) found that sex determination in mammals depended on the presence of the Y chromosome, with chromosomally XO mice being normal fertile females. The final piece of the jigsaw was the variegation seen in female mice heterozygous for X-linked genes affecting coat color or texture, with patches of mutant and of normal color, together with similar effects in females heterozygous for X-autosome translocations (Lyon, 1962). When these facts were put together, the hypothesis was formulated that one X chromosome in somatic cells of female mammals is genetically inactive and condensed during interphase to form the sex chromatin body (Lyon, 1961). The inactivation occurs early in embryogeny and is stably inherited through subsequent mitotic divisions. Thus, in the adult there are large clumps of cells descended from a single cell at the time of inactivation, all cells having the same X chromosome active, resulting in the variegated patches seen in heterozygotes for X-linked genes. The effect of this X-inactivation is that chromosomally XY males and XX females each have a single gene dosage of X-linked gene products, and hence the phenomenon is seen as providing dosage compensation.

In the years following the discovery of X-inactivation the idea was consolidated in several ways. Having been originally put forward for the mouse, the theory was soon extended to all mammals (Lyon, 1962). In the human, many sex chromosome aneuploids, e.g., XXX, XXY, XXXX, were known, and in each case the maximum number of sex chromatin bodies per cell was one less than the number of X chromosomes. Hence, the hypothesis was modified to postulate that a single X chromosome remained active in diploid cells, rather than that one became inactive. A problem lay in the abnormal phenotypes of some human sex chromosome aneuploids, particularly XO females. In contrast to the normal phenotype of XO female mice, the majority of human XO females die prenatally, and the remainder are malformed with Turner's syndrome. It was suggested that there might be a small pairing region of homology between X and Y chromosomes, and that genes in this region would not undergo inactivation, as dosage compensation would not be required. In XO's and other aneuploids there would be an incorrect dosage of these genes (Lyon, 1962).

In the germ cells a different picture was seen. Ohno (1967) and colleagues showed that although mouse oogonia had a condensed X-chromosome, oocytes had none, suggesting that both X chromosomes were active in these cells. This was confirmed when Gartler *et al.* (1972, 1973) showed that in human females heterozygous for electrophoretic variants of the X-linked enzyme glucose-6-phosphate dehydrogenase (G6PD), the oocytes formed a heterodimer of the two forms of G6PD. This could happen only if both alleles of the *G6PD* gene were active in the same cell. Conversely, in male germ cells the single X-chromosome was apparently inactive in that it did not make RNA when the autosomes were doing so (Monesi, 1965). Thus, single X-chromosome activity is a feature of somatic cells only (Fig. 1), and the germ cells show differential X-chromosome activity in the two sexes. Some types of sex chromosome aneuploids show sterility and this may be due to incorrect X-chromosome dosage in germ cells. However, it is not clear whether differing amounts of gene products are required in the germ cells, or whether there is a need for particular activity states of the X chromosome during meiotic pairing.

Inactivation of the sex chromosomes at male meiosis is seen in other animal groups with an XX : XY sex-determining mechanism. Miklos (1974) [and see recent discussion by Jablonka and Lamb (1988)] suggested that the active euchromatic state of the chromosomes is required for meiotic pairing, and also that the presence of unpaired active pairing sites is deleterious to a meiotic cell. Thus, when the X and Y chromosomes become differentiated in evolution there are potential unpaired pairing sites remaining on the X chromosome. The inactivation of the X chromosome in male meiosis, it is postulated, protects these unpaired sites from deleterious effects. Conversely, in the female, reactivation of the inactive X chromosome at meiosis ensures full meiotic pairing of the two X chromosomes. These ideas are attractive, as they could account for the effects of various types of chromosome translocations and insertions on male fertility. All known reciprocal X-

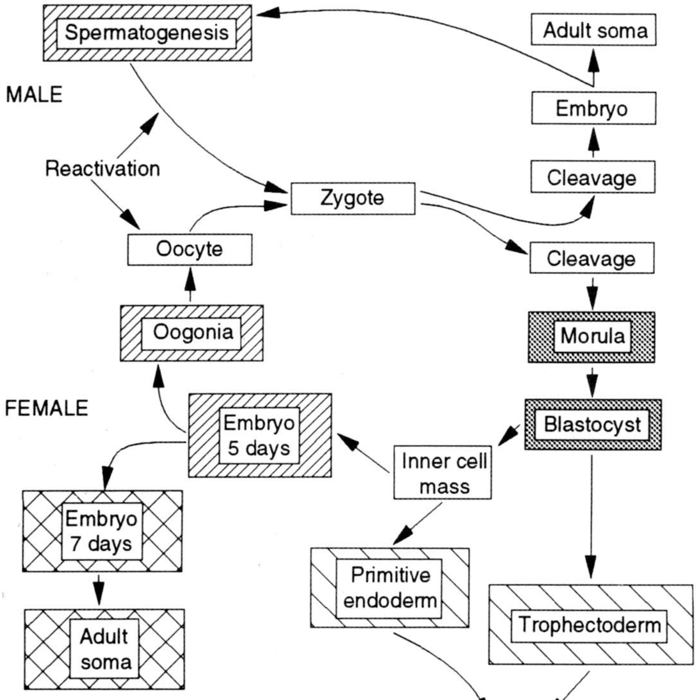

	Xist expression	Inact-ivation	Imprint (patenal)	Me
(open)	-	-	-	-
(stippled)	+	-	+	-
(wide hatch)	+	+	+	-
(fine hatch)	+	+	-	-
(cross-hatch)	+	+	-	+

Figure 1 Cycle of X-chromosome activity in the mouse, showing the relation of *Xist* expression, imprinting, and methylation (Me) to inactivation and reactivation. Methylation is shown as not occurring in the extraembryonic membranes, but there is some doubt about this (see text). Reprinted with permission from Lyon (1993a).

autosome translocations in the mouse cause male sterility, whereas autosomal insertions into the X-chromosome and X-autosome Robertsonian translocations permit fertility. This could result if X-inactivation had more adverse effects on pairing in reciprocal translocations than in the other types of aberrations (reviewed in Lyon, 1989).

II. Early Ideas on Mechanisms

A. X-Inactivation Center

A very early concept about the mechanism of inactivation was that of an X-inactivation center, lying on the X chromosome, which determined whether the chromosome was active or inactive. This idea arose from studies of X-autosome translocation in mice. Coat color genes in the translocated autosomal segments could exhibit variegation, similar to that shown by X-linked genes, indicating that inactivation could spread into the attached autosomal segment. However, genes in only one of the two segments into which the autosome was broken showed this variegation. The interpretation was that there was a center on the X chromosome from which the inactivation spread, and that segments of X chromosome separated from the center by a translocation could not undergo inactivation (Lyon, 1963; Russell, 1963) (Fig. 2). Support for this interpretation came from recent studies of the translocations T37H and T38H, which both have X-chromosome breakpoints near the locus of ornithine transcarbamylase (*Otc*). In females heterozygous for T38H and for a mutant deficient in Otc there was a typical mosaic pattern of *Otc*-positive and -negative cells in the liver, whereas in similar females heterozygous for T37H, all cells stained positively for Otc. *In situ* hybridization to the chromosome showed that in T38H the *Otc* locus lay on the same X-chromosome fragment as the putative X-inactivation center, whereas in T37H the translocation break lay between the *Otc* locus and the inactivation center. The interpretation was that in T37H the segment carrying *Otc* did not undergo inactivation (Lyon *et al.*, 1986).

B. Properties of the Inactive X Chromosome

In addition to its property of condensation during interphase and mitotic prophase, the inactive X replicates its DNA late during mitotic S phase (Taylor, 1960). The combination of late replication and condensation is a general property of heterochromatin. Heterochromatin can be divided into two types: constitutive heterochromatin including centromeric and other heterochromatic blocks, and facultative heterochromatin, such as the inactive X chromosome, in which the same chromatin can adopt the euchromatic active, or heterochromatic inactive, state in different cells. Another well-studied example of facultative heterochromatin is seen in mealy bugs, in which the entire paternally derived

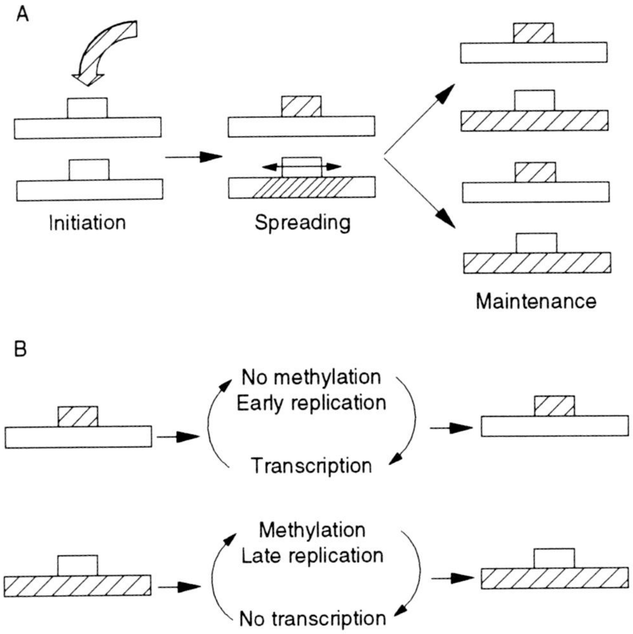

Figure 2 Initiation and maintenance of X-chromosome inactivation. The X-inactivation center is shown as a box. At initiation (A), a signal blocks one inactivation center, and the other center initiates the spread of inactivation in both directions along the chromosome. In maintenance (B), the active or inactive state of the chromosome is maintained through successive cell generations, probably by a feedback loop, which may involve late replication and in some cases methylation. Reprinted with permission from Lyon (1993b).

genome set becomes heterochromatic and inactive in embryos destined to be male. Yet another property of the inactive X is dark staining after hot hypotonic treatment, so-called "Kanda staining" (Kanda, 1973). All of these properties, particularly late replication and Kanda staining, have been highly valuable in studies of X-inactivation. For instance, in mice with X-autosome translocations, late replication (Takagi, 1980) and Kanda staining (Rastan, 1983) are only seen in one of the X-chromosomal segments, and these properties can extend into the attached autosomal material, in line with the results on gene activity. An unresolved challenge is to find the causal relations among the properties, and the involvement of the X-inactivation center in bringing them about.

III. Further Developments

A. Ohno's Law

In 1967, Ohno put forward the idea that has become known as Ohno's Law, namely that a gene X-linked in one mammalian species is X-linked in all (see Chapter 6, this volume). The rationale behind this was that owing to the effectively single gene dosage of X-linked genes, translocations, either of X-linked genes to autosomes or the reverse, would result in imbalance of autosomal and X-linked gene products, and would hence be detrimental and selected against. There is now strong evidence in favor of Ohno's Law. It has been highly valuable in studying X-inactivation by making it possible to identify X-linked genes and hence to study the phenomenon in species in which relatively little genetic variation is known.

B. X-Inactivation in Marsupials and Imprinting

In particular, Ohno's Law enabled the study of X-inactivation in marsupials. The first unexpected finding was in the choice of X chromosome for inactivation. In eutherians, either the maternally or the paternally derived X chromosome becomes inactive in different cells. However, in marsupials the paternally derived X chromosome is preferentially inactivated in all cells (Sharman, 1971; Cooper *et al.*, 1977). This is an example of genetic imprinting, in which the chromosomes or genes retain some cell memory or imprint of their gametic origin (Table I). Preferential paternal X-inactivation was later found in the trophectoderm and primitive endoderm cell lineages of mouse and rat embryos (Takagi, 1978) (Fig. 1 and Table I) and also, although somewhat less clearly, in human extraembryonic tissues (Harrison, 1989). Experimental work in mice showed that the imprint is not an essential part of the mechanism in the lineages concerned. In diploid parthenogenotes, with both X chromosomes of maternal origin, X-inactivation can still occur, and in XO embryos with only a paternal X chromosome this can remain active. Thus, the imprint appears to affect the choice of X chromosome for inactivation (reviewed in Lyon, 1988, 1989). Similar imprinting is seen in mealy bugs, in which the paternal genome set is inactivated in male embryos, and in other cases of inactivation or elimination of one homolog of a pair (Lyon and Rastan, 1984). It is possible that such an imprint, resulting in the two homologs having differing susceptibilities

Table I
Comparison of X-Chromosome Inactivation in Monotremes, Marsupials, and Eutheria

Extent on X		Tissues	Imprint	Complete or stable	Methylation
	Monotremes	Some	?	?	?
	Marsupials				
	Embryonic tissues	Some	Yes	No	?
	Adult	Most or all	Yes	No	No
	Eutheria				
	Extraembryonic tissues	Most or all	Yes	No	No
	Embryo and adult	All	No	Yes	Yes

Note: In monotremes inactivation affects only part of the X-chromosome in some tissues. The extraembryonic tissues of eutherians resemble adult tissues of marsupials in showing imprinted, paternal, and relatively unstable inactivation, and lack differential methylation. From Lyon (1993b), with permission.

to inactivation or loss, may reduce the risk of error in which both homologs behave similarly, presumably a lethal event. Thus, the imprint may confer a selective advantage.

In eutherians, X-inactivation occurs later in the lineage in which inactivation occurs randomly, namely the primitive ectoderm, which gives rise to the embryo proper. Thus, the imprint may already have been lost before inactivation, or else the mechanism is different and ignores the imprint.

In view of the fact that the X chromosome is inactivated in male germ cells, it has been suggested that paternal X-inactivation was the more primitive type and evolved from an earlier mechanism of inactivation in male germ cells. A further unexpected finding in marsupials was that X-inactivation was less complete or stable than in eutherians (Table I). Particular X-linked genes could be reactivated either in cultured cells (*G6pd*), or in certain tissues *in vivo* (e.g., *Pgk-1*) (Cooper *et al.*, 1977). By contrast, X-inactivation is highly stable in eutherians. A low level of reactivation, increasing with age, has been detected in mice for the *Otc* gene (Wareham *et al.*, 1987), and for some autosomal genes translocated to the X chromosome, but the levels are much lower than in marsupials and with other genes no such effect is detectable (Moore *et al.*, 1992). It may be that the selective advantage that led to the evolution of random inactivation in eutherians was the greater stability and therefore more complete dosage compensation that accompanied it.

C. X-Inactivation in Monotremes

Recently, evidence has been obtained on X chromosome inactivation in the remaining subclass of mammals, the monotremes. Graves and Watson (1991) showed that in the platypus the X and Y chromosomes were divided into cytogenetically similar, putatively homologous, and visibly different putative differential segments. The differential segment of the X chromosome showed late replication in some tissues (Table I). Thus, the platypus may be showing an evolutionarily early form of X-chromosome inactivation, in which only part of the X chromosome is involved. It may be that X-inactivation first began at a time when the X and Y chromosomes differed in only a small region. There would have been an inactivation center, but the extent of spreading of the signal from the inactivation center would have been much less. As the Y chromosome became smaller by loss of genetical material homologous with the X chromosome, there will have been selection for increased spread of X chromosome inactivation, giving better dosage compensation.

IV. Initiation and Maintenance of X-Inactivation

An advance in concepts of the mechanism of X-inactivation involved the separate consideration of initiation and maintenance of inactivation (Lyon, 1971, 1974). Initiation meant the first differentiation of the two X chromosomes in the cells of the early female embryo. Maintenance was the faithful replication of the active or inactive state of each chromosome through subsequent mitotic divisions (Fig. 2). It might be brought about by some form of feedback loop, originally suggested to involve chromosome proteins. Initiation was postulated to involve the inactivation center and was an event occurring at a specific developmental stage. An important question was whether the X chromosomes were active in early embryos and thus whether differentiation of the X chromosomes entailed the activation of a previously inactive chromosome or the reverse. In 1978, Epstein *et al.* showed that both X chromosomes of early female mouse embryos were active and thus the phenomenon was indeed one of inactivation. It appeared to involve a counting mechanism establishing a 2A : 1X ratio, where A is a haploid autosome set, since there was evidence that in contrast to the single X-activity of diploids, in triploids either one or two X chromosomes (Endo *et al.*, 1982; Vogel *et al.*, 1983) are active and in tetraploid mouse embryos two X chromosomes are active (Webb *et al.,* 1992). In mouse embryos X-inactivation was first seen in trophectoderm cells in blastocysts of around 4 days postcoitum and occurred last in the primitive ectoderm lineage, giving rise to the embryo itself, at around 5 days (Fig. 1) (Takagi *et al.,* 1982; Sugawara *et al.,* 1985).

In the 1970s, cell lines of embryonal carcinoma (EC) cells and embryonic stem (ES) cells became available. Some such XX cell lines had both X chromosomes active, whereas in others an inactive X chromosome was present, as though the X-chromosome state depended on the embryonic stage, before or after X-inactivation, from which the foundation cells had been derived. Martin *et al.* (1978) showed that when an EC cell line with both X chromosomes active was allowed to differentiate, X-inactivation took place. Similar results were obtained by McBurney and Strutt (1980). Thus, such cells provide a useful system for studying the factors involved in initiating X-inactivation. They also provide a clue to the mechanism, as the factors involved in initiating X-inactivation must be included in those bringing about differentiation. As an example of work with EC cells, Takagi *et al.* (1983) were able to induce reactivation of the X chromosome in adult female mouse somatic cells after fusing them with EC cells. This result suggests that the necessary factors for the reactivation which occurs in female germ cells are still present in the early embryo, and may be responsible for the reactivation of the paternally derived X chromosome after fertilization (Fig. 1).

Clues to the factors involved in inactivation or reactivation have been very difficult to obtain, as cells in general have resisted attempts to alter the activity state of their X chromosomes, and thus EC and ES cells are highly valuable.

V. Summary of Current Knowledge of X-Inactivation

Thus, the general picture is that inactivation is a positive process, not a spontaneous loss of activity, initiated in early embryogeny from an inactivation center and spreading in a cis-limited way in both directions. There is then some mechanism for maintaining or stabilizing the inactivation, probably by a feedback loop. Eutherians and marsupials differ in the stability of inactivation. There is a counting mechanism determining a 2A : 1X gene dosage ratio, and in marsupials and some eutherian cell lineages a parental imprint is involved in choice of X for activity. Changes of state of X chromosomes occur also in male and female germ cells.

VI. Recent Developments

A. X-Inactivation Center and *Xist* Gene

There have been highly promising recent developments concerning the X-inactivation center. By study of different translocations and deletions the center has been precisely mapped, to band Xq13 in the human and band XD in the mouse. A gene termed *Xist* (*X* *i*nactive *s*pecific *t*ranscript), cloned from the relevant region, first in human (Brown *et al.*, 1991) and then in mouse (Borsani *et al.*, 1991; Brockdorff *et al* , 1991), has the unique property of being expressed from the inactive but not the active X chromosome. The combination of the location and the pattern of expression of *Xist* makes it a strong candidate for a role in the inactivation center. Several kilobases of the gene have been sequenced, 15 kb in mouse (Brockdorff *et al.*, 1992) and 17 kb in man (Brown *et al.*, 1992), and in both species no substantial open reading frame has been found, and the RNA is confined to the nucleus. Thus, there is no evidence that any protein product is formed, and the gene may act through RNA. In both species there are several blocks of repeats, whose significance is not yet known. Brockdorff *et al.* (1992) speculate that the gene may be involved in altering the conformation of chromatin in the region.

An unexpected feature of the gene, as a candidate for the X-inactivation center, is that it is expressed in all tissues of the adult female (Fig. 1). The inactivation center is postulated to be involved in the initiation of inactivation in the early embryo, rather than the later maintenance of inactivation. As *Xist* is so widely expressed, the factor involved in the initiation of inactivation at a specific stage remains to be discovered. However, it is known that *Xist* is expressed at appropriate stages for it to have a causal role in inactivation, rather than a consequential one. Kay *et al.* (1993) found that, in the mouse embryo, *Xist* was not expressed at the two-cell stage, but was expressed in the morula and blastocyst, before the initiation of inactivation (Fig. 1). Furthermore, in the lineages (trophectoderm and primitive endoderm) showing imprinting, only the paternal allele of *Xist* was expressed, and in ES cells *Xist* expression was triggered by differentiation (known to be accompanied by inactivation). An important early finding, relevant to the X-inactivation center, was that of the *Xce* (*X-c*ontrolling *e*lement) locus, studied by Johnston and Cattanach (1981), Rastan (1982), and colleagues. Various strains of mice carry different alleles at the *Xce* locus and these alleles affect the probability of the X chromosome on which they lie becoming the active one. Thus, females heterozygous for different *Xce* alleles show unequal ratios of activity of their two X chromosomes, such as 60 : 40 or 70 : 30. The *Xce* locus maps to the same region as the X-inactivation center, and may be a part of it. Kay *et al.* (1993) found that the expression of *Xist* in heterozygotes for *Xce* was inversely proportional to the "strength" of the *Xce* allele if a strong allele is one that preferentially remains active.

In the male, *Xist* is expressed only in germ cells of the testis (Richler *et al.*, 1992; Salido *et al.*, 1992; McCarrey and Dilworth, 1992), suggesting that somatic X-inactivation in females and germ cell X-inactivation in males may share common features. In female germ cells *Xist* expression disappears at around the time of reactivation of the X chromosome, again suggesting a similar role for the gene in germ and somatic cells (McCarrey and Dilworth, 1992).

The inactivation center would be expected to be involved in the 2A : 1X counting mechanism, but so far it is not clear whether the *Xist* gene takes part. Numerous speculations concerning the mechanism that limits activity to a single X chromosome have fallen into two categories, either postulating a limited quantity of some substance or suggesting cooperative binding of an agent at the center (Gartler and Riggs, 1983; Grant and Chapman, 1988; McBurney, 1988). In either case, some binding site for an agent would be expected and has not yet been identified. Furthermore, no relevant autosomal genes that could take part in the numerator element have yet been found. Jacobs and Migeon (1989) looked at human trisomic fetuses, for 18 of the 22 autosomes, seeking anomalies of X-inactivation, but found no such anomalies. Thus, it is still possible that the numerator A component depends on some general effect of the whole genome, such as a change in nuclear volume.

B. Spreading and Maintenance of X-Inactivation

The spreading of inactivation from the center in both directions along the chromosome entails the passage of some signal for long distances, in a cis-limited way, without skipping to other chromosomes. Riggs (1975) and Holliday and Pugh (1975) suggested that a DNA methylase might have suitable properties to be the agent involved. A *de novo* methylase that ran along the chromosome would provide the spreading of inactivation at initiation, and a maintenance methylase that methylated already half-methylated sites would provide a suitable feedback loop for maintenance of the inactive state (Fig. 2). There is now good evidence for differential methylation of active and inactive X chromosomes. The CpG islands in 5′ promoter regions of genes on the inactive X chromosome are heavily methylated, whereas such CpG islands are unmethylated on the active X chromosome and elsewhere in the genome. Treatment with the demethylating agent 5-azacytidine induced reactivation of some X-linked genes, and the CpG islands of these reactivated genes had a reduced level of methylation (reviewed in Riggs, 1990a,b; Bird, 1992). Thus, the methylation appears to have a functional effect. However, in marsupials there is no methylation of CpG islands on the inactive X chromosome (Kaslow and Migeon, 1987) (Table I), indicating that methylation is not the primary spreading signal. In view of the lower stability of X-inactivation in marsupials compared with that in eutherians, another possibility is that methylation is involved in stabilizing inactivation. In addition, there is evidence that methylation may occur later than inactivation, or not at all, in some eutherian cells. Bartlett *et al.* (1991) induced inactivation in an EC cell line by allowing it to differentiate, and methylation did not occur. Similarly, in the mouse embryo Lock *et al.* (1987) reported that methylation of the *Hprt* gene did not occur until some days after the time of inactivation. However, Grant *et al.* (1992), using a sensitive PCR technique, claimed that methylation was present as early as 3.5 days postcoitum for *Pgk-1* and 5.5 days for *G6pd,* which is very soon after initiation of inactivation. Possibly methylation may travel slowly along the chromosome away from the inactivation center. By contrast Grant *et al.* (1992) found no methylation in embryonic female germ cells at any stage even though an inactive X chromosome is present in these cells from their earliest detection to the onset of meiosis. There is also evidence that cells of the extraembryonic endoderm may escape DNA modification, as DNA from such cells could transfect $HPRT^-$ cells, whereas DNA from the embryo proper could not (Kratzer *et al.,* 1983). Grant *et al.* (1992), however, found evidence of some methylation in yolk sac endoderm cells and hence there is some doubt on this point. Thus, methylation of CpG islands on the inactive X chromosome seems a feature of some but not all eutherian cell lineages (Fig. 1 and Table I). The lineages in which it is not found tend to be those in which reversal of inactivation occurs naturally, as in marsupials or eutherian female germ cells, or can be induced, as in the extraembryonic tissues of eutherians, where Migeon *et al.* (1986) found complete reactivation of the inactive X chromosome in a cell line derived from human chorionic villus cells. This association between instability and lack of methylation, taken together with the appearance of methylation after other signs of inactivation, has led to the view that methylation is part of the stabilizing mechanism of inactivation.

Thus, one must look further for the spreading mechanism, and consider the other properties of the inactive X chromosome, including late replication and condensation. Riggs (1990a,b), Riggs and Pfeifer (1992), and Grant and Chapman (1988) have suggested that late replication, like methylation, could form a feedback loop that would stabilize inactivation. This could come about if late replication prevented transcription and transcription were required for early replication (Fig. 2). Unlike methylation, late replication seems to be a feature of X-inactivation in all cell types, and appears at an early stage in embryogeny (Issa *et al.*, 1969; Sugawara *et al.*, 1985; Takagi *et al.*, 1982). Thus, it could be part of the primary spreading mechanism. The questions then are how the asynchronous replication is brought about, and what is the connection between the late replication and chromosome condensation. At present the control of replication origins in higher organisms is not understood (Fangman and Brewer, 1992) and hence the cause of late replication is unknown. For the condensation, Riggs (1990a,b) has suggested the possibility of a mechanism involving type I DNA reeling, in which an enzyme at a scaffold attachment site reels in DNA toward itself in both directions until it meets another such attachment site. This would provide the type of cis-limited mechanism that is required. Like methylation, condensation at the visible level, to form the sex chromatin body, is not found in all cells having an inactive X chromosome, and appears later in embryogeny than late replication does. Hence, visible condensation is likely to be a secondary feature, but it probably depends on some underlying change in the chromatin fiber. Binding of proteins to DNA is obviously of importance. Methylated DNA binds two proteins: MeCP1 and MeCP2 (Bird, 1992). Pfeifer and Riggs (1991) and Pfeifer *et al.* (1990) studied DNA footprinting of the 5′ region of the human *PGK* gene on active and inactive chromosomes and showed that the promoter region of the active gene was much occupied by protein factors, whereas the inactive gene was not. The results suggested binding of nucleosomes to the 5′ region of the inactive gene. However, these differences are probably the result of methylation and hence not primary.

C. Gene and Chromosome Responses to the Spreading Signal

Although the inactivation center is of major importance in the initiation of inactivation, it appears that other factors along the chromosome may also be involved. In mouse X-autosome translocations, inactivation travels into the attached autosome for large distances, over several G-bands and many megabases of DNA. However, it appears not to travel as far as in X-chromosome material itself. This has led Riggs (1990a,b) to suggest that there may be "booster elements" along the X chromosome that enhance the spread. Perhaps these elements would be certain features of the interspersed repetitive DNA, with different features either enhancing or obstructing the spread. It should be noted that they would need to be spaced at much larger distances than are the scaffold attachment sites suggested by Riggs as involved in DNA reeling. A further point of interest is the apparent difference between mammals and *Drosophila* in the distances over which spreading effects operate. In *Drosophila* the phenomenon of position effect variegation is regarded as due to effects of heterochromatin spreading to normally euchromatic genes translocated nearby. However, in *Drosophila* the distance of spread is limited to around 1.0 to 1.5 Mb (Eissenberg and Elgin, 1991; Henikoff, 1990), whereas in mice the spread of X-inactivation into attached autosomal material may be up to 20 cM or about 40 Mb. In *Drosophila* a specific protein, heterochromatin protein 1 (HP1) is believed to be involved in position effect variegation (Eissenberg and Elgin, 1991; Eissenberg *et al.*, 1992). Homologs of the *HP1* gene found in mammals should repay investigation.

There is also evidence of response of individual genes to the spreading signal. In humans, there are various genes known to escape inactivation, and the number is increasing (Ellison *et al.*, 1992). It was originally predicted that genes in the pairing segment would not undergo inactivation, and indeed some of the human noninactivated genes, such as *MIC2,* do indeed lie in this segment. However, others, including *STS, ZFX, UBE1* and *RPS4X,* that lie outside this region have been found, and some have normally inactivated genes on both sides (Wang *et al.*, 1992). This implies that the spreading signal must have passed through these genes. Either they have resisted the inactivating signal itself or they have been inactivated and have resisted the stabilizing signal, so that they have been immediately reactivated. Sex chromosome aneuploids, such as XO females, will have an incorrect dosage of gene products of these genes, and this could account for the abnormal phenotype of these aneuploids. Homologous genes in the mouse undergo inactivation normally, and no mouse genes that escape inactivation are yet known (Ashworth *et al.*, 1991; Kay *et al.*, 1991). Thus, mice have more effective and complete dosage compensation than humans have. This could explain the difference in phenotype of XO females, who die prenatally or are malformed in humans but are normal in mice.

This difference in noninactivated genes also provides another clue to the evolution of X-inactivation. Since X-inactivation can spread for long distances into autosomes and many autosomal genes can undergo inactivation, it seems unlikely that specific X-chromosomal DNA sequences are of major importance either in travel of the spreading signal or in response of individual genes to it. Rather it seems that in the course of evolution of the X chromosome there has been selection against interspersed sequences that inhibit spread and against regulatory sequences of individual genes that cause resistance to the spreading signal or to stabilization of inactivation. Work in *Drosophila* is again of interest. Here "boundary elements" have been found in the regulatory regions of genes, which can protect the genes from the inactivating influence of nearby heterochromatin (Eissenberg and Elgin, 1991). Thus, one may speculate that similar protective elements might exist for mammalian genes, and there might have been loss of such elements from X-chromosomal genes during evolution.

VII. Conclusion

Research into molecular mechanisms of X-chromosome inactivation is thus at present poised for rapid progress. Studies of the *Xist* gene should provide important insight into the nature of the X-inactivation center and its role in initiation of inactivation. New methods of study of chromatin structure and function may lead to unravelling of the causal relations among late replication, condensation, and methylation. Comparison of human genes that escape inactivation with their mouse homologs should reveal the features important in the response of genes to the inactivating or stabilizing signals.

References

Ashworth, A., Rastan, S., Lovell-Badge, R., and Kay, G. (1991). X-chromosome inactivation may explain the difference in viability of XO humans and mice. *Nature (London)* **351,** 406–408.

Barr, M. L., and Bertram, L. F. (1949). A morphological distinction between neurones of the male and female and the behavior of the nucleolar satellite during accelerated nucleoprotein synthesis. *Nature (London)* **163,** 676–677.

Bartlett, M. H., Adra, C. N., Park, J., Chapman, V. M., and McBurney, M. W. (1991). DNA methylation of two X chromosome genes in female somatic and embryonal carcinoma cells. *Somatic Cell Mol. Genet.* **17,** 35–47.

Bird, A. P. (1992). The essentials of DNA methylation. *Cell (Cambridge, Mass.)* **70,** 5–8.

Borsani, G., Tonlorenzi, R., Simmler, M. C., Dandolo, L., Arnaud, D., Capra, V., Grompe, M., Pizzuti, A., Muzny, D., Lawrence, C., Willard, H. F., Avner, P., and Ballabio, A. (1991). Characterization of a murine gene expressed from the inactive X chromosome. *Nature (London)* **351,** 325–329.

Brockdorff, N., Ashworth, A., Kay, G. F., Cooper, P., Smith, S., McCabe, V. M., Norris, D. P., Penny, G. D., Patel, D., and Rastan, S. (1991). Conservation of position and exclusive expression of mouse *Xist* from the inactive X chromosome. *Nature* (*London*) **351,** 329–331.

Brockdorff, N., Ashworth, A., Kay, G. F., McCabe, V. M., Norris, D. P., Cooper, P. J., Swift, S., and Rastan, S. (1992). The product of the mouse *Xist* gene is a 15 kb inactive X-specific transcript containing no conserved ORF and located in the nucleus. *Cell* (*Cambridge, Mass.*) **71,** 515–526.

Brown, C. J., Ballabio, A., Rupert, J. L., Lafreniere, R. G., Grompe, M., Tonlorenzi, R., and Willard, H. F. (1991). A gene from the region of the human X inactivation centre is expressed exclusively from the inactive X chromosome. *Nature* (*London*) **349,** 38–44.

Brown, C. J., Hendrich, B. D., Rupert, J. L., Lafreniere, R. G., Xing, Y., Lawrence, J., and Willard, H. F. (1992). The human *XIST* gene: Analysis of a 17 kb inactive X-specific RNA that contains conserved repeats and is highly localized within the nucleus. *Cell* (*Cambridge, Mass.*) **71,** 527–542.

Cooper, D. W., Johnston, P. G., Sharman, G. B., and Vandeberg, J. L. (1977). The control of gene activity on eutherian and metatherian X chromosomes: A comparison. *In* "Reproduction and Evolution" (J. H. Calaby and C. H. Tyndale-Biscoe, eds.), pp. 81–87. Australian Academy of Sciences, Canberra.

Eissenberg, J. C., and Elgin, S. C. R. (1991). Boundary functions in the control of gene expression. *Trends Genet.* **7,** 335–340.

Eissenberg, J. C., Morris, G. D., Reuter, G., and Hartnett, T. (1992). The heterochromatin-associated protein HP-1 is an essential protein in Drosophila with dosage-dependent effects on position-effect variegation. *Genetics* **131,** 345–352.

Ellison, J., Passage, M., Yu, L.-C., Yen, P., Mohandas, T. K., and Shapiro, L. (1992). Directed isolation of human genes that escape X inactivation. *Somatic Cell Mol. Genet.* **18,** 254–268.

Endo, S., Takagi, N., and Sasaki, M. (1982). The late-replicating X chromosome in digynous mouse triploid embryos. *Dev. Genet.* **6,** 137–143.

Epstein, C. J., Smith, S., Travis, B., and Tucker, G. (1978). Both X chromosomes function before visible X chromosome inactivation in female mouse embryos. *Nature* (*London*) **274,** 500–503.

Fangman, W. L., and Brewer, B. J. (1992). A question of time: Replication origins of eukaryotic chromosomes. *Cell* (*Cambridge, Mass.*) **71,** 363–366.

Gartler, S. M., and Riggs, A. D. (1983). Mammalian X-chromosome inactivation. *Annu. Rev. Genet.* **17,** 155–190.

Gartler, S. M., Liskay, R. M., Campbell, B. K., Sparkes, R., and Gant, N. (1972). Evidence for two functional X chromosomes in human oocytes. *Cell Differ.* **1,** 215–218.

Gartler, S. M., Liskay, R. M., and Gant, N. (1973). Two functional X-chromosomes in human fetal oocytes. *Exp. Cell Res.* **82,** 464–466.

Grant, M., Zuccotti, M., and Monk, M. (1992). Methylation of CpG sites of two X-linked genes coincides with X-inactivation in the female mouse embryo but not in the germ line. *Nat. Genet.* **2,** 161–166.

Grant, S. G., and Chapman, V. M. (1988). Mechanisms of X-chromosome regulation. *Annu. Rev. Genet.* **22,** 199–233.

Graves, J. A. M., and Watson, J. M. (1991). Mammalian sex chromosomes: Evolution of organization and function. *Chromosoma* **101,** 63–68.

Harrison, K. B. (1989). X-chromosome inactivation in the human cytotrophoblast. *Cytogenet. Cell Genet.* **52,** 37–41.

Henikoff, S. (1990). Position-effect variegation after 60 years. *Trends Genetics* **6,** 422–426.
Holliday, R., and Pugh, J. E. (1975). DNA modification mechanisms and gene activity during development. *Science* **187,** 226–232.
Issa, M., Blank, C. E., and Atherton, G. W. (1969). The temporal appearance of sex chromatin and of the late-replicating X chromosome in blastocysts of the domestic rabbit. *Cytogenetics* **8,** 219–237.
Jablonka, E., and Lamb, M. J. (1988). Meiotic pairing constraints and the activity of sex chromosomes. *J. Theor. Biol.* **133,** 23–36.
Jacobs, P. A., and Migeon, B. R. (1989). Studies of X-chromosome inactivation in trisomies. *Cytogenet. Cell Genet.* **50,** 75–77.
Johnston, P. G., and Cattanach, B. M. (1981). Controlling elements in the mouse. IV. Evidence of non-random X-inactivation. *Genet. Res.* **37,** 151–160.
Kanda, N. (1973). A new differential technique for staining the heteropycnotic X-chromosome in female mice. *Exp. Cell Res.* **80,** 463–467.
Kaslow, D. C., and Migeon, B. R. (1987). DNA methylation stabilizes X chromosome inactivation in eutherians but not in marsupials: Evidence for multistep maintenance of mammalian X dosage compensation. *Proc. Natl. Acad. Sci. U.S.A.* **84,** 6210–6214.
Kay, G. F., Ashworth, A., Penny, G. D., Dunlop, M., Swift, S., Brockdorff, N., and Rastan, S. (1991). A candidate spermatogenesis gene on the mouse Y chromosome is homologous to ubiquitin-activating enzyme E1. *Nature (London)* **354,** 486–489.
Kay, G. F., Penny, G. D., Patel, D., Ashworth, A., Brockdorff, N., and Rastan, S. (1993). Expression of Xist during mouse development suggests a role in the initiation of X chromosome inactivation. *Cell (Cambridge, Mass.)* **72,** 171–182.
Kratzer, P. G., Chapman, V. M., Lambert, H., Evans, R. E., and Liskay, R. M. (1983). Differences in the DNA of the inactive X chromosomes of fetal and extraembryonic tissues of mice. *Cell (Cambridge, Mass.)* **33,** 37–42.
Lock, L. F., Takagi, N., and Martin, G. R. (1987). Methylation of the *Hprt* gene on the inactive X occurs after chromosome inactivation. *Cell (Cambridge, Mass.)* **48,** 39–46.
Lyon, M. F. (1961). Gene action in the X-chromosome of the mouse (*Mus musculus* L). *Nature (London)* **190,** 372–373.
Lyon, M. F. (1962). Sex chromatin and gene action in the mammalian X-chromosome. *Am. J. Hum. Genet.* **14,** 135–148.
Lyon, M. F. (1963). *In* "Second International Conference on Congenital Malformations," pp. 67–68. International Medical Congress, Ltd., New York.
Lyon, M. F. (1971). Possible mechanisms of X chromosome inactivation. *Nature (London), New Biol.* **232,** 229–232.
Lyon, M. F. (1974). Mechanisms and evolutionary origins of variable X-chromosome activity in mammals. *Proc. R. Soc. London, Ser. B* **187,** 243–268.
Lyon, M. F. (1988). The William Allan memorial award address: X-chromosome inactivation and the location and expression of X-linked genes. *Am. J. Hum. Genet.* **42,** 8–16.
Lyon, M. F. (1989). X-chromosome inactivation as a system of gene dosage compensation to regulate gene expression. *Prog. Nucleic Acid Res. Mol. Biol.* **36,** 119–130.
Lyon, M. F. (1992). Some milestones in the history of X-chromosome inactivation. *Annu. Rev. Genet.* **26,** 15–27.
Lyon, M. F. (1993a). Epigenetic inheritance in mammals. *Trends Genet.* **9,** 123–128.
Lyon, M. F. (1993b). Evolution of mammalian sex chromosomes. *In* "The Differences Between the Sexes" (R. V. Short, ed.). Cambridge Univ. Press, Cambridge (in press).
Lyon, M. F., and Rastan, S. (1984). Parental source of chromosome imprinting and its relevance for X chromosome inactivation. *Differentiation (Berlin)* **26,** 63–67.
Lyon, M. F., Zenthon, J., Evans, E. P., Burtenshaw, M. D., Wareham, K. A., and Williams,

E. D. (1986). Lack of inactivation of a mouse X-linked gene physically separated from the inactivation centre. *J. Embryol. Exp. Morphol.* **97,** 75–85.

Martin, G. R., Epstein, C. J., Travis, B., Tucker, G., Yatziv, S., Martin, D. W., Clift, S., and Cohen, S. (1978). X-chromosome inactivation during differentiation of female teratocarcinoma stem cells *in vitro. Nature (London)* **271,** 329–333.

McBurney, M. W. (1988). X chromosome inactivation: A hypothesis. *BioEssays* **9,** 85–88.

McBurney, M. W., and Strutt, B. J. (1980). Genetic activity of X chromosomes in pluripotent female teratocarcinoma cells and their differentiated progeny. *Cell (Cambridge, Mass.)* **21,** 357–364.

McCarrey, J. R., and Dilworth, D. D. (1992). Expression of *Xist* in mouse germ cells correlates with X-chromosome inactivation. *Nat. Genet.* **2,** 200–203.

Migeon, B. R., Schmidt, M., Axelman, J., and Cullen, C. R. (1986). Complete reactivation of X-chromosomes from human chorionic villus cells with a switch to early DNA replication. *Proc. Natl. Acad. Sci. U.S.A.* **83,** 2182–2186.

Miklos, G. L. G. (1974). Sex-chromosome pairing and male fertility. *Cytogenet. Cell Genet.* **13,** 558–577.

Monesi, V. (1965). Differential rate of ribonucleic acid synthesis in the autosomes and sex chromosomes during male meiosis in the mouse. *Chromosoma* **17,** 11–21.

Moore, T. F., Sheardown, S. A., and Whittingham, D. G. (1992). Failure to detect age-related reactivation at the *Hprt* locus. *Mouse Genome* **90,** 687–689.

Ohno, S. (1967). "Sex Chromosomes and Sex-Linked Genes." Springer-Verlag, Berlin.

Ohno, S., Kaplan, W. D., and Kinosita, R. (1959). Formation of the sex chromatin by a single X-chromosome in liver cells of *Rattus norvegicus. Exp. Cell Res.* **18,** 415–418.

Pfeifer, G. P., and Riggs, A. D. (1991). Chromatin differences between active and inactive X chromosomes revealed by genomic footprinting of permeabilized cells using DNase I and ligation-mediated PCR. *Genes Dev.* **5,** 1102–1113.

Pfeifer, G. P., Tanguay, R. L., Steigerwald, S. D., and Riggs, A. D. (1990). In vivo footprint and methylation analysis by PCR-aided genomic sequencing: Comparison of active and inactive X chromosomal DNA at the CpG island and promoter of human PGK-1. *Genes Dev.* **4,** 1277–1287.

Rastan, S. (1982). Primary non-random X-inactivation caused by controlling elements in the mouse demonstrated at the cellular level. *Genet. Res.* **40,** 139–147.

Rastan, S. (1983). Non-random X-chromosome inactivation in mouse X-autosome translocations—location of the inactivation centre. *J. Embryol. Exp. Morphol.* **78,** 1–22.

Richler, C., Soreq, H., and Wahrman, J. (1992). X inactivation in mammalian testis is correlated with inactive X-specific transcription. *Nat. Genet.* **2,** 192–195.

Riggs, A. D. (1975). X inactivation, differentiation and DNA methylation. *Cytogenet. Cell Genet.* **14,** 9–25.

Riggs, A. D. (1990a). Marsupials and mechanisms of X chromosome inactivation. *Aust. J. Zool.* **37,** 419–441.

Riggs, A. D. (1990b). DNA methylation and late replication probably aid cell memory, and type I DNA reeling could aid chromosome folding and enhanced function. *Philos. Trans. R. Soc. London, Ser. B* **326,** 285–297.

Riggs, A. D., and Pfeifer, G. P. (1992). X-chromosome inactivation and cell memory. *Trends Genet.* **8,** 169–174.

Russell, L. B. (1963). Mammalian X-chromosome action: Inactivation limited in spread and in region of origin. *Science* **140,** 976–978.

Salido, E. C., Yen, P. H., Mohandas, T. K., and Shapiro, L. J. (1992). Expression of the X-inactivation-associated gene *XIST* during spermatogenesis. *Nat. Genet.* **2,** 196–199.

Sharman, G. B. (1971). Late DNA replication in the paternally derived X chromosome of female kangaroos. *Nature (London)* **230,** 231–232.

Sugawara, O., Takagi, N., and Sasaki, M. (1985). Correlation between X-chromosome inactivation and cell differentiation in female preimplantation mouse embryos. *Cytogenet. Cell Genet.* **39,** 210–219.

Takagi, N. (1978). Preferential inactivation of the paternally derived X chromosome in mice. *In* "Genetic Mosaics and Chimeras in Mammals" (L. B. Russell, ed.), pp. 341–360. Plenum, New York.

Takagi, N. (1980). Primary and secondary nonrandom X chromosome inactivation in early female mouse embryos carrying Searle's translocation T(X;16)16H. *Chromosoma* **81,** 439–459.

Takagi, N., Sugawara, O., and Sasaki, M. (1982). Regional and temporal changes in the pattern of X-chromosome replication during the early post-implantation development of the female mouse. *Chromosoma* **85,** 275–286.

Takagi, N., Yoshida, M. A., Sugawara, O., and Sasaki, M. (1983). Reversal of X-inactivation in female mouse somatic cells hybridized with murine teratocarcinoma stem cells in vitro. *Cell (Cambridge, Mass.)* **34,** 1053–1062.

Taylor, J. H. (1960). Asynchronous duplication of chromosomes in cultured cells of Chinese hamster. *J. Biophys. Biochem. Cytol.* **7,** 455–464.

Vogel, W., Trautmann, T., Hörler, H., and Pentz, S. (1983). Cytogenetic and biochemical investigations on fibroblast cultures and clones with one and two active X chromosomes of a 69,XXY triploidy. *Hum. Genet.* **64,** 246–248.

Wang, J. C., Passage, M. B., Ellison, J., Becker, M. A., Yen, P. H., Shapiro, L. J., and Mohandas, T. K. (1992). Physical mapping of loci in the distal half of the short arm of the human X chromosome: Implications for the spreading of X-chromosome inactivation. *Somatic Cell Mol. Genet.* **18,** 195–200.

Wareham, K. A., Lyon, M. F., Glenister, P. H., and Williams, E. D. (1987). Age related reactivation of an X-linked gene. *Nature (London)* **327,** 725–727.

Webb, S., De Vries, T. J., and Kaufman, M. H. (1992). The differential staining pattern of the X chromosome in the embryonic and extraembryonic tissues of postimplantation homozygous tetraploid mouse embryos. *Genet. Res.* **59,** 205–214.

Welshons, W. J., and Russell, L. B. (1959). The Y-chromosome as the bearer of male-determining factors in the mouse. *Proc. Natl. Acad. Sci. U.S.A.* **45,** 560–566.

8

Sex Determination in Marsupials and Monotremes

Gregory K. Hampikian
Worcester Foundation for Experimental Biology
Shrewsbury, Massachusetts 01545

Desmond W. Cooper
School of Biological Sciences
Macquarie University,
Sydney, New South Wales 2109, Australia

Jennifer A. Marshall Graves
Department of Genetics and Human Variation
La Trobe University
Melbourne, Victoria 3083, Australia

Molecular Genetics of Sex Determination

I. Introduction

Woven into the diversity of mammalian forms are those common threads borne by our ancestors. In addition to possessing fur and suckling the young, mammals also share a Y-chromosome-based system of sex determination. Recently, we showed the importance of comparative mammalian genetics by using marsupials to verify or reject candidates for the mammalian testis-determining gene on the Y chromosome (*TDF,* for testis-determining factor). Our studies proved the efficacy of the so-called "marsupial test," a strategy that examined the conservation in sequence and Y-chromosome location of candidates for *TDF.* In this chapter we review current research in marsupial and monotreme sex chromosomes and sex determination, and we demonstrate how comparative studies are being used to unravel the mysteries of mammalian sex determination.

It is believed that mammals first appeared some 200 MYBP (million years before present) (Kemp, 1982). Within 30–50 million years after its genesis (150–170 MYBP) the class had split (Fig. 1). One branch led to the egg-laying monotremes (subclass Protheria), and the other gave rise to all other mammals (subclass Theria). The therian line itself split about 120–150 MYBP, producing the two infraclasses Eutheria (all nonmarsupials, including the common laboratory and agricultural mammals) and Metatheria (the marsupials). We are now taking advantage of the temporal distance between these groups of mammals to resolve the genetics and evolution of sex determination. Conserved and disparate elements are more easily distinguished when compared across these three distinct lineages, and the base sequence of their genes reflects the history of the class Mammalia.

II. Marsupial Development and the Experimental Potential of Australian Mammals

Marsupials are unique among mammals in the live birth of remarkably undeveloped offspring. These "pouch young" are born at a stage comparable to a eutherian embryo. Immediately after birth, the tiny offspring manage to climb from the genital opening to a nipple (usually in the pouch), using only their well-developed forelimbs equipped with sharp curved claws; the hindlimbs remain underdeveloped and nonfunctional. It is precisely this early birth that makes marsupials a uniquely attractive system to study. In the tammar wallaby, for example, testis differentiation does not occur until 3 days after birth (O *et al.*, 1988; Renfree *et al.*, 1987;

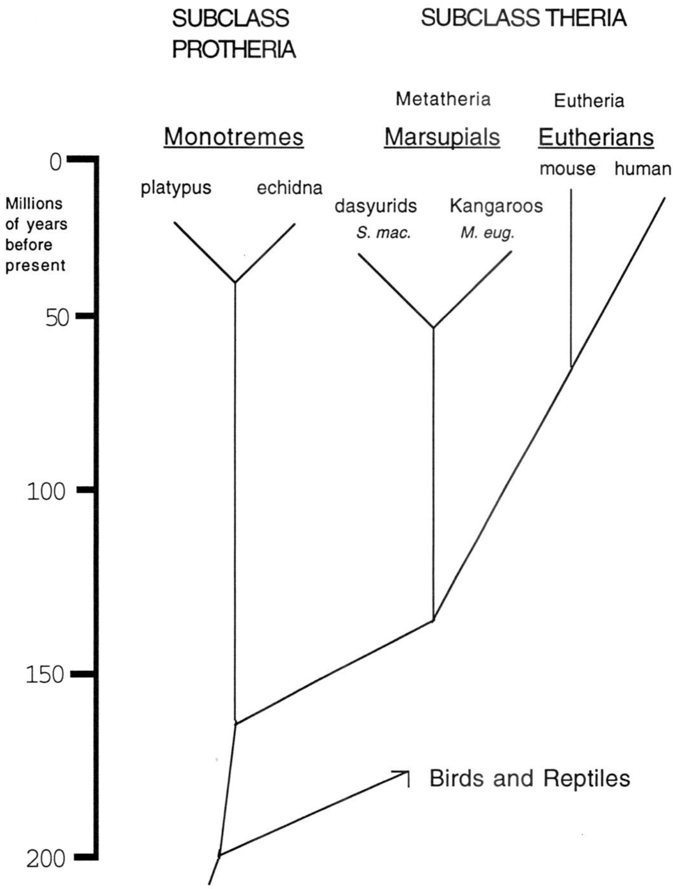

Figure 1 Mammalian family tree, illustrating approximate divergence dates of the subclasses Protheria and Theria; the therian infraclasses Metatheria and Eutheria; and several species mentioned in this chapter.

Renfree and Short, 1988); in contrast, this stage is generally reached by about the midpoint of eutherian gestation. Marsupials are the only mammals to provide such an open window on development.

Several distinctly marsupial features are linked to the birth of embryo-like young, the most obvious being the pouch. This structure, which can have an anterior or posterior opening (helpful for burrowing animals), appears to have evolved several times within the marsupial lineages (Wood Jones, 1923-1925; Sharman, 1976). For centuries, naturalists considered the pouch a second womb; thus the generic term for the common American opossum *Didelphis* means literally two wombs. Actually, the pouch is a well-developed skin fold that covers the mammary glands. Although the term marsupial comes from the Latin word *marsupium,* meaning pouch, several species of small opossums and some Australian carnivorous marsupials are pouchless, and other species have "temporary" pouches that appear only during the breeding season (Strahan, 1983). Some nonmarsupials also have pouches. An example is the monotreme echidna, which has a temporary egg-pouch where hatchlings develop.

Internally, female marsupials retain the ancestral reptile-like double vagina, a consequence of the ureters developing between the sex ducts (see Chapter 3 in Tyndale-Biscoe and Renfree, 1987). In eutherians, the ureters pass outside of the sex ducts, which allows posterior fusion of the oviducts, and thus formation of a single large vagina—a necessary adaptation for the passage of well-developed neonates.

A. Exceptional Monotremes

Mammals in the order Monotremata lay eggs which are incubated against the mother's belly and held in place by either the mother's tail (platypus) or an egg pouch (echidna). Although they lack nipples, monotremes can nurse their young with milk from ducts lining the mother's abdomen. The name Monotremata means "one hole," in reference to the single opening through which urine, feces and eggs are passed (the male has a separate penile duct for ejaculation). Monotremes do not have scrota and their testes are intraabdominal (Griffiths, 1978). This fascinating order is represented by only three extant species—the platypus and two species of echidna—all of which are found only in Australia and New Guinea.

Research on monotremes has been hampered by lack of a successful breeding program and the protected status of animals in the wild. Recently a single platypus was captive-bred, leading to the hope that monotreme genetics will soon become feasible. Cell culture from platypus and echidna has been successful, and this has allowed intensive study of the monotreme chromosomes. One difficulty has been the large number of chromosomes that remain indistinguishable from one another (Watson, 1990). However, the sex chromosomes are distinguishable, and monotremes are being used increasingly for comparative genetic studies, especially in the area of sex chromosome evolution and sex determination.

III. Mammalian Sex Differentiation: Conservation and Divergence

In therian mammals at least, the possession of a Y chromosome leads to testis development. Since 1959, it has been known that in human and mouse, animals with a XX or XO sex chromosome constitution are female, and those with a XY or XXY complement are male (Ford *et al.*, 1959; Welshons and Russell, 1959; Jacobs and Strong, 1959). In eutherians, the primordial cells of the testis are the first sexually dimorphic structures (Magre, 1985). All subsequent male characteristics, including the formation of a scrotum, are induced by hormones produced in the testis (Jost, 1970; Wilson *et al.*, 1981). Thus, the simple paradigm first formulated by Jost (Y chromosome → testis → androgen → all male attributes) works well for eutherians.

In marsupials, however, there are important exceptions to the eutherian model. Whereas most male marsupial characteristics are probably attributable to the presence of a Y chromosome (Sharman *et al.*, 1970, 1990), the decision to form a scrotum or a pouch is influenced by the X chromosome and not the Y. The evidence for this has accumulated over the last few decades, but only recently have the outlines of a coherent picture begun to emerge (Tyndale-Biscoe and Renfree, 1987; Renfree and Short, 1988; Renfree, 1992).

The first line of evidence concerns the timing of differentiation. Unlike eutherians, marsupial embryos develop several sex-specific structures before gonadal differentiation is detectable. This was first noted in the tammar wallaby by Alcorn (1975), and followed up in a series of investigations in the same species by O *et al.* (1988; summarized in Renfree and Short, 1988). They showed that somatic sexual dimorphisms are evident at 5 days before birth, whereas morphological evidence of testis formation is not seen until 3 days after birth. Sexing in these investigations was done by karyotyping, which is straightforward, given the small number of chromosomes in this species ($2n = 16$) and the distinctive size difference of the X and Y. The four male characteristics that precede testis differentiation in the tammar wallaby are a well-developed gubernaculum and processus vaginalis (structures concerned with testis function), paired scrotal anlagen, and the complete absence of mammary anlagen.

The second group of studies involves the administration of exogenous hormones. Burns (1961) was the first to show that testosterone treatment of early female pouch young does not inhibit pouch or mammary gland development in the Virginia opossum, *Didelphis viginiana*. Conversely, he demonstrated that estradiol treatment of males does not inhibit development of the scrotum or induce pouch and mammary glands, although penis development is suppressed. As Renfree and Short (1988) discuss in their detailed analysis of these early experiments, the results were duly reported but their significance was not realized. Instead, an attempt was made to explain the results according to the Jost paradigm.

Recognition that the data did not support this model led to several more recent experiments that classified developmental processes in terms of their response to steroid treatment (Fadem and Tesoriero, 1986; Shaw *et al.*, 1988; Moore and Thurstan, 1990; reviewed by Renfree, 1992). On the basis of these studies it was concluded that the scrotum, mammary gland, and female gonad were unaffected by steroid hormones, whereas the testis, sex-specific behavior, and prostate were affected by steroids.

A third set of experiments examined the effect of testis grafts on 10-day-old female pouch young; these grafts remained palpable for 3–6 months. A parallel study examined castrated males; sham operations on both sexes were performed as controls. The females with testis grafts exhibited a mixture of male and female traits. They possessed a prostate, penis, and Cowper's gland indistinguishable from those of a normal male. On the other hand, the pouch, mammary glands, vaginal complex, uteri, and oviducts were those of a normal female. All grafted females were sterile and had ovaries that lacked oocytes or follicles. The castrated males had no penis or Cowper's gland, and Müllerian duct derivatives were absent. However, these animals did develop a normal scrotum with gubernaculum and vas deferens. Curiously, their weights were the same as those of the control males. The results supported the conclusion that scrotum, pouch, and mammary glands are steroid-independent, and suggested that growth in males may be as well.

The last group of data concerns the karyotypes of intersexes. There are published accounts of eight Australian intersexes, and a brief preliminary description of one individual of the American species, *Monodelphis domestica* (Robinson in Renfree and Short, 1988). A further seven Australian animals have been studied (Hughes *et al.*, 1992; Cooper *et al.*, 1993). Three categories of phenotype have been documented in the published accounts of intersexual marsupials. They are: (*a*) male without scrotum, but with a pouch containing nipples (XXY or XY); (*b*) female with empty scrotum but no pouch or mammary gland (XO or XO/XX); and (*c*) bilaterally asymmetric female, with empty hemiscrotum on one side and hemipouch on the other (XO or XO/XX/XXX).

In all cases, the presence of a Y is associated with testes (which are abdominal in the absence of a scrotum). Lack of a Y is associated with ovaries. In contrast, the pouch can be found in the presence of a Y, and a scrotum can develop without a Y. Whereas the pouch and scrotum are not considered homologous structures, they are antithetical, as demonstrated by the bilaterally asymmetric females (category c above). The genetic key to the pouch–scrotum decision is most clearly associated with the number of X chromosomes and is independent of the Y. It is intriguing that the two features that appear to be influenced by the X rather than Y are sexually dimorphic before any gonadal differentiation is detectable. The possibility that gene dosage is involved in the pouch–scrotum decision is consistent with evidence that X-inactivation is incomplete in marsupials (discussed in Section IV).

It will be necessary to survey more intersexual marsupials before the underlying mechanisms can be understood. Such a survey will be facilitated by the sexual dimorphism in size of many marsupials, especially kangaroos and wallabies. A small female-sized animal with a scrotum and no pouch is the phenotype predicted for XO, whereas a large male-sized animal with a pouch but no scrotum may be XXY.

It is important to keep in mind that marsupials are not a monolithic group, but rather a lineage as old and diverse as the eutherians. Even among closely related marsupials, differences in sexual development have been described. For example, the American opossums and their Australian cousins have been separated for at least 40 million years. This separation is evidenced by a number of physiological differences. A notable example is the presence of mammary gland rudiments in young males of the American opossums, whereas there is no evidence of mammary anlagen in the males of any Australian species (Renfree, 1992). Although such differences should be appreciated, a basic plan of marsupial sex determination is emerging.

In summary, marsupials display several sexual dimorphisms that precede testis differentiation, are not affected by steroid treatment, and are associated with the number of X chromosomes, independent of the Y. These dimorphisms have been studied extensively in male embryos of the tammar wallaby and include well-developed gubernaculum and process vaginalis; lack of mammary anlagen, which are evident in female embryos; and clearly developed paired scrotal anlagen. Since direct genetic control of these male features in marsupials precedes testis differentiation and androgen production, it is clear that some aspects of sex differentiation are not conserved among mammals.

IV. Evolution of the Mammalian Sex Chromosomes

Almost certainly the mammalian X and Y chromosomes are derived from an ancestral homomorphic pair (Ohno, 1967; also see Chapter 6 this volume). Remnants of this ancient homology still exist, although much of the modern X–Y identity is the result of more recent translocations with autosomes (Graves and Schmidt, 1992). In eutherians, the sex chromosomes pair at male meiosis, and recombination occurs along the pseudoautosomal region (PAR). Apart from this region, and other vestiges of homology, the X and Y have become genetically differentiated by some combination of translocation, deletion, repetition, mutation, and the trapping of mobile elements and pseudogenes. The question is the order and relative importance of each of these processes. It is not just a question of how, but why these particular chromosomes have evolved in such a unique manner.

Gene mapping in contemporary species is the best available method to trace chromosome evolution. The ultimate history of mammalian sex chromosomes will require data from the mammals, birds and reptiles, all of which shared a common ancestor about 200 million years ago. In birds, sex determination is chromosomal but the homomorphic sex is male (ZZ) and the heteromorphic sex is female (ZW). Reptiles use a variety of sex determinating methods, including some that are based on environmental rather than genetic factors. Y-chromosome-based sex determination is a system apparently limited to mammals, suggesting that it developed early in mammalian evolution, before the division of the two extant mammalian subclasses. In order to reconstruct the ancestral homomorphic X–Y pair, it is necessary to analyze their distant offspring, comparing the sex chromosomes of each mammalian lineage. Monotremes, marsupials and eutherians have in common the following features: a large gene-rich X, some form of X-inactivation, and testis determination associated with the Y chromosome. However, recent studies have revealed characteristics unique to each group, which challenge many of the basic assumptions of sex chromosome evolution.

A. Conservation of the X Chromosome

The conservation in size of the eutherian X chromosome led Ohno (1967) to propose that its genetic constitution would be preserved in all mammals (see Chapter 6, this volume). In eutherians, the X chromosome constitutes uniformly about 5% of the haploid genome. Genetic mapping in eutherians bolstered Ohno's Law, by showing that the same set of genes is X-linked in all species (see O'Brien and Graves, 1990).

It was believed that X-chromosome conservation was essential because of X-inactivation, since translocations interfering with this system were expected to result in dosage disorders. Experimental support for this idea came from X–autosome translocations (X;A), which induced sterility in mice and were associated with cases of human sterility. Cytologically, X-inactivation appeared complete in eutherians, with one entire X chromosome becoming condensed and visible as a sex chromatin body at interphase. The model of complete X-inactivation was supported by early expression studies, which showed that alleles on the inactive-X are not expressed, and that this silencing is stably passed on to somatic daughter cells. All of these observations on eutherian X chromosomes indicated that the X was conserved *in toto,* therby confirming Ohno's Law.

When comparisons were broadened to include marsupials, however, some unexpected results were obtained. The most obvious difference was the relative size of sex chromosomes. The X chromosomes of many marsupial species are smaller (3% of the haploid genome) than are the X chromosomes of the typical eutherian (5%) (Hayman *et al.,* 1982; Young *et al.,* 1982); and the Y chromosome, which is 2–3% of the eutherian genome, can be minute in marsupials (less than 1%). Another difference concerned the PAR. In fact, there is no obvious PAR in marsupials, as judged by the absence of chiasmata between the marsupial X and Y in studies of male meiosis from 30 Australian and South America species (reviewed by Hayman, 1990). Certainly, these differences indicated that not all of the X chromosome is conserved.

Marsupial X-inactivation also differs from that of the eutherian. Rather than being random, marsupial X-inactivation is restricted to the paternally derived X chromosome. It is less stable through cell divisions than eutherian X-inactivation is, and it is tissue-specific and incomplete. Expression studies on several X-linked isozymes illustrate this last point. The enzymes glucose-6-phosphate dehydrogenase (G6PD), phosphoglycerate kinase 1 (PGK1), α-galactosidase (GLA), and hypoxanthine phosphoribosyl transferase (HPRT) are all subject to X-inactivation in eutherians. However, marsupial tissue and fibroblast culture experiments (Cooper *et al.,* 1990) show a complex pattern of partial expression of the paternally derived X, which may be related to the position of these genes on the X (Dawson and Graves, 1988). These studies helped to explain why sex chromatin bodies are not common in marsupial cells, and expanded debate on how X-inactivation spreads. In order to trace the evolution of sex chromosomes and the spread of X-inactivation, genetic maps of the marsupial and monotreme X chromosomes were required.

B. Marsupial Gene Mapping

Classical genetic studies have been performed on the captive breeding colonies of three marsupial species: *Sminthopsis crassicaudata* (Bennett *et al.,* 1990), *Macropus eugenii* (Hinds *et al.,* 1990), and *Monodelphis domestica*

(VandeBerg, 1990). These three species, representing three separate orders that shared a common ancestor about 80 million years ago, now supply researchers with a pedigreed source of material for molecular mapping. Lynne Selwood at La Trobe University, Melbourne, has also established a colony of *Sminthopsis macroura,* which we have used extensively for our research on sex determination.

Initial mapping of sex-linked genes in marsupials relied on the expression of isozymes from homologs of human X-linked genes (reviewed by VandeBerg *et al.,* 1987; Cooper *et al.,* 1990). These experiments, first performed in the early 1970s for *PGK1* and *G6PD* and later for *GLA,* revealed the surprising pattern of marsupial X-inactivation: paternal, incomplete, and tissue specific.

A refinement in mapping genes on marsupial sex chromosomes came with the advent of rodent–marsupial cell hybrids (Hope and Graves, 1978). Hybrids that retained only the X chromosome (or a part thereof) from the marsupial could be analyzed for expression of suspected X-linked genes. The first gene products examined—HPRT, PGK, G6PD, and GLA—indicated that the genes were on the marsupial X, supporting Ohno's Law (Graves, 1979; Donald and Hope, 1981; Dobrovic and Graves, 1986; Dawson and Graves, 1988; reviewed by Graves *et al.,* 1990). However, steroid sulphatase (STS), which is X-linked in eutherians, was not expressed in cell hybrids retaining the marsupial X (Cooper *et al.,* 1984; Dawson and Graves, 1988). Thus *STS* and marsupial cell hybrids provided the first data suggesting a deviation from Ohno's Law, but without precise localization of this gene, the question of its position could not be resolved.

Mapping to a chromosomal location required the development of *in situ* hybridization to marsupial chromosomes. This was achieved by using homologous probes obtained from cloned marsupial sequences (Wainwright and Hope, 1985). But marsupial mapping really came of age with the advent of heterologous *in situ* hybridization, painstakingly developed by Andrew Sinclair in the laboratory of Jennifer Graves (Sinclair *et al.* 1987). This technique is now routinely performed and has provided a number of surprising insights concerning sex chromosome evolution and sex determination.

A major discovery was the autosomal location in marsupials of several genes that map to the short arm of the human X chromosome (Xp): *POLA, OTC, DMD, CYBB, MAOA, SYN1* (Sinclair *et al.,* 1987; Spencer *et al.,* 1991). As more genes were mapped, it became clear that genes from the human Xp were clustered primarily on two marsupial autosomes, indicating that at some point in mammalian history, large rearrangements had led to the present assortment of sex chromosomes. This conclusion generated a question: Which arrangement came first? Did the common ancestor of all mammalian X chromosomes possess the genes present on the human Xp, indicating that the marsupial X had lost these genes in recombination with autosomes; or, did the ancestral X lack the genes on human Xp, gaining them in the eutherian line as a result of translocation with autosomes? This question could not be settled without looking at the even more distant relatives of the eutherians, the monotremes.

C. Monotreme Gene Mapping and the Pedigree of the X Chromosome

In many respects, monotreme sex chromosomes are unique among mammals. The Y chromosome is unusually big, about 4% of the haploid genome, and large regions of the X and Y, which pair during meiosis, are G-band-homologous (Murtagh, 1977; Wrigley and Graves, 1988). This suggests that among extant mammals, monotreme sex chromosomes most closely resemble the ancestral homomorphic pair (Graves, 1987; Wrigley and Graves, 1988). The presumed X chromosome is larger than that in other mammals (6% of the haploid genome), and X-inactivation, as indicated by replication asynchrony, is tissue-specific and restricted to the unpaired region. There are also several small unpaired chromosomal elements, which, together with the sex chromosomes, form a translocation chain at meiosis (Murtagh, 1977).

To answer the question of whether the eutherian Xp gained the region containing *POLA, OTC, DMD, CYBB, MAOA,* and *SYN1,* or the marsupial X lost it, these genes were mapped in monotremes. The results show that all of the eutherian Xp genes that are autosomal in marsupials are also autosomal in monotremes. Even the clustering of human Xp genes on two marsupial autosomes is paralleled more or less on two monotreme autosomes. This supports the hypothesis that large autosome-to-X translocations occurred in the eutherian line after the marsupial/eutherian split 120–150 million years ago (Fig. 2).

Curiously, four genes conserved on the X chromosome in all mammals map to the region of the monotreme X that pairs with the Y chromosome and is apparently not inactivated (Watson *et al.*, 1990). Conservation without inactivation would seem to contradict Ohno's notion that X-chromosome conservation is related to inactivation. Perhaps, as suggested by Graves and Watson (1991), the X chromosome is just a "particularly good example of the conservation of the mammalian genome as a whole."

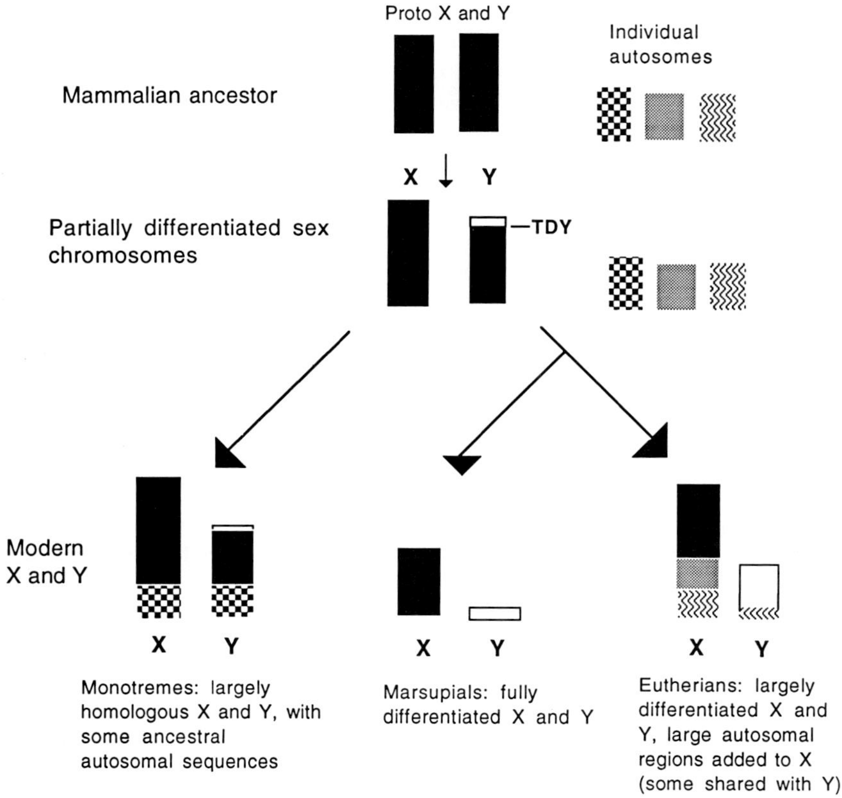

Figure 2 Proposed evolution of mammalian sex chromosomes from an ancestral homomorphic pair. The first step involves the acquisition of the sex-determining function in a newly differentiated region of the Y (open boxes). Later steps include recombination with autosomes, as well as heterochromatinization and deletions along the Y. Depictions of the modern X and Y illustrate the different degrees of sex chromosome differentiation seen in extant mammals. Drawings are not to scale.

V. Marsupials and the Quest for the Testis Determinant

Testes are obviously critical to the survival of mammalian species. Therefore conservation of the *TDF* gene, which controls their differentiation, seems likely. Through the years several candidates for *TDF* have been put forward and tested. Marsupials have played a key role in assessing the two most recent candidate genes.

Page *et al.* (1987) cloned a promising candidate that mapped to a region of the Y associated with human sex reversal. The gene, *ZFY,* was present on the Y chromosome of all eutherian mammals assayed by Southern blotting, and a related sequence (*ZFX*) was detected on human Xp. Page provided us with a *ZFY* probe to determine whether the gene was conserved on the marsupial Y chromosome. Surprisingly, *ZFY* homologs were found, but were located on marsupial *autosomes* in the same region as other human Xp genes (Sinclair *et al.*, 1988). This was an early indication that *ZFY* is not the mammalian *TDF*, and it demonstrated the importance of information provided by marsupial genetics. Since this time, *ZFY* homologs have also been found to be antosomal in monotremes (Watson *et al.*, 1993).

These findings launched the careers of several key players in the testis-determination story. Andrew Sinclair, who performed the *ZFY in situ* hybridization, was then a student in Graves' laboratory. He went on to do postdoctoral work with Peter Goodfellow in London, where he and colleagues cloned the *SRY* gene (sex-determining region Y). The *SRY* gene was isolated from a 35-kb region of the human Y chromosome present in XX males possessing only a fragment of the Y chromosome. It appeared to be a good candidate for *TDF,* because it encoded a 223 amino acid protein with homology to a DNA-binding domain found in high mobility group (HMG) proteins (Sinclair *et al.*, 1990). This gene, like *ZFY,* was conserved on the Y chromosomes of all eutherians tested. Moreover, in Southern blots, the *SRY* probe hybridized to several fragments that were not Y-specific. It was believed that these bands represented genes with homology to the conserved DNA-binding region of *SRY*. They were thus called SOX (*SRY*-like HMG box) genes. Experimental verification that *Sry* is the mouse *Tdy* (corresponding to the human *TDF*) was obtained in the laboratory of Robin Lovell-Badge, who, in collaboration with Australian Peter Koopman, found that *Sry* can reverse sex in transgenic XX mice (Koopman *et al.*, 1991; see Chapter 3).

Jamie Foster, another student in Graves' laboratory, subsequently characterized marsupial *SRY* homologs for his Ph.D. dissertation (see Foster *et al.*, 1992). Among the homologs characterized by Foster and colleagues is the first X-chromosome gene with homology to *SRY* (unpublished results). The advantages of marsupial genetics, as demonstrated by the *TDF* story, also attracted the attention of this chapter's first author, who worked closely with Jamie Foster and Francine Brennan, in Graves' laboratory. Brennan constructed the first marsupial testis cDNA library and isolated *SRY* cDNA from *Sminthopsis macroura.*

A. The Search for Marsupial *SRY*

As soon as the new *TDF* candidate was isolated by Sinclair *et al.* (1990), the question of a marsupial homolog was raised. Would *SRY* pass the marsupial test? Together with Sinclair and colleagues, Foster, on loan from the Graves' laboratory, began a frustrating series of Southern blots, using labeled human *SRY* to probe marsupial male and female genomic DNA. The species used represented the two major orders of Australian marsupials, the tammar wallaby, a member of the kangaroo family (*Macropus eugenii,* order Diprotodonta, family Macropodidae), and *Sminthopsis macroura* (order Polyprotodonta, family Dasyuridae), a mouse-like carnivorous marsupial. Unfortunately, the probe bound nonspecifically or washed completely off the blot. In a final attempt before returning to Australia, Foster left a Southern blot exposing in the freezer while he went on a last vacation in Europe. When the film was processed, 6 weeks later, a crisp autoradiograph revealed several bands with homology to the human *SRY* probe, and for the first time, a male-specific band was seen on a marsupial Southern blot. The Y-chromosome homologs of *SRY* were detected as a male-specific 4.3-kb *Eco*RI fragment in the tammar wallaby, and a male-specific 2.6-kb fragment in *S. macroura.* (Fig. 3a). This was the first evidence of a Y-chromosome gene conserved between the infraclasses; it demonstrated a common origin for these distant and cytologically different Y chromosomes.

Obtaining a second such autoradiograph proved as difficult as obtaining the first. Three of us tried dozens of times under every set of conditions we could generate. Only twice were satisfactory results obtained. What followed was 18 months of intensive effort to clone the marsupial homologs of *SRY*. In addition to the male-specific *SRY* band, in each species there were several other bands on our Southern blots. These presumably represented SOX genes, which contain the conserved HMG motif.

In our effort to clone the unique Y-chromosome gene, we took three approaches: (*a*) PCR amplification using *SRY* primers from the HMG-box region conserved between mouse and humans; (*b*) isolation of *SRY* homologous sequences from male total-genomic libraries, and (*c*) cDNA cloning from an adult testis library. This last approach was based on the curious detection of *Sry* transcripts in mouse adult testis (Sinclair *et al.,* 1990; Koopman *et al.,* 1990). While Foster constructed and screened several total genomic libraries, Brennan and Hampikian worked together on the cDNA and PCR strategies. Fourteen of the 18 months were spent pursuing SOX genes from genomic libraries, sequencing artifacts from PCR reactions, and trying to construct a good testis cDNA library. Finally, with Hampikian due to return to the United States in 4 months, and Foster expected to begin a fellowship in Goodfellow's laboratory, we decided to refocus our efforts.

Brennan and Hampikian continued to work on a cDNA library, we abandoned PCR, and the three of us worked together to construct size-selected libraries from gel slices of restricted genomic DNA. The slices corresponded to the approximate molecular weight of the male-specific bands seen on Southerns from *S. macroura* and the tammar wallaby. Several lambda libraries were made from the appropriate gel slices of each species; in all, over 10 million recombinants were screened. The first good news came on Christmas Eve 1991, when Foster detected three positive plaques, all from tammar wallaby libraries. One clone had a 4.8-kb insert that corresponded to a SOX gene seen in both males and females, but two of the clones had 4.3-kb inserts corresponding to the molecular weight (MW) of the male-specific *SRY* band. When the 4.3 kb insert was used as a probe on Southern blots of *Eco*RI-digested male and female DNA from both species, only the male-specific 4.3-kb band in the tammar wallaby hybridized to the probe. This confirmed that we had cloned *SRY* from the tammar wallaby.

New Year's Eve brought even better tidings. Brennan isolated a single positive recombinant from the *S. macroura* cDNA library. When we reprobed our Southern blots from both species with a labeled fragment from this cDNA, it hybridized strongly to the male-specific *SRY* bands in both mansupial species (Fig. 3b). This confirmed that our *S. macroura* cDNA was expressed from the Y-chromosome fragment that had hybridized to human *SRY* on Southern blots (Fig. 3a).

At this point we had cloned an *SRY* cDNA from *S. macroura* and a genomic clone from the tammar wallaby. Although elated at the prospect having the first nonhuman *SRY* cDNA clone, we were concerned about the possibilities of rearrangements akin to those that have plagued all attempts to clone a mouse *Sry* cDNA. A genomic clone from *S. macroura* was needed to verify our cDNA's sequence integrity. Therefore, in a final effort to obtain a *S. macroura* genomic clone, we used the cDNA to rescreen size-selected genomic libraries. This time we obtained an *S. macroura* partial genomic clone identical to the final 464 bp of the cDNA. We were relieved to find that there were no 3′ rearrangements in our *SRY* cDNA.

Final confirmation that marsupial *SRY* is indeed on the Y chromosome was obtained by *in situ* hybridization (Graves *et al.*, 1992). Although the Y chromosome of *S. macroura* is too small (smaller than a single silver grain) for this procedure, clear results were obtained for the tammar wallaby. Metaphase spreads of tissue-culture cells were probed with various dilutions of a tritium-labeled probe from the tammar wallaby *SRY* clone. Fifty-one spreads with a near-zero level of cytoplasmic background were scored. The chromosomal locations of 204 individual silver grains were analyzed using the Z_{max} algorithm (Ewens *et al.*, 1992), and the results

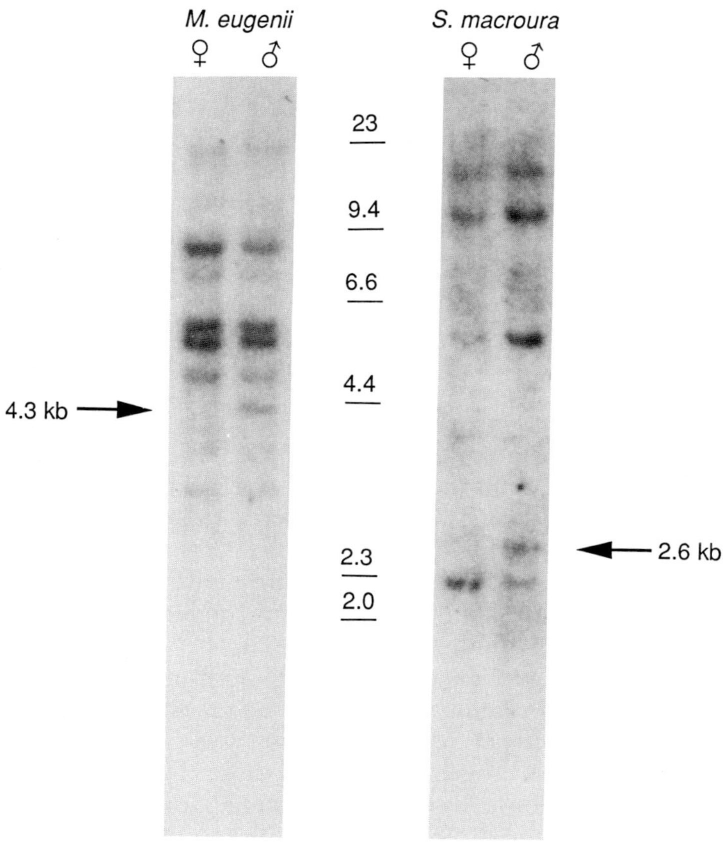

Figure 3 Identification of SRY homologs in two marsupial species. Southern blots of *Eco*R1 digested genomic DNA from male and female pairs of *S. macroura* and *M. eugenii* were hybridized with (a) ^{32}P-labeled human SRY probe or (b) ^{32}P-labeled probe derived from the cloned *S. macroura* SRY c-DNA. The male-specific band in each species is indicated by an arrow, at 2.6 kb for *S. macroura*, and 4.3 kb for *M. eugenii.*

confirmed our Southern blot data. A significant site ($P < .001$) of hybridization was detected on the Y chromosome, and a significant secondary site ($P < .01$) was mapped to the X chromosome. This latter site is detected as a band that shows X-chromosome dosage on Southern blots probed with *SRY*.

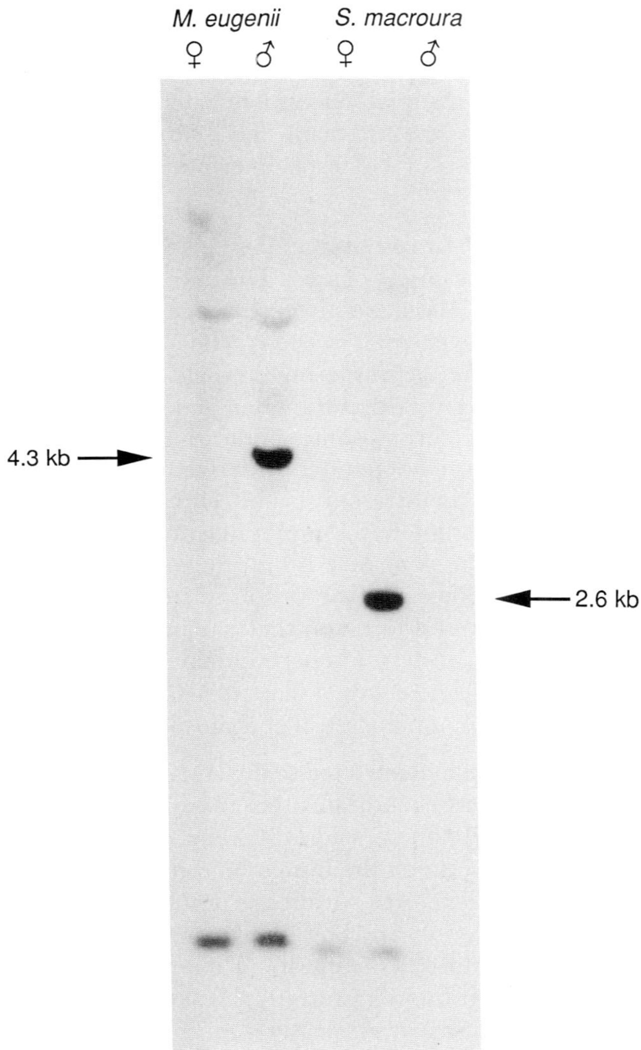

VI. Analysis of the Marsupial *SRY* Gene

Our mapping data, and sequence comparisons between the *SRY* genes of the two marsupials, human, mouse, and rabbit, led to the following conclusions:

***SRY* Has Passed the Marsupial Test.** It is conserved on the Y chromosome across the therian–metatherian boundary representing 120–150 million years of mammalian evolution. *SRY* is transcribed in the adult testis of at least one marsupial, *Sminthopsis macroura,* consistent with its expression pattern in mouse and man. These results support the hypothesis that *SRY* is the conserved testis-determining gene, at least in all therian mammals.

The Mammalian Y Chromosome Has a Monophyletic Origin. Y chromosomes come in a variety of sizes. The eutherian Y is typically about 2–3% of the haploid genome. This percentage is approximated in the tammar wallaby. However, in *S. macroura.* the Y chromosome is minute (less than 1%). Since the Y chromosome is male-determining in all these species, it was possible either that the same gene in the sex-determining path is resident on all Y chromosomes, indicating a monophyletic origin, or that different genes on independently derived "Y" chromosomes act as the "master sex-switch" in different species. Our results establish that the mammalian Y chromosome has a monophyletic origin. We have shown that the sex determining gene *SRY* is indeed conserved between marsupials and eutherians. Recently, the presumptive spermatogenesis gene *Ube1Y* has also been found to lie on the marsupial Y, as well as on the mouse Y (Mitchell *et al.,* 1992). These reports provide the first molecular evidence that the Y chromosomes of marsupials and eutherian mammals are derived from a common ancestral chromosome.

***SRY* Conservation is Restricted to the HMG-Box.** A comparison of mammalian *SRY* sequences indicates that there is no detectable conservation outside of the 79 amino acid HMG-box. Within the box, there is significant homology between the human and marsupial *SRY* sequences (Fig. 4). At the amino acid level, within this domain, tammar wallaby and humans show a 70% identity and 86% similarity; *S. macroura* and humans have a 69% identity and 82% similarity (Table I). This is similar to the mouse/human comparison. Significantly, all seven amino acid residues associated with human sex reversal (Berta *et al.,* 1990; Jager *et al.,* 1990; Harley *et al.,* 1992; Hawkins *et al.,* 1992) are conserved in every species examined.

VII. Discussion

A. *SRY* Is the Mammalian *TDF*

The quest for the mammalian testis-determining gene has ended. Several lines of evidence confirm that *SRY* is the mammalian *TDF*: *Sry* and its

```
S. macroura SRVKRPMNAFMVWSQTQRRKVALQNPKMHNSEISKQLGVTWKLLSDSEKRPFIDEAKRLRDKHKQVS-DYKYQPRRKTK
M. eugenii  DRVKRPMNAFMIWSRSQRRKVALENPKMHNSEISKHLGFTWKMLPDNEKQPFIDEAERLRAKHREEFPDYKYQPRRKKF
Rabbit      ERVKRPMNAFMVWSQHQRRQVALENPKMRNSDISKQLGHQWKMLSEAEKWPFFQEAQRLQAMHKEKYPDYKYRPRRKVK
Human       DRVKRPMNAFIVWSRDQRRKMALENPRMRNSEISKQLGYQWKMLTEAEKWPFFQEAQKLQAMHREKYPNYKYRPRRKAK
Mouse       GHVKRPMNAFMVWSRGERHKLAQQNPSMQNTEISKQLGCRWKSLTEAEKRPFFQEAQRLKILHREKYPNYKYQPHRRAK
              *|||||||**||  *|* *| *|| | |**||| ||  || |** || || *|| *|    |**   *||| |*|*
```

Figure 4 Comparison of the amino acid sequence of the HMG-box of *SRY* in five species. A dash was introduced in the *S. macroura* sequence to maintain alignment. Invariant amino acids at positions associated with human sex-reversal mutations (XY females) are boxed. Other invariant amino acids are indicated with an asterisk (*), and conservative changes with a vertical line (|). Conserved amino acids are grouped as follows: HRK, GASTP, QEDN, MLVI, and FYW. (Adapted from Foster *et al.*, 1992.)

adjacent sequences are sufficient to induce testis differentiation in XX transgenic mice (Koopman *et al.*, 1991); mutations in the HMG-box of this gene are associated with human sex reversal (Berta *et al.*, 1990; Jager *et al.*, 1990; Harley *et al.*, 1992; Hawkins *et al.*, 1992); and *SRY* is conserved over the 120- to 150-million-year divergence of marsupials and eutherians (Foster *et al.*, 1992). Even the tiny Y chromosome of *S. macroura* possesses an *SRY* gene, which, like its homologs in mouse and human, is transcribed in adult testis.

B. *SRY* Sequence Diversity and Function

Like other HMG genes, *SRY* probably regulates gene activity by coding for a DNA-binding protein. Recombinant *SRY* protein has been shown to bind specific DNA sequences, and mutant peptides derived from sex-reversed patients show decreased binding capabilities (Harley *et al.*, 1992;

Table I
Interspecies Homology of the *SRY*-Conserved Region

	Human	Rabbit	Mouse	*S. macroura*
Human				
Rabbit	82/94 (50)			
Mouse	71/84 (80)	65/80 (80)		
S. macroura	63/84 (145)	71/81 (145)	62/84 (145)	
M. eugenii	68/85 (145)	70/82 (145)	60/80 (145)	79/85 (50)

Note. Comparison of the amino acid sequences of the HMG-box region of *SRY* for the five species shown in Fig. 4. Percentage identity/similarity are presented, followed by (in parentheses) the estimated divergence times (millions of years before present). (Adapted from Graves *et al.*, 1992.)

Nasrin *et al.*, 1991). Giese and colleagues (1992) have shown that *SRY* can induce DNA bending *in vitro,* and suggest that this bending mediates transcription of remote loci. The *SRY* protein has not yet been observed in early embryos, but it is expected to exert its effects at the genital ridge just prior to testis differentiation. Koopman and colleagues (1991) reported that a burst of *Sry* transcripts appear 10.5–11.5 days post coitum (dpc) in the mouse, just before testis differentiation is cytologically evident (12.5 dpc.) All of these data point to a finely regulated system, in which *SRY* either enhances or diminishes transcription from other specific loci. In fact, *SRY* may do both since Haqq and colleagues (1993) have shown that *SRY* protects specific residues of the MIS and P-450 aromatase genes. The inference is that *SRY* enhances MIS and inhibits P-450 aromatase expression (see Chapter 19, this volume).

Since *SRY* is the first mammalian gene to induce organogenesis experimentally, we might expect it to be highly conserved. Yet, despite its conservation of function, expression, and location, *SRY* shows remarkable interspecies heterogeneity at the amino acid level. It appears that *SRY* is evolving rapidly and is conserved only within the HMG-box. Even within this "conserved" motif, the tammar wallaby and mouse show only a 60% identity and 80% similarity; *S. macroura* and mouse show a 62% identity and 84% similarity at the amino acid level (Table 1). In comparison, we have identified a marsupial SOX gene (J. W. Foster, unpublished results) with an HMG-box that is 99% similar to a mouse SOX gene that has been implicated in neurogenesis. Other SOX genes also show more conservation than *SRY,* for example SOX2 is 98% identical in human and mouse, whereas *SRY* is only 71% identical in these species.

The divergence of *SRY* genes in different species is consistent with either of two opposed scenarios. Either the functional constraints on the *SRY* sequence are lax, or conversely, each species has a specific and functionally unique *SRY.* In support of the former, some have proposed that *SRY* is a repressor or competitor of an ovary determinant, rather than a transcriptional activator of testis differentiation. The assumption is that an interfering agent need not be as precise as an enhancer; any spanner in the works will do. Repression could be carried out at either the nucleic acid or the protein level. For example, the *SRY* gene could titrate out binding factors destined for ovary inducing genes, or the *SRY* peptide could bind and repress elements required for female determination; these activities might require that only small parts of *SRY* be conserved. However, although such a flexible role for *SRY* is possible, the evidence to date depicts a more constrained gene whose species identity is functionally important. Several lines of evidence suggest that the nonconserved and even the noncoding portions of *SRY* are functionally significant. For example, when Koopman *et al.* (1991) used the human *SRY* gene to construct transgenic mice, they did not obtain sex reversal even though the transgene was expressed at the correct developmental stage. Also, transgenic mice that did develop as XX males required not only the putative coding region of mouse *Sry*, but also several kilobases of adjacent sequence. It appears that *SRY* is species-specific and requires more than coding regions to function.

C. Is *SRY* Driving Speciation?

The divergence of *SRY* sequences could be a major factor in reproductive isolation and speciation. For example, it may explain the sex reversal and fertility seen in crosses between two mouse subspecies, first reported by Eicher and colleagues in 1982. When a *Mus domesticus* Y chromosome is crossed into a *Mus musculus* background, the XY offspring do not develop testes (Eicher *et al.*, 1982). Eicher and Washburn (1983) attributed this to a later-acting testis determinant coded by Y^{DOM}, and this was confirmed several years later (Palmer and Burgoyne, 1991; Taketo *et al.*, 1991). In the original paper reporting the phenomenon, Eicher and colleagues (1982) suggested that mutations in *TDF* might be involved in mammalian speciation. Recently, a single amino acid difference (Thr for Ile at position 61) in the conserved portion of *Sry* was reported for these two subspecies (Nasrin *et al.*, 1991). Although the significance of this finding was not noted by the authors, it has been proposed that this difference, and other changes in *Sry*, are responsible for the gonadal dysgenesis seen in the *M. musculus* × *M. domesticus* crosses (see Chapter 5, this volume).

Is *SRY* the agent of speciation imagined by Eicher and colleagues? Graves and Erickson (1992) have analyzed several mouse strains and report that all *M. domesticus* strains have the *Sry* threonine substitution, and all *M. musculus* lack it. However, they point out that this change alone does not explain the gradation of sex reversal seen when different *M. domesticus* strains are bred to C57BL/6 females. Biddle and Nishioka (1988) have used a mouse Y-sequence probe to identify a polymorphism that correlates well with the degree of sex reversal caused by Y chromosomes from different *M. domesticus* strains. However, it is not known whether this polymorphism represents a difference in the *Sry* sequence. The effect of single base changes in *Sry* will be best addressed by constructing *Sry* transgenic mice. If it is found that changes in *Sry* are responsible for the incompatibility of *M. domesticus* and *M. musculus*, the theory that *SRY* drives mammalian speciation will receive experimental support.

D. What Is Driving *SRY* Diversity?

Comparisons of *SRY* sequences support the hypothesis that genes unique to the Y chromosome evolve rapidly. The reasons for this are not clear, though there is no shortage of possible causes, e.g., the lack of a pairing partner, or the heterochromatic environment and late replication of the Y. One factor that has not been experimentally examined is the effect on the Y chromosome of passage through sperm in every generation (Graves and Schmidt, 1992). Sperm-borne chromosomes go through more divisions than those in the egg, they are subjected to compaction after having their histones replaced with sperm-specific protamines, and they are transferred to a new cellular environment where their nucleosomes are again restructured. Whereas all other chromosomes pass through sperm half of the time, the Y chromosome is restricted to this gamete, and whereas the X and autosomes are met by homologous partners in the egg, genes unique to the Y chromosome have no partner and therefore homologous recombination and gene conversion are not possible. This is true for all genes that are unique to the Y chromosome; however, in the case of *SRY*, mutations may be critical to both the survival of the lineage and the creation of new species. A rapidly evolving *SRY* is an ideal candidate for speciation, since mutations in *SRY* can greatly affect reproduction.

Implicit in the theory that *SRY* is driving speciation, is the requirement that the X-borne and autosomal genes that interact with *SRY* adapt to its changes. It is easiest to imagine that functional mutations in *SRY* originate in the male germline. Sperm that carry these mutations will produce fertile offspring only if the conceptus has cooperative or compensating alleles for factors that interact with *SRY*. Otherwise, the offspring will not develop testes, and the Y chromosome is effectively dead. The cooperative alleles on autosomes might be maintained in the population as recessives that produce XX and XY females, until they are rescued by the mutant *SRY (SRY*)*. If the *SRY**/cooperative allele combination confers selective advantage, these chromosomes will quickly spread. Spreading will be achieved by the selective advantage of the modified chromosomes and by the infertility of *SRY** offspring without the cooperative alleles. If *SRY** effectively splits the gene pool through reproductive isolation, the result is two diverging species that compete for the same niche. Thus, it is possible to view *SRY* not only as a source of speciation but also as a source of intense competition and even extinction.

E. The Future of Sex Determination Research on Australian Mammals

With two marsupial *SRY* genes now sequenced, the obvious next step is to map and sequence a monotreme *SRY*. This work is in progress in Graves' laboratory. However, without XO and XXY platypuses or echidnas, we cannot be sure that the Y chromosome is sex-determining in these species. Even if they do conform to the mammalian rule of Y-chromosome-based sex determination, it is conceivable that monotremes do not share the *SRY* = *TDF* equation. On their giant Y chromosome it is possible that another gene regulates sex, a gene that has lost its function, changed location, or been usurped in the therian line of descent (Graves and Schmidt, 1992). By including monotremes in comparative studies, the mystery of sex determination is being pushed back to the earliest days of mammalian life.

Australian mammals will also play a key role in explaining how the scrotum arose and testicles descended. As noted above, monotremes have no scrotum and their testes are intraabdominal (Griffiths, 1978). The eutherian scrotum is associated with the Y chromosome, develops only in response to androgen, and is homologous with the female labia majora. In contrast, the marsupial scrotum develops before the testis, is independent of the Y chromosome and hormonal stimulation, is anterior to the penis, and has no known female homolog [marsupials lack labia, and the urogenital opening is identical in both sexes (Hughes *et al.*, 1992)]. The question is, did the scrotum arise twice independently in the therian subclass, or are the same genes at the root of scrotal development in all species? Identifying the determinants or inhibitors of the pouch–scrotum switch on the marsupial X chromosome will help to answer this question and resolve the genetics of scrotal development.

Finally, the next phase in marsupial research will require the production of transgenic animals. When this is achieved, researchers will have an unprecedented opportunity to study mammalian development. In eutherians, embryonic development cannot be easily observed without inducing trauma or death. In marsupials, developing neonates, at stages comparable to eutherian embryos, can be examined in the pouch and manipulated without injury. An obvious candidate for early transgenic experiments will be *SRY*. But in a wider sense transgenics would provide a new marsupial test for any conserved gene. Sequence conservation alone does not prove conserved function, especially in complex systems like sex determination. With transgenics, knockout and ectopic expression experiments will lead to a marsupial test based not only on conserved sequence but also on conserved function.

Acknowledgments

The authors thank Francine Brennan for help in producing the illustrations, and Jamie Foster for permission to cite unpublished data. The work on marsupial and monotreme gene mapping was supported by the Australian Grants Council, and the work on marsupial *SRY* by the National Health and Medical Research Council. Gregory K. Hampikian was supported by a National Science Foundation Fellowship.

References

Alcorn, G. T. (1975). Development of the ovary and urogenital ducts in the tammar wallaby *Macropus eugenii* (Desamrest, 1817). Ph.D. Thesis, Macquarie University, Sydney.

Bennett, J. H., Hayman, D. L., Hope, R. M., and Breed, W. (1990). Reproductive and genetical studies with a laboratory colony of the dasyurid marsupial *Sminthopsis crassicaudata*. *In* "Mammals from Pouches and Eggs" (J. A. M. Graves, R. M. Hope, and D. W. Cooper, eds.), pp. 65–81. CSIRO, Melbourne.

Berta, P., Hawkins, J. R., Sinclair, A. H., Griffiths, B. L., Goodfellow, P. N., and Fellous, M. (1990). Genetic evidence equating *SRY* and the testis-determining factor. *Nature (London)* **348,** 448–450.

Biddle, F. G., and Nishioka, Y. (1988). A DNA polymorphism from five inbred strains of the mouse identifies a functional class of *domesticus*-type Y chromosomes that produce the same phenotypic distribution of gonadal hermaphrodites. *Genome* **30,** 870–878.

Burns, R. K. (1961). Role of hormones in the differentiation of sex. *In* "Sex and Internal Secretions " (W. G. Young, ed.), Vol. 1, pp. 76–158. Williams & Wilkins, London.

Cooper, D. W., McAllan, B. M., Donald, J. A., Dawson, J. A., Dobrovic A., and Graves, J. A. M. (1984). Steroid sulphatase is not detected on the X chromosome of Australian marsupials. *Hum. Gene Mapping* **7,** 439.

Cooper, D. W., Johnston, P. G., VandeBerg, J. L., and Robinson, E. S. (1990). X-chromosome inactivation in marsupials. *In* "Mammals from Pouches and Eggs" (J. A. M. Graves, R. M. Hope, and D. W. Cooper, eds.), pp. 270–275. CSIRO, Melbourne.

Cooper, D. W. *et al.* (1993). In preparation.

Dawson, G. W., and Graves, J. A. M. (1986). Gene mapping in marsupials and monotremes. II. Assignment to the *X* chromosome in four species of dasyurid marsupial. *Cytogenet. Cell Genet.* **42,** 80–84.

Dobrovic, A., and Graves J. A. M. (1986). Gene mapping in marsupials and monotremes. II. Assignment of four genes to the X chromosome in four species of dasyarid marsupial. *Cytogenet. Cell Genet.* **42,** 9–13.

Donald, J. A., and Hope, R. M. (1981). an autosomal assignment in a marsupial: The gene for LDHA is on chromosome 5 in the red kangaroo. *Biochem. Genet.* **19,** 901–908.

Eicher, E. M., and Washburn, L. L. (1983). Inherited sex reversal in mice: Identification of a new primary sex-determining gene. *J. Exp. Zool.* **228,** 297–304.

Eicher, E. M., Washburn, L. L., Whitney, J. B., III, and Morrow, K. E. (1982). *Mus poschiavinus* Y chromosome in the C57BL/6J murine genome causes sex reversal. *Science* **217,** 535–537.

Ewens, W. J., Griffiths, R. C., Ethier, S. N., Wilcox, S. A., and Graves, J. A. M. (1992). Statistical analysis of *in situ* hybridization data: Derivation and use of the Z_{max} test. *Genomics* **12,** 675–682.

Fadem, B. H., and Tesoriero, J. V. (1986). Inhibition of testicular development and feminization of the male genitals by neonatal oestrogen treatment in a marsupial. *Biol. Reprod.* **34,** 771–776.

Ford, C. E., Jones, K. W., Polani, P., de Almeida, J. C., and Briggs, J. H. (1959). A sex chromosome anomaly in a case of gonadal dysgenesis (Turner's syndrome). *Lancet* **1,** 711–713.

Foster, J. W., Brennan, F. E., Greg Hampikian, G. K., Goodfellow, P. N., Sinclair, A. H., Lovell-Badge, R., Selwood, L., Renfree, M. B., Cooper, D. W., and Graves, J. A. M. (1992). Evolution of sex determination and the Y chromosome: *SRY*-related sequences in marsupials. *Nature (London)* **359,** 531–533.

Giese, K., Cox, J., and Grosschedl, R. (1992). The HMG domain of lymphoid enhancer factor 1 bends DNA and facilitates assembly of functional nucleoprotein structures.

Graves, J. A. M. (1979). Mouse–marsupial cell hybrids containing fragments of the marsupial X chromosome. *Somatic Cell Genet.* **5,** 481–489.

Graves, J. A. M. (1987). The evolution of mammalian sex chromosomes and dosage compensation-clues from marsupials and monotremes. *Trends Genet.* **3,** 252–256.

Graves, J. A. M., and Schmidt, M. M. (1992). Mammalian sex chromosomes: Design or accident? *Curr. Opin. Gen. Dev.* **2,** 890–901.

Graves, J. A. M., and Watson, J. M. (1991). Mammalian sex chromosomes: Evolution of organization and function. *Chromosoma* **101,** 63–68.

Graves, J. A. M., Sinclair, A. H., and Spencer, J. A. (1990). Marsupial gene mapping and the evolution of mammalian sex chromosome form and function. *In* "Mammals from Pouches and Eggs" (J. A. M. Graves, R. M. Hope, and D. W. Cooper, eds.), pp. 224–241. CSIRO, Melbourne.

Graves, J. A. M., Foster, J. W., Hampikian, G. K., and Brennan, F. E. (1993). Sex-determination in marsupial mammals. *In* "Sex Chromosomes and Sex-Determining Genes" (K. C. Reed, and J. A. M. Graves, eds.), pp. 325–336. Harwood Academic Publishers, Melbourne.

Graves, P. E., and Erickson, R. P. (1992). An amino acid change in the DNA-binding region of *Sry,* found in *Mus musculus domesticus* and other species, does not explain C57BL/6J-Y^{DOM} sex-reversal. *Biochem. Biophys. Res. Commun.* **185,** 310–316.

Griffiths, M. (1978). "The Biology of the Monotremes." Academic Press, New York.

Haqq, C. M., King, C., Donahue, P. K., and Neiss, M. A. (1993). SRY recognizes conserved DNA sites in sex-specific promoters. *Proc. Natl. Acad. Sci. U.S.A.* **90,** 1097–1101.

Harley, V. R., Jackson, D. I., Hextall, P. J., Hawkins, J. R., Berkovitz, G. D., Sockanathan, S., Lovell-Badge, R., and Goodfellow, P. N. (1992). DNA binding activity of recombinant SRY from normal males and XY females. *Science* **255,** 453–456.

Hawkins, J. R., Taylor, A., Berta, P., Levilliers, J., Van der Auwera, B., and Goodfellow, P. N. (1992). Mutational analysis of SRY: nonsense and missense mutations in XY sex reversal. *Hum. Genet.* **88,** 471–474.

Hayman, D. L. (1990). Marsupial cytogenetics. *In* "Mammals from Pouches and Eggs" (J. A. M. Graves, R. M. Hope, and D. W. Cooper, eds.), pp. 189–209. CSIRO, Melbourne.

Hayman, D. L., Ashworth, L. K., and Carrano, A. V. (1982). The relative DNA contents of the eutherian and marsupial X chromosome. *Cytogenet. Cell Genet.* **34,** 256–270.

Hinds, L. A., Poole, W. E., Tyndale-Biscoe, C. H., van Oorschot, R. A. H., and Cooper, D.W. (1990). Reproductive biology and the potential for genetic studies in the tammar wallaby. *In* "Mammals from Pouches and Eggs" (J. A. M. Graves, R. M. Hope, and D. W. Cooper, eds.), pp. 81–92. CSIRO, Melbourne.

Hope, R. M., and Graves, J. A. M. (1978). Fusion and hybridization of marsupial and eutherian cells. VI. Hybridization. *Aust. J. Biol. Sci.* **31,** 527–543.

Hughes, R. L., Pearse, A.-M., Cooper, D.W., Joss, J. M. P., Johnson, P. G., and Jones, M.P. (1993). The genetic basis of marsupial gonadogenesis and sexual phenotype. *In* "Sex Chromosomes and Sex-Determining Genes" (K. C. Reed, and J. A. M. Graves, eds.), pp. 17–48. Harwood Academic Publishers, Melbourne.

Jacobs, P. A., and Strong, J.A. (1959). A case of human intersexuality having a possible XXY sex-determining mechanism. *Nature* (*London*) **183,** 302–303.

Jäger, R. J., Anvret, M., Hall, K., and Scherer, G. (1990). A human XY female with a frameshift mutation in SRY, a candidate testis determining gene. *Nature* (*London*) **348,** 452–454.

Jost, A. (1970). Hormonal factors in the sex differentiation of the mammalian foetus. *Philos. Trans. R. Soc. London B Ser.* **259,** 119–130.

Kemp, T. S. (1982). "Mammal-like reptiles and the origin of mammals." Academic Press, New York.

Koopman, P., Münsterberg, A., Capel, B., Vivian, N., and Lovell-Badge, R. (1990). Expression of a candidate sex-determining gene during mouse testis differentiation. *Nature* (*London*) **348,** 450–452.

Koopman, P., Gubbay, J., Vivian, N., Goodfellow, P., and Lovell-Badge, R. (1991). Male development of chromosomally female mice transgenic for *Sry. Nature* (*London*) **351,** 117–21.

Magre, S. (1985). Différentiationdes cellules de Sertoli et morphogénèse testiculaire chez le foetus de rat. *Arch. Anat. Microsc. Morphol. Exp.* **74,** 64–68.

Mitchell, M. J., Woods, D. R., Wilcox, S. A., Graves, J. A. M., and Bishop, C. E. (1992). The marsupial Y chromosome encodes a homologue of the mouse Y-linked candidate spermatogenesis gene *Ube. Nature* (*London*) **359,** 528–530.

Moore, H. D. M., and Thurstan, S. M. (1990). Sexual differentiation in the grey short tailed opossum, *Monodelphis domestica,* and the effect of oestradiol benzoate on development in the male. *J. Zool.* **221,** 639–658.

Murtagh, C. E. (1977). A unique cytogenetic system in monotremes. *Chromosoma* **65,** 37–57.

Nasrin, N., Buggs, C., Kong, X. F., Carnazza, J., Goebl, M., and Alexander-Bridges, M. (1991). DNA-binding properties of the testis-determining gene and a related protein. *Nature* (*London*) **354,** 317–320.

O, W.-S., Short, R. V., Renfree, M.B., and Shaw, G. (1988). Primary genetic control of sexual differentiation in a mammal. *Nature* (*London*) **331,** 716–717.

O'Brien, S. J., and Graves, J. A. M. (1990). Geneticists converge on divergent mammals: An overview of comparative mammalian genetics. *In* "Mammals from Pouches and Eggs" (J. A. M. Graves, R. M. Hope, and D. W. Cooper, eds.), pp. 5–15. CSIRO, Melbourne.

Ohno, S. (1967). "Sex Chromosomes and Sex-linked Genes." Springer-Verlag, Berlin.

Page, D. C., Mosher, R., Simpson, E. M., Fisher, E. M. C., Mardon, G., Pollack, J., McGillivray, B., de la Chapelle, A., and Brown, L.C. (1987). The sex-determining region of the Y chromosome encodes a finger protein. *Cell* (*Cambridge, Mass.*) **51,** 1091–1104.

Palmer, S. J., and Burgoyne, P. S. (1991). The *Mus domesticus Tdy* allele acts later than the

Mus musculus musculus Tdy allele: A basis for XY sex-reversal in C57BL/6-Y^{POS} mice. *Development, (Cambridge, UK)* **113,** 709–714.

Renfree, M. (1992). The role of genes and hormones in marsupial sexual differentiation. *J. Zool. Lond.* **226,** 165–173.

Renfree, M. B., and Short, R. V. (1988). Sex determination in marsupials: Evidence for a marsupial-eutherian dichotomy. *Philos. Trans. Soc. London, Ser. B.* **322,** 41–53.

Renfree, M. B., Shaw, G., and Short, R. V. (1987). Sexual differentiation in marsupials. *In* "Genetic Markers of Sex Differentiation" (F. P. Haseltine, M.E. McClure, and E. H. Goldberg, eds.), pp. 27–41. Plenum, New York.

Sharman, G. B. (1976). Evolution of viviparity in mammals. *In* "Reproduction in Mammals 6" (C. R. Austin and R. V. Short, eds.), pp. 32–70. Cambridge Univ. Press, Cambridge.

Sharman, G. B., Robinson, E. S., Walton, S. M., and Berger, P. J. (1970). Sex chromosomes and reproductive anatomy of some intersexual marsupials. *J. Reprod. Fertil.* **21,** 57–68.

Sharman, G. B., Hughes, R. L., and Cooper, D. W. (1990). The chromosomal basis of sex differentiation in marsupials. *Aust. J. Zool.* **37,** 45–66.

Shaw, G., Renfree, M. B., Short, R. V., and O, W.-S. (1988). Experimental manipulation of sexual differentiation in wallaby pouch young with exogenous steroids. *Development, (Cambridge, UK)* **104,** 689–701.

Sinclair, A. H., Wrigley, J. M., and Graves, J. A. M. (1987). Autosomal assignment of OTC in marsupials and monotremes: Implications for the evolution of sex chromosomes. *Genet. Res.* **50,** 131–136.

Sinclair, A. H., Foster, J. W., Spencer, J. A., Page, D. C., Palmer, M., Goodfellow, P. N., and Graves, J. A. M. (1988). Sequences homologous to *ZFY,* a candidate human sex-determining gene, are autosomal in marsupials. *Nature (London)* **336,** 780–783.

Sinclair, A. H., Berta, P., Palmer, M. S., Hawkins, J. R., Griffiths, B. L., Smith, M. J., Foster, J. W., Frischauf, A.-M., Lovell-Badge, R., and Goodfellow, P. N. (1990). A gene from the human sex-determining region encodes a protein with homology to a conserved DNA-binding motif. *Nature (London)* **346,** 240–244.

Spencer, J. A., Watson, J. M., Lubahn, D. B., Joseph, D. R., French, F. S., Wilson, E. M., and Graves, J. A. M. (1991). The androgen receptor gene is located on a highly conserved region of the X chromosomes of marsupial and monotreme as well as eutherian mammals. *J. Hered.* **82,** 134–139.

Strahan, R. (1983). Australian mammal fauna. *In* "The Complete Book of Australian Mammals" (R. Strahan, ed.), pp. xvii-xx. Angus & Robertson, London.

Taketo, T., Saeed, J., Nishioka, Y., and Donahoe, P. K. (1991). Delay of testicular differentiation in the B6.Y^{DOM} ovotestis demonstrated by immunocytochemical staining for Müllerian Inhibiting Substance. *Dev. Biol.* **146,** 386–395.

Tyndale-Biscoe, C. H., and Renfree, M. B. (1987). "Reproductive Physiology of Marsupials." Cambridge Univ. Press, Cambridge.

VandeBerg, J. L. (1990). The grey-short tailed opossum (*Monodelphis domestica*) as a model didelphid species for marsupial genetic research. *In* "Mammals from Pouches and Eggs" (J. A. M. Graves, R. M. Hope, and D. W. Cooper, eds.), pp. 93–107. CSIRO, Melbourne.

VandeBerg, J. L., Robinson, E. S., Samollow, P. B., and Johnston, P. G. (1987). X linked expression and X chromosome inactivation: Marsupials, mouse and man compared. *Curr. Top. Biol. Med. Res.* **15,** 225–253.

Wainwright, B. J., and Hope, R. M. (1985). Cloning and chromosomal location of the alpha and β globin genes from a marsupial. *Proc. Natl. Acad. Sci. U.S.A.* **82,** 8105–8108.

Watson, J. M. (1990). Monotreme genetics and cytology, and a model for mammalian sex chromosome evolution. *In* "Mammals from Pouches and Eggs" (J. A. M. Graves, R. M. Hope, and D. W. Cooper, eds.), pp. 243–267. CSIRO, Melbourne.

Watson, J. M., Spencer, J. A., Riggs A. D., and Graves, J. A. M. (1990). The X Chromosome of monotremes shares a highly conserved region with the marsupial and eutherian and marsupial X despite the absence of X inactivation. *Proc. Natl. Acad. Sci. U.S.A.* **87,** 7125–7129.

Welshons, W. J., and Russell, L. B. (1959). The Y chromosome as the bearer of male determining factors in the mouse. *Proc. Natl. Acad. Sci. U.S.A.* **45,** 560–566.

Wilson, J. D., George, F. W., and Griffin, J. E. (1981). The hormonal control of sexual development. *Science* **211,** 1278–1284.

Wood Jones, F. (1923-1925). "The Mammals of South Australia." Government Printer, Adelaide, South Australia.

Wrigley, J. M., and Graves, J. A. M. (1988). Sex chromosome homology and incomplete tissue-specific *X* inactivation suggest that monotremes represent an intermediate stage of mammalian sex chromosome evolution. *J. Hered.* **79,** 115–118.

Young, G. J., Graves, J. A. M., Barbieri, I., Wooley, P. A., Cooper, D. W., and Westerman, M. (1982). The chromosomes of dasyurids (Marsupalia). *In* "Carnivorous Marsupials" (M. Archer, ed.), pp. 783-795. R. Zool. Soc. N.S.W., Sydney.

9

The Feminine Mystique: The Initiation of Sex Determination in *Drosophila*

Claire Cronmiller
Department of Biology
University of Virginia
Charlottesville, Virginia 22901

Helen K. Salz
Department of Genetics
Case Western Reserve School of Medicine
Cleveland, Ohio 44106

Molecular Genetics of Sex Determination

I. Introduction

In organisms where the genetic basis for sex determination has been extensively investigated, a small number of genes appear to act as binary switches to bring about the sexual differentiation that is appropriate to the chromosomal sex of the individual. For example, in mammals the Y chromosome, acting as a "male determining" signal, turns on a genetic switch that diverts sexual development from its default pathway, i.e., female differentiation, in order to produce males (reviewed in Goodfellow and Lovell-Badge, 1991; also see Chapters 1–3, this volume). In contrast, the default pathway for sex determination in *Drosophila* is male. Here the chromosomal signal (XX) for female development must activate a genetic switch to impose a *female* identity on an embryo that would otherwise develop as a male. Thus, when the genetic switch is ON, the female developmental pathway is selected; if the switch defaults to the OFF mode, male development ensues.

The primary switch that initiates sex determination in *Drosophila* controls both major aspects of sexual cell fate: sexual differentiation and X-chromosomal dosage compensation (Fig. 1). Through sexual differentiation the sexes diverge with respect to structural, behavioral, and tissue-specific physical characteristics. Conversely, the process of X-chromosome dosage compensation maintains a molecular similarity between the sexes at the cellular level: Transcriptional hyperactivation of the single X chromosome in the male ensures that both sexes are provided with the same levels of most X-linked gene products.

Identified as both necessary and sufficient to control female sexual differentiation and dosage compensation, the *Sex-lethal* gene (*Sxl*) qualifies as the primary binary switch of fly sex determination. The alternative activity states of *Sxl* (functionally ON/OFF) determine the sexual pathway choice. Moreover, *Sxl* appears to coordinate the sexual identity

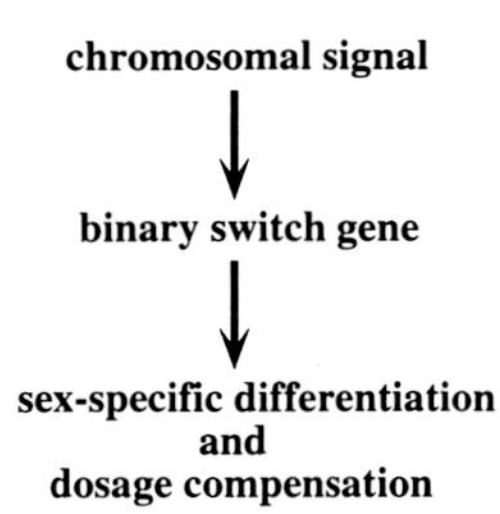

Figure 1 The simplified view. The somatic sex determination pathway consists of three major regulatory stages.

of somatic and germline cell lineages. In this review we summarize the genetic and molecular mechanisms that signal and activate the *Sxl* switch, thereby leading to the establishment of a female identity. And, because these mechanisms appear to be mediated by substantially different sets of genes in the soma versus the germline, we consider these two lineages independently.

II. Regulation of Somatic Sex Determination

A. Introducing the Master (*Mistress?*) Switch

A terse view of the somatic sex determination pathway consists of three stages: the chromosomal signal, the genetic switch whose "position" is set by the signal, and a collection of regulatory genes that subsequently control the expression of sexual differentiation and dosage compensation (Fig. 1). The first advances toward elaborating details of the genetic regulation of sex determination came at either end of this pathway. As early as 1916, Bridges recognized that, although the karyotypes of male and female flies normally could be distinguished by the presence or absence of a distinct Y chromosome, the Y chromosome itself played no role in determining the individual's sex. Subsequent analysis of polyploids and chromosomal aneuploids demonstrated that the primary determinant of sex determination in *Drosophila* is a chromosomal parameter referred to as X/A, i.e., the number of X chromosomes relative to the number of sets of autosomes (Bridges, 1925). XXAA (X/A = 1.0) embryos develop as females, whereas XAA (X/A = 0.5) embryos develop as males (we will come back to the genetic explanation of this chromosomal signal). Meanwhile, at the other end of the pathway, numerous sex-specific mutations readily identified genes that have since been found to control sexual differentiation (Morgan *et al.*, 1943; Sturtevant, 1945; Hildreth, 1965; Watanabe, 1975) or X chromosome dosage compensation (Fukunaga *et al.*, 1975; Belote and Lucchesi, 1980; Lucchesi *et al.*, 1982). For details of the current understanding of how these "downstream" genes function to effect the expression of sexual cell fate, see reviews by Belote (1992) and Burtis and Wolfner (1992).

Strategically located in the middle of this abbreviated pathway is the binary switch whose operation is initially mandated by the primary sex-determining signal. Thereafter, the position of the switch determines, either actively or by default, the expression patterns of the downstream genes that regulate sexual differentiation and dosage compensation. The gene, *Sex-lethal* (*Sxl*), acts as this central regulator of sex determination in the fly. The identification of *Sex-lethal* as the switch followed from the recognition that loss-of-function and gain-of-function alleles of the locus produce reciprocal, sex-specific effects, and that these effects encompass dosage compensation-related lethality as well as phenotypic transformations of sexually dimorphic structures (Cline, 1978, 1979a,b, 1983; Lucchesi and Skripsky, 1981; Sánchez and Nöthiger, 1982; Gergen, 1987; Gorman *et al.*, 1993). Indeed, the gene's name cleverly alludes to the contrasting lethality of both kinds of *Sxl* mutant alleles. (This derivation of the name contradicts the misconception of an anonymous reviewer who once facetiously described the wild-type function of the gene as being required to prevent death during sex. This deduction would almost have been an understandable error since genes in *Drosophila* generally are named after their mutant phenotypes. It might have been supposed that *Sex-lethal* mutants die when they engage in sex. However, a gene with such a function would more appropriately have been named, *lethal-sex*.)

Considerable genetic evidence has established that the functional state of *Sex-lethal,* as the switch, becomes set in response to the X/A signal and is subsequently maintained independently of that signal in a heritable fashion. Changing the X/A signal from 1.0 to 0.5 by removing one of the X chromosomes from diplo-X cells results in the eventual loss of these cells and their descendents if this change is made after the blastoderm stage of development (Sánchez and Nöthiger, 1983; Bachiller and Sánchez, 1991). XO cells thus derived from postblastoderm XX cells seem incapable of "resetting" their mode of dosage compensation to accomplish the hyperactivation of their single X chromosome; such an adjustment would be necessary for normal viability and growth. Similarly, XO clones that are generated in this way and do survive (e.g., in the "less competitive" environment of abdominal differentiation) exhibit female sexual differentiation. Despite the change in the signal, sexual differentiation and dosage compensation remain in the female mode. Thus, the preblastoderm commitment to female versus male development, including sex and dosage compensation, is normally stably inherited. That this unalterable commitment to sexual cell fate is mediated through the functional state of *Sex-lethal* is apparent in the phenotype of *Sxl*$^{-}$ clones generated in XX individuals: They exhibit male differentiation, as well as growth handicaps indicative of dosage compensation defects (Cline, 1979a; Sánchez and Nöthiger, 1982).

Simply stated, *Sxl* is the genetic ON/OFF switch of fly sex determination. This apparently straightforward functional role was first supposed from the opposite sex-specific effects of complementary classes of mutations in the gene. Loss-of-function mutations affect only XXAA individuals: Null alleles, such as Sxl^{f1}, are recessive, female-specific lethals (Cline, 1979a). Conversely, gain-of-function mutations, such as Sxl^{M1}, affect only XAA individuals: Constitutive alleles are dominant, male-specific lethals (Cline, 1978, 1979a). Thus, the *Sxl* wild-type function is essential to chromosomal females: Turning the gene ON directs embryonic development to the female pathway. And, *Sxl* must remain OFF for the survival of chromosomal males: Inappropriate expression of the gene in otherwise wild-type XAA individuals is sufficient to promote the female developmental pathway. The characterization of *Sxl* function in terms of a binary genetic switch has been substantiated further by the demonstration that induced expression of a regulatable *Sxl* transgene in chromosomal males mimics the male-lethal effects of Sxl^{M} mutations (Bell *et al.*, 1991).

A number of molecular mechanisms can be envisioned to underlie the functional basis of an ON/OFF genetic regulatory switch. Prior to the molecular isolation of the *Sxl* locus, the easiest expectation for the mode of this gene's regulation was transcriptional ON/OFF regulation. It came as a surprise, therefore, when the first Northern blots of *Sxl* RNA expression were developed. The *Sxl* gene produces a complex set of transcripts, showing stage-, tissue-, and sex-specificity (male as well as female) (Fig. 2) (Bell *et al.*, 1988; Salz *et al.*, 1989; Samuels *et al.*, 1991). With the determination of the *Sxl* expression pattern, it became painfully obvious to some that the mystery of *Sxl* regulation could not be solved with a dissecting microscope and a fly-sorting paintbrush alone!

The assortment of *Sxl* mRNA products, although mostly overlapping, can be distinguished by a number of structural differences (Bell *et al.*, 1988; Salz *et al.*, 1989; Samuels *et al.*, 1991). Seven RNA species, four female-specific and three male-specific, are expressed continuously from mid-embryo through adult stage (Fig. 2). The sex-specific difference between these two *late* RNA types derives from the inclusion in the male-specific varieties of an extra exon (exon E-male: Fig. 3). Since the presence of this small exon in the male-specific RNAs results in premature termination of translation of these messages, *Sxl* activity in males is effectively rendered nonfunctional by this splicing alternative (see below for an elaboration of *Sxl* sex-specific splicing regulation). Although the late RNAs all appear to use a common 5′ end, additional structural differences among the late transcripts (in both sexes) appear to arise from 3′ end variations, including added or alternative terminal exons and/or alternative polyadenylation (Salz *et al.*, 1989; Samuels *et al.*, 1991). Included among the late female-specific RNAs are two whose expression is germline-dependent. However, internal structural differences have not yet been found that would functionally distinguish these transcripts from the other female-specific messages (Salz *et al.*, 1989; Samuels *et al.*, 1991) (see below for discussion of *Sxl* germline regulation).

Three additional *Sxl* RNAs are expressed transiently during early embryogenesis (Fig. 2). These *early* RNAs are not only embryo-specific, but *female*-specific as well (Keyes *et al.*, 1992) (see below for an elaboration of *Sxl* sex-specific transcriptional regulation). Like the late RNAs, the assorted early RNAs appear to differ from each other by variations in their 3′ ends, while originating from a common 5′ start site. The 5′ end

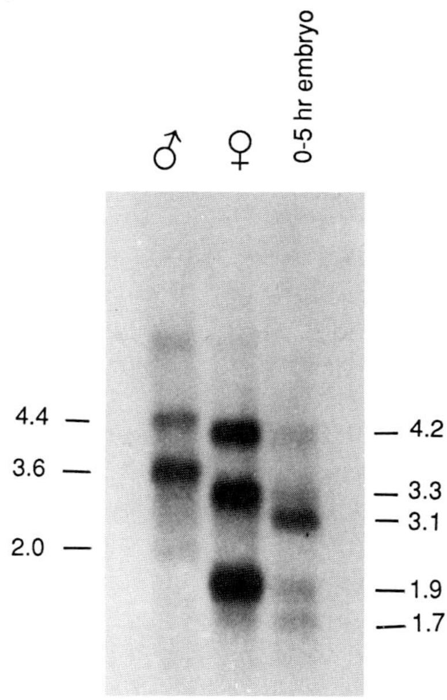

Figure 2 The pattern of *Sxl* transcription. Northern blot of poly(A)$^+$ RNA from adult males, adult females, and 0- to 5-hour embryos, probed with a genomic fragment from exon "E-common." Three major size classes of RNA are present in adults of both sexes. In females the RNAs are approximately 200 bp smaller than each corresponding class in males. As described in the text, differences *between* the sexes are due to the inclusion in male transcripts of the male-specific exon. Sex *nonspecific* differences arise through nonspecific splicing variations and alternative polyadenylation site usage. Also visible is a large (~7 kb) RNA that may be an incompletely processed message. Female-specific adult RNAs include two size classes that are expressed only in the germline: 1.9 and 3.3 kb. (A second 3.3-kb RNA class is *not* germline dependent; it is not distinguishable from the germline-dependent 3.3-kb RNA on standard Northern blots.) Early embryo RNA includes two embryo-specific messages (3.1 and 1.7 kb), as well as the female-specific (adult-type) RNAs. Additionally, a larger (~4.0 kb) embryo-specific RNA class can be resolved on higher percentage gels. Reprinted by permission from Salz *et al.* (1989) and Cold Spring Harbor Laboratory Press.

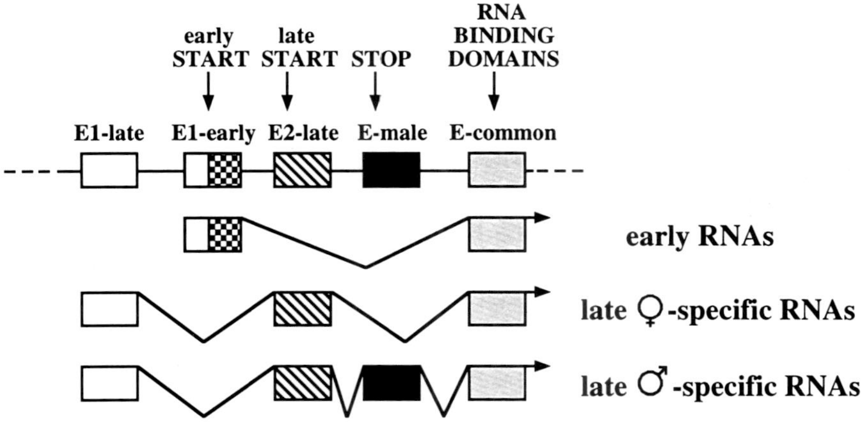

Figure 3 A summary of the different processing events that produce the female-, male-, and embryo-specific RNAs. The 10 different *Sxl* RNA classes can be divided into three groups: early, late female-specific, and late male-specific. The female and male RNAs differ by the inclusion or exclusion of the male-specific exon (E-male). The early RNAs are not only embryo-specific, but also female-specific. They differ in two ways from the late RNAs: (*a*) They have a unique, internal transcription start site (upstream of exon E1-early), and (*b*) a unique splicing event joins "E1-early" directly to "E-common," skipping both "E2-late"and "E-male."

of the early messages, however, is different (downstream) from that used to transcribe the late messages. Moreover, the 5′ exon that encodes the translation start site in the early RNAs is different from the corresponding exon in the late RNAs (exon E1-early vs E2-late: Fig. 3).

The structural variation observed for *Sxl* mRNAs results in the coding potential for as many as six slightly different polypeptides (Samuels *et al.*, 1991). How many of these possible polypeptides are synthesized *in vivo* is not clear. Western immunoblots containing extracts from animals at various stages of the life cycle, including embryos and adults, show multiple forms of SXL protein (Bopp *et al.*, 1991). The demonstration of SXL protein diversity *in vivo* was welcomed because it could be used to account hypothetically for the several distinguishable *Sxl* functions that have been documented genetically, including positive autoregulation (implementation and maintenance of the female-specific late expression), control of sexual differentiation (through *tra:* see below), and control of dosage compensation (target unknown). However, strict specialization of protein activities does not appear to be the scenario *in vivo*. At least in somatic tissues, a single *Sxl* cDNA, with the coding potential of a single polypeptide, has been shown to retain some amount of all three somatic functions (Bell *et al.*, 1991; Keyes *et al.*, 1992). In order to simplify our further discussion of *Sxl* regulation, we will refer to any functional polypeptide products of the gene collectively as "SXL protein."

So, what kind of molecule is the SXL protein and exactly what is its biochemical activity? In controlling downstream sexual differentiation, *Sxl* is required as a positive regulator of the *transformer* (*tra*) gene (Cline, 1979a; McKeown *et al.*, 1988; Nagoshi *et al.*, 1988). Biochemically, this regulation takes the form of RNA processing: *tra* gene function is controlled by sex-specific alternative splicing (Boggs *et al.*, 1987). By this mechanism, a mature *tra* mRNA with a long open reading frame (ORF) is made only in females. Increasing evidence endorses a model in which the SXL protein interacts directly with the *tra* pre-mRNA to bring about the female-specific splicing pattern of this message (Sosnowski *et al.*, 1989; Inoue *et al.*, 1990; Valcárcel *et al.*, 1993). Fortunately, the predicted amino acid sequence of the SXL protein includes two copies of a conserved functional motif that is consistent with this proposed role in RNA processing (Bell *et al.*, 1988). This duplicated domain in the SXL protein is a conserved sequence feature (*R*NA *r*ecognition *m*otif: RRM) of numerous ribonucleoproteins (RNPs), many of which have been implicated in RNA processing and can bind RNA *in vitro* (Kenan *et al.*, 1991). All of the SXL polypeptide variants synthesized from the female-specific transcripts, both early and late, are predicted to contain these RRM domains (Samuels *et al.*, 1991). Consequently, SXL protein is imagined to be involved in a single biochemical process, namely RNA processing, for all of its different genetic functions. Although it seems likely that SXL is involved similarly in alternative splicing reactions to control not only sexual differentiation in the female germline, but also dosage compensation in diplo-X embryos, confirmation of such a hypothesis awaits the identification and molecular isolation of the gene targets of *Sxl* regulation in these two processes.

B. *Sex-Lethal* Regulation: Four Phases

The molecular process of selecting and maintaining the activity state of *Sxl* can be subdivided into four phases (Fig. 4). The first two phases involve the transcriptional regulation of *Sxl* expression; the next two phases involve the control of the gene's female-specific function by RNA splicing regulation.

1. Phase 1: Promoter Choice

The *Sxl* gene contains two functionally distinct promoters that direct the synthesis of two temporally separable sets of RNAs: early (P_E) and late (P_L). The complication of two *Sxl* promoters was first suggested from a genetic dissection of the locus (Salz *et al.*, 1987). In this study, numerous rearrangements of the *Sxl* gene were generated by mobilization of a transposable *P* element that was inserted at the 5′ end of the gene. The deletions that were recovered in these screens included several that, although extending into the RNA coding region of the gene, did not totally disrupt Sxl^+ function. Deletions that remove the 5′ end of the *Sxl* transcription unit eliminate only some of the gene's functions. Subsequent molecular characterization of all classes of *Sxl* mRNA has not only confirmed the existence of two promoters, but mapped them as well (Salz *et al.*, 1987; Samuels *et al.*, 1991; Keyes *et al.*, 1992).

Transcriptional activation of the *Sxl* early promoter is the first step toward turning this developmental switch ON permanently and specifically in XX embryos (Fig. 4,I). Not only is the transcriptional activation of P_E responsible for the first zygotic expression of *Sxl,* but this expression is female-specific (Keyes *et al.*, 1992). Using a stage-specific probe made from the embryo-specific E1-early exon, Keyes *et al.* (1992) were able to visualize the activity of the early promoter by *in situ* RNA hybridization experiments on whole-mount embryos. These experiments detailed the developmental timing of the early transcript expression and clearly demonstrated that only XX embryos synthesize these *Sxl* transcripts: In female embryos the early transcripts first appear during nuclear cycle 9, with peak levels midway through the syncytial blastoderm stage. Their expression subsequently diminishes, trailing off to low levels during gastrulation. Early *Sxl* mRNAs (i.e., transcripts containing the specific E1-early exon) are not synthesized in male embryos. Basically, the sex-specificity of P_E results from its functional dependence on the X/A signal; we will consider the molecular genetic nature of that dependence in a later section.

From a functional perspective, the structure of the female-specific early *Sxl* mRNAs is related most significantly to the structure of the late female-specific mRNAs. That is, just like the processing of the late *Sxl* transcripts in females, splicing of the early female-specific transcripts *excludes* the "E-male" exon, which is present in the late male-specific transcripts and thereby renders those mRNAs functionally useless (Keyes *et al.*, 1992). The predicted SXL protein product(s) of the early female-specific mRNAs is largely the same as those of the late female-

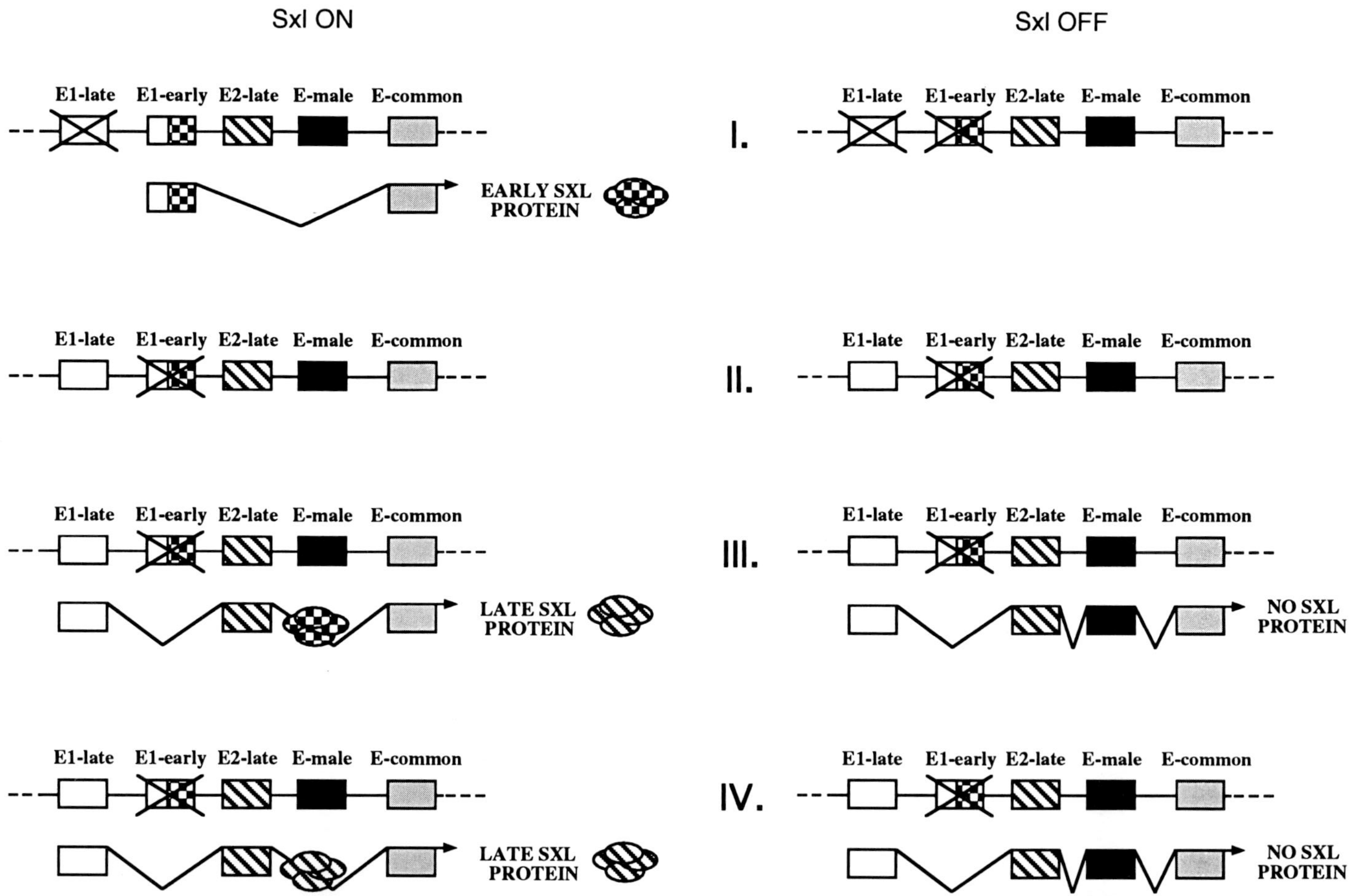

Sxl ON
Sxl OFF
I.
II.
III.
IV.
E1-late
E1-early
E2-late
E-male
E-common
EARLY SXL PROTEIN
LATE SXL PROTEIN
NO SXL PROTEIN

specific mRNAs (Keyes *et al.,* 1992). Aside from small (24–26 aa) differences at their N termini, the predicted amino acid sequences of the early and late SXL proteins are identical, including the two conserved RNA binding domains. Consequently, the early SXL protein functionally imitates the late SXL protein as a splicing factor.

The mechanism by which the male-specific exon, E-male, is excluded from the early female-specific transcripts is not known. Although it is clear that SXL protein itself is required to accomplish the sex-specific splicing of the gene's own late transcripts (see Phase 3, this chapter), splicing of the early *Sxl* pre-mRNAs to their functional messages seems to occur without SXL protein. Indeed, the first SXL protein found in the embryo appears to be that synthesized from the early transcripts. Prior to nuclear cycle 12, when the expression of the early transcripts peaks, no maternally encoded SXL protein has been detected in the embryo (Bopp *et al.,* 1991). It is conceivable that splicing factors other than SXL, whether generic or specific, are provided through the maternal genome for early *Sxl* splicing. No candidate molecules have been identified. Alternatively, some inherent structural characteristic distinct to the early pre-mRNAs, such as size or secondary structure, may facilitate their processing to bypass exons, E2-late and E-male. Whatever the actual molecular mechanism of the early *Sxl* splicing, however, the utilization of a separate promoter to produce transcripts with a 5′ exon different from the 5′ exon of the late transcripts has somehow obviated the *usual* requirement for SXL protein in the splicing of its own message (Keyes *et al.,* 1992).

Figure 4 Model for sex-specific regulation of *Sxl:* Four phases. *Sxl* regulation can be divided into four phases. I. Promoter choice: The early, female-specific promoter (P_E) directs the transcription of the early RNAs, whereas the late, sex nonspecific promoter (P_L) remains inactive. Consequently, the early RNAs are synthesized and processed to encode protein only in females. II. Promoter swap: At or around the cellular blastoderm stage, P_E becomes inactive, whereas P_L is activated in both sexes. III. Early self-splicing: Initially, the early SXL protein (present only in females) promotes the splicing of the late sex nonspecific *Sxl* transcripts from P_L such that exon E-male is skipped. As a result, the late SXL protein is produced only in females. Since early SXL protein is absent in males, the late male-specific transcripts are processed to include exon E-male, and no functional late SXL protein is made. IV. Late self-splicing: The late SXL protein promotes the female-specific splicing of the late *Sxl* transcripts from P_L, thus ensuring that *Sxl* pre-mRNAs will continue to be processed into a functional, protein-encoding form in females. In males, the absence of SXL protein guarantees that the late transcripts will include exon E-male and therefore be nonfunctional.

2. Phase 2: Promoter Swap

The second phase of *Sxl* activation involves a transcriptional transition from the use of the early to the late promoter (Fig. 4,II). In fact, the temporal expression of P_E and P_L must, at least in females, be closely coupled, such that transcripts from P_L are synthesized before the products of P_E disappear; after all, early SXL protein is required to process the late transcripts into functional messages (see Phase 3 below). Late *Sxl* transcription can first be detected (with a specific probe prepared from exon, E2-late) at nuclear cycle 13, the stage at which the accumulation of early RNAs reaches maximum levels (Keyes, 1992). Since SXL protein *is* present in female embryos by nuclear cycle 12, accessibility of this splicing factor to the late transcripts is assured. Conversely, the absence of early SXL protein in male embryos at this time ensures that the feminizing switch remains OFF.

Although P_L functions in both sexes, there is some evidence that the activation of this promoter may be sensitive to, though not require, the same female-specific signal that activates P_E (Keyes, 1992). First, the *Sxl* late transcription begins at slightly different times in male and female embryos. As noted above, late transcripts first appear in female embryos at nuclear cycle 13. Late transcripts in male embryos, however, do not appear until after cellularization, late in cycle 14. Moreover, whereas mutations in genes that contribute to the X/A signaling system (see Sections II,C,1 and II,C,2, below) do not interfere with the normal expression of P_L in male embryos (Cline, 1984; Maine *et al.*, 1985; Keyes *et al.*, 1992; Keyes, 1992), such signal alterations do perturb the normal onset of P_L in female embryos. In such mutant female embryos, activation of P_L appears to be delayed relative to wild-type. Since P_L expression is not eliminated in these mutants, the X/A signaling system is clearly not required for this promoter's activation in either sex. However, when normally present, the components of the X/A signal seem to facilitate the activation of P_L, resulting in slightly earlier expression in females than in males. Alternatively, the X/A signal may alleviate in female embryos a temporary block of the expression of P_L, a block that is soon afterward removed or bypassed in all embryos. In any case, the earlier activation of P_L in female embryos may serve to ensure the temporal overlap of early SXL protein with late *Sxl* pre-mRNA.

3. Phase 3: Early Self-Splicing

The switch in promoters from the early female-specific to the late sex nonspecific promoter necessitates a post-transcriptional mode of regulation in order to maintain the ON/OFF regulation (Fig. 4,III). An examination of the structures of the late RNAs originally suggested that this regulation takes place at the level of splicing. In males the inclusion of a male-specific exon results in a truncated, apparently nonfunctional protein; a number of stop codons contained within this exon abort translation of the message (Bell *et al.*, 1988). In females, this exon is excluded from the mature transcript, and protein is produced (Bell *et al.*, 1988; Bopp *et al.*, 1991).

Although it is now clear that the post-transcriptional regulation of *Sxl* function depends upon a sex-specific splicing process that involves the SXL protein itself, early clues of the autoregulatory nature of this process came from a series of elegant classical genetic experiments by Cline (1984). He demonstrated that one *Sxl* allele has the ability to activate another *in trans;* that is, the *Sxl* gene autoregulates. Thus, it is possible that the protein product of the early RNAs, which is available in female embryos when transcription from the late promoter is initiated, can direct the splicing of the late pre-mRNAs such that the male-specific exon is skipped. Keyes *et al.* (1992) provided experimental support for this model by showing that ectopic expression of one of the early proteins is capable of transactivating an otherwise silent allele of *Sxl:* A brief pulse of ectopic *Sxl* expression kills males. However, Sxl^{-} males survive. Since a mutant *Sxl* genotype rescues the male lethal phenotype produced by transient ectopic Sxl^{+} expression, death must result from the inappropriate activation of the endogenous *Sxl* allele, not from the pulse of ectopic expression. These data not only suggest that the early proteins are capable of activating an endogenous allele of *Sxl,* but also establish that activation of the endogenous *Sxl* allele is irreversible, in support of the early genetic evidence that *Sxl* activity is maintained by an autoregulatory feedback loop (Cline, 1984).

4. Phase 4: Late Self-Splicing

Maintenance of the female-specific splicing pattern is likely to be carried out by the late SXL protein (Fig. 4,IV). For one thing, although the early SXL protein is capable of directing the splicing of the late pre-mRNAs, this *Sxl* product is present only briefly during embryonic development. Moreover, Bell *et al.* (1991) have shown that ectopic expression of late SXL protein, like that of early SXL protein, is also capable of transactivating an otherwise silent allele of *Sxl.* It appears that once SXL protein is made, it ensures the continual production of more SXL by establishing an autoregulatory loop (Cline, 1984). As a result, the female-specific splicing pattern is maintained throughout development.

Although the mechanism by which SXL regulates exon skipping of its own pre-mRNA is not yet understood, it is likely that it is similar to the mechanism by which SXL directs the female-specific splicing of *tra*. During *tra* pre-mRNA splicing, SXL protein has been shown to influence the 3′ splice site choice by binding to the sex nonspecific intron at or near the polypyrimidine tract (Sosnowski *et al.,* 1989; Inoue *et al.,* 1990; Valcárcel *et al.,* 1993). Similarly, the splicing of its own pre-mRNA appears to involve the binding of SXL to analogous sequences within the intron between exon E2-late and exon E-male (Figure 4). A complete blockage of the inclusion of E-male in the mature *Sxl* message, however, additionally requires sequences that are present in the intron between exon E-male and the first common exon (Sakamoto *et al.,* 1992; Horabin and Schedl, 1993). Thus, it may be necessary for SXL to block both the 3′ and 5′ splice site acceptors of the male-specific exon in order to force the joining of exon E2-late to exon E-common.

These data suggest a model by which SXL protein prevents recognition of the male-specific exon by preventing formation of a spliceosome. Consistent with this model is the finding that, in mammalian systems, the crucial first step in spliceosome assembly is the juxtaposition of the 5′ and 3′ portions of the intron (Seraphin and Rosbash, 1989, 1991; Michaud and Reed, 1991; Bennett *et al.,* 1992). This process requires a number of non-snRNP spliceosome-associated proteins, as well as the U1 snRNP. One of these proteins, U2AF, is similar to SXL in that it binds the polypyrimidine tract in the 3′ portion of the intron (Zamore and Green, 1991; Zamore *et al.,* 1992). Indeed, excess amounts of SXL protein can prevent U2AF binding *in vitro* (Valcárcel *et al.,* 1993); perhaps SXL functions *in vivo* to prevent binding of U2AF.

The biochemical mechanism of *Sxl* splicing could be the same whether the SXL protein itself is produced from early or late transcripts. Even the temporal distinction made here between *Early* versus *Late Self-Splicing* might be less explicit *in vivo.* What might distinguish these two stages in the establishment of *Sxl*'s splicing pattern are batteries of splicing cofactors that cooperate with either early SXL or late SXL. Potential candidate gene products are discussed below.

C. *Sex-Lethal* Regulation: "With a Little Help from My Friends"

The molecular mechanisms that underlie the initiation and subsequent performance of *Sxl* functions in defining a female identity are complex. Clearly, there must be an assortment of other molecules involved in these biochemical processes. Indeed, a number of gene products can be named as essential "helpers" in sex determination, having been identified by genetic and molecular approaches.

1. The Signal

The quantitative nature of the X/A signal was first recognized by Bridges (1925), who discovered that XX flies with three rather than the normal two sets of autosomes exhibit a mosaic sexual morphology; XXAAA flies, called triploid intersexes, display a mixture of distinct male and female tissue characteristics. Bridges' resulting "genetic balance" theory of sex determination stated that whereas each X chromosome has a "net female tendency," the autosomes are "male-determining in their action." According to this model, the course of sexual development would follow from the relative balance of female-determining (predominantly X-linked) versus male-determining (primarily autosomal) genes. Studies that followed Bridges' original observation attempted to locate such "sex tendency" genes by examining the effects of duplications or deletions of chromosomal fragments of the X (Dobzhansky and Schultz, 1931), or of an autosome (Pipkin, 1947) on the sexual phenotype of triploid intersexes. The experiments of Dobzhansky and Schultz (1931) added support for Bridges' genic balance theory by demonstrating that duplications of several regions of the X chromosome have feminizing effects on triploid intersexes; the postulate of a single female-determining factor on the X could not account for the observed results. The analogous study carried out for chromosome 2, however, failed to detect any sex-modifying regions (Pipkin, 1947). In retrospect, it is likely that even the sex determination of triploid intersexes is not sensitive enough to detect the subtle effects of small changes in the denominator of X/A, given the polygenic nature of the system as we now know it (see below for more sensitive assays).

Implicit in Bridges' model for the genetic nature of X/A is the notion that such female- or male-determining genes exert their effects in a dose-sensitive manner, hence the modifying effects of X-chromosomal duplications on the sex of XXAAA flies. Indeed, once the link between somatic sexual differentiation and dosage compensation was recognized—through identification of the switch gene (*Sxl*) that controls both processes—it became possible to predict more precisely the phenotypes of mutations in the genetic elements whose doses contribute directly to the measurement of X/A (Cline, 1988). For example, specific loss of signal elements on the X (numerator elements) should result in a reduction of the X/A value that is perceived by *Sxl*, even though the cell carries two X chromosomes. The consequent failure to activate *Sxl* upsets dosage compensation as the cell cannot cope with two hyperactivated X chromosomes. Thus, loss-of-function mutations in numerator elements of X/A should be female-specific lethals. Conversely, an increased dose of the same numerator element should be lethal to chromosomal males. Again, lethality results from dosage compensation defects as a consequence of inappropriate *Sxl* expression.

The application of these dose-related criteria as a test for X/A signal elements has led to the identification of three numerator components of X/A: *sisterless-a* (*sis-a*), *sisterless-b* (*sis-b*), and *runt* (*run*) (Cline, 1986, 1988; Duffy and Gergen, 1991; Torres and Sánchez, 1992). As predicted, reduced dosages of these genes result in female-specific lethal effects, whereas increased dosages lead to male-specific lethality. Both sex-specific lethal phenotypes derive from the establishment of activity states of *Sxl* that are inappropriate for the gross chromosomal sex of the individual. This is concluded from the demonstration that the gain-of-function mutation, Sxl^{M1}, suppresses the female lethal effects, and the null allele, Sxl^{f1}, suppresses the male lethal effects. Consistent with the role of these X-linked signal elements as positive regulators of *Sxl* is the observation that mutations in these genes exhibit synergistic mutant effects with Sxl^{-}, as well as with each other. Finally, since the effects of adding or deleting $sis\text{-}a^{+}$, $sis\text{-}b^{+}$, and run^{+} are only semiadditive, the explicit parts of X/A contributed by these genes are not functionally identical. Nevertheless, increased activity of one gene can partially compensate for reduced activity in others: Duplication of $sis\text{-}b^{+}$ can partially suppress $sis\text{-}a^{-}/sis\text{-}a^{-}$ female lethality (Cline, 1988). This, too, is predicted behavior for constituents of the quantitative X/A.

Although the classical triploid intersex approach failed to detect autosomal regions containing putative denominator constituents of X/A, an autosomal signal element has, nevertheless, been found: *deadpan* (*dpn*) (Younger-Shepherd *et al.*, 1992). Like numerator elements, denominator elements of X/A should show sex-specific dose-related effects, only in the reciprocal direction. Increased denominator dose will *decrease* the perceived X/A and, therefore, exert its effects in females, whereas decreased denominator dose will *increase* the perceived X/A and exert its effects in males. Located on chromosome 2, *dpn* has been shown to exhibit the expected sex-specific dosage effects when combined with altered doses of *sis-b*. The relative viability of males with one extra copy of $sis\text{-}b^{+}$ and also one less copy of dpn^{+} is less than 50%. And, this lethality has been shown to result from the misexpression of *Sxl*, since it is suppressed by Sxl^{f1}. Conversely, the relative viability of females with one less copy of $sis\text{-}b^{+}$ and also one extra copy of dpn^{+} is less than 10%; presumably, the perceived X/A signal in these XXAA individuals is just too low to ensure *Sxl* activation. Thus, in contrast to the positive regulators of *Sxl* that comprise the numerator of X/A, *dpn* acts as a negative regulator of *Sxl* in its contribution to the "male tendency" of the autosomes.

2. Interpreting the Signal

Through its quantitative constituents, the X/A signal acts zygotically to dictate the sex-specific expression state of *Sxl;* yet, activation of *Sxl* in XX embryos requires at least one additional positive regulator that is not itself a part of the initiating signal. That regulator is the maternally supplied product of the *daughterless* (*da*) gene (Cline, 1980, 1983, 1984, 1988). Numerous genetic experiments have demonstrated the essential role of the maternal *da* gene in activating *Sxl.* Daughters (X/A = 1) from mothers homozygous for the original temperature-sensitive, hypomorphic allele, da^1, do not live (Sandler, 1972; Mange and Sandler, 1973). In the total absence of maternal germline da^+ function, all progeny embryos, even those with two X chromosomes, fail to activate *Sxl;* only sons survive (Cronmiller and Cline, 1987). Thus deprived of da^+ function, diplo-X embryos die as a consequence of upsets in X chromosome dosage compensation, just as with fatal reductions in numerator elements (Gergen, 1987). Also, as is the case with the female lethal effects due to loss-of-function zygotic numerator gene mutations, the gain-of-function mutation, Sxl^{M1}, suppresses the da^- female-specific maternal effect. And, loss-of-function *da* mutations exhibit dominant, synergistic interactions with zygotic *Sxl* and numerator loss-of-function mutations. Thus, loss of maternal da^+ function exerts its deleterious effects on XX zygotes as a result of misregulation of *Sxl.* The zygotic genotype with respect to *da* appears normally not to affect *Sxl* activation (Cline, 1980; Cronmiller and Cline, 1986).

Maternally supplied da^+ product and zygotic X/A signal numerator gene products all collaborate to accomplish the transcriptional activation of the *Sxl* early promoter in chromosomally female embryos (see Phase 1, this chapter). A number of experimental approaches have provided molecular evidence for the involvement of these positive regulators in the initial transcriptional activation of P_E. Both the maternal *da* genotype and the doses of the numerator elements have been shown to affect *Sxl* P_E transcription. No embryos from homozygous da^1 mothers (at restrictive temperatures) synthesize *Sxl* early transcripts, as detected by either primer extension or *in situ* hybridization assays (Keyes, 1992; Keyes *et al.*, 1992). Reduced zygotic doses of *sis-a, sis-b,* and/or *run* result in abnormal expression of the early (sex specific) transcripts. *In situ* hybridization experiments, with the specific E1-early exon as a probe, have shown that synthesis of the *Sxl* early transcripts is not only substantially reduced, but nonuniformly distributed throughout numerator-reduced embryos (Keyes *et al.*, 1992). Similarly, P_E expression has been detected by means of a reporter gene constructed as a fusion transgene in which transcription of a *lacZ* gene is directed by *Sxl* P_E (Keyes, 1992). Here visualization of P_E activation by expression of *lacZ* reiterates the effects observed on the *Sxl* early transcript pattern: Embryos from da^1/da^1 mothers exhibit no β-galactosidase staining, and decreased zygotic *sis-a, sis-b,* and *run* doses result in low levels of nonuniform β-galactosidase staining in XX embryos. Conversely, increased *sis-a* and *sis-b* doses result in significant, though nonuniform, expression of the *lacZ* reporter gene in XY embryos.

Although *da, sis-a, sis-b,* and *run* are all required as positive regulators of *Sxl* early in XX embryos, their biochemical activities must not all be equally discriminating with respect to their sex-specificity. By virtue of their X-chromosomal locations, *sis-a, sis-b,* and *run* vary sex-specifically in their doses, since their functioning in *Sxl* P_E activation must take place prior to the establishment of dosage compensation, which will obliterate most X-linked gene dosage differences between the sexes. However, the *da* gene product, since it is provided maternally, will be present in equal amounts in all zygotes, whether diplo-X or haplo-X. In this cooperative scheme, the *da* gene product acts as a generic partner in a molecular effort that is rendered female-specific by the numerator gene products.

The molecular characterizations of the Da, Sis-b and Runt proteins do indeed identify these *Sxl* regulators as transcription factors; the most extensive experimental documentation has been accumulated for Da and Sis-b. On the basis of the genetic behavior of deletions and duplications, Cline (1988) originally mapped the *sis-b* locus to the *achaete–scute* complex (AS-C), only to rediscover an "old" mutant allele of the gene, long-forgotten among the *scute* literature as sc^{3-1}. [The female-specific allele from which sc^{3-1} is derived, sc^3, had actually been considered much earlier for its significance to the study of sex determination when Sturtevant (1945) examined its effects on the phenotype of *transformer.* It had none.] Within AS-C *sis-b* corresponds to *scute-α*, whose function is provided by the AS-C transcript *T4* (Torres and Sánchez, 1989; Parkhurst *et al.*, 1990; Erickson and Cline, 1991). The protein product of *sis-b* (AS-C T4) is a member of the basic-helix–loop–helix (bHLH) family of transcriptional regulators (Villares and Cabrera, 1987; Alonso and Cabrera, 1988; Murre *et al.*, 1989a). Moreover, the Da protein is also a member of this family (Caudy *et al.*, 1988b; Cronmiller *et al.*, 1988;

Murre *et al.*, 1989a). Since the HLH domain of these proteins mediates the homo- or heterodimer formation that is a prerequisite to their function (reviewed in Garrell and Campuzano, 1991), it appears that the genetic collaboration of the *da* and *sis-b* genes actually reflects a direct biochemical collaboration between their products. Indeed, Da and Sis-b (AS-C T4) proteins do form heterodimers and bind DNA target sequences *in vitro* (Murre *et al.*, 1989b; Cabrera and Alonso, 1991; Van Doren *et al.*, 1991). Furthermore, heterodimeric complexes consisting of Da and another AS-C HLH protein, AS-C T3, whose bHLH domain is very similar to that of AS-C T4, not only bind DNA, but also activate transcription of a reporter gene in a yeast assay system (Cabrera and Alonso, 1991). In all likelihood, Da and Sis-b form critical heterodimeric complexes to activate transcription from *Sxl* P_E in chromosomally female embryos.

A model for the transcriptional activation of *Sxl* P_E (promoter choice) is illustrated in Fig. 5. Functional activation complexes must include Da protein, which is available nonspecifically to XX and XY embryos. Additionally, they must include numerator proteins (Sis-a, Sis-b, Runt, etc.), whose levels are twofold greater in XX embryos than they are in XY embryos. Only in XX embryos will there be sufficient quantities of numerator proteins to ensure transcriptional activation of P_E. The delicate balance of this molecular measure of the number of X chromosomes has been demonstrated *in vivo* by interfering with the "counting" process. The addition of a transgene encoding the AS-C T4 HLH (Sis-b) protein increases the calculated assessment of the X chromosome number; such an increase can lead to ectopic *Sxl* expression and its consequent lethality in chromosomal (haplo-X) males (Parkhurst *et al.*, 1990; Erickson and Cline, 1991).

Further *in vivo* evidence for HLH protein involvement in the numerical value of the X chromosome count comes from misexpression of another HLH protein that is *not* normally involved in the sex-determination process, namely the *Hairy* (*h*) protein. Early ectopic expression of *h* inhibits *Sxl* expression, thereby resulting in female (XX)-specific lethality (Parkhurst *et al.*, 1990). Since *h* acts negatively in the regulation of segmentation gene expression, one possible account of Hairy protein function is that it acts by sequestering positively acting HLH proteins. Perhaps the Hairy protein ectopically expressed in coincidence with X-chromosome counting in the early embryo asserts a competitive, nonproductive affinity for the HLH proteins involved in this process, e.g., Sis-b. According to this explanation, the removal of "free" Sis-b protein would result in levels too low to contribute productively to *Sxl* transcriptional activation, i.e., levels more like those en-

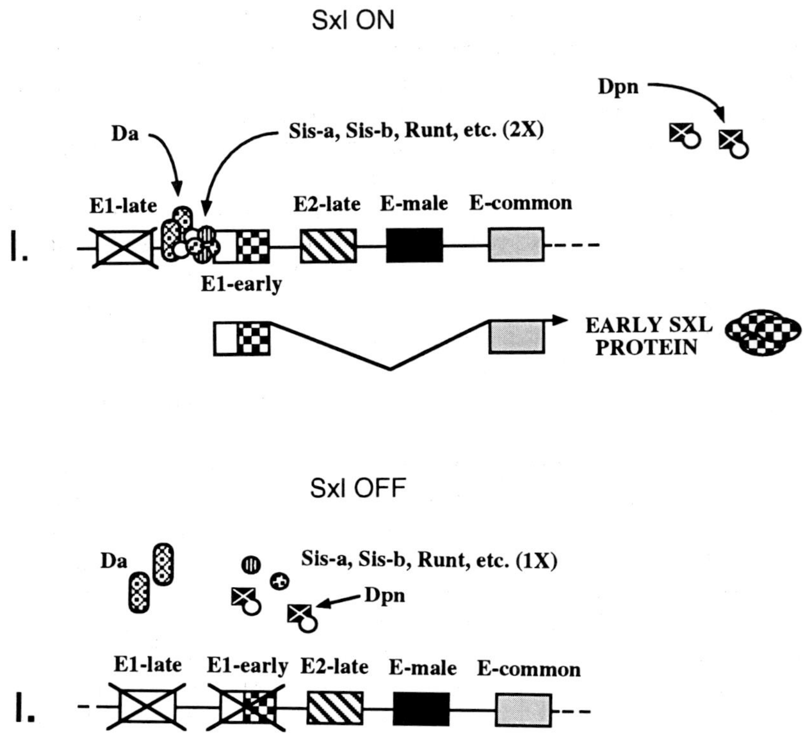

Figure 5 Hypothetical scheme for *Sxl* transcriptional regulation. I. *Promoter choice.* The female-specific P_E promoter of *Sxl* is only active if X/A = 1. The protein products of the X-linked (positive regulating) numerator genes (Sis-a, Sis-b, Runt, and others?) are twice as abundant in females (diplo-X) as they are in males (haplo-X). The protein products of the autosomal (negative regulating) denominator genes (Dpn and others?) are present in equal amounts in males and females. Only in females is the dose of numerator element proteins high enough to guarantee that sufficient functional transcription complexes are produced to activate expression of P_E. In males, the lower dose of numerator element proteins is insufficient to overcome negative regulation by denominator element proteins. Without the requisite transcriptional activation complexes, P_E remains inactive. In addition to the components of the X/A signal, transcription from P_E requires the participation of maternally encoded Da protein.

countered in haplo-X embryos. Da protein is not likely to be the molecule whose availability is limited by association with ectopic Hairy protein, because duplications of da^+ do not suppress the observed female lethality. However, duplications of $sis\text{-}a^+$ and/or $sis\text{-}b^+$ do suppress the lethal effects of ectopic *h* expression, leading to the conclusion that Hairy protein in this contrived scenario is interfering with the numerator component of the X/A signal (Parkhurst *et al.*, 1990). This serendipitous situation brought about by ectopic *h* expression may have anticipated the biochemical nature of Bridges' autosomal male-tendency factors in his genic balance theory.

In an apparent direct counterbalance to the bHLH numerator proteins that act as positive regulators of *Sxl* P_E, at least one denominator protein is an HLH protein that acts as a negative regulator: *dpn* encodes a Hairy-like HLH protein (Bier *et al.*, 1992). As noted above, *dpn* exhibits the expected genetic characteristics of an autosomal component of X/A. The sequence similarity between Hairy and Dpn, especially in the DNA-binding basic domain, encourages the speculation that Dpn exerts its male-determining effects through an inhibitory binding of HLH-positive regulators. And, since X chromosomes are counted by the doses of numerator gene products, Dpn probably interacts with these proteins, e.g., Sis-b (Fig. 5). In other regulatory pathways, transcriptional repression by HLH proteins has been documented; the inhibiting proteins lack the basic domain, which is essential for DNA binding, but retain the HLH motif, which mediates protein homo- or heterodimerization (Benezra *et al.*, 1990; Ellis *et al.*, 1990; Garrell and Modolell, 1990; Van Doren *et al.*, 1991). In these instances, repression apparently follows from a failure of the regulatory complexes to bind DNA target sites. In the case of Dpn (and Hairy in its role in segmentation?), since the protein sequence contains a significantly conserved basic domain, Dpn-containing complexes might exhibit altered site-specificity for, rather than elimination of, DNA binding. It is even possible that a more complex, indirect mechanism, involving additional denominator elements, leads to the overall male-determining nature of the autosomes.

Although the HLH proteins have been the most extensively characterized components of the X/A signaling system, they are certainly not the only molecular constituents. The *run* gene has been characterized molecularly. Although the Runt protein sequence contains no obvious homology to recognized DNA-binding motifs, Runt is localized to the nucleus, consistent with the gene's supposed role in segmentation (Kania *et al.*, 1990). Additionally, the Runt protein sequence contains a putative ATP-binding site. Rather than a direct (DNA contacting) role in transcriptional regulation, *run*'s function may involve modulating the regulatory activities of DNA-binding factors, as through variable phosphorylation states (Duffy and Gergen, 1991; Torres and Sánchez, 1992).

In reality, the biochemical scheme required to activate *Sxl* P_E is likely to be considerably more complex than is illustrated in Fig. 5. For one thing, *run* affects *Sxl* activation only in the central region of the embryo

(Duffy and Gergen, 1991; Keyes, 1992). This may mean that numerator elements make spatially nonuniform contributions to *Sxl* activation. Alternatively, since pole-biased heterogeneity in *Sxl* activation has been observed as a consequence of both increases and decreases of *sis* gene dosage as well, this pattern could reflect an inherent facility in activating *Sxl* at the poles, relative to the center of the embryo (Keyes, 1992). The basis for such differential responsiveness of *Sxl* P_E could be the involvement of minor regulators whose peripheral roles in sex determination have been masked by other vital functions. Possible examples include two other AS-C proteins, T3 and T5, which contain bHLH domains that are closely related to T4 (Sis-b) (Parkhurst *et al.,* 1993), and also the *extramacrochaetae* (*emc*) protein, a negative regulator of sensory organ development that encodes an HLH protein (Younger-Shepherd *et al.,* 1992).

Given the growing number of recognized regulatory factors that act *in trans* to control the activation of *Sxl,* where might they all function? A preliminary analysis of the *Sxl* promoter(s) has been made by using a *lacZ* reporter gene fusion and a series of deletions of the putative enhancer region of the early promoter (Keyes, 1992). Although this enhancer region contains numerous minimal consensus HLH binding sites [CANNTG: Blackwell and Weintraub (1990)], no single site appears to be absolutely essential for enhancer activity. Rather, multiple *cis*-acting regulatory elements, located within 1.5 kb of the early transcription start site, may be required for complete P_E expression.

3. Establishing and/or Maintaining Positive Autoregulation

SXL protein participates in the regulated splicing of the gene's own primary message. Yet, SXL protein is apparently not sufficient to accomplish the necessary processing of its RNAs. Mutations in three genes have been identified that prevent the transition to, and/or the maintenance of, the *Sxl* autoregulatory feedback loop. Analysis of these genes has been complicated, however, because they are required for additional functions during development. Like *daughterless,* hints of their roles in sex determination have come from the analysis of weak (nonnull) mutations.

One putative regulator of *Sxl* splicing is *fl(2)d,* whose role in sex determination was discovered because a weak mutation in the locus causes a female-specific lethal phenotype, hence the name, *female-lethal(2)d* (Granadino *et al.,* 1990). However, *fl(2)d* has at least two functions, only one of which is female-specific. Complete loss-of-function alleles are sex nonspecific lethals, indicating that *fl(2)d* is also required for a process that is essential to both sexes (Granadino *et al.*, 1990, 1992). But since *fl(2)d*$^{-}$ clones survive, that process must not be a cell vital one.

The female-specific *fl(2)d* function is required for female sexual differentiation and dosage compensation (Granadino *et al.*, 1990, 1991). By clonal analysis it is clear that *fl(2)d*$^{-}$ cells do survive, but in females (XX) they show sexual transformations, developing male structures. And, salivary glands from homozygous *fl(2)d*^{1} female larvae exhibit increased levels of transcription of the X-linked gene *sgs-4* relative to those measured for the heterozygous sibling females. The sex nonspecific roles that are provided by *fl(2)d* during development have not been identified.

Since sexual differentiation and dosage compensation are both regulated through the activity state of the switch, *Sxl,* and since *fl(2)d* is required throughout development, one plausible hypothesis is that *fl(2)d* functions to maintain the female-specific expression of *Sxl.* That is, *fl(2)d* may be an accessory to *Sxl* female-specific splicing. Indeed, chromosomally female larvae that are homozygous for *fl(2)d* mutations express *Sxl* in the male mode (Granadino *et al.*, 1990). Conceivably, the *fl(2)d* product acts throughout the life of the fly to assist SXL protein in female-specific splicing regulation. Because the developmental profile of *fl(2)d* expression is not yet known, it is impossible to predict whether the gene product also functions in the early self-splicing stage of *Sxl* regulation.

Another apparent regulator of *Sxl* is *virilizer* (*vir*). As with *fl(2)d,* weak *vir* mutations affect only females, whereas the most severe mutations are sex nonspecific lethal (Hilfiker, 1990; Hilfiker and Nöthiger, 1991). Like *fl(2)d, vir* is required for somatic sexual differentiation: At restrictive temperatures, females homozygous for *vir*^{ts} develop as intersexes. Preliminary genetic studies suggest that *vir* is also required for proper dosage compensation (Hilfiker, 1990). Although the genetic characterization of *vir* suggests that this gene is required for expression of *Sxl,* there are currently no molecular data to elaborate its role. The strong phenotypic similarities between *fl(2)d* and *vir* might imply that the *vir* gene product is also involved in the regulation of *Sxl* female-specific splicing.

A third putative regulator of *Sxl* splicing is the product of the gene *snf* [a.k.a. *sans-fille, liz,* or *fs(1)1621*]. Unlike *fl(2)d* and *vir, snf* has no apparent vital function in either males or females. In fact, *snf*'s role in somatic sex determination is only evident when the probability of *Sxl* activation has been otherwise reduced (Cline, 1988; Oliver *et al.*, 1988; Steinmann-Zwicky, 1988; Salz, 1992). In an otherwise wild-type genetic background, females homozygous for a weak *snf* mutation produce both female and male progeny. When these females are mated to *Sxl*$^{-}$/*Y* males, however, they produce only male progeny; i.e., heterozygous *Sxl*$^{-}$ females cannot survive a weak *snf* maternal effect (Salz, 1992). Female-specific lethality in this case is due to the lack of maternal *snf* function, suggesting that, like *daughterless, snf* makes a product that is deposited in the egg for subsequent activation of *Sxl* (Albrecht and Salz, 1993). Unlike the defective *Sxl* activation in XX embryos suffering from the *da*$^{-}$ maternal effect, however, *Sxl* early expression in embryos from *snf* mutant mothers *is* initiated in response to a female X/A signal. But, the transition to stable *Sxl* expression is not made (Albrecht and Salz, 1993). In such mutant embryos, the early *Sxl* transcription pattern and protein accumulation appear to be normal; subsequently there is little accumulation of the late female-specific spliced RNAs and protein. Certainly, *snf* is required to establish female-specific splicing under some conditions, although it is not clear whether *snf* is an essential factor for this process, or rather a "facilitator" of splicing whose loss is not critical *except* when there is insufficient SXL protein present. Furthermore, similar genetic studies have suggested that zygotic expression of *snf* can also be required for *Sxl* activation. Perhaps *snf* is required throughout development for the maintenance of *Sxl* function (Oliver *et al.*, 1988; Steinmann-Zwicky, 1988; Salz, 1992). Identification and characterization of a *snf* null mutation, and the biochemical analysis of the *snf* gene product, should clarify this gene's role in establishing the female-specific splicing pattern of *Sxl*.

D. Are There Accessory Genes Yet to Be Identified?

Although many of the molecular details of the regulation of somatic sex determination in *Drosophila* are within reach experimentally, not all of the participating genes have been identified. How can they be found? Although systematic screens, based on increasingly varied criteria, should continue to turn up candidate genes, serendipitous observations will no doubt remain an important source of discovery.

Historically, the simplest approach to identifying genetic components of a regulatory pathway has been to search for mutants with specific phenotypes, sex-specific phenotypes in the case of sex determination. However, this strategy owes much of its success to exceptionally good luck, since pleiotropy is widespread among these genes' functions. A perfect example of both luck and pleiotropy is the *daughterless* gene. The role of *da*$^{+}$ in sex determination was evident from the female-specific maternal effect phenotype of the spontaneous mutation, *da*^{1}:

Homozygous da^1 mothers produce only sons (Bell, 1954; Sandler, 1972). Null alleles of *da,* however, are sex nonspecific lethals as a result of the gene's involvement in other vital developmental processes (Cronmiller and Cline, 1987; Caudy *et al.*, 1988a; Cline, 1989). Recognition of *da* as a sex-determination gene would have been considerably delayed without the fortuitous discovery of this special allele, da^1, which disrupts the wild-type sex-specific (germline) function more severely than it disrupts the sex nonspecific (somatic) function. Other pleiotropic sex-determination regulatory genes whose null phenotypes show little, if any, sex specificity are *sis-b* (*scute-α*), *dpn, vir, fl(2)d,* and *run.* The sex-determining roles of these genes have been assigned on the basis of either sex-specific phenotypes of non-null alleles [*vir, fl(2)d*] or dominant interactions with known sex determination mutants (*sis-b, dpn, run*). Mutagenesis screens based on dominant synergistic interactions with known sex determination mutants largely avoid bias against the recovery of pleiotropic genes. Such a screen was used to isolate new *da* alleles, and since the crossing scheme never required the mutagenized chromosomes to be homozygosed, null (lethal) alleles were readily recovered (Cronmiller and Cline, 1987).

In the future it is likely that so-called "reverse genetics" will contribute significantly to the identification of additional sex determination loci. Since many of the principal molecules in both the transcriptional and the splicing regulation of *Sxl* have been described, biochemical searches can be carried out *in vitro.* Enhancer binding, for example, has been used as an assay to screen expression libraries for specific DNA-binding regulatory proteins. Or, immunoprecipitation will search for protein-binding partners. Molecules identified by any biochemical route, however, must ultimately have their functions validated genetically, a reasonable task in *Drosophila.*

III. Regulation of Germline Sex Determination

As in the soma, the establishment of the female identity in the germline requires *Sxl* activity (Schüpbach, 1985; Salz *et al.*, 1987). A comparison of normal and agametic adult females first identified two of the four female-specific *Sxl* RNAs as germline-dependent. The differences among the female-specific RNAs appear to arise primarily from the use of alternative polyadenylation signals, suggesting a mechanism for tissue-specific regulation (Samuels *et al.*, 1991). Since no differences have been found in the 5′ ends or internal splice junctions among these RNAs, it is likely that their sex-specific regulation is the same in the soma and germline. Although there is no direct evidence that *Sxl* autoregulates in the germline, the fact that *fl(2)d, snf,* and *vir* are all required for *Sxl* germline expression strongly suggests that it does (Steinmann-Zwicky, 1988; Hilfiker, 1990; Granadino *et al.,* 1992; Salz, 1992).

Although the mechanisms that maintain *Sxl* germline activity may be very similar to those that maintain the gene's somatic activity, the mechanisms that signal and activate the *Sxl* switch may be very different in the germline and soma. To understand how these differences may originate necessitates a review of the embryological divergence of these two cell types. In *Drosophila,* the first nuclear divisions take place in the absence of cytokinesis, resulting in a syncytium of embryonic nuclei. Following the 10th nuclear division (nuclear cycle 10), the nuclei migrate to the periphery of the embryo. The pole cells are the germline primordia and are set aside at this early stage in development at the posterior end of the embryo: A small subpopulation of nuclei that migrate to the posterior pole of the embryo cellularize to form the pole cells, while the remaining (somatic) nuclei undergo approximately four more acellular divisions before general cellularization takes place. Whereas SXL protein can be detected in all somatic nuclei by nuclear cycle 12, *no* SXL protein is evident in the pole cells at this time (Bopp *et al.,* 1991). Moreover, SXL protein is still undetectable in these cells during gastrulation, at which time the elongation of the germ band carries them into the embryo interior where they eventually colonize the presumptive gonadal mesoderm. *Sxl* activation is likely to occur at some point after pole cell invagination, but prior to the end of embryogenesis. First-instar larvae already show signs of sexual dimorphism: The number of germ cells in immature larval gonads is greater in males than in females, indicating that sex-specific regulation of germ cell division has occurred (Sonnenblick, 1941).

These temporal differences in the tissue-specific activation of *Sxl* suggest that the mechanisms that turn ON the female-specific switch are functionally different in the germline and soma. Indeed, *Sxl* activation in the germline has been shown to require not only a female-specific chromosomal signal (X/A = 1) but also an inductive signal from the somatic gonad (Steinmann-Zwicky *et al.,* 1989). Experiments in which pole cells were transplanted into agametic hosts suggest that female development of the germline occurs only when gonad and germline are both female (XXAA). Switching the sex of the soma or germline, by mutating *Sxl* or by changing the X/A signal, results in male-like development. Thus, XX germ cells may make a product that receives and/or interprets a signal from the female soma in order to effect *Sxl* activation.

How is *Sxl* activated in the germline? Is activation of *Sxl* in the germline analogous to activation of *Sxl* in the soma? Possibly. Recall that a genetic analysis of a series of deletions of the *Sxl* gene originally predicted the existence of a distinct, internal promoter that is embryo-specific (Salz *et al.,* 1987). The same analysis found evidence for an internal germline-specific promoter. Unfortunately, although sought, no RNA has yet been found that contains a germline-specific 5′ end. Such an RNA is likely to be rare *in vivo,* however, since its expression would be restricted to germ cells during a very short period of germline development. The existence of a distinct germline promoter is an attractive idea because the activation of *Sxl* in this tissue might depend on an activating signal that is qualitatively different from that defined in the soma: Germ cells that are mutant for the somatic numerator gene, *sis-b,* can enter the female-specific developmental pathway (Steinmann-Zwicky, 1993). If female differentiation of the germline is similarly independent of the wild-type functions of the other known somatic X/A components (*sis-a, run,* and *dpn*), a separate signaling system must be operating in the germline. Consistent with this hypothesis, *daughterless* function is not required in the germline for the differentiation of eggs (Cronmiller and Cline, 1987). There may exist a set of dose-sensitive elements that are germline-specific transcription factors analogous to the X/A somatic lot. Alternatively, given the importance of signal transduction in the initiation of germline *Sxl* expression, it is possible that X-linked dose-sensitive elements are receptors and/or ligands in this signaling pathway.

The genetic approach to dissecting the regulatory mechanisms of germline sex determination would be to isolate mutants in which the process is disrupted. But, what phenotype should one expect from such mutations? Mutations in all of the genes known to be required for both germline and somatic sex determination, including *Sxl, snf, fl(2)d,* and *vir,* give rise to a single mutant phenotype in the ovary. Loss-of-function mutations in these genes develop egg chambers in which neither the oocyte nor the nurse cells differentiate. Instead, the germarial stem cells continue to divide; this results in the production of "ovarian tumors." In some cases, these tumors have been shown to express male-specific RNAs, suggesting that this phenotype really does result from sexual transformation (unpublished results cited in Pauli and Mahowald, 1990; Gönczy *et al.,* 1992). Conceivably, additional genes that are required solely for germline sex determination could be identified by characterizing mutants with this sort of ovarian tumor phenotype. However, these analyses should be made with caution; it has been suggested that the tumorous ovary phenotype could also arise from abnormal mitotic proliferation of germ cells (McKearin and Spradling, 1990; Szabad *et al.*, 1991; Lantz *et al.*, 1992). For example, mutations in genes involved in germ cell entry into meiosis are likely to cause continued mitotic proliferation. Among the mutants with tumorous ovary phenotypes, *ovo* (*ovo*) and *ovarian tumor* (*otu*) seem most likely to reflect wild-type gene functions in some aspect of germline sex determination. However, specifically what those functions might be remains unclear (Oliver *et al.*, 1990; Pauli and Mahowald, 1990; Geyer *et al.*, 1992; Steinmann-Zwicky, 1992a,b, and references cited therein).

IV. Conclusion

Although the regulation of sex determination in *Drosophila* appears superficially to show little similarity to the regulation of sex determination in mammals, this system nevertheless provides an excellent model for elucidating basic molecular genetic control mechanisms that lead to determination events during development. The analysis of *Sex-lethal* regulation has established one way in which a single gene can not only respond to a transiently expressed determinative signal, but also subsequently maintain its new activity state independently of that signal.

Acknowledgments

We are grateful to Yuh Nung Jan, Susan Parkhurst, and Monica Steinmann-Zwicky for sharing results prior to publication. We thank Patricia Hunt for comments on the manuscript.

References

Albrecht, E. B., and Salz, H. K. (1993). The *Drosophila* sex determination gene *snf* is utilized to establish the female-specific splicing pattern of *Sex-lethal*. *Genetics*, **134,** 801–807.

Alonso, M. C., and Cabrera, C. V. (1988). The *achaete-scute* gene complex of *Drosophila melanogaster* comprises four homologous genes. *EMBO J.* **7,** 2585–2591.

Bachiller, D., and Sánchez, L. (1991). Production of XO clones in XX females of *Drosophila*. *Genet. Res.* **57,** 23–28.

Bell, A. E. (1954). A gene in *Drosophila melanogaster* that produces all male progeny. *Genetics* **39,** 958–959.

Bell, L. R., Maine, E. M., Schedl, P., and Cline, T. W. (1988). *Sex-lethal,* a *Drosophila* sex determination switch gene, exhibits sex-specific RNA splicing and sequence similarity to RNA binding proteins. *Cell (Cambridge, Mass.)* **55,** 1037–1046.

Bell, L. R., Horabin, J. I., Schedl, P., and Cline, T. W. (1991). Positive autoregulation of *Sex-lethal* by alternative splicing maintains the female determined state in *Drosophila. Cell (Cambridge, Mass.)* **65,** 229–239.

Belote, J. M. (1992). Sex determination in *Drosophila melanogaster:* From the X : A ratio to *doublesex. Semin. Dev. Biol.* **3,** 319–330.

Belote, J. M., and Lucchesi, J. C. (1980). Male-specific lethal mutations of *Drosophila melanogaster. Genetics* **96,** 165–185.

Benezra, R., Davis, R. L., Lockshon, D., Turner, D. L., and Weintraub, H. (1990). The protein Id: A negative regulator of helix-loop-helix DNA binding proteins. *Cell (Cambridge, Mass.)* **61,** 49–59.

Bennett, M., Michaud, S., Kingston, J., and Reed, R. (1992). Protein components specifically associated with prespliceosome and spliceosome complexes. *Genes Dev.* **6,** 1986–2000.

Bier, E., Vaessin, H., Younger-Shepherd, S., Jan, L. Y., and Jan, Y. N. (1992). *deadpan,* an essential pan-neural gene in *Drosophila,* encodes a helix-loop-helix protein similar to the *hairy* gene product. *Genes Dev.* **6,** 2137–2151.

Blackwell, T. K., and Weintraub, H. (1990). Differences and similarities in DNA-binding preferences of MyoD and E2A protein complexes revealed by binding site selection. *Science* **250,** 1104–1110.

Boggs, R. T., Gregor, P., Idriss, S., Belote, J. M., and McKeown, M. (1987). Regulation of sexual differentiation in *D. melanogaster* via alternative splicing of RNA from the *transformer* gene. *Cell (Cambridge, Mass.)* **50,** 739–747.

Bopp, D., Bell, L. R., Cline, T. W., and Schedl, P. (1991). Developmental distribution of female-specific *Sex-lethal* proteins in *Drosophila melanogaster.. Genes Dev.* **5,** 403–415.

Bridges, C. B. (1916). Non-disjunction as proof of the chromosome theory of heredity. *Genetics* **1,** 1–52.

Bridges, C. B. (1925). Sex in relation to chromosomes and genes. *Am. Nat.* **59,** 127–137.

Burtis, K. C., and Wolfner, M. F. (1992). The view from the bottom: Sex-specific traits and their control in *Drosophila. Semin. Dev. Biol.* **3,** 331–340.

Cabrera, C. V., and Alonso, M. C. (1991). Transcriptional activation by heterodimers of the *achaete-scute* and *daughterless* gene products of *Drosophila. EMBO J.* **10,** 2965–2973.

Caudy, M., Grell, E. H., Dambly-Chaudière, C., Ghysen, A., Jan, L. Y., and Jan, Y. N. (1988a). The maternal sex determination gene *daughterless* has zygotic activity necessary for the formation of peripheral neurons in *Drosophila. Genes Dev.* **2,** 843–852.

Caudy, M., Vässin, H., Brand, M., Tuma, R., Jan, L. Y., and Jan, Y. N. (1988b). *daughterless,* a *Drosophila* gene essential for both neurogenesis and sex determination has sequence similarities to *myc* and the *achaete-scute* complex. *Cell (Cambridge, Mass.)* **55,** 1061–1067.

Cline, T. W. (1978). Two closely linked mutations in *Drosophila melanogaster* that are lethal to opposite sexes and interact with *daughterless. Genetics* **90,** 683–698.

Cline, T. W. (1979a). A male-specific lethal mutation in *Drosophila melanogaster* that transforms sex. *Dev. Biol.* **72,** 266–275.

Cline, T. W. (1979b). A product of the maternally-influenced *Sex-lethal* gene determines sex in *Drosophila melanogaster. Genetics* **91,** Suppl., s22.

Cline, T. W. (1980). Maternal and zygotic sex-specific gene interactions in *Drosophila melanogaster. Genetics* **96,** 903–926.

Cline, T. W. (1983). The interaction between *daughterless* and *Sex-lethal* in triploids: A lethal sex transforming maternal effect linking sex determination and dosage compensation in *Drosophila melanogaster. Dev. Biol.* **95,** 260–274.

Cline, T. W. (1984). Autoregulatory functioning of a *Drosophila* gene product that establishes and maintains the sexually determined state. *Genetics* **107,** 231–277.

Cline, T. W. (1986). A female-specific lesion in the X-linked positive regulator of the *Drosophila* sex determination gene, *Sex-lethal. Genetics* **113,** 641–663; corrigendum: **114,** 345.

Cline, T. W. (1988). Evidence that *sisterless-a* and *sisterless-b* are two of several discrete "numerator elements" of the X/A sex determination signal in *Drosophila* that switch *Sxl* between two alternative stable expression states. *Genetics* **119,** 829–862.

Cline, T. W. (1989). The affairs of *daughterless* and the promiscuity of developmental regulators. *Cell (Cambridge, Mass.)* **59,** 231–234.

Cronmiller, C., and Cline, T. W. (1986). The relationship of relative gene dose to the complex phenotype of the *daughterless* locus in *Drosophila. Dev. Genet.* **7,** 205–221.

Cronmiller, C., and Cline, T. W. (1987). The *Drosophila* sex determination gene *daughterless* has different functions in the germline versus the soma. *Cell (Cambridge, Mass.)* **48,** 479–487.

Cronmiller, C., Schedl, P., and Cline, T. W. (1988). Molecular characterization of *daughterless,* a *Drosophila* sex determination gene with multiple roles in development. *Genes Dev.* **2,** 1666–1676.

Dobzhansky, T., and Schultz, J. (1931). Evidence for multiple sex factors in the X-chromosome of *Drosophila melanogaster. Proc. Natl. Acad. Sci. U.S.A.* **17,** 513–518.

Duffy, J. B., and Gergen, J. P. (1991). The Drosophila segmentation gene *runt* acts as a position-specific numerator element necessary for the uniform expression of the sex-determining gene *Sex-lethal. Genes Dev.* **5,** 2176–2187.

Ellis, H. M., Spann, D. R., and Posakony, J. W. (1990). *extramacrochaetae,* a negative regulator of sensory organ development in *Drosophila,* defines a new class of helix-loop-helix proteins. *Cell (Cambridge, Mass.)* **61,** 27–38.

Erickson, J. W., and Cline, T. W. (1991). Molecular nature of the *Drosophila* sex determination signal and its link to neurogenesis. *Science* **251,** 1071–1074.

Fukunaga, A., Tanaka, A., and Oishi, K. (1975). *maleless,* a recessive autosomal mutant of *Drosophila melanogaster* that specifically kills male zygotes. *Genetics* **81,** 135–141.

Garrell, J., and Campuzano, S. (1991). The helix-loop-helix domain: A common motif for bristles, muscles and sex. *BioEssays* **13,** 493–498.

Garrell, J., and Modolell, J. (1990). The *Drosophila extramacrochaetae* locus, an antagonist of proneural genes that, like these genes, encodes a helix-loop-helix protein. *Cell (Cambridge, Mass.)* **61,** 39–48.

Gergen, J. P. (1987). Dosage compensation in *Drosophila:* Evidence that *daughterless* and *Sex-lethal* control X chromosome activity at the blastoderm stage of embryogenesis. *Genetics* **117,** 477–485.

Geyer, P. K., Patton, S., Rodesch, C., and Nagoshi, R. N. (1993). Genetic and molecular characterization of P element-induced mutations reveal that the *Drosophila ovarian tumor* gene has maternal activity and a variable null phenotype. *Genetics* **133,** 265–278.

Gönczy, P., Viswanathan, S., and DiNardo, S. (1992). Probing spermatogenesis in *Drosophila* with P-element enhancer detectors. *Development (Cambridge, UK)* **114,** 89–98.

Goodfellow, P., and Lovell-Badge, R., eds. (1991). "Sex Determination and the Mammalian Y Chromosome," Semin. Dev. Biol., Vol. 2, No. 4. Saunders, Philadelphia.

Gorman, M., Kuroda, M. I., and Baker, B. S. (1993). Regulation of the sex-specific binding of the *maleless* dosage compensation protein to the male X chromosome in *Drosophila. Cell (Cambridge, Mass.)* **72,** 39–49.

Granadino, B., Campuzano, S., and Sánchez, L. (1990). The *Drosophila melanogaster fl(2)d* gene is needed for the female-specific splicing of *Sex-lethal* RNA. *EMBO J.* **9,** 2597–2602.

Granadino, B., San Juán, A. B., and Sánchez, L. (1991). The gene *fl(2)d* is required for various *Sxl*-controlled processes in *Drosophila* females. *Roux's Arch. Dev. Biol.* **200,** 172–176.

Granadino, B., San Juán, A. B., Santamaria, P., and Sánchez, L. (1992). Evidence of a dual function in *fl(2)d,* a gene needed for *Sex-lethal* expression in *Drosophila melanogaster. Genetics* **130,** 597–612.

Hildreth, P. E. (1965). *doublesex,* a recessive gene that transforms both males and females of *Drosophila* into intersexes. *Genetics* **51,** 659–678.

Hilfiker, A. (1990). Genetic and developmental analysis of *virilizer,* a gene involved in sex determination and dosage compensation of *Drosophila melanogaster.* Ph.D. Thesis, University of Zürich, Switzerland.

Hilfiker, A., and Nöthiger, R. (1991). The temperature sensitive mutation vir^{ts} (*virilizer*) identifies a new gene involved in sex determination of *Drosophila. Roux's Arch. Dev. Biol.* **200,** 240–248.

Horabin, J. I., and Schedl, P. (1993). Regulated splicing of the *Drosophila Sex-lethal* male exon involves a blockage mechanism. *Mol. Cell. Biol.* **13,** 1408–1414.

Inoue, K., Hoshijima, K., Sakamoto, H., and Shimura, Y. (1990). Binding of the *Drosophila Sex-lethal* gene product to the alternative splice site of *transformer* primary transcript. *Nature* (*London*) **344,** 461–463.

Kania, M. A., Bonner, A. S., Duffy, J. B., and Gergen, J. P. (1990). The *Drosophila* segmentation gene *runt* encodes a novel nuclear regulatory protein that is also expressed in the developing nervous system. *Genes Dev.* **4,** 1701–1713.

Kenan, D. J., Query, C. C., and Keene, J. D. (1991). RNA recognition: Towards identifying determinants of specificity. *Trends in Biochem. Sci.* **16,** 214–220.

Keyes, L. N. (1992). Sex-specific activation of the *Sex-lethal* gene in *Drosophila melanogaster.* Ph.D. Thesis, Princeton University, Princeton, NJ.

Keyes, L. N., Cline, T. W., and Schedl, P. (1992). The primary sex-determination signal of *Drosophila* acts at the level of transcription. *Cell* (*Cambridge, Mass.*) **68,** 933–943.

Lantz, V., Ambrosio, L., and Schedl, P. (1992). The Drosophila *orb* gene is predicted to encode sex-specific germline RNA-binding proteins and has localized transcripts in the ovaries and early embryos. *Development* (*Cambridge, UK*) **115,** 75–88.

Lucchesi, J. C., and Skripsky, T. (1981). The link between dosage compensation and sex differentiation in *Drosophila melanogaster. Chromosoma* **82,** 217–227.

Lucchesi, J. C., Skripsky, T., and Tax, F. E. (1982). A new male-specific lethal mutation in *Drosophila melanogaster. Genetics* **100,** Suppl., s42.

Maine, E. M., Salz, H. K., Schedl, P., and Cline, T. W. (1985). *Sex-lethal,* a link between sex determination and sexual differentiation in *Drosophila melanogaster. Cold Spring Harbor Symp. Quant. Biol.* **50,** 595–604.

Mange, A. P., and Sandler, L. (1973). A note on the maternal effect mutants *daughterless* and *abnormal oocyte* in *Drosophila melanogaster. Genetics* **73,** 73–86.

McKearin, D. M., and Spradling, A. C. (1990). *bag of marbles:* A *Drosophila* gene required to initiate both male and female gametogenesis. *Genes Dev.* **4,** 2242–2251.

McKeown, M., Belote, J. M., and Boggs, R. T. (1988). Ectopic expression of the female *transformer* gene product leads to female differentiation of chromosomally male *Drosophila. Cell* (*Cambridge, Mass.*) **53,** 887–895.

Michaud, S., and Reed, R. (1991). An ATP-independent complex commits pre-mRNA to the mammalian spliceosome assembly pathway. *Genes Dev.* **5,** 2534–2546.

Morgan, T. H., Redfield, H., and Morgan, L. V. (1943). Maintenance of a *Drosophila* stock centre in connection with investigations on the constitution of germinal material in relation to heredity. *Year Book—Carnegie Inst. Washington* **42,** 171–174.

Murre, C., McCaw, P. S., and Baltimore, D. (1989a). A new DNA binding and dimerization motif in immunoglobulin enhancer binding, *daughterless, MyoD,* and *myc* proteins. *Cell (Cambridge, Mass.)* **56,** 777–783.

Murre, C., McCaw, S. P., Vaessin, H., Caudy, M., Jan, L. Y., Jan, Y. N., Cabrera, C. V., Buskin, J. N., Hauschka, S. D., Lasser, A. B., Weintraub, H., and Baltimore, D. (1989b). Interactions between heterologous helix-loop-helix proteins generate complexes that bind specifically to a common DNA sequence. *Cell (Cambridge, Mass.)* **58,** 537–544.

Nagoshi, R. N., McKeown, M., Burtis, K. C., Belote, J. M., and Baker, B. S. (1988). The control of alternative splicing at genes regulating sexual differentiation in *D. melanogaster. Cell (Cambridge, Mass.)* **53,** 229–236.

Oliver, B., Perrimon, N., and Mahowald, A.P. (1988). Genetic evidence that the *sans fille* locus is involved in *Drosophila* sex determination. *Genetics* **120,** 159–171.

Oliver, B., Pauli, D., and Mahowald, A. P. (1990). Genetic evidence that the *ovo* locus is involved in *Drosophila* germ line sex determination. *Genetics* **125,** 535–550.

Parkhurst, S. M., Bopp, D., and Ish-Horowicz, D. (1990). X : A ratio, the primary sex-determining signal in *Drosophila,* is transduced by helix-loop-helix proteins. *Cell (Cambridge, Mass.)* **63,** 1179–1191.

Parkhurst, S. M., Lipshitz, H. D., and Ish-Horowicz, D. (1993). *achaete-scute* feminizing activities and *Drosophila* sex determination. *Development (Cambridge, UK)* **117,** 737–749.

Pauli, D., and Mahowald, A. P. (1990). Germ-line sex determination in *Drosophila melanogaster. Trends Genet.* **6,** 259–264.

Pipkin, S. B. (1947). A search for sex genes in the second chromosome of *Drosophila melanogaster* using the triploid method. *Genetics* **32,** 592–607.

Sakamoto, H., Inoue, K., Higuchi, I., Ono, Y., and Shimura, Y. (1992). Control of *Drosophila Sex-lethal* pre-mRNA splicing by its own female-specific product. *Nucleic Acids Res.* **20,** 5533–5540.

Salz, H. K. (1992). The genetic analysis of *snf:* A *Drosophila* sex determination gene required for activation of *Sex-lethal* in both the germline and the soma. *Genetics* **130,** 547–554.

Salz, H. K., Cline, T. W., and Schedl, P. (1987). Functional changes associated with structural alterations induced by mobilization of a P element inserted in the *Sex-lethal* gene of *Drosophila. Genetics* **117,** 221–231.

Salz, H. K., Maine, E. M., Keyes, L. N., Samuels, M. E., Cline, T. W., and Schedl, P. (1989). The *Drosophila* female-specific sex determination gene, *Sex-lethal,* has stage-, tissue-, and sex-specific RNAs suggesting multiple modes of regulation. *Genes Dev.* **3,** 708–719.

Samuels, M. E., Schedl, P., and Cline, T. W. (1991). The complex set of late transcripts from the *Drosophila* sex determination gene *Sex-lethal* encodes multiple related polypeptides. *Mol. Cell. Biol.* **11,** 3584–3602.

Sánchez, L., and Nöthiger, R. (1982). Clonal analysis of *Sex-lethal,* a gene needed for female sexual development in *Drosophila melanogaster. Roux's Arch. Dev. Biol.* **191,** 211–214.

Sánchez, L., and Nöthiger, R. (1983). Sex determination and dosage compensation in *Drosophila melanogaster:* Production of male clones in XX females. *EMBO J.* **2,** 485–491.

Sandler, L. (1972). On the genetic control of genes located in the sex-chromosome heterochromatin of *Drosophila melanogaster. Genetics* **70,** 261–274.

Schüpbach, T. (1985). Normal female germ cell differentiation requires the female X chromosome to autosome ratio and expression of *Sex-lethal* in *Drosophila melanogaster. Genetics* **109,** 529–548.

Seraphin, B., and Rosbash, M. (1989). Identification of functional U1 snRNA-pre-mRNA complexes committed to spliceosome assembly and splicing. *Cell (Cambridge, Mass.)* **59,** 349–358.

Seraphin, B., and Rosbash, M. (1991). The yeast branchpoint sequence is not required for the formation of a stable U1 snRNA-pre-mRNA complex and is recognized in the absence of U2 snRNA. *EMBO J.* **10,** 1209–1216.

Sonnenblick, B. P. (1941). Germ cell movements and sex differentiation of the gonads in the *Drosophila* embryo. *Proc. Natl. Acad. Sci. U.S.A.* **27,** 484–489.

Sosnowski, B. A., Belote, J. M., and McKeown, M. (1989). Sex-specific alternative splicing of RNA from the *transformer* gene results from sequence-dependent splice site blockage. *Cell (Cambridge, Mass.)* **3,** 449–459.

Steinmann-Zwicky, M. (1988). Sex determination in *Drosophila:* The X-chromosomal gene *liz* is required for *Sxl* activity. *EMBO J.* **7,** 3889–3898.

Steinmann-Zwicky, M. (1992a). How do germ cells choose their sex? *Drosophila* as a paradigm. *BioEssays* **14,** 513–518.

Steinmann-Zwicky, M. (1992b). Sex determination of *Drosophila* germ cells. *Semin. Dev. Biol.* **3,** 341–347.

Steinmann-Zwicky, M. (1993). Sex determination in *Drosophila: sis-b,* a major numerator element of the X : A ratio in the soma, does not contribute to the X : A ratio in the germ line. *Development (Cambridge, UK)* **117,** 763–767.

Steinmann-Zwicky, M., Schmid, H., and Nöthiger, R. (1989). Cell-autonomous and inductive signals can determine the sex of the germ line of *Drosophila* by regulating the gene *Sxl. Cell (Cambridge, Mass.)* **57,** 157–166.

Sturtevant, A. H. (1945). A gene in *Drosophila melanogaster* that transforms females into males. *Genetics* **30,** 297–299.

Szabad, J., Jursnich, V. A., and Bryant, P. J. (1991). Requirement for cell-proliferation control genes in *Drosophila* oogenesis. *Genetics* **127,** 525–533.

Torres, M., and Sánchez, L. (1989). The *scute* (T4) gene acts as a numerator element of the X : A signal that determines the state of activity of *Sex-lethal* in *Drosophila. EMBO J.* **8,** 3079–3086.

Torres, M., and Sánchez, L. (1992). The segmentation gene *runt* is needed to activate *Sex-lethal,* a gene that controls sex determination and dosage compensation in *Drosophila. Genet. Res.* **59,** 189–198.

Valcárcel, J., Singh, R., Zamore, P. D., and Green, M. R. (1993). The protein Sex-lethal antagonizes the splicing factor U2AF to regulate alternative splicing of *transformer* pre-mRNA. *Nature (London)* **362,** 171–175.

Van Doren, M., Ellis, H. M., and Posakony, J. W. (1991). The *Drosophila extramacrochaetae* protein antagonizes sequence-specific DNA binding by *daughterless/achaete-scute* protein complexes. *Development (Cambridge, UK)* **113,** 245–255.

Villares, R., and Cabrera, C. V. (1987). The *achaete-scute* gene complex of *D. melanogaster:* Conserved domains in a subset of genes required for neurogenesis and their homology to *myc. Cell (Cambridge, Mass.)* **50,** 415–424.

Watanabe, T. K. (1975). A new sex transforming gene on the second chromosome of *Drosophila melanogaster. Jpn. J. Genet.* **50,** 269–271.

Younger-Shepherd, S., Vaessin, H., Bier, E., Jan, L. Y., and Jan, Y. N. (1992). *deadpan,* an essential pan-neural gene encoding an HLH protein, acts as a denominator in *Drosophila* sex determination. *Cell (Cambridge, Mass.)* **70,** 1–20.

Zamore, P. D., and Green, M. R. (1991). Biochemical characterization of U2 snRNP auxiliary factor: An essential pre-mRNA splicing factor with a novel intranuclear distribution. *EMBO J.* **10,** 207–214.

Zamore, P. D., Patton, J., and Green, M. R. (1992). Cloning and domain structure of the mammalian splicing factor U2AF. *Nature (London)* **355,** 609–614.

10

The Human Y Chromosome

Ulrich Müller
Institut für Humangenetik
Justus-Liebig-Universität Giessen
D-35392 Giessen, Germany

Molecular Genetics of Sex Determination

I. Introduction

The Y chromosome constitutes approximately 1% of the DNA content of the diploid human genome. This amounts to 60×10^6 bp of DNA, given an estimated DNA content of the human genome of 6×10^9 bp. Much of the Y chromosome consists of repeated satellite DNA, the bulk of which is located in the heterochromatin of the distal long arm (primarily but not entirely Yq12) (Cooke, 1976). The satellite DNA of Yq is composed of the 3.4- and 2.5-kb repeat families DYZ1 and DYZ2 (Cooke *et al.*, 1983). DYZ1 is a subset of human satellite III and consists of 5-bp subunits organized into the 3.4-kb higher-order unit (Nakahori *et al.*, 1986). DYZ2, a subset of human satellite I, contains AT-rich sequences including one Alu sequence per repeat (Frommer *et al.*, 1984). Together DYZ1 and DYZ2 may account for 50–70% of the DNA content of the Y chromosome. No function of these repeats is apparent. Males lacking Yq12 are normal.

A block of alphoid satellite DNA called DYZ3 is present in the centromeric region of the Y (Wolfe *et al.*, 1985; Tyler-Smith and Brown, 1987). The centromeric alphoid satellite DNA is organized into 5.7 or 6.0 kb units, which contain subunits of 170 bp. DYZ3 makes up approximately 1% of the Y chromosome. A role of DYZ3 in centromere function is likely. This notion is supported by the detection of alphoid satellite DNA in the centromeres of all higher eukaryotes. Alphoid DNA may be required for binding of nucleosome-positioning proteins (Strauss and Varshavsky, 1984) and/or for attachment of proteins involved in spindle insertion (Gall and Atherton, 1974). The exact role of alphoid satellite DNAs and thus of DYZ3 is not yet understood, however. The remaining $20–30 \times 10^6$ bp of Y-chromosomal DNA can be subdivided into the pseudoautosomal region of distal Yp and the more proximal short and long arms of the Y.

The pseudoautosomal region makes up the terminal 2.5×10^6 bp of DNA of Yp (Yp11.32) and displays sequence homology with the distal short arm of the X chromosome (Xp22.3) (Brown, 1988; Petit *et al.*, 1988; Rappold and Lehrach, 1988). This region undergoes obligatory recombination during male meiosis. The term "pseudoautosomal" (Burgoyne, 1982) refers to the seemingly autosomal mode of inheritance of DNA sequences from this region (Cooke *et al.*, 1985; Simmler *et al.*, 1985). Several "minisatellites" have been detected in the pseudoautosomal region (Simmler *et al.*, 1987; Page *et al.*, 1987a). These minisatellites account for the hypervariability of DNA sequences in distal Xp and Yp. A gene, *MIC 2,* has been assigned to the proximal pseudoautosomal region (Darling *et al.*, 1986). *MIC 2* codes for the cell surface antigen

12E7, a glycoprotein of 32 kd that appears to be involved in T-cell adhesion processes (Gelin *et al.*, 1989). Another gene in this region codes for the receptor of the granulocyte—macrophage colony-stimulating factor, *GM-CSF*. (Gough *et al.*, 1990).

The following review focuses on the area of the Y chromosome that covers Yp from the proximal pseudoautosomal region to the centromere, and Yq from the centromere to the heterochromatic region (Yq12). This area is of special interest because it contains Y-specific genes including the testis-determining gene *SRY;* X/Y homologous genes such as *ZFY/ZFX, RPS4Y/RPS4X,* and *AMGY/AMGX;* and perhaps sequences coding for a spermatogenesis factor.

II. Deletion Mapping of the Testis-Determining Locus (*TDF*)

The close correlation between the presence of a Y chromosome and a male phenotype suggests that a gene(s) on the Y governs male sex differentiation in mammals. One Y chromosome is sufficient for testicular development, even in the presence of supernumerary X chromosomes, as in human individuals with 47,XXY, 48,XXXY, or 49,XXXXY karyotypes (Jacobs and Strong, 1959; Barr *et al.*, 1959; Fraccaro *et al.*, 1960).

The mapping of the testis-determining gene (or genes) on the Y chromosome is facilitated by the occurrence of mutations in man that interfere with normal gonadal differentiation, and reverse sex, as in 46,XX males and 46,XY females. The XX males are frequently indistinguishable phenotypically from normal 46,XY males. Yet they have a female karyotype and are always infertile. Clinical findings in XX males are listed in Table I. The 46,XY female condition called "Swyer syndrome" or XY gonadal dysgenesis is characterized by a lack of testis differentiation in patients with a 46,XY karyotype and a female phenotype. Table II summarizes the clinical findings in Swyer syndrome.

The differentiation of testes in XX males signals the presence of the testis-determining function in these individuals. In fact, Y-chromosomal DNA is found in about 80% of XX males (Guellaen *et al.*, 1984; Page *et al.*, 1985; Müller *et al.*, 1986b; Affara *et al.*, 1986, 1987; Vergnaud *et al.*, 1986). These may be referred to as (Y+) XX males, and those lacking Y-chromosomal DNA, as (Y−) XX males. Whereas Y-chromosomal genes are likely to account for testis differentiation in (Y+) XX males and may be considered primary testis-determining genes, the mutations resulting in (Y−) XX males must be different. Such mutations may affect the promoter region of an X or autosomal structural gene con-

Table I
Clinical Findings in 46,XX Males

Testicular atrophy with Leydig cell hyperplasia, tubular sclerosis, impaired spermatogenesis
Gynecomastia
Female hair pattern including female-type pubic hair, pectoral alopecia, poor growth of facial hair
Infertility
Low testosterone levels
Elevated or normal gonadotropine levels

Note. Although XX males are always infertile, some of the features listed above may vary considerably, so that XX males may be indistinguishable from normal 46,XY males (from Müller, 1990).

trolled by the Y. Alternatively they may affect genes acting secondarily to a primary gene(s) in the testis-determining pathway. The Y-chromosomal DNA in (Y+) XX males derives from Yp and was localized to the short arm of one of the two X chromosomes in all cases tested (Magenis *et al.*, 1982; Andersson *et al.*, 1986; Müller *et al.*, 1987; Schempp *et al.*, 1989; Mohandas *et al.*, 1990). These findings are in agreement with the hypothesis that an aberrant X–Y interchange during paternal meiosis is the underlying cause of the XX male syndrome (Ferguson-Smith, 1966). The X–Y interchange hypothesis has been substantiated by the analysis in several XX males of DNA sequences derived from the distal short arms of their fathers' sex chromosomes. These analyses demonstrated that one X chromosome in XX males had lost X-chromosomal DNA and gained Y-chromosomal DNA (Page *et al.*, 1987b; Petit *et al.*, 1987).

The amount of Y-chromosomal DNA varies in different XX males. This allows the construction of a deletion map of Yp with respect to the testis-determining locus (*TDF*). Sequences detected in most XX males

Table II
Clinical Findings in 46,XY Females (Swyer Syndrom)

Streak gonads with increased risk of gonadoblastoma or dysgerminoma
Infantile or normal appearing female internal and external genitalia
Poor breast development
Primary amenorrhea
Low estrogen levels
Elevated gonadotropin levels

Note. Some 46,XY females with Swyer syndrome have well-developed breasts (from Müller, 1990).

are thought to be closest to *TDF* and those sequences that are present in a few XX males only are more distal from this locus. Assuming a linear interchange of Y-chromosomal material in XX males, this approach places *TDF* to a more distal region of the Y chromosome, yet proximal to the pseudoautosomal region (Affara *et al.*, 1986; Müller *et al.*, 1986a; Vergnaud *et al.*, 1986). A deletion map of Yp is shown in Fig. 1.

Assignment of *TDF* to a distal region of Yp is further substantiated by findings in XY females. Loss of Y-chromosomal DNA is detected in some of these individuals and this DNA derives from the region of Yp in close proximity to *TDF* (Müller *et al.*, 1986a,b; Disteche *et al.*, 1986). For example, Y-chromosomal DNA absent in some 46,XY females includes DNA sequences Y-280 and Y-286, which were assigned to a region close to *TDF* (Fig. 1).

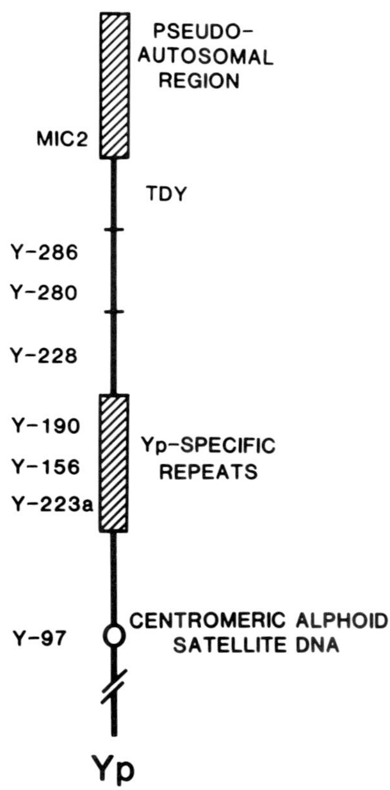

Figure 1 Deletion map of the short arm of the human Y chromosome (TDY = TDF). Distances are arbitrary and do not reflect the physical lengths of deletion intervals. The map was based on findings in XX males and XY females. See text. From Müller (1989).

III. Testis-Determining Region of Yp

Analysis of the testis-determining region by pulsed-field gel electrophoresis places sequences Y-280 and Y-286 of Fig. 1 within less than 0.8 Mbp of each other and adjacent to the pseudoautosomal region (Fig. 2; Müller and Lalande, 1990). A gene coding for a zinc binding finger protein was isolated from this region (Page *et al.,* 1987c). This gene, referred to as *ZFY,* lies between sequences Y-280 and Y-286 (Fig. 2).

ZFY was thought by some to be identical to *TDF.* This notion was supported by findings in an XX male and in an XY female who carried a Y;22 translocation. Part of the X-chromosomal DNA present in the XX male was missing in the unusual XY female, thus seemingly delineating 160 kb of DNA including *ZFY* as the region critical for testis determination (Page *et al.,* 1987c).

It has now been proven that *ZFY* and *TDF* are not identical. *ZFY,* which has an X chromosomal homolog (*ZFX*), was assigned to an autosome in marsupials, as was *ZFX* (Sinclair *et al.,* 1988). Given that the Y chromosome has a male-determining function in marsupials as well as in eutherians, this finding is hardly compatible with a testis-determining function of *ZFY* in mammals. *ZFY* is expressed in the germ cells but not somatic cells of the murine testis (Koopman *et al.,* 1989). Given that somatic cells but not germ cells are required for testicular differentiation, this further argues against a role for *ZFY* in gonadal differentiation. In addition, several XX males and XX true hermaphrodites were shown to lack *ZFY* but to have sequences more distal to *ZFY* (Palmer *et al.,* 1989). This places the *TDF* gene in a region of Yp closer to Y-286 and the pseudoautosomal region (Fig. 2). Significantly, the region of Yp

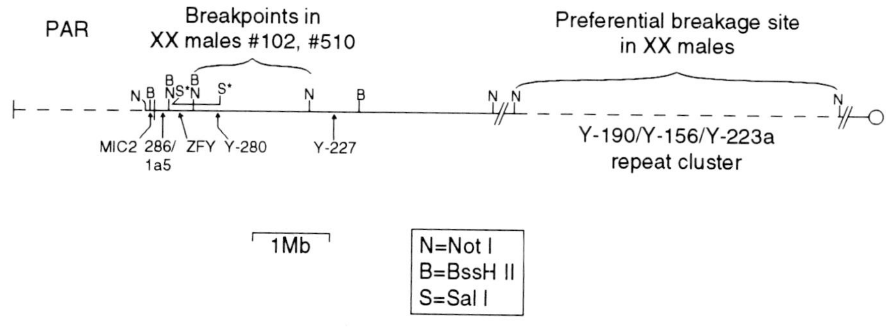

Figure 2 Physical map of the short arm of the human Y chromosome. PAR, pseudoautosomal region. See text. From Müller and Lalande (1990).

surrounding Y-286 contains Y-specific DNA sequences, unlike the region surrounding *ZFY* (which displays X homology). Thus any gene isolated from the Y-286 region that has been conserved during evolution of the mammalian Y chromosome would be a candidate for *TDF*.

Such a gene has now been detected. Rare XX males that lack *ZFY* but contain more distal Y chromosomal DNA sequences enabled the delineation of a small segment of approximately 35 kb of Y-chromosomal DNA close to the pseudoautosomal region. Assuming that *TDF* is located within this small portion of the Y chromosome, Sinclair *et al.* (1990) set out to clone the entire region and test for evolutionarily conserved DNA sequences. One cloned DNA segment, pY53.3, turned out to be conserved in all male, but not female, mammals tested. This segment was sequenced and found to contain two open reading frames corresponding to 99 and 223 amino acids reading toward the telomere. The deduced polypeptide sequence coded for by the larger open reading frame shows significant similarity to the mating-type protein, Mc, of the yeast *Schizosaccharomyces pombe* and to a DNA-binding motif present in high mobility group (HMG) proteins 1 and 2. The DNA-binding properties suggest that the gene, termed sex-determining region Y (*SRY*) in humans and *Sry* in the mouse, is a transcription factor. This notion is supported by the finding that recombinant *SRY* protein binds to sequences recognized by some transcription factors (Harley *et al.*, 1992).

There is now much evidence that *SRY* actually is *TDF*. First, in the mouse embryo, expression of *Sry* is restricted to the somatic cells of the male gonad (Koopman *et al.*, 1990). Expression is first detectable in the genital ridge at embryonic day 10.5, the time when morphologic differentiation of the testis starts. *Sry* continues to be expressed in the gonad until Day 12.5 and disappears thereafter. (In the adult mouse testis, however, *Sry* is transcribed again. Unlike the situation in embryonic gonads, in which expression is independent of the presence of germ cells, adult expression seems to depend on this cell type. At present, it is not clear whether the embryonic and adult transcripts are physically and functionally distinct.)

Second, several point mutations of *SRY* have been found in some 46,XY females without cytogenetic deletions of Yp (Berta *et al.*, 1990; Jäger *et al.*, 1990; Hawkins *et al.*, 1992). Of these, one resulted in a frameshift that rendered the *SRY* protein nonfunctional (Jäger *et al.*, 1990). Two additional mutations were detected in the HMG box of *SRY*, its DNA binding domain (Hawkins *et al.*, 1992). Using recombinant mutated *SRY* protein, binding was abolished to those target sequences to which wild-type *SRY* binds (Harley *et al.*, 1992). At present, *de novo* mutations have been described in 7 of 63 46,XY females tested. It is assumed that the nonfunctional *SRY* is the cause of sex reversal in these individuals. Sex reversal in the remaining 46,XY females might be due to mutations in the regulatory region of *SRY* or to mutations in genes acting downstream of *SRY* in the testis differentiation pathway.

Third, a DNA sequence containing *Sry* was shown to cause sex reversal of chromosomally female (40,XX) mice. The transgenic mice were produced by use of a 14-kb fragment of murine Y-chromosomal DNA containing the conserved gene *Sry* (Koopman *et al.*, 1991; see also Chapter 3, this volume). Of a total of 158 embryos obtained, 2 male embryos were found to have a 40,XX karyotype and to express *Sry*. This indicates, that the 14-kb fragment containing *Sry* is sufficient to induce testis differentiation. The additional 40,XX embryos that were found to be transgenic for *Sry*, however, were not sex reversed. Lack of sex reversal in these mice may be explained by low copy number of *Sry*, position effects, and (or) mosaicism. Koopman *et al.* (1991) also succeeded in studying the adult phenotype of one 40,XX *Sry*-sex-reversed mouse. This male was infertile, had male genitals, and displayed normal male mating behavior. Histological examination of the testis revealed apparently normal Sertoli, Leydig, and myoid cells but no germ cells.

Taken together, these findings strongly suggest a role for *SRY*/*Sry* in mammalian testis induction. It appears that *SRY*/*Sry* acts by regulating the expression of morphogenetic genes that may not be located on the Y chromosome. One candidate for such a gene is the gene coding for Müllerian inhibiting substance (MIS), which is required for the repression of female genital differentiation in males and perhaps for some aspects of testicular differentiation as well. It is probably identical to testis-secreted serologically detectable H-Y antigen (Müller and Wachtel, 1991; Müller *et al.*, 1993). Its gene has been assigned to the short arm of chromosome 19.

IV. Preferential Breakage Site on Yp Involved in the Origin of XX Males

Although the amount of Y-chromosomal DNA detected in (Y+) XX males varies among individuals, about 70% of the XX males investigated by the author are positive for the same set of sequences from Yp. These sequences include Y-specific repeated DNA assigned to a more proximal region of Yp (in addition to the more distal sequences of Yp shown in Fig. 1). The repeats designated DYZ5 are detected by probes Y-156, Y-190, and Y-223a (Müller *et al.*, 1986a).

Analysis of the DYZ5 repeats by pulsed-field gel electrophoresis and phage and cosmid mapping showed them to be organized into two blocks of 20 kb of tandemly repeated Y-chromosomal DNA. The larger block is hypervariable in size and may contain 20 to 40 repeat units; i.e., it may vary in size between 400 and 800 kb. The smaller block has a constant size of about 60 kb and is composed of three repeat units. In most Y chromosomes, the larger block is distal to the smaller one (Tyler-Smith *et al.*, 1988). The two blocks are located on a *Not*I fragment of approximately 5.3 Mbp (Müller *et al.*, 1989).

The structure of DYZ5 was investigated by pulsed-field gel electrophoresis in six XX males positive for probes Y-156, Y-190, and Y-223a. Although the large repeat block was detected in DNA from all six XX males, the smaller block was missing in the DNA from five of them (83%). This indicates that chromosome breaks had occurred between the two blocks. The large *Not*I fragment of 5.3 Mbp to which both repeat blocks are assigned in normal males was found to be rearranged in the XX males. In all five cases it had a size of about 4.4 Mbp. Evidently the breaks occurred within a small region between the two blocks. This region appears to be a focus of structural lability in the human Y chromosome.

Structural lability in the region between the two repeat blocks is further supported by findings in 46,XY females. In two of three individuals with a deletion of Yp, the larger repeat block was missing and the smaller was present. Evidently, the breakpoints lay within the two blocks of DYZ5. The rearranged *Not*I fragments differed in size in the two 46,XY females. Unlike the situation in the XX males, the breaks in the two XY females thus do not appear to have occurred at similar sites. A larger number of 46,XYp- females needs to be investigated to prove preferential breakage in proximal Yp as a common cause of the 46,XYp- condition.

V. Sequences from Yq in an XX Male (No. 102)

Although DNA sequences from Yp are found in all (Y+) XX males, there is one in whom sequences from Yq are detected. This XX male, No. 102, is also positive for sequences from distal Yp (Y-280 and Y-286; Fig. 1) (Müller *et al.*, 1986a) but not from a more proximal region of Yp. Two of 12 sequences from Yq tested were present in this patient's DNA (Table III). These sequences, designated Y-198 and Y-253, were assigned to a proximal region of Yq by deletion mapping (Fig. 3) and by *in situ* hybridization (Donlon and Müller, 1991). Y-198 and Y-253 are located on two adjacent *Eag*I fragments of 620 and 480 kb, respectively,

Table III
Hybridization of Y-Chromosomal Cloned DNA Sequences with DNA from Various Cell Lines

	DNA Sequence												
Cell Line	Y-198	Y-253	Y-157a	Y-221	Y-294	Y-202	Y-460	Y-214	Y-286 (Yq)	Y-402	Y-216a	Y-418	Y-286 (Yp)
45, X/46, X, del(Y)(pter-q11.21:) (#F1)	–	–	–	–	–	–	–	–	–	–	–	–	+
45, X/46, i(Yp) (pter-q11::q11-pter) (#1184)	+	+	+	+	+	–	–	–	–	–	–	–	+
46, XYq- (#L1078)	+	+	+	+	+	+	–	–	–	–	–	–	+
46, X, del(Y)(pter-q11.23:) (#TL44)	+	+	+	+	+	+	+	+	+	+	+	+	+
#102	+	+	–	–	–	–	–	–	–	–	–	–	+

Note. Y-286 hybridizes with two HindIII restriction fragments of 5.2 and 1.7 kb, respectively. The 5.2-kb fragment maps to the testis-determining region of Yp and the 1.7 kb fragment to proximal Yq.

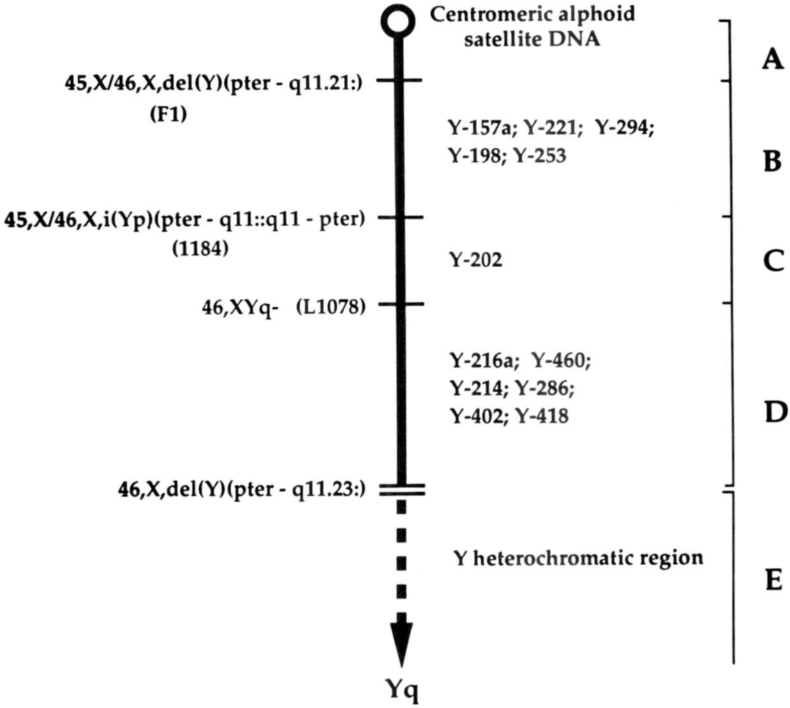

Figure 3 Deletion map of the euchromatic portion of the long arm of the Y chromosome. The map is based on the findings summarized in Table III. Deletion intervals are marked as A–E. Length of deletion intervals A–E do not reflect physical distances.

and on a *Bss*HII fragment of 1100 kb. Probes Y-294, Y-157, and Y-221, assigned to the same deletion interval as Y-198 and Y-253, detect different *Eag*I and *Bss*HII restriction fragments, indicating that they do not lie within 1 Mbp of DNA with which Y-198 and Y-253 hybridize.

In situ hybridization with Y-253 and Y-198 demonstrated that both sequences are on the short arm of one X chromosome in this patient, as are the sequences from distal Yp. This finding is most readily explained by the assumption that this XX male arose from an interchange between an X and an *inverted* Y chromosome. Inversions of parts of the Y are common and may result in the juxtaposition of sequences from distal Yp and Yq, as indicated in Fig. 4. Accordingly, the interchange resulted in the transfer of the testis-determining region of distal Yp and parts of proximal Yq onto one X chromosome of the patient. This would place sequences Y-198 and Y-253 distal to Y-294, Y-157, and Y-221 in deletion interval B of Fig. 3 (Donlon and Müller, 1991). De-

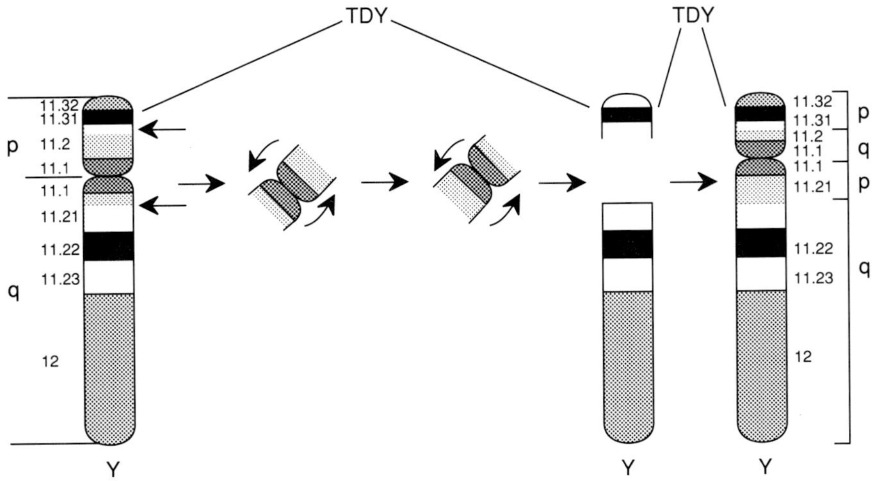

Figure 4 Origin of an inverted Y chromosome that might have been present in the father of XX male No. 102. The inversion event places sequences of Yq close to the TDF region. During an aberrant X/Y interchange, sequences from Yq and the TDF region are thought to have been translocated to an X of the patient (TDY = TDF).

letion mapping now suggests that Y-198 and Y-253 are proximal to Y-294, Y-157, and Y-221 (Cheng *et al.*, 1992). Therefore, the rearrangement observed in XX male, No. 102 is more complex than a simple inversion.

VI. The Spermatogenesis Locus of Yq

Many chromosome aberrations, including balanced and unbalanced translocations, interfere with normal spermatogenesis and cause oligozoospermia or azoospermia (reviewed by Fraccaro, 1983). These aberrations may include the autosomes only, e.g., Robertsonian translocations or inversions (Batanian and Hulten, 1987; Gabriel-Robez *et al.*, 1986), or may involve the sex chromosomes as well. Although Robertsonian translocations usually result in oligozoospermia (Fraccaro, 1983), Y-autosome translocations have in many cases been associated with azoospermia (Zuffardi and Tiepolo, 1982). Furthermore, karyotype–phenotype correlations in patients with deletions of Yq suggest that there is a special gene locus on Yq, the presence of which is a prerequisite for normal spermatogenesis to occur. Although deletions of most if not all of the euchromatic portion of Yq (Yq12) do not interfere with normal spermatogenesis and male function (Borgaonkar and Hollander, 1971; Verma *et al.*, 1978; Genest, 1981; also, see Introduction), deletions including the distal euchromatic portion of Yq are associated with azoospermia (Neu *et al.*, 1973; Tiepolo and Zuffardi, 1976; Yunis *et al.*, 1977; Fitch *et al.*, 1985). This prompted Tiepolo and Zuffardi (1976) to postulate a gene locus in the distal euchromatic portion of Yq that is required in order for normal spermatogenesis to occur. The existence of a Y-chromosomal "spermatogenesis locus," however, has been questioned. Münke *et al.* (1985) argue that azoospermia in patients with deletions of most of Yq is due to meiotic arrest as a result of impeded pairing and synaptonemal complex formation of the sex chromosomes and not to the deletion of a particular gene. This might indeed apply to individuals with azoospermia and complicated aberrations involving the Y chromosome. Such aberrations include Y-autosome translocations with various breakpoints on the Y (Dutrillaux and Gueguen, 1975; Smith *et al.*, 1979; Gonzales *et al.*, 1981; Petit *et al.*, 1982), monocentric and dicentric Y chromosomes with duplications of Yq (Butler *et al.*, 1981; Ataya *et al.*, 1983), and Y chromosomal rings (Chandley *et al.*, 1976). On the other hand, this might not be the case in individuals with simple terminal deletions of Yq. Deletions of the entire heterochromatic portion of Yq do not seem to interfere with fertility and can be traced through many generations (Genest *et al.*, 1970). Yet deletions of an additional small fragment of Yq that might involve an additional few megabases only result in complete absence of spermatozoa. If this were due to meiotic arrest rather than a gene deletion, one might expect oligozoospermia and not azoospermia in infertile Yq- individuals. In other words, one would expect a gradual increase in the frequency of meiotic arrest that is correlated with the size of Yq deletions.

Molecular genetic methods now allow the distinction between the two possible mechanisms underlying sterility in Yq- individuals. Provided there is a spermatogenesis locus on Yq, it will be possible to isolate DNA sequences in the vicinity of such a locus. In a first step toward this goal, a deletion map of Yq with respect to the putative spermatogenesis locus was constructed (Fig. 5) (Cheng *et al.*, 1992). Several sequences from euchromatic Yq are missing in infertile patients with a 46,XYq − karyotype. This region is now being further analyzed by pulsed-field gel electrophoresis for the presence of HTF islands, which are frequently associated with functional genes (Bird, 1986). Furthermore, analysis of yeast artificial chromosomes (YACs) derived from this region (Burke *et al.*, 1987) will aid in the isolation of evolutionarily conserved sequences, which are frequently parts of functional genes. Deletion maps of Yq with respect to the spermatogenesis region similar to the one given in

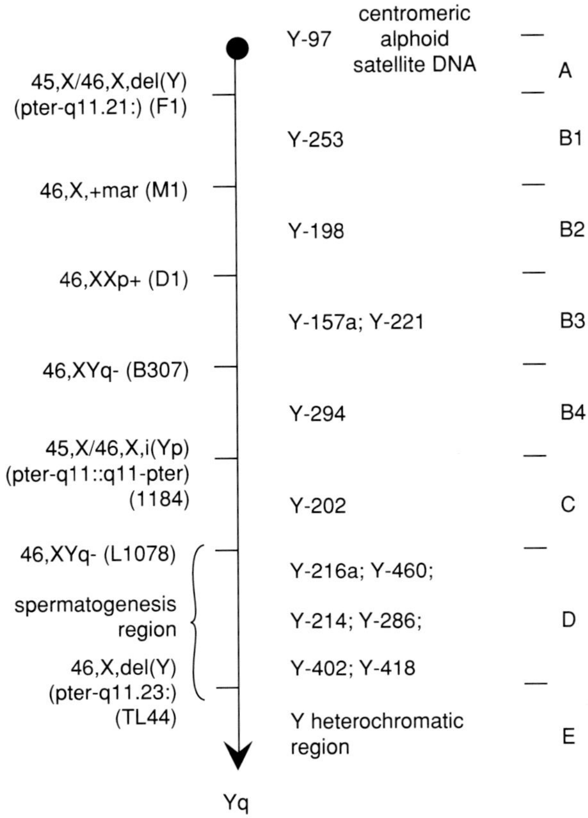

Figure 5 Detailed deletion map of Yq with respect to the putative spermatogenesis region. This map is an extension of the one given in Fig. 3. For details see text; adapted from Cheng *et al.* (1992).

Fig. 5 have been constructed by others. Thus Ma *et al.* (1992) present a detailed map of that region of the euchromatic portion of Yq that is close to the heterochromatic portion. Studying infertile patients with probes from this region, the authors detected interstitial deletions in two patients. These deletions, however, did not overlap, indicating that there might be two loci involved in spermatogenesis or that an inversion of Yq had occurred in one patient. Therefore, more infertile individuals with deletions of Yq need to be tested until "positional cloning" of a putative spermatogenesis locus becomes feasible. If such genes are expressed during male germ cell differentiation, they are likely candidates for genes required for spermatogenesis on Yq.

VII. X–Y Homologous Genes

SRY is currently the only Y-chromosomal gene known to be Y-specific. The remaining Y-chromosomal genes isolated display X–Y homology. These genes are *ZFY, RPS4Y,* and *AMGY,* and their respective X-chromosomal homologs are *ZFX, RPS4X,* and *AMGX.*

ZFY and its homolog in Xp, *ZFX,* code for a zinc finger protein (Page *et al.,* 1987c). *ZFY/ZFX* appear to be ubiquitously expressed and *ZFX* escapes inactivation in the female. The function of this gene (genes) is unknown.

RPS4Y, which has been mapped to Yp and its homolog in proximal Xq, *RPS4X,* code for isoforms of the ribosomal protein S4 (RPS4) (Fisher *et al.,* 1990). *RPS4Y/RPS4X* are ubiquitously expressed. *RPS4X* escapes inactivation in the female. It has been argued that *RPS4Y/RPS4X* are genes that prevent the expression of Turner stigmata (Watanabe *et al.,* 1993). This now appears to be unlikely. Patients with X-chromosomal aberrations, such as 46,X,i(Xq), were shown to express *RPS4X* in greater than double dosage, yet they have Turner stigmata (Just *et al.,* 1992). Conversely, no Turner stigmata were detected in a 46,XYp− female that had a large deletion of the Y short arm including *RPS4Y* (Müller *et al.,* 1992; see also Chapter 11, this volume).

AMGY, which has been assigned to Yp and its homolog in Xp, *AMGX,* code for a matrix protein of the dental enamel. They appear to regulate the formation of enamel and are expressed transiently by ameloblasts during tooth development. Mutations of *AMGX* appear to result in X-linked amelogenesis imperfecta (AI) (Lagerström *et al.,* 1991) thus implying an important role for this gene in normal tooth development. A physiological role of *AMGY* is not certain at present. Although *AMGY* is expressed, its level of expression is only 10% of that of its X chromosomal homolog (Salido *et al.,* 1992).

VIII. Summary

This chapter covers findings on the structure and function of the human Y chromosome. Data on the location and organization of repeated DNA sequences of the Y are summarized. Results obtained by deletion mapping and pulsed-field gel electrophoresis are reviewed and the assignment of the testis-determining locus (*TDF*) to the distal short arm of the Y chromosome is deduced. Experimental findings are reviewed. These findings suggest that *SRY* is the testis-determining gene in man and its mouse homolog *Sry* is pivotal in murine testicular induction.

Evidence is given for a focus of structural lability in Yp demarcated by two blocks of the tandem repeat DYZ5. Breaks within this region frequently result in the origin of (Y+) XX males. Findings are reported in one exceptional (Y+) XX male of DNA sequences from both distal Yp and proximal Yq. The findings in this unusual XX male are explained by an aberrant interchange between an X and a rearranged Y chromosome during paternal meiosis. Finally, observations are discussed that indicate the existence of a spermatogenesis locus in the long arm of the Y chromosome, and characteristics of X–Y homologous genes are given.

References

Affara, N. A., Florentin, L., Morrison, N., Kwok, K., Mitchell, M., Cooke, A., Jamieson, D., Glasgow, L., Meredith, L., Boyd, E., and Ferguson-Smith, M. A. (1986). Regional assignment of Y-linked DNA probes by deletion mapping and their homology with X chromosome and autosomal sequences. *Nucleic Acids Res.* **14,** 5353–5373.

Affara, N. A., Ferguson-Smith, M. A., Magenis, R. E., Tolmie, J. L., Boyd, E., Cooke, A., Jamieson, D., Kwok, K., Mitchell, M., and Snadden, L. (1987). Mapping the testis determinants by an analysis of Y-specific sequences in males with apparent XX and XO karyotypes and females with XY karyotypes. *Nucleic Acids Res.* **15,** 7325–7342.

Andersson, M., Page, D. C., and de la Chapelle, A. (1986). Chromosome Y-specific DNA is transferred to the short arm of the X chromosome in human XX males. *Science* **233,** 786–788.

Ataya, K. M., Dudin, G., and Mroueh, A. (1983). Dicentric i(Yq) chromosome and azoospermia. *Am. J. Med. Genet.* **14,** 583–590.

Barr, M. L., Shaver, E. L., and Carr, D. H. (1959). An unusual sex chromatin pattern in three mentally deficient subjects. *J. Ment. Def. Res.* **3,** 78.

Batanian, J., and Hulten, M. A. (1987). Electron microscopic investigations of synaptonemal complexes in an infertile human male carrier of a pericentric inversion inv(1)(p32q42). Regular loop formation but defective synapsis including a possible interchromosomal effect. *Hum. Gen.* **76,** 81–89.

Berta, P., Hawkins, J. R., Sinclair, A. H., Taylor, A., Griffiths, B. L., Goodfellow, P. N., and Fellous, M. (1990). Genetic evidence equating SRY and the testis determining factor. *Nature (London)* **348,** 448–450.

Bird, A. P. (1986). CpG-rich islands and the function of DNA methylation. *Nature (London)* **321,** 209–213.

Borgaonkar, D. S., and Hollander, D. H. (1971). Quinacrine fluorescence of the Y chromosome. *Nature (London)* **230,** 52.

Brown, W. R. A. (1988). A physical map of the human pseudoautosomal region. *Eur. Mol. Biol. Organ. J.* **7,** 2377–2385.

Burgoyne, P. S. (1982). Genetic homology and crossing-over in the X and Y chromosomes of mammals. *Hum. Genet.* **61,** 85–90.

Burke, D. T., Carle, G. F., and Olson, M. V. (1987). Cloning of large segments of exogenous DNA into yeast by means of artificial chromosome vectors. *Science* **236,** 806–812.

Butler, M. G., Sanger, W. G., and Walzak, M. P. (1981). A unique Y/Y translocation in an infertile male. *Cytogenet. Cell Genet.* **31,** 175–177.

Chandley, A. C., MacLean, N., Edmond, P., Fletcher, J., and Watson, G. S. (1976). Cytogenetics and infertility in man. II. Testicular histology and meiosis. *Ann. Hum. Genet.* **40,** 165–176.

Cheng, S. D., Gasparini, R., and Müller, U. (1992). Molecular analysis of aberrations of Xp and Yq. *Hum. Genet.* **88,** 379–382.

Cooke, H. J. (1976). Repeated sequences specific to human males. *Nature* (*London*) **262,** 182–186.

Cooke, H. J., Fantes, J., and Green, D. (1983). Structure and evolution of human Y chromosome DNA. *Differentiation* (*Berlin*) **23,** Suppl., S48.

Cooke, H. J., Brown, W. R. A., and Rappold, G. A. (1985). Hypervariable telomeric sequences from the human sex chromosomes are pseudoautosomal. *Nature* (*London*) **317,** 687–692.

Darling, S. M., Banting, G. S., Pym, B., Wolff, J., and Goodfellow, P. N. (1986). Cloning an expressed gene shared by the human sex chromosomes. *Proc. Natl. Acad. Sci. U.S.A.* **83,** 135–139.

Disteche, C. M., Casanova, M., Saal, H., Friedman, C., Sybert, V., Graham, J., Thuline, H., Page, D., and Fellous, M. (1986). Small deletions of the short arm of the Y chromosome in 46,XY females. *Proc. Natl. Acad. Sci. U.S.A.* **83,** 7841–7844.

Donlon, T. A., and Müller, U. (1991). Deletion mapping of DNA segments from the Y chromosome long arm and their analysis in an XX male. *Genomics* **10,** 51–56.

Dutrillaux, B., and Gueguen, J. (1975). Etude méiotique et mitotique dans un cas de translocation t (5;Y). *Hum. Genet.* **27,** 241–245.

Ferguson-Smith, M. A. (1966). X-Y chromosomal interchange in the aetiology of true hermaphroditism and of XX Klinefelter's syndrome. *Lancet* **2,** 475–476.

Fisher, E. M. C., Beer-Romero, P., Brown, L. G., Ridley, A., McNeil, J. A., Bentley, L., Willard, H. F., Bieber, F. R., and Page, D. C. (1990). Homologous ribosomal protein genes on the human X and Y chromosomes: Escape from X inactivation and possible implications for Turner syndrome. *Cell* (*Cambridge, Mass.*) **63,** 1205–1218.

Fitch, N., Richer, C. L., Pinsky, L., and Kahn, A. (1985). Deletion of the long arm of the Y chromosome and review of Y chromosome abnormalities. *Am. J. Med. Genet.* **20,** 31–42.

Fraccaro, M. (1983). Chromosome abnormalities and gamete production in man. *Differentiation* (*Berlin*) **23,** Suppl., S40–S43.

Fraccaro, M., Kaijser, K., and Lindsten, J. (1960). A child with 49 chromosomes. *Lancet* **2,** 899–902.

Frommer, M., Prosser, J., and Vincent, P. C. (1984). Human satellite I sequences include a male-specific 2.47 kb tandemly repeated unit containing one Alu family member per repeat. *Nucleic Acids Res.* **12,** 2887–2900.

Gabriel-Robez, O., Ratomponirina, C., Rumpler, Y., Le Marec, B., Luciani, J. M., and Guichaoua, M. R. (1986). Synapsis and synaptic adjustment in an infertile human male heterozygous for a pericentric inversion in chromosome 1. *Hum. Genet.* **72,** 148–152.

Gall, J. G., and Atherton, D. D. (1974). Satellite DNA sequences in drosophila virilin. *J. Mol. Biol.* **835,** 633–664.

Gelin, C., Aubrit, F., Phalipon, A., Raynal, B., Cole, S., Kaczorek, M., and Bernard, A. (1989). The E2 antigen, a 32 kD glycoprotein involved in T-cell adhesion processes, is the MIC2 gene product. *Eur. Mol. Biol. Organ. J.* **8,** 3253–3259.

Genest, P. (1981). Étude complémentaire sur la nature d'un petit Y multicentenaire. *Ann. Génét.* **24,** 165–166.

Genest, P., Laberge, C., Poty, J., Gagne, R., and Bouchard, M. (1970). Transmission d'un petit "Y"durant onze générations dans une lignée familiale. *Ann. Génét.* **13,** 233–238.

Gonzales, J., Lesourd, S., and Dutrillaux, B. (1981). Mitotic and meiotic analysis of a reciprocal translocation t(Y;3) in an azoospermic male. *Hum. Genet.* **57,** 111–114.

Gough, N. M., Gearing, D. P., Nicola, N. A., Baker, E., Pritchard, M., Callen, D. F., and Sutherland, G. R. (1990). Localization of the human GM-CSF receptor gene to the X-Y pseudoautosomal region. *Nature (London)* **345,** 734–736.

Guellaen, G., Casanova, M., Bishop, C., Geldwerth, D., Andre, G., Fellous, M., and Weissenbach, J. (1984). Human XX males with Y single-copy DNA fragments. *Nature (London)* **307,** 172–173.

Harley, V. R., Jackson, D. I., Hextall, P. J., Hawkins, J. R., Berkovitz, G. D., Sockanathan, S., Lovell-Badge, R., and Goodfellow, P. N. (1992). DNA binding activity of recombinant SRY from normal males and XY females. *Science* **255,** 453–456.

Hawkins, J. R., Taylor, A., Berta, P., Levilliers, J., van der Auwera, B., and Goodfellow, P. N. (1992). Mutational analysis of SRY: Nonsense and missence mutations in XY sex reversal. *Hum. Genet.* **88,** 471–474.

Jacobs, P. A., and Strong, J. A. (1959). A case of human intersexuality having a possible XXY sex-determining mechanism. *Nature (London)* **183,** 302–303.

Jäger, R. J., Anvret, M., Hall, K., and Scherer, G. (1990). A human XY female with a frame shift mutation in the candidate testis-determining gene SRY. *Nature (London)* **348,** 452–454.

Just, W., Geerkens, C., Held, K., and Vogel, W. (1992). Expression of RPS4X in fibroblasts from patients with structural aberrations of the X-chromosome. *Hum. Genet.* **89,** 240–242.

Koopman, P., Gubbay, J., Collignon, J., and Lovell-Badge, R. (1989). Zfy gene expression patterns are not compatible with a primary role in mouse sex determination. *Nature (London)* **342,** 940–942.

Koopman, P., Münsterberg, A., Blanche, C., Vivian, N., and Lovell-Badge, R. (1990). Expression of a candidate sex-determining gene during mouse testis differentiation. *Nature (London)* **348,** 450–452.

Koopman, P., Gubbay, J., Vivian, N., Goodfellow, P., and Lovell-Badge, R. (1991). Male development of chromosomally female mice transgenic for SRY. *Nature (London)* **351,** 117–121.

Lagerström, M., Dahl, N., Nakahori, Y., Nakagome, Y., Bäckman, B., Landergren, U., and Pettersson, U. (1991). A deletion in the amelogenin gene (AMG) causes X-linked amelogenesis imperfecta (AIH1). *Genomics* **10,** 971–975.

Ma, K., Sharkey, A., Kirsch, S., Vogt, P., Keil, R., Hargreave, T. B., McBeath, S., and Chandley, A. C. (1992). Towards the molecular localisation of the AZF locus: Mapping of microdeletions in azoospermic men with subintervals of interval 6 of the human Y chromosome. *Hum. Mol. Genet.* **1,** 29–33.

Magenis, R. E., Webb, M. J., McKean, R. S., Tomar, D., Allen, L. J., Kammer, H., van Dyke, K. L., and Louvrien, E. (1982). Translocation (X;Y)(p22.3;p11.2) in XX males: Etiology of male phenotype. *Hum. Genet.* **62,** 271–276.

Mohandas, T. K., Stern, H. J., Mecker, C. A., Passage, M. B., Müller, U., Page, D. C., Yen, P. H., and Shapiro, L. J. (1990). Steroid sulfatase gene in XX males. *Am. J. Hum. Genet.* **46,** 369–376.

Müller, U. (1989). Molecular genetic approaches to testis differentiation. *In* "Pediatric Endocrinology" (R. Collu, J. R. Ducharme, and H. Guyda, eds.), 2nd ed. pp. 295–305. Raven Press, New York.

Müller, U. (1990). The search for the testis determining factor (TDF). *Contemp. Obstet. Gynecol.* **35,** 22–32.

Müller, U., and Lalande, M. (1990). A physical map of the human Y chromosome short arm. *Genomics* **7,** 517–523.

Müller, U., and Wachtel, S. (1991). Are testis-secreted H-Y serological antigen and Müllerian inhibiting substance the same? *Am. J. Hum. Genet.* **49,** Suppl. 22 (abstr.).

Müller, U., Lalande, M., Donlon, T. A., and Latt, S. A. (1986a). Moderately repeated DNA sequences specific for the short arm of the human Y chromosome are present in XX males and reduced in copy number in an XY female. *Nucleic Acids Res.* **14,** 1325–1340.

Müller, U., Donlon, T. A., Schmid, M., Fitch, N., Richer, C. L., Lalande, M., and Latt, S. A. (1986b). Deletion mapping of the testis determining locus with DNA probes in 46,XX males and in 46,XY and 46,XY,dic(Y) females. *Nucleic Acids Res.* **14,** 6489–6505.

Müller, U., Latt, S. A., and Donlon, T. A. (1987). Y-specific DNA sequences in male patients with 46,XX and 47,XXX karyotypes. *Am. J. Med. Genet.* **28,** 393–401.

Müller, U., Lalande, M., Donlon, T. A., and Heartlein, M. W. (1989). Breakage of the human Y chromosome short arm between two blocks of tandemly repeated DNA sequences. *Genomics* **5,** 153–156.

Müller, U., Kirkels, V. G. H. J., and Schleres, J. M. J. (1992). Absence of Turner stigmata in a 46,XYp- female. *Hum. Genet.* **90,** 239–242.

Müller, U., Wachtel, S. S., Jaswaney, V. L., and Goldberg, E. H. (1993). H-Y (SDM) antibody specifically binds Müllerian inhibiting substance. *Hum. Genet.* **91,** 515–518.

Münke, M., de Martinville, B., Lieber, E., and Francke, U. (1985). Minute chromosomes replacing the Y chromosome carry Y-specific sequences by restriction fragment analysis and in situ hybridization. *Am. J. Med. Genet.* **22,** 361–374.

Nakahori, Y., Mitani, K., Yamada, M., and Nakagome, Y. (1986). A human Y-chromosome-specific DNA family (DYZ1) consists of a tandem array of pentanucleotides. *Nucleic Acids Res.* **14,** 7569–7580.

Neu, R. L., Barlow, M. J., and Gardner, L. I. (1973). A 46,XYq- male with aspermia. *Fertil. Steril.* **24,** 811–813.

Page, D. C., de la Chapelle, A., and Weissenbach, J. (1985). Chromosome Y-specific DNA in related human XX males. *Nature (London)* **315,** 224–226.

Page, D. C., Bieker, K., Brown, L. G., Hinton, S., Leppert, M., Lalouel, J. L., and Lathrop M. (1987a). Linkage, physical mapping and DNA sequence analysis of pseudoautosomal loci on the human X and Y chromosomes. *Genomics* **1,** 243–256.

Page, D. C., Brown, L. G., and de la Chapelle, A. (1987b). Exchange of terminal portions of X- and Y-chromosomal short arms in human XX males. *Nature (London)* **328,** 437–440.

Page, D. C., Mosher, R., Simpson, E. M., Fisher, E. M. C., Mardon, G., Pollack, J., McGillivray, B., de la Chapelle, A., and Brown, L. G. (1987c). The sex-determining region of the human Y chromosome encodes a finger protein. *Cell (Cambridge, Mass.)* **51,** 1091–1104.

Palmer, M. S., Sinclair, A. H., Berta, P., Ellis, N. A., Goodfellow, P. N., Abbas, N. E., and Fellous, M. (1989). Genetic evidence that ZFY is not the testis-determining factor. *Nature (London)* **342,** 937–939.

Petit, C., de la Chapelle, A., Levilliers, J., Castillo, S., Noel, B., and Weissenbach, J. (1987). An abnormal terminal X-Y interchange accounts for most but not all cases of human XX maleness. *Cell (Cambridge, Mass.)* **49,** 595–602.

Petit, C., Levilliers, J., and Weissenbach, J. (1988). Physical mapping of the human pseudoautosomal region; comparison with genetic linkage map. *Eur. Mol. Biol. Organ. J.* **7,** 2369–2376.

Petit, P., Unglik, A., and Fryns, J. P. (1982). Translocation 46,X,t(Y;14)(q122;q111) in a case of sterility in the male. *Ann. Génét.* **25,** 63–64.

Rappold, G. A., and Lehrach, H. (1988). A long range restriction map of the pseudoautosomal region by partial digest PFGE analysis from the telomere. *Nucleic Acids Res.* **16,** 5361–5377.

Salido, E. C., Yen, P. H., Koprivnikar, K., Loh-Chung, Y., and Shapiro, L. J. (1992). The human enamel protein gene amelogenin is expressed from both the X and the Y chromosomes. *Am. J. Hum. Genet.* **50,** 303–316.

Schempp, W., Müller, G., Scherer, G., Bohlander, S. K., Rommerskirch, W., Fraccaro, M., and Wolf, U. (1989). Localization of Y chromosome sequences and X chromosomal replication studies in XX males. *Hum. Genet.* **81,** 144–148.

Simmler, M. C., Rouyer, F., Vergnaud, G., Nyström-Lahti, M., Ngo, K. Y., de la Chapelle, A., and Weissenbach, J. (1985). Pseudoautosomal DNA sequences in the pairing region of the human sex chromosomes. *Nature (London)* **317,** 692–697.

Simmler, M. C., Johnsson, C., Petit, C., Rouyer, F., Vergnaud, G., and Weissenbach, J. (1987). Two highly polymorphic minisatellites from the pseudoautosomal region of the human sex chromosomes. *Eur. Mol. Biol. Organ. J.* **6,** 963–969.

Sinclair, A. H., Foster, J. W., Spencer, J. A., Page, D. C., Palmer, M., Goodfellow, P. N., and Graves, J. A. M. (1988). Sequences homologous to ZFY, a candidate human sex determining gene, are autosomal in marsupials. *Nature (London)* **336,** 780–782.

Sinclair, A. H., Berta, P., Palmer, M. S., Hawkins, J. R., Griffiths, B. L., Smith, M. J., Foster, J. W., Frischauf, A. M., Lovell-Badge, R., and Goodfellow, P. N. (1990). A gene from the human sex-determining region encodes a protein with homology to a conserved DNA-binding motif. *Nature (London)* **346,** 240–244.

Smith, A., Fraser, I. S., and Elliot, G. (1979). An infertile male with balanced Y;19 translocation. Review of Y; autosome translocations. *Ann. Génét.* **22,** 189–194.

Strauss, F., and Varshavsky, A. (1984). A protein binds to a satellite DNA repeat at three specific sites that would be brought into mutual proximity by DNA folding in the nucleosome. *Cell (Cambridge, Mass.)* **37,** 889–901.

Tiepolo, L., and Zuffardi, O. (1976). Localization of factors controlling spermatogenesis in the nonfluorescent portion of the human Y chromosome long arm. *Hum. Genet.* **34,** 119–124.

Tyler-Smith, C., and Brown, W. R. A. (1987). Structure of the major block of alphoid satellite DNA on the human Y chromosome. *J. Mol. Biol.* **195,** 457–470.

Tyler-Smith, C., Taylor, L., and Müller, U. (1988). Structure of a hypervariable tandemly repeated DNA sequence on the short arm of the human Y chromosome. *J. Mol. Biol.* **203,** 837–848.

Vergnaud, G., Page, D. C., Simmler, M. C., Brown, L., Rouyer, F., Noel, B., Botstein, D., de la Chapelle, A., and Weissenbach, J. (1986). A deletion map of the human Y chromosome based on DNA hybridization. *Am. J. Hum. Genet.* **38,** 109–124.

Verma, R. S., Dosik, H., Scharf, T., and Lubs, H. A. (1978). Length heteromorphism of fluorescent (f) and non-fluorescent (nf) segments of human Y chromosome: Classification, frequencies, and incidence in normal Caucasians. *J. Med. Genet.* **15,** 277–281.

Watanabe, M., Zinn, A. R., Page, D. C., and Nishimoto, T. (1993). Functional equivalence of human X- and Y-encoded isoforms of ribosomal protein S4 consistent with a role in Turner syndrome. *Nat. Genet.* **4,** 268–271.

Wolfe, J., Darling, S. M., Erickson, R. P., Craig, I. W., Buckle, V. J., Rigby, P. W. J., Willard, H. F., and Goodfellow, P. N. (1985). Isolation and characterization of an alphoid centromeric repeat family from the human Y chromosome. *J. Mol. Biol.* **182,** 477–485.

Yunis, E., Garcia-Conti, F. L., Torres de Caballero, O. M., and Giraldo, A. (1977). Yq deletion, aspermia, and short stature. *Hum. Genet.* **39,** 117–122.

Zuffardi, O., and Tiepolo, L. (1982). Frequencies and types of chromosome abnormalities associated with male fertility. *In* "Genetic Control of Gamete Production and Function" (P. G. Crosignani and B. L. Rubin, eds.), p. 261. Academic Press, London.

11

DNA Sequence Homology between the Human Sex Chromosomes

Nabeel A. Affara and Malcolm A. Ferguson-Smith

Department of Pathology
University of Cambridge
Cambridge CB2 1QP
United Kingdom

Molecular Genetics of Sex Determination

I. Introduction

The genetic functions carried by the sex chromosomes have become the subject of intense investigation largely because of the opportunity afforded by the functional monosomy of the X and Y chromosomes in males. The classical view of the mammalian sex chromosomes perceives their evolution from an ancestral pair of homologs (Ohno, 1967). Morphological differentiation has accompanied the emergence of a sex-determining role on the Y and a mechanism of dosage compensation for X-linked genes in females, mediated by random X inactivation. It might be expected that sequence homology from ancestral times may have been retained in functionally important domains of the sex chromosomes. For example, despite their great disparity in size and morphology, the X and Y chromosomes pair at meiosis, thus ensuring correct segregation during gametogenesis. This pairing region can encompass almost the whole of the short arm of the Y and involves up to 27% of Xp from the telomere (Polani, 1982; Chandley *et al.*, 1984). It is not unreasonable to expect that sequence homology is important for correct pairing and may extend over some of this region.

The failure of the sex chromosomes to participate in recombination beyond the pairing region has precluded a conventional genetic analysis for most of the Y. Suppression of recombination is likely to be an essential requirement for preserving the sex-determining role of the differential part of the Y (Goodfellow and Darling, 1988). It is, therefore, not surprising that models of the Y have relied heavily on phenotype–karyotype correlations through detailed cytogenetic analysis of abnormal Y chromosomes (for review, see Davis, 1981). Models of the Y derived from such studies necessarily reflect the level of resolution that cytogenetics can provide. However, in recent years, molecular analysis using cloned DNA sequences (isolated primarily from flow-sorted Y chromosome specific libraries) has been combined with analysis of structural abnormalities of the Y to enable the construction of comprehensive deletion maps covering Yp and Yq (Vergnaud *et al.*, 1986; Affara *et al.*, 1986b, 1987; Müller *et al.*, 1986; Page, 1986; Bardoni *et al.*, 1991; O'Reilly *et al.*, 1992; Ma *et al.*, 1992; Nakahori *et al.*, 1991b; Vollrath *et al.*, 1992). Our understanding of Y-chromosome structure and patterns of X–Y homology has developed considerably from such studies. In this review, we consider what has been learned about the patterns of sequence homology between the sex chromosomes, how these are distributed on the X and Y, and how they have evolved.

II. The Pseudoautosomal Region

The pseudoautosomal region represents the most comprehensively studied segment of X–Y homology. This is, in large measure, because genetic and molecular analyses have been possible since the Y recombines with the X within the pairing segment (for review, see Ellis and Goodfellow, 1989).

A. Genetic Analysis

The electronmicrographs of Chandley *et al.* (1984) depicting the human X–Y synaptonemal complex are striking, implying that the X and Y share sequence homology in this region. Measurements of the degree of synapsis from these studies indicate that pairing can involve from 11.4 to 56.3% of the Y and 3.3 to 17.3% of Xp. If, as suggested by Goodfellow *et al.* (1985), the stringency of pairing decreases as meiosis proceeds, the lower limit of synapsis observed by Chandley *et al.* (1984) may reflect the extent of strict sequence homology. On the basis of a size of approximately 50 Mb for the Y, one can estimate a size of approximately 5 Mb for the human pairing region.

Such X–Y pairing is well documented for several mammalian species (Koller and Darlington, 1934; Solari, 1974). Darlington (1937) and, more recently, Burgoyne (1982) proposed that an obligatory crossover occurs in the pairing region during male meiosis and that, in consequence, at least this segment of the Y should be amenable to genetic analysis. It follows that loci in this part of the X and Y will show partial sex-linkage or pseudoautosomal behavior at meiosis. There is considerable evidence for recombination in the mouse pairing segment where inheritance of the Sxr mutation (Singh and Jones, 1982; Evans *et al.*, 1982; Bishop *et al.*, 1987), the steroid sulfatase (*STS*) gene (Keitges *et al.*, 1985, 1987), and an inserted proviral sequence (Harbers *et al.*, 1986) has been shown to be pseudoautosomal in character.

As the collection of DNA sequences mapping to the human Y chromosome has expanded, various patterns of X–Y homology have emerged. Many of these do not map to the pairing region and include Yp–Xq, Yq–Xp, Yp–Xp and Yq–Xq associations. These will be discussed further in subsequent sections. However, many DNA sequences have been isolated (primarily from cosmid clones) that map to the pairing segment; these have been pivotal in defining the human pseudoautosomal region.

Among these is the *MIC2* gene, which encodes a surface antigen recognized by the 12E7 monoclonal antibody (see below); some of the sequences are summarized in Table I, listed by their human gene mapping (HGM) D numbers where these have been assigned (Cooke *et al.*, 1985; Simmler *et al.*, 1985; Rouyer *et al.*, 1986; Darling *et al.*, 1986) (for detailed listings and references of all X–Y homologous sequences, see Weissenbach and Goodfellow, 1991; Davies *et al.*, 1991). Of particular importance is the sequence DXYS14 described by Cooke *et al.* (1985). These authors have shown by exonuclease digestion with *Bal*31 from the Yp telomere, that DXYS14 lies within 10 kb of the chromosome end and hence defines the limit of Yp. It would appear that the human pairing segment is rich in hypervariable minisatellite sequences, as many of the sequences mapping to this region are associated with repeated DNA and are highly polymorphic (Cooke *et al.*, 1985; Simmler *et al.*, 1985; Cooke and Smith, 1986; Rouyer *et al.*, 1986; Weissenbach *et al.*, 1987); frequent unequal recombination within the pairing region may be responsible for generating a spectrum of length polymorphisms of the VNTR type. All these sequences are strictly homologous with the X and have been mapped to the telomeric region of Xp between Xp22.3-pter by the use of somatic cell hybrids or *in situ* hybridization (Buckle *et al.*, 1985; Cooke *et al.*, 1985; Simmler *et al.*, 1985).

Data from a number of segregation studies have furnished formal proof that there is exchange between the X and the Y in the human pseudoautosomal region. The loci DXYS14 (Cooke *et al.*, 1985), DXYS15, and DXYS17 (Simmler *et al.*, 1985; Rouyer *et al.*, 1986; Weissenbach *et al.*, 1987) and a polymorphic sequence from the *MIC2* gene (Goodfellow *et al.*, 1986) have been used to elaborate a genetic map based on the recombination fractions measured between these loci and that of each locus to sex, the latter being used as a marker to delineate the proximal limit of the pseudoautosomal region on the Y. Figure 1 summarizes the map and recombination fractions that have been measured between each locus in male meiosis.

Several important points emerge from these studies. First, DXYS14 shows almost 50% recombination with sex, thus confirming that an obligatory crossover occurs in this region during male meiosis. Second, the sum of the recombination fractions between each locus is consistent with the recombination fractions of each locus to sex. This suggests that the pseudoautosomal region does not contain recombination hotspots that distort the genetic map. However, a more rigorous assessment of this finding will require the use of further markers to develop a more detailed genetic map. Third, the *MIC2* locus shows only 2% recombination with

Table I
Xp–Yp Homologous Sequences in the Pseudoautosomal Region

HGM locus symbol	Probe name at locus
DXYZ2	708, 113F, 68B
DXYS14	29C1
DXYS15	113B, 113D
DXYS17	601, 602
DXYS20	pDP320, 362A, 29A24
DXYS28	pDP411a
DXYS41	pMR24/1, cY24
DXYS59	68D2
DXYS60	U7A
DXYS62	GMGXY1
DXYS75	P9
DXYF36F1	P117
DXS406, DXYS85	P12
DXS415, DXYS87	P99
DXS419, DXYS86	P131
MIC2 (gene for 12E7 antigen)	Several
CSF2RA (gene for GMCSF receptor α subunit)	
XE7 (gene of unknown function)	

sex, suggesting that *MIC2* is very close to the pseudoautosomal boundary. This accords with theoretical arguments that *TDF* (the testis-determining gene) is adjacent to the boundary (Bengtsson and Goodfellow, 1987). This prediction has been borne out by the studies describing the *SRY* gene (the current candidate for *TDF*), which place it some 5 kb from the pseudoautosomal boundary into the differential part of the Y (Sinclair *et al.*, 1990). Fourth, a comparison to recombination over the pseudoautosomal region in female meiosis suggests that recombination in male meiosis occurs at a rate that is approximately 10-fold greater (Goodfellow *et al.*, 1986; Weissenbach *et al.*, 1987). This is perhaps not surprising as recombination between the sex chromosomes in female meiosis can occur at any point along the X, whereas in males it is normally confined to the pairing region. Alternatively, specific factors may be responsible for enhancing pairing region recombination in male meiosis to guarantee formation of an obligatory crossover. Thus the 50 cm between *MIC2* and DXYS14 on the Y measures some 5 cm on the X. This latter distance is consistent with the lower estimate of 5 Mb for the size of the pseudoautosomal region based on the degree of X–Y synapsis visualized in electron micrographs.

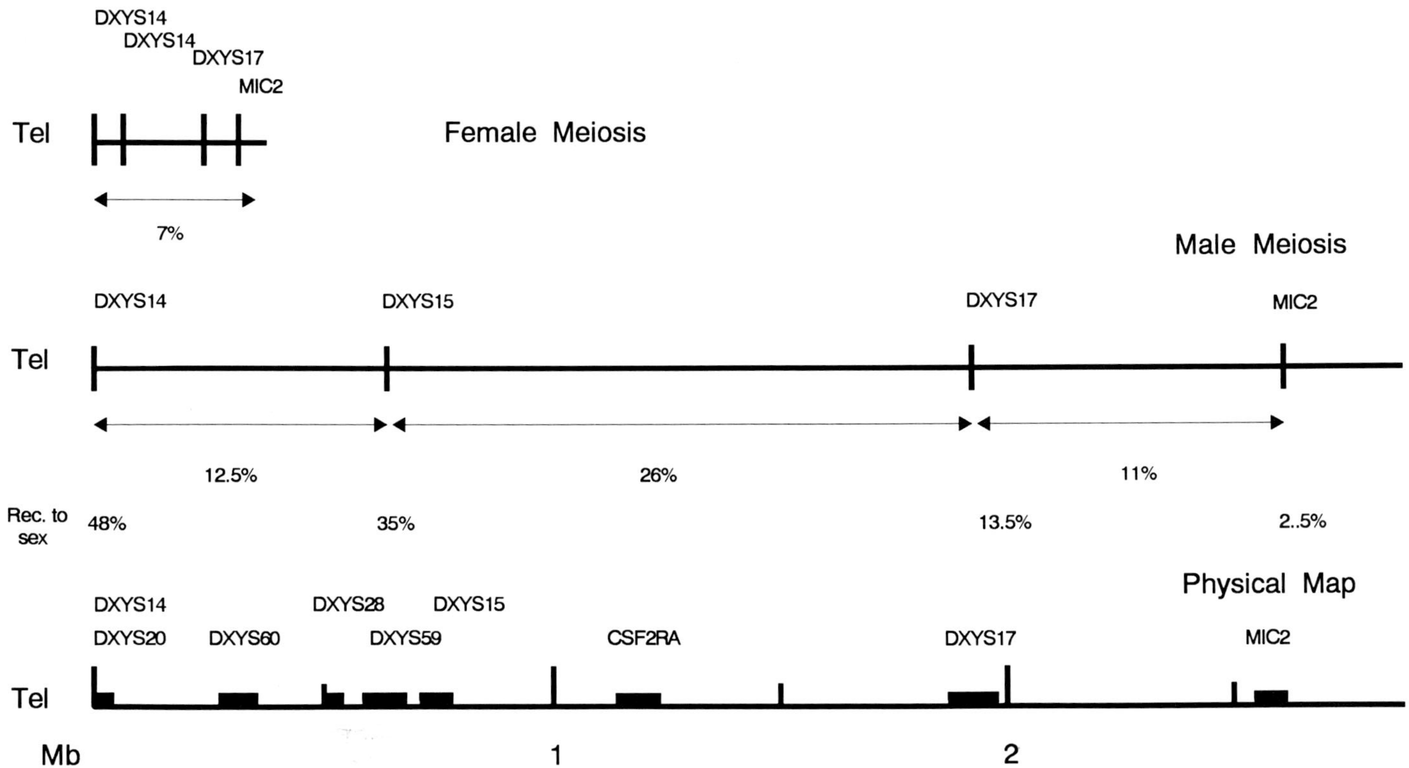

Figure 1 Genetic and physical maps of the human pseudoautosomal region. These maps were compiled from the data presented in Cook *et al.* (1985), Simmler *et al.* (1985), Rouyer *et al.* (1986), Goodfellow *et al.* (1986), Weissenbach *et al.*, (1987), Brown (1988), Petit *et al.* (1988), and Rappold and Lehrach (1988).

B. Physical Analysis

Direct physical analysis has resolved the debate over the size of the pseudoautosomal region. Long-range restriction maps of the pseudoautosomal region have been produced by a number of investigators using the technique of pulsed-field gel electrophoresis and digestion of high-molecular-wieght DNA with a range of rare cutting restriction enzymes. Such studies have enabled not only measurement of the size of the pseudoautosomal region, but also mapping of the location of CpG islands (clusters of undermethylated GC rich rare restriction enzyme sites that occur frequently at the 5′ ends of genes) (see Bird, 1986), which may flag the position of pseudoautosomal genes. All these studies were conducted on DNA prepared from peripheral blood lymphocytes from normal males and females, individuals with multiple X or Y chromosomes, XO Turner patients, or somatic cell hybrids containing only the X or Y chromosome. Thus any differences in methylation at CpG islands on active and inactive X chromosomes could be scored.

The telomeric location of sequences DXYS14 and DXYS20 has proved valuable in mapping the region by partial digestion from the natural end of the chromosome, exploiting these probes as end labels for the fragments produced. Brown (1988), Rappold and Lehrach (1988) and Petit *et al.* (1988) have all used this approach to measure the size of the pseudoautosomal region and to give detailed information on the organization of the subtelomeric region of Xp and Yp. Brown (1988) also employed probes DXYS15 (Simmler *et al.*, 1985), DXYS17 (Rouyer *et al.*, 1986), and p19B proximal to the CpG island at the *MIC2* locus (Pritchard *et al.*, 1987) to map the rest of the region extending to the pseudoautosomal boundary close to *MIC2*. In a very similar analysis, Petit *et al.* (1988) used a larger number of markers including DXYS15, DXYS17, DXYS60 (Rouyer *et al.*, 1986), DXYS28 (Page *et al.*, 1987), DXYS59, p19B, and a second marker, p44A, which lies distal to the island at the *MIC2* locus (Pritchard *et al.*, 1987).

The results from all three studies are highly congruent and a consensus map (aligned to the genetic map) is shown in Fig. 1. The size of the pseudoautosomal region from the telomere to *MIC2* is in the range 2.5–3 Mb, consistent with the minimal region of X–Y pairing observed by Chandley *et al.* (1984). This is indicated by the divergence of the X and Y maps just proximal to the *MIC2* locus. Further, the physical distances between markers reflect the order and recombination distances of the genetic map. Somewhat surprising, however, is the high density of CpG islands found in the subtelomeric region of both the X and Y, covering about 500 kb pairs. The islands in this region are very closely clustered, implying that there may be a high density of X–Y homologous genes close to the telomere. However, Petit *et al.* (1988) suggest that this could reflect sequence organization and composition close to the telomere and that many of these islands may not be associated with genes but hypomethylated DNA, which is perhaps a property of telomeric domains. Ellis and Goodfellow (1989) advanced the proposal that the CpG rich subtelomeric domain may function in X–Y pairing or perhaps nuclear membrane attachment. Clustering of CpG islands is also evident in more proximal locations in the central segment of the pseudoautosomal region and around the boundary near the *MIC2* locus. The spacing between these islands is closer to that expected (Bird, 1986; Brown and Bird, 1986) for the genome as a whole and hence are more likely to be associated with genes. The similarity in distribution of pseudoautosomal islands in Turner patients and individuals with multiple X chromosomes indicates that their methylation status on active and inactive X chromosomes is the same over the pairing segment. It has been shown that CpG islands of genes that undergo X inactivation are methylated on the inactive X, but not on the active X (Keith *et al.*, 1986; Yen et al., 1984). Hypomethylation on the inactive X at the *MIC2* CpG island (Goodfellow *et al.*, 1987a and b) correlates with escape from X inactivation. Therefore, it is likely that genes associated with other pseudoautosomal islands showing hypomethylation also escape X inactivation. This line of reasoning has fuelled speculations by the authors responsible for these studies that pseudoautosomal CpG islands may map the locations of genes involved in suppressing the Turner phenotype (discussed in the next section; see also Chapters 10 and 14, this volume).

Definitive positioning of the pseudoautosomal boundary has been achieved by chomosome walking from the *MIC2* locus and Southern analysis of male, female, and X and Y somatic cell hybrid genomic DNA with walk clones (Ellis *et al.*, 1989). The boundary occurs some 85–90 kb proximal to the *MIC2* CpG island. Sequencing of DNA fragments spanning the boundary has revealed that it is abrupt, being defined by a 300-bp Alu sequence present on the Y but absent from the X, where strict X–Y sequence homology breaks down. Proximal to this point the X and Y have diverged by 23% over a 225-bp region, beyond which all sequence homology is lost. Ellis *et al.* (1989) argue that this indicates the existence of an earlier boundary on the Y and calculate that the present boundary was created some 25 million years ago. A study of polymorphisms on the X and Y at the boundary region has shown far greater variation on the X between and within races than on the Y, thus indicating that the present X boundary is of considerably earlier origin (Ellis *et al.*, 1990).

C. Pseudoautosomal Genes

Observations on karyotype–phenotype correlations in patients with Turner syndrome have stimulated speculations that genes involved in this condition reside in homologous regions in or adjacent to the pairing segment. Deletions of the X chromosome short arm leading to monosomy of part or all of Xp are associated with Turner stigmata. It has been suggested by Ferguson-Smith (1965) that escape from X inactivation of one or more key X-linked genes ensures the necessary double dose of expression required to suppress the Turner phenotype. Ferguson-Smith (1965) developed this line of thought further by proposing that in normal males the second dose is provided by functional homologous genes on the Y chromosome. Burgoyne (1982) argued subsequently that these loci are located in the pseudoautosomal region precisely because it escapes X inactivation and is a region of X–Y homology.

It is tempting to accept the suggestion that CpG islands at homologous pseudoautosomal locations on the X and Y mark the positions of genes involved in Turner's syndrome. There is, however, strong evidence that anti-Turner loci do not lie in the pseudoautosomal region (see Chapter 10 and discussion below).

First, several reports exist in the literature of 46,Y,t(X;Y)(p22.3;q11) males deleted for either the *Xg* or the *STS* loci (Akesson *et al.*, 1980; Alderdice *et al.*, 1983; Boyd *et al.*, 1981; Ferguson-Smith *et al.*, 1982; Metaxotou *et al.*, 1983; Tiepolo *et al.*, 1977; reviewed in Shapiro, 1985). Both *Xg* and *STS* show X-linked inheritance in humans and hence lie outside the pseudoautosomal region. Both genes have been mapped to distal Xp between Xp22.3-pter (Ferguson-Smith *et al.*, 1982; Tiepolo *et al.*, 1980) and hence their loss in these patients with terminal deletions is indicative of deletion of the entire X pseudoautosomal region. These patients do not have Turner stigmata and in consequence, this argues against a role for the pseudoautosomal region in suppressing the Turner phenotype. The association of these deletions with short stature could indicate the existence of genes concerned with height in the pseudoautosomal segment or the region proximal to it on the X. In this respect, Ogata *et al.* (1992) and Henke *et al.* (1991) have provided evidence from karyotype–phenotype correlations for a stature locus or loci in the mid to distal region of the pairing segment.

Second, a number of cases have been reported of XYp− females with Turner's stigmata (Rosenfeld *et al.*, 1979; Magenis *et al.*, 1984: Disteche *et al.*, 1986), thus indicating that loci on Yp are implicated in the Turner phenotype. Affara *et al.* (1987) demonstrated that the deletion in the case reported by Magenis *et al.* (1984) was caused by an abnormal X–Y interchange and that the patient carries two X pseudoautosomal regions—one of paternal origin on her Y chromosome, and one on the maternal X. Levilliers *et al.* (1989) have also demonstrated this in a further two XYp− females. These cases are important because they demonstrate that loss of loci on the Y proximal to the pseudoautosomal region results in Turner stigmata despite the presence of two pairing segments. One would therefore predict that X–Y homologous genes are present in Yp proximal to the pseudoautosomal region. Escape from X inactivation is not solely the preserve of the pseudoautosomal region. The *Xg* locus (Race and Sanger, 1975), the X–Y homologous genes (*a*) *STS* (Shapiro *et al.*, 1979), (*b*) *KAL* (Kallmann syndrome) (Franco *et al.*, 1991; Legouis *et al.*, 1991), (*c*) *ZFX* (Schneider-Gadicke *et al.*, 1989), (*d*) the ribosomal protein S4 gene *RPS4X* (Fisher *et al.*, 1990), and (*e*) the gene *GS1* whose function is unknown (Yen *et al.*, 1992) and the human equivalent (A1S9T) on Xp of a mouse cell cycle gene (Brown and Willard, 1989) have been shown to escape X inactivation. These loci all lie outside the pseudoautosomal region, thus illustrating that candidate loci for the Turner phenotype on the X that are not subject to dosage compensation could equally well be positioned in regions of Xp proximal to the pseudoautosomal segment.

Genetic linkage studies with DXYS14 have provided weak evidence for a pseudoautosomal gene implicated in predisposition to schizophrenia (Collinge *et al.*, 1989; see also Crow, 1988). At present this assignment is tentative and requires confirmation by more extensive linkage analysis.

The first functional locus shown to escape X inactivation that mapped unequivocally to the pseudoautosomal region was the *MIC2* gene whose product (a surface antigen) is detected by the 12E7 monoclonal antibody (reviewed by P. J. Goodfellow *et al.*, 1987a; P. N. Goodfellow *et al.*, 1985). This was achieved by showing recombination of *MIC2* with *TDF*, through the use of somatic cell hybrids to demonstrate escape from X inactivation (Goodfellow and Tippet, 1981; Goodfellow *et al.*, 1983, 1984), cloning of the cDNA sequence (Darling *et al.*, 1986), and its use in *in situ* hybridization (Buckle *et al.*, 1985). Four other cloned genes have been assigned to the pseudoautosomal region: the gene *CFS2RA*, which encodes the α subunit of the GM-CFS receptor (Gough *et al.*, 1990); a gene of unknown function designated *XE7* (Ellison *et al.*, 1992a,b); the gene *ANT3* (an ADP/ATP translocase), which also escapes X inactivation (Schiebel *et al.*, 1993); the gene *HIOMT* (hydroxyindole-O-methyltransferase) (Yi *et al.*, 1993) which catalizes the final reaction in the synthesis of melatonin and is a possible candidate gene for neuropsychiatric diseases that appear to be associated with pseudoautosomal loci. Physical mapping has placed

CFS2RA at about 1200 kb from the telomere (see Fig. 1) and adjacent to a CpG island (Rappold *et al.*, 1992). The physical location of *XE7* is unknown. However *ANT3* and *HOIMT* have been mapped proximal to CFS2RA and distal to *MIC2*.

A further pseudoautosomal locus has been proposed based on the quantitative polymorphism in the level of expression of the *MIC2* gene associated with alleles at the *Xg* locus (Goodfellow and Tippett, 1981; Tippett *et al.*, 1986). Goodfellow *et al.* (1987a,) have called this locus *XGR* and postulate that it is polymorphic with different alleles acting in cis to up or down regulate coordinately *XGa* and *MIC2*. No DNA sequences corresponding to such a locus have been isolated.

III. X–Y Homology in the Differential Region of the Y Chromosome

A. Deletion Analysis of the Y Chromosome

Models of the structure of the differential portion of the Y have been elaborated from deletion maps developed by analyses with cloned Y DNA sequences. As mentioned earlier, this has relied heavily on the exploitation of abnormal Y chromosomes defined by cytogenetic analysis. The greater resolution offered by molecular analysis has enabled a more precise assessment of the distribution of X–Y homologous sequences outside the pseudoautosomal region. In this section we illustrate the type of map of Yp and Yq we have derived as a framework for examining the locations of X–Y homologies elucidated by ourselves and others (Affara *et al.*, 1986a,b, 1987, 1989; Bickmore and Cooke, 1987; Bishop *et al.*, 1984; Cooke *et al.*, 1984; Geldwerth *et al.*, 1985; Koenig *et al.*, 1984, 1985; Kunkel *et al.*, 1983; Page *et al.*, 1982, 1983; Wolfe *et al.*, 1984; Fraser *et al.*, 1987; Yen *et al.*, 1988; Petit *et al.*, 1990; Bailey *et al.*, 1992; Lambson *et al.*, 1992).

Mapping of the short arm is based largely on the variable extent of Yp found transferred to the paternal X in XX males as a consequence of abnormal X–Y interchange during male meiosis (Ferguson-Smith, 1966). Several laboratories have used this approach (Vergnaud *et al.*, 1986; Affara *et al.*, 1986b, 1987; Koenig *et al.*, 1985; Müller *et al.*, 1986; Page, 1986; Pritchard *et al.*, 1987; Vollrath *et al.*, 1992). In our own work we have supplemented the XX male panel with structural abnormalities of the Y that provide additional breakpoints on Yp (Ferguson-Smith *et al.*, 1987). These are summarized schematically in Fig. 2 and consist of a male

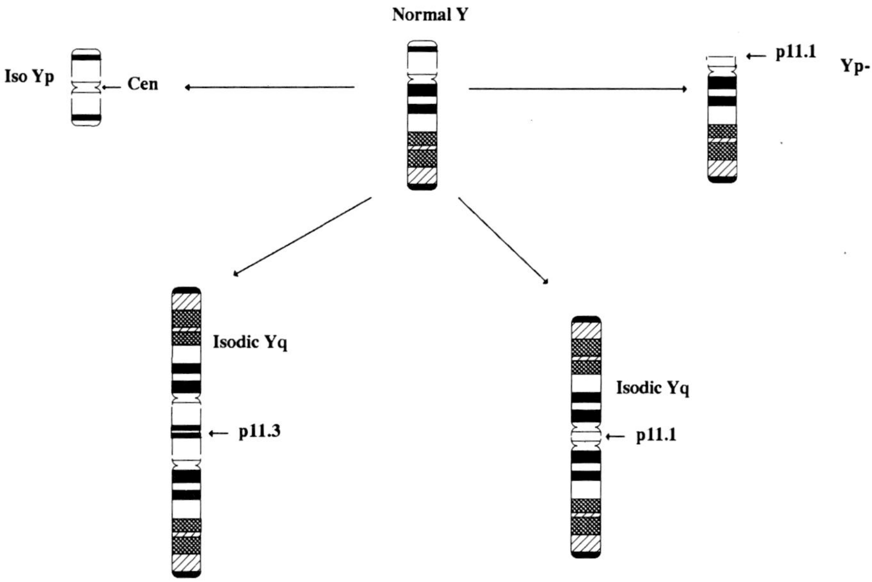

Figure 2 Schematic representation of chromosomes contributing to the Yp deletion panel used to derive the map presented in Fig. 3.

possessing a monocentric i(Yp) (isochromosome Yp), a female with two dicentric i(Yq) chromosomes (breakpoint in Yp11.1), a male with a dicentric i(Yq) chromosome (breakpoint in Yp11.3), the XYp− female described by Magenis *et al.* (1984) (breakpoint in Yp11.1), and an XO male with a fragment of Y chromosome origin. Figure 3 summarizes an updated map of Yp that we developed from our own panel (including DNA probes isolated by other groups) and illustrates the division of the short arm into 20 intervals; X–Y homologous sequences are boxed and sequences detecting Yp–Yq homologies are marked with a dot.

An analogous approach has proved valuable in defining a deletion map for Yq (Ferguson-Smith *et al.*, 1987; Bardoni *et al.*, 1991; Ma *et al.*, 1992; O'Reilly *et al.*, 1992; Vollrath *et al.*, 1992). Figure 4 shows a schematic representation of the Y chromosomes with breakpoints on Yq contributing to our panel. They consist of three categories of cytogenetically distinguishable dicentric i(Yp) chromosomes with breakpoints in Yq11.21, Yq11.22, and Yq11.3; an X–Y translocation where the derivative X carries the region Yq11.21-qter; a Y-5 translocation where the derivative 5 carries Ypter-Yq11.21; a Y-14 translocation where the derivative 14 carries Yq11.22-qter; the monocentric i(Yp) chromosome described above. Figure 5 illustrates the current map of Yq and demon-

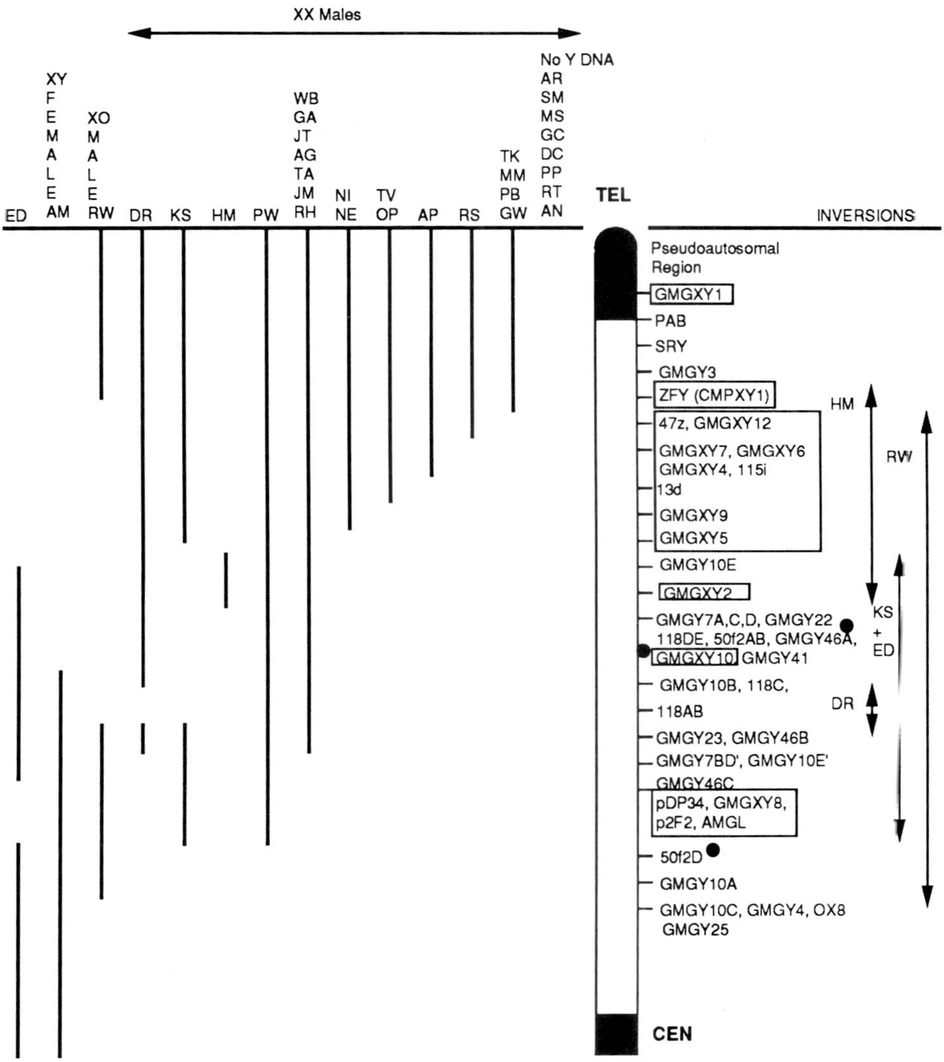

Figure 3 Deletion map of the Y short arm. The boxed sequences are X–Y homologous sequences and the dots mark Yp–Yq homologuous regions.

strates that 14 intervals can be resolved. Though not indicated on the figure, several of the Yq sequences detect homologies with autosomes. Because various maps of the Y chromosome developed by different groups have been based on different combinations of probes and aberrant Y chromosomes, it has not been possible to create a completely

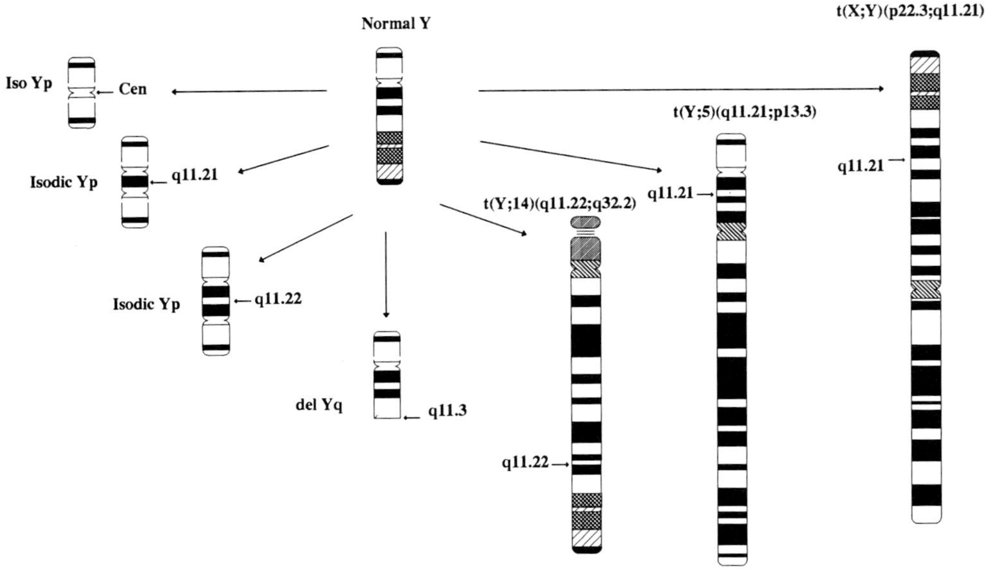

Figure 4 Schematic representation of chromosomes contributing to the Yq deletion panel used to derive the map presented in Fig. 5.

integrated deletion map of the Y. However, the overlapping clone map of the euchromatic portion of the Y based on yeast artificial chromosomes (YACs) ordered by a series of *STS* markers (Foote *et al.*, 1992) will provide a means of cross-referencing the different maps. This will aid the further mapping of X–Y homologous regions on the Y chromosome.

As in other studies (Affara *et al.*, 1986a; Bishop *et al.*, 1984; Koenig *et al.*, 1985; Geldwerth *et al.*, 1985) we used a panel of somatic cell hybrids and lymphoblastoid cell lines carrying deleted X chromosomes to delineate Y homologous regions on the X. Table II summarizes the mapping data on the X for X–Y homologous probes shown in Figs. 3 and 5 and sequences mapped by other studies. Excluding the pseudoautosomal region, the patterns of homology point to 10 further distinct sequence relationships between the X and the Y chromosomes. These are (*a*) distal Yp11.1–Xp21.2-p22.1 (*ZFY*/*ZFX*); (*b*) distal Yp11.1–Xq13 (*RPS4X*/*RPS4Y*); (*c*) distal Yp11.1–Xq21.1-q21.3; (*d*) proximal Yp11.1–Xp22.3-pter; (*e*) proximal Yp11.1–Xp21.2-22.1 (*amelogenin*); (*f*) proximal Yp11.1–Xq13; (*g*) proximal Yp11.1–Xq21.1-q21.3; (*h*) Ycen-q11.21–Xp22.3-pter; (*i*) Yq11.22-q11.3–Xq28; (*j*) Yqter–Xq28. In the following sections we discuss each block of homology and the data contributed by other groups.

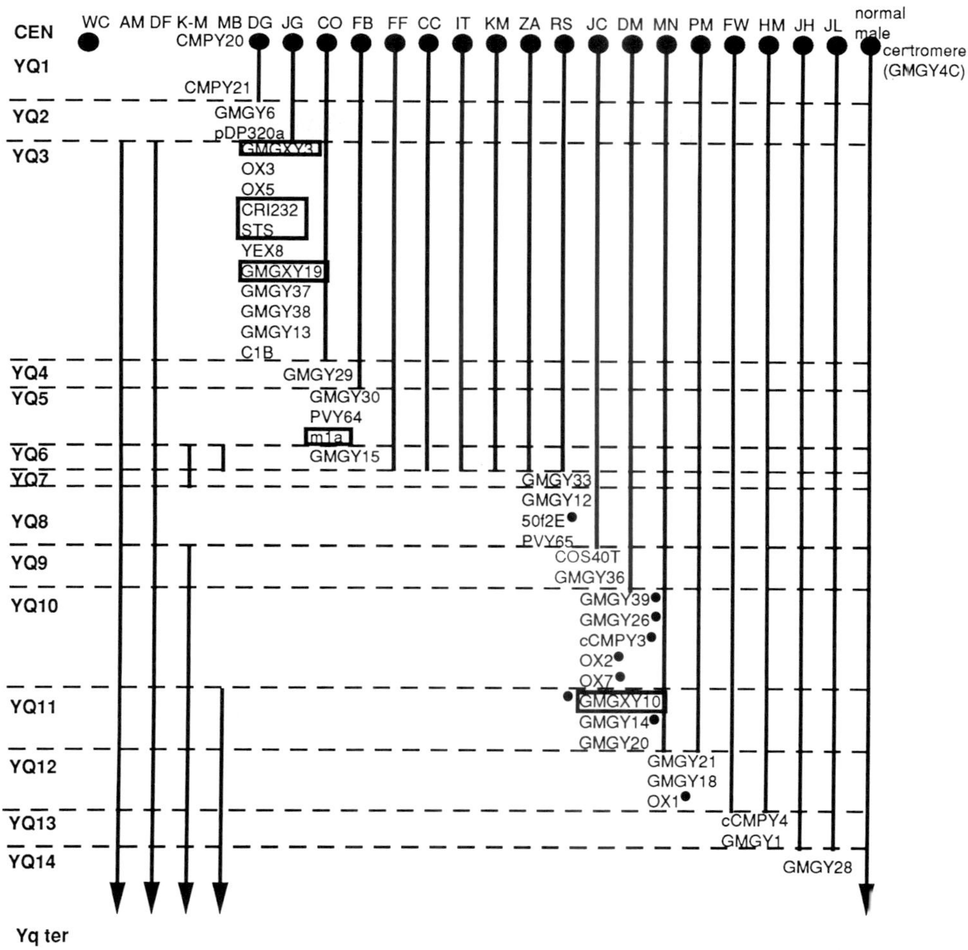

Figure 5 Deletion map of the Y long arm. X–Y homologous sequences are boxed and Yp–Yq homologs are marked with a dot.

B. The Yp–Xq Homologies

These homologies account for the majority of X–Y homologous sequences described. Table III lists some of the sequences contained in this category that have been useful in defining these homologies (see also Weissenbach and Goodfellow, 1991; Davies *et al.*, 1991). The archetypical

Table II
Distribution of X–Y Homologous Sequences on the X Excluding the Pseudoautosomal Region

	Region on the X				
	Xq22.3-pter	Xp21.2-p22.1	Xq13	Xq21.1-q21.3	Xq28
	DXYS29	ZFX	RPS4X	DXYS5	DXYS37
	DYS74	AMG	DXYS22	DXYS21	DXYS61
Probe	DXS31			DXYS30	DXYS64
	STS			DXYS31	
	DXS278			DXYS32	
	DXYS62			DXYS33	
	DXS69			DXYS34	
	DXF28S1			DXYS35	
	KAL			DXYS7	
	GS1			DXYS8	
				DXYS1	
				DXYS49	
				DXYS38	
				DXYS39	
				DXYS42	
				DXYS3	
				DXYS4	
				DXYS6	
				DXYS7	
				DXYS8	
				DXYS9	
				DXYS46	
				DXYS67	
				DXYS68	
				DXYS12	
				DXYS13	
				DXYS25	
				DXYS27	

sequence DXYS1 defining this pattern of homology was first described by Page *et al.* (1982) and mapped to Yp and Xq13-q21 by *in situ* hybridization and somatic cell hybrid studies (Page *et al.*, 1983). Comparative restriction analysis on the X and Y over a distance of 36 kb indicated 99% homology over the region (Page *et al.*, 1983). Subsequently other investigators have described many more sequences belonging to this group and have been able to subdivide the sequences into further classes of homology (Affara *et al.*, 1986a, 1987; Bishop *et al.*, 1983, 1984; Geldwerth *et al.*, 1985; Koenig *et al.*, 1985; Wolfe *et al.*, 1984; Fisher *et al.*, 1990; Lambson *et al.*, 1992).

Table III
X–Y Homologous sequences Mapping between Yp and Xq

HGM locus symbol	Probe name at locus
DXYS1	pDP34, DP31
DXYS2	7b
DXYS3	22b
DXYS4	p1
DXYS5	47z, 47a
DXYS6	p16
DXYS7	13d
DXYS8	p115, pDP61
DXYS9	p8j
DXYS12	st25-1, st25-2
DXYS13	pTAK2
DXYS21	GMGXY2
DXYS22, DYS25	p2f(2)
DXYS23	pDP132
DXYS25	p75/78
DXYS27	cY59γ
DXYS30	GMGXY4
DXYS31	GMGXY5
DXYS32	GMGXY6
DXYS33	GMGXY7
DXYS34	GMGXY8
DXYS35	GMGXY9
DXYS38	pMR3/2
DXYS39	pMR2/1
DXYS36	GMGXY12
DXYS42	pDP307
DXYS46	MJO-23
DXYS49	FR80-II
DXYS67	QST-54
DXYS68	RX-50
DXYS73	87-27

The majority of the sequences fall into the first class. These map to distal Yp11.1 and to the region Xq21.1-q21.3 on the X as defined by somatic cell hybrid studies and exhibit the same high degree of conservation between the X and Y as DXYS1 (Lambson *et al.*, 1992). In many cases, presence in XX males confirms their location on Yp (Affara *et al.*, 1986b,1987; Koenig *et al.*, 1985; Vergnaud *et al.*, 1986). Some of these sequences detect polymorphisms on the X and may therefore be useful in linkage studies of proximal Xq (Page *et al.*, 1982; Koenig *et al.*, 1985; Mitchell, 1989).

The second class of Yp–Xq homology is categorized by sequences such as DXYS1 and DXYS34 (GMGXY8), which map to proximal Yp11.1 and Xq21.1-q21.3. Our deletion map of Yp illustrates the division of the Yp–Xq homologous sequences into two blocks: the small block located in proximal Yp and a second larger block in distal Yp just proximal to *ZFY*. The blocks are separated by a region rich in Y-specific probes, several of which detect multiple loci on the Y including the long arm. It is not clear whether these sequences originally formed one contiguous block on Yp and have subsequently become separated as a result of paracentric inversions in the Y short arm. Anomalous sequence order in some XX males (see Fig. 3) supports this possibility. This may be an explanation for the location in proximal Yp11.1 of the sequences DXYS1 and GMGXY8, which share a common evolutionary history with the main group of sequences in distal Yp11.1 (see below). The independent evolutionary history (and pattern of homology) of the other two sequences forming the proximal Yp11.1 block (*amelogenin* and DXYS22-p2F2) suggests that it has been created by at least two further evolutionary events (see below).

The third class is defined by the sequence DXYS22, which maps to the proximal block in Yp11.1 but detects a homology in Xq13.

The fourth class of homology is defined by the conserved X–Y homologous ribosomal protein gene sequence *RPS4Y(X)* (Fisher *et al.*, 1990). On the Y this gene maps just distal to *ZFY* and is therefore separated from the main block of Yp–Xq homology found in distal Yp11.1. On the X, the gene *(RPS4X)* maps to Xq13. Both X and Y homologs of this gene are functional and the X locus has been shown to escape X inactivation. Fisher *et al.* (1990) have suggested that *RPS4Y* is a candidate gene for the anti-Turner function exhibited by Yp (see also Watanabe *et al.*, 1993). However, this is inconsistent with its location on Xq since the Turner phenotype is associated primarily with deletions of Xp.

From the yeast artificial chromosome contig of Foote *et al.* (1992) the size of the distal block on Yp can be estimated to cover approximately 5 Mb. On the X, the size is unknown and at present we have no indication that any further genes are encoded by these X–Y homologous sequences.

C. The Yp–Xp Homologies

Outside the pseudoautosomal region only three further Yp–Xp homologies have been reported.

The first of these (Page *et al.*, 1987) is a highly conserved, X–Y homologous sequence that encodes a zinc finger transcriptional activator protein. DNA sequencing indicates >95% homology between the X

(*ZFX*) and the Y (*ZFY*) sequences (Schneider-Gadicke *et al.*, 1989; Lau and Chan, 1989; Palmer *et al.*, 1990; North *et al.*, 1991). The two genes exhibit an extraordinary degree of evolutionary conservation, being found on both sex chromosomes in all eutherians examined. This suggests an important function. Somatic cell hybrids carrying X deletions and *in situ* hybridization have placed the X gene in Xp21.2-p22.1 (Affara *et al.*, 1989; Müller and Schempp, 1989; Schneider-Gadicke *et al.*, 1989) and Schneider-Gadicke *et al.* (1989) have demonstrated that it escapes inactivation. On the Y, the gene has been physically mapped to a position approximately 260 kb proximal to the *MIC2* CpG island (Verga and Erickson, 1989).

Early hopes that *ZFY* was the pivotal testis-determining gene have receded in the light of strong evidence to the contrary and the isolation of the *SRY* gene (Sinclair *et al.*, 1988, 1990; Koopman *et al.*,1989, 1990, 1991; Gubbay *et al.*, 1990). The function of *ZFY* and *ZFX* is unclear, but a number of possibilities exist: they may have an adjuct but secondary role in sex determination; they could be involved in germ cell proliferation. However, the appropriate map location of *ZFY* on Yp and escape from X inactivation by *ZFX* have prompted the obvious speculation that *ZFX* and *ZFY* may be involved in suppressing the Turner phenotype (Burgoyne, 1989). However, the XY female patient studied by by Fisher *et al.* (1990) rather counters this possibility as this patient is deleted for *ZFY* but does not exhibit Turner stigmata.

The second Yp–Xp homology is defined by the *amelogenin* gene, which is involved in forming tooth enamel and is expressed in the developing tooth bud (Lau *et al.*, 1989; Nakahori and Nakagome, 1989). The Y-linked gene for amelogenin appears to be intact and probably functional, although it may be expressed at only one-tenth of the level of the X-linked gene (Nakahori and Nakagome, 1989; Nakahori *et al.*, 1991a; Salido *et al.*, 1992). Comparison of the exon sequences between the Y- and the X-linked genes shows that the extent of homology varies between 76 and 100%, with most exons showing 90–95% homology (Salido *et al.*, 1992). Thus the X and Y sequences are highly conserved but to a lesser extent than observed for *ZFX*/*ZFY*.

By deletion mapping, Petit *et al.* (1989) placed *amelogenin* proximal to *STS* and analysis of somatic cell hybrids carrying deleted X chromosomes has placed the X homolog in the interval carrying *ZFX* between Xp22.1 and Xp21.2 (Bailey *et al.*, 1992). Several reports have shown the Y homolog in proximal Yp (Fincham, 1991; Nakahori *et al.*, 1991b; Bailey *et al.*, 1992; Salido *et al.*, 1992) and from the deletion map shown in Fig. 3, it can be seen that the Y homolog maps into the same interval as the proximal block of Yp–Xq homologous sequences described above. A number of YACs linking *amelogenin* on the Y to these Yp–Xq homologous sequences (DXYS1 and GMGXY8) have been isolated, indicating that they are very closely associated (N. A. Affara, unpublished results).

The third group of Yp–Xp homology lies between sequences just proximal to the X-situated *MIC2* gene (DXS431 and DXS432) and Y sequences (DYS137 and DYS138) present in proximal Yp (Levilliers *et al.*, 1989; Petit *et al.*, 1989, 1990). These are partial homologies and may have originated from the pseudoautosomal region by means of a paracentric inversion on Yp.

D. The Xp–Yq Homologies

Sequences conforming to this pattern of homology are summarized in Table IV. They are far fewer in number than the Xq–Yp sequences but have been studied in considerable detail. Kunkel *et al.* (1983) and Koenig *et al.* (1984) reported the first sequences (DXS69 and DXS31 (or M1a) respectively) showing homology between Xp22.3-pter and Yq. These were judged to show 80% homology between the X and the Y, suggesting that the chromosomes have diverged to a greater extent in this region.

A number of other sequences fitting this pattern of homology have been described. The steroid sulfatase gene (Yen *et al.*, 1987, 1988; Fraser *et al.*, 1987), the Kallmann syndrome gene *KAL* (an adhesion molecule involved in neuronal migration) (Franco *et al.*, 1991; Legouis *et al.*, 1991), DXYS29 (GMGXY3), DYS74 (GMGXY19) (Affara *et al.*, 1986a, 1987), DXF28S1 (Petit *et al.*, 1990), and DXS278 (Knowlton *et al.*, 1989) all have homologous sequences in proximal Yq. Another gene sequence (*GS1*, of unknown function) has been mapped to Xp22.3–pter and has a homolog

Table IV
X–Y Homologous Sequences Mapping between Xp and Yq

HGM locus symbol	Probe name at locus
DXYS29	GMGXY3
DXS31, DYS22	M1A
DXS69	p71-7a
DYS74	GMGXY19
DXS278	CRI-S232
DXF28S1	IP415
STS, STSP (steroid sulfatase gene and pseudogene)	Several probes
KAL, KALP (pseudogene on the Y)	
GS1 (not yet mapped on the Y—function unknown)	

on the Y chromosome (Yen *et al.*, 1992). However, it has not been shown whether the Y homolog is located in proximal Yq. In the case of *STS*, only the 5′ half of the gene is present on the Y and hence the Y sequences form a nonfunctional pseudogene (Frazer *et al.*, 1987; Yen *et al.*, 1987, 1988). This locus has been termed *STSP*. Similarly, the Y homolog of the *KAL* gene (*KALP*) has suffered exon deletions and frameshift mutations, thus rendering it a nonfunctional, nonprocessed pseudogene (Castillo *et al.*, 1992; Incerti *et al.*, 1992). Preliminary findings indicate that the Y homolog of *GS1* (like *STS*) has lost the 3′ half of the gene and it too is nonfunctional (Yen *et al.*, 1992). Sequence comparison between exons of the X and Y copies of the *STS* and *KAL* genes indicates that the homology varies between 85 and 95%, consistent with the degree of divergence observed for DXS31.

DXYS29, DYS74, DXS278, and *GS1* are commonly deleted in the majority of boys with *STS* deficiency caused by interstitial deletions of the X that remove the *STS* gene (Gillard *et al.*, 1987; Ballabio *et al.*, 1987, 1989a; Newman *et al.*, 1990). This implies that the sequences are closely clustered on the X. The absence of any additional clinical manifestations in deletion patients with *STS* deficiency compared to nondeletion cases suggests that the *GS1* gene is not important clinically.

Physical mapping around the steroid sulfatase gene by pulsed-field gel electrophoresis (Ross *et al.*, 1990; Newman *et al.*, 1990; Petit *et al.*, 1989) and deletion mapping (using DNA from patients with STS deficiency, Kallmann's syndrome, chondrodysplasia punctata, and X-linked mental retardation) (Ballabio *et al.*, 1989b; Petit *et al.*, 1989) has provided information about the organization of these sequences on the X and the probable location of the genes responsible for these syndromes. *STS*, DXYS29, DYS74, and DXS278 are linked on the same 750-kb *Sfi*I fragment therefore confirming the clustering of these loci (Ross *et al.*, 1990; Newman *et al.*, 1990). Figure 6 shows the X physical map. DXS31 is separated from this cluster in a distal location and has not yet been linked to the main group of X–Y homologous sequences. From the physical mapping studies of Petit *et al.* (1989), DXS31 is approximately 3 Mb from the pseudoautosomal boundary on the X and *STS* between 3 and 7 Mb. Thus DXS31 could be as much as 4 Mb distal to the main block of X–Y homologous probes.

The map of the X block of homology shown in Fig. 6, which incorporates a number of other X probes in the region, covers 3.5 Mb and shows that the X–Y homologous sequences do not form a contiguous block but are interspersed with X-specific sequences such as DXS237 (GMGX9). DXS278 detects at least three loci over this distance, showing that this sequence is an interspersed repeat. The map indicates the location of at

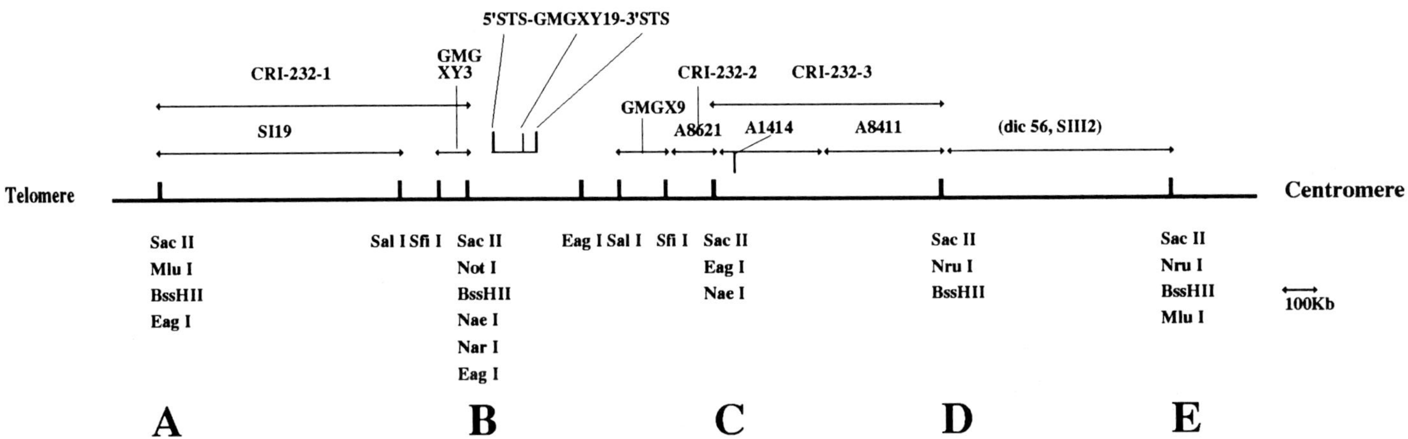

Figure 6 Physical map on the X of the Xp22.3-pter/Yq homology.

least five CpG islands. The *KAL* gene has been shown to be associated with island D, and the *GS1* gene with island B, with transcription being from the island toward the telomere. It is not known whether the remaining CpG islands define the position of further gene sequences.

The deletion map of Yq presented in Fig. 5 shows that these sequences are clustered together on the Y chromosome. *STSP*, DXS278 (CRI232), and DYS74 map to the same interval and DXYS29 is in the immediately adjacent proximal interval. On our map, DXS31 (M1A) lies two intervals distal to this block, correlating with its physical separation from these sequences on the X. These findings are confirmed by the YAC map of Foote *et al.* (1992), which additionally indicates the association of the *KALP* locus with the Y cluster. The precise location of Y sequences homologous to *GS1* is still to be established.

Two points are noteworthy. First, the frequent class of X and Y translocation (reviewed by Shapiro, 1985), involving breakpoints in Xp22.3-pter and proximal Yq, often deletes the X block of homologous sequences but retains the Y homologs (Ballabio *et al.*, 1989c). Ballabio *et al.* (1989c) have proposed that this may be provoked by an illegitimate recombination between Xp and Yq promoted by sequence homology. Second, all three genes (*KAL, STS,* and *GS1*) escape X inactivation and this may indicate a common evolutionary history. With the exception of *amelogenin,* the general observation that every X-linked gene having a Y homolog escapes X inactivation (irrespective of its Y functional status) may reflect the origin of the sex chromosomes from a common ancestral pair of homologs (see below).

E. The Yq–Xq Homologies

A small number of sequences demonstrate Yq–Xq homology. These appear to fall into two groups. Cooke *et al.* (1984) isolated the sequence DXYS61Y from a Y flow-sorted library that mapped to distal Xq and Yq. Comparison of the X (DXYS61X) and Y sequences over a 1-kb distance and restriction maps over a 47-kb distance has revealed complete identity (Bickmore and Cooke, 1987). This high degree of conservation suggests that this homology is of recent origin, perhaps later than the Yp–Xq homology characterized by DXYS1 (see below). The localization on the X has been narrowed down to Xq28 and shown to be very closely linked to a second sequence (DXYS64X) exhibiting this pattern of homology (Arveiler *et al.*, 1989). In both cases the Y homology has been shown to be close to the Yq telomere.

A further sequence DXYS37Y (GMGXY10) (Affara *et al.*, 1987) has been isolated which maps to Xq28 and to Yq. The sequences detected on the X by DXYS37, DXYS61, and DXYS64 are linked on a YAC contig and hence are very close together in Xq28 (Jon Warner, personal communication). From the deletion map of Yq in Fig. 5 it can be seen that DXYS37 is located in the distal portion of the Yq euchromatin and is thus distinct from DXYS61 and DXYS64 in its pattern of homology. This is supported by studies on primates (discussed below). From the deletion maps in Figs. 3 and 5 it can be seen that DXYS37 (GMGXY10) maps to the region of homology between the long and short arms of the Y chromosome.

Figure 7 summarizes the patterns of X–Y homology documented in the above sections.

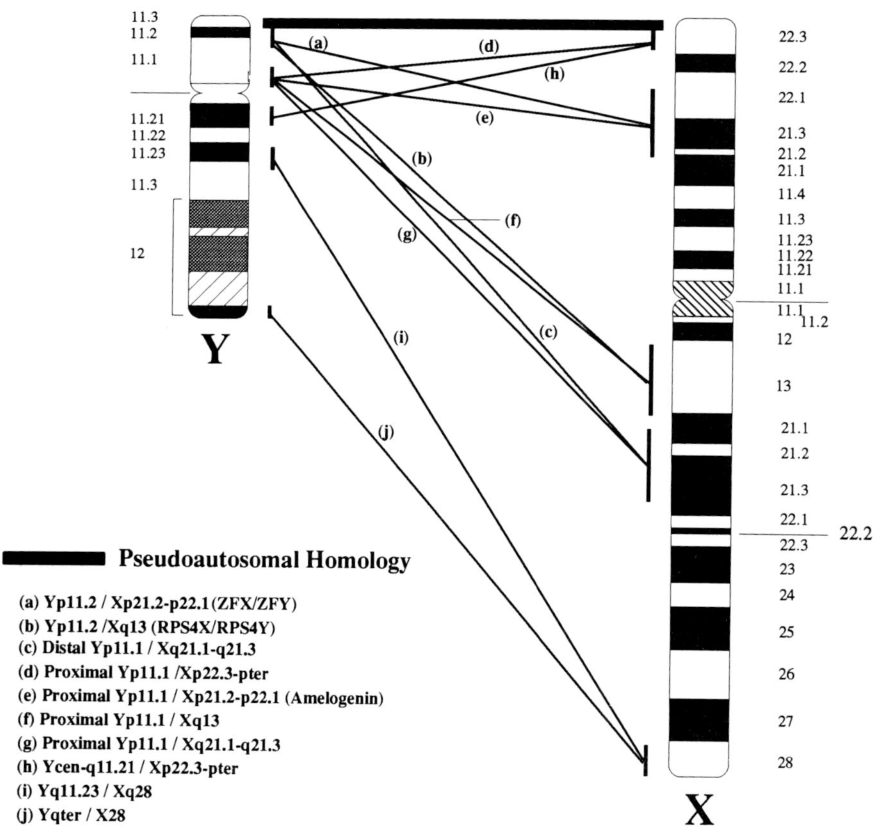

Figure 7 Summary diagram of X–Y homologies.

IV. Evolution of Sex Chromosome Homologies

It is widely accepted that the X and Y have evolved from ancestral homologs (Ohno, 1967; also see Chapter 6) and, therefore, it is reasonable to expect that X–Y homologous sequences reflect remnants of this past. The investigation of the pseudoautosomal region in primates other than humans has been very limited but confirms this prediction. By use of human pseudoautosomal probes, Weber *et al.* (1989) demonstrated X–Y exchange in the chimpanzee. *In situ* hybridization and Southern analysis have shown conservation of the interspersed human pseudoautosomal repeat DXYZ2 (probe 113F) in the great apes and as far back as the prosimians (Weber *et al.*, 1987; Schempp *et al.*, 1989). This is perhaps not surprising, as one would expect pseudoautosomal sequences to represent the most highly conserved group of X–Y homologous sequences because they undergo frequent recombination and are presumably instrumental in the essential function of X–Y pairing and segregation at meiosis.

Most primate studies have concentrated on homologies outside the pseudoautosomal region and on the observation that homologies exist that occur between inappropriate arms of the X and Y. In the following section we discuss what light these homologies, as studied in other primates, can throw on our understanding of sex chromosome evolution. Table V summarizes the evolution of each class of X–Y homology as it is currently understood on the basis of those X–Y homologous probes that have been applied to the analysis of primate species.

A. X–Y Homologies Showing Limited Divergence

By examining the pattern of X–Y linkage in different primate (and other mammalian) species, the first appearance of a sequence on both the X and the Y chromosomes can be monitored. On this basis, it is possible to differentiate between the various patterns of homology in relation to their evolutionary history. The *ZFY* and *ZFX* genes show >95% homology between the X and the Y sequences and are X–Y linked in all eutherian mammals tested (Page *et al.*, 1987). The pattern of X–Y linkage of *RPS4X*/*RPS4Y* has not been examined in a wide range of mammals, though it does appear that a Y homolog cannot be found on the mouse Y chromosome. This could be indicative of deletion from the rodent Y chromosome or divergence to the point where cross-hybridization no longer occurs. However, the human *RPS4X*/*RPS4Y* sequences share 82% homology in coding regions (Fisher *et al.*, 1990), indicating that they have been diverging over a long period of time. Accepting the rate of diver-

Table V
Distribution of X–Y Homologous Regions in Primates

	Mouse	Prosimians	New World	Old World	Orangutan	Gorilla	Chimpanzee	Human
Homology								
Yp11.2–Xp21.2-p22.1								
ZFY/ZFX	XY	XY	XY	XY	XY	XY	XY	XY
Dist.Yp11.1–Xq21.1-q21.3								
All sequences tested	–	X	X	X	X	X	X	XY
except DXYS36	–	–	X	X,XY in baboon	XY	XY	X	XY
Prox.Yp11.1–Xq21.1-q21.3								
DXYS1(DP34)	–							
DXYS34(GMGXY8)		X	X	X	X	X	X	XY
Prox.Yp11.1–Xq13								
DXYS22(p2F2)	–	–	X	X	X	XY	XY	XY
Prox.Yp11.1–Xp21.2-p22.1								
AMG	X	–	XY or X	XY or X	XY	XY	XY	XY
Ycen-q11.21–Xp22.3-pter								
STS, KAL,DYS74(GMGXY19)	XY	XY(PAR)	X	XY or X	XY	X	XY	XY
DXS31(M1a),DXS274	–	XY(PAR)	X	X	XY	XY	XY	XY
DXYS29(GMGXY3)	–	–	X	X	XY	X	XY	XY
Yq11.22-q11.3–Xq28								
DXYS37(GMGXY10)	–	–	X	X	XY	XY	XY	XY
Yqter–Xq28								
DXYS61,DXYS64	–	–	–	–	–	–	X	XY

Note. –, not tested; PAR, pseudoautosomal region.

gence of 0.24% per million years for two genetically isolated loci (Sibley and Alquist, 1987), this would suggest that the two genes began to diverge at least 75 million years ago. Thus both the *ZFX/ZFY* and the *RPS4X/RPS4Y* homologies are clearly of ancient origin and are likely to date from the ancestral pair of homologs that gave rise to the sex chromosomes. The Yp–Xq pattern of homology shown by *RPS4X* and *RPS4Y* may reflect rearrangements on the sex chromosomes as they have diverged. This is suggested by the location of *RPS4X* in Xq within a strongly conserved homology between the mouse and the human X chromosomes (Rastan and Brown, 1990; Hamvas *et al.*, 1992).

The X and Y homologs of *amelogenin* are also likely to be of ancient origin. Nakahori and Nakagome (1989), Nakahori *et al.* (1991b), and Bailey *et al.* (1992) have demonstrated X–Y linkage of *amelogenin* in the higher primates and in some but not all species of old and new world monkeys investigated. However, as with *RPS4X/RPS4Y,* a Y-linked copy of the *amelogenin* gene cannot be detected on the mouse and rat Y chromosomes (Lau *et al.*, 1989; Nakahori *et al.*, 1991b), suggesting either deletion of the gene from the rodent Y or extensive divergence. X–Y linkage of *amelogenin* has been demonstrated in cattle (Nakahori *et al.*, 1991b) and indicates that its presence on both sex chromosomes dates from very early times.

In contrast, the near 100% homology between the X and the Y for the majority of Yp–Xq sequences and the Yq–Xq homology detected by DXYS61 and DXYS64 does not signal a high degree of conservation and homologies of ancient origin, but rather recent homologies that have not yet diverged extensively. Studies with DNA from males and females of various primate species have confirmed this conclusion. The majority of the members of the Yp–Xq homology tested thus far detect only X-linked sequences in chimpanzee and gorilla (Bishop *et al.*, 1983, 1984; Geldwerth *et al.*, 1985; Koenig *et al.*, 1985; Page *et al.*, 1982, 1983; Bickmore and Cooke, 1987; Mitchell, 1989). Lambson *et al.* (1992) extended the examination to orangutan, various old world monkeys, and marmoset as representative of new world monkeys. The majority of the probes tested from this category of homology show dosage between males and females and no Y-specific fragment in males (except in the human), indicating that they are only on the X. Therefore, these homologies represent a recent colonization of the Y by X sequences rather than a remnant of an ancient homology.

Page *et al.* (1982, 1983) proposed a duplicative transposition event from the X to the Y to account for homologies of this nature. This must have occurred at some point after the divergence of the chimpanzee and human lineages as defined by the accepted phylogeny of human evolution (Sibley and Alquist, 1987). This would date the event at some 8 million years ago, which is consistent with the estimated degree of X–Y divergence found in these homologous regions in humans, as judged by restriction analysis (Lambson *et al.*, 1992). Not all the sequences in the Yp–Xq and Yq–Xq homologies have a homogeneous evolutionary history. The pattern of X–Y linkage for three probes, DXYS22 (p2F2) (proximal Yp11.2–Xq), DXYS36 (GMGXY12) (distal Yp11.2–Xq21.1-21.3), and DXYS37 (GMGXY10) (DXYS37Y) (proximal Yq–Xq28), reveals independent lines of evolution, thus distinguishing them from the general pattern established by the majority of sequences in this group (Lambson *et al.*, 1992). The patterns are summarized in Table V. It is clear these homologies arose before the divergence of chimpanzee and human; between orangutan and gorilla for p2F2 (10–15 million years ago); between old world monkeys and orangutan for DXYS36 (GMGXY12) (15–34 million years ago); and probably between the divergence of new world and old world monkeys for DXYS37 (GMGXY10) since this sequence is X–Y linked in baboon. Thus these homologies may represent independent transposition events from the X to the Y. Further, the homolog of DXYS36 appears to have been lost from chimpanzee. Thus there are additional layers of complexity in the evolution of these broad blocks of homology, which detailed analysis with a more comprehensive range of markers will help to elucidate.

DXYS37Y (GMGXY10) lies within two tightly clustered blocks of sequences that show Yp–Yq homology (see Figs. 3 and 5), suggesting there has been a duplication on the Y at some point in the past. It is difficult to estimate when this occurred, but on the basis that DXYS37Y detects multiple loci on the Y in earlier primates as well as humans, Mitchell (1989) has suggested this may have occurred prior to the branch leading to the orangutan. A transposition event may be responsible for the current Yp–Yq distribution seen in humans.

B. Xp22.3–Yq Homology: A Pseudoautosomal Remnant

In contrast to the homologies discussed above, DXS31 shows 80% homology between X and Y (Koenig *et al.*, 1984), and *STS* and *KAL* show about 85–95% (Yen *et al.*, 1988; Castillo *et al.*, 1992; Incerti *et al.*, 1992). Different wash stringencies in Southern blots and restriction analysis on male and female human DNA (to detect X–Y differences) indicate that this is common to other sequences constituting this group (Mitchell, 1989; Lambson *et al.*, 1992), suggesting a greater degree of X–Y evolutionary divergence. From this, one can conclude that these sequences represent an older homology.

Studies with primates support this conclusion. Members of this homology group have been employed to examine primate species as far back as prosimians. DXS31, various sequences from the *STS* pseudogene, and exon probes from the *KALP* gene show no Y-specific sequences or dosage between males and females in lemurs and other prosimians (Koenig *et al.*, 1984; Yen *et al.*, 1988; Castillo *et al.*, 1992), thus suggesting an autosomal or pseudoautosomal location 70 million years ago. Since *STS* is pseudoautosomal in the mouse (which branched off from a common anscestor to the prosimians 10 million years earlier), the latter is likely to be the case (Keitges *et al.*, 1987). DXS31 and DXS278 detect only X-linked sequences in the new and old world monkeys so far examined but show X and Y sequences in orangutan, gorilla, chimpanzee, and human (Koenig *et al.*, 1984, 1985; Lambson *et al.*, 1992). However, *STS* sequences detect X and Y fragments in old world monkeys, orangutan, chimpanzee, and humans, but only X-linked sequences in gorilla and the range of new world monkeys thus far tested (Yen *et al.*, 1988). Similarly, *KALP* appears to be pseudoautosomal or autosomal in prosimians, X-linked in new world monkeys, and X–Y linked in old world monkeys (Castillo *et al.*, 1992). Finally, Mitchell (1989) and Lambson *et al.* (1992) have demonstrated that DXYS29 (GMGXY3) is XY homologous in chimpanzee and orangutan, but only X linked in gorilla and the old and new world monkeys tested. DYS74 (GMGXY19) shows the same pattern of X–Y linkage but extends this to the old world monkeys (prosimians have not been examined with these sequences). Table V summarizes the findings.

Several features point to a common evolutionary history for this group of sequences: (*a*) close physical linkage of these sequences on X and Y; (*b*) correspondence of locus order on the X and Y with respect to the Xp and Yq telomeres; (*c*) escape from X inactivation for the *STS, KAL,* and *GS1* genes (although the map location of *GS1* on the Y has not been established); (*d*) the degree of divergence between X and Y is similar and all three genes appear to have functionless Y pseudogenes; (*e*) they share very similar patterns of X–Y linkage in primates; (*f*) with the exception of DXS31 and DXS278 all these sequences have been deleted from or have diverged extensively on the gorilla Y chromosome. The checkered pattern of X–Y linkage for *STS* in old and new world monkeys implies X–Y linkage in the earliest ancestor giving rise to these lineages. Subsequent deletion from the Y in later ancestors of various sublineages can account for their unique X linkage in some species. This may explain the small deviations exhibited by the markers DXYS29, DX31, and DXS278 from the general pattern of X–Y linkage. A systematic study with all the probes from this region on a wide range of primates will elucidate the point at which such deletions are likely to have occurred and may help to map the position of these evolutionary deletion breakpoints.

Yen *et al.* (1988) postulated that this group of sequences was once in the pseudoautosomal segment and was removed by a pericentric inversion on the Y. If *STS* is taken as the benchmark sequence, it is clear from these findings that this inversion must have occurred between the prosimians and the split creating the new world monkeys. On the evolutionary time scale this would place it between 45 and 70 million years ago and is consistent with the degree of divergence between the X and Y estimated by Mitchell (1989) and Lambson *et al.* (1992) for DYS74 and the observed sequence differences between the X and Y homologs of *STS* and *KAL* (Yen *et al.*, 1988; Castillo *et al.*, 1992; Incerti *et al.*, 1992). Such an inversion would also account for the homologs on Yq. The hypervariable minisatellite sequence DXS278 (Knowlton *et al.*, 1989) further supports the idea of a previous pseudoautosomal location, as it shares the characteristics of several highly polymorphic sequences isolated from the exchange region. It follows that a pericentric inversion on the Y would result in the creation of a new pseudoautosomal boundary distinct from that on the X and is therefore entirely compatible with the findings of Ellis *et al.* (1989, 1990) in this respect.

The proposal that the Y is prone to chromosomal rearrangements has some attraction in considering mechanisms for preserving its sex-determining role. Bengtsson and Goodfellow (1987) have argued cogently that a small number of illegitimate recombination events outside the pseudoautosomal region could lead to an expansion of the boundary limit. Ellis *et al.* (1989) presented evidence for a previous more proximal boundary defining the primate X–Y exchange region. Clearly where such exchanges result in XX males, the expansion will be self-limiting, as the new chromosomes created will not be contributed to the next generation since these individuals are infertile. Thus *TDF* will define the limit of the tendency of the pseudoautosomal region to expand. For obvious reasons, to maintain a chromosomal basis of sex determination, it is important that the frequency of recombination events including *TDF* be minimized. Therefore, there is great advantage in minimizing the degree of X–Y homology beyond *TDF* as a means of suppressing recombination. Rearrangements of the Y chromosome mediated by pericentric and paracentric inversions could achieve this effect and prevent the progressive homogenization of the X and Y by recombination beyond the sex-determining locus. It is also essential that such rearrangements not compromise the function of the pairing segment in gametogenesis.

There is considerable evidence that such rearrangements are relatively common on the Y: the anomalous sequence order on Yp in some XX males, the proximal location on Yp of homologs of the probes DXS431 and DXS432, which are adjacent to the pseudoautosomal region on the X

(Levilliers *et al.*, 1989; Petit *et al.*, 1989); the splitting of the Yp–Xq homology on the short arm of the Y (Affara *et al.*, 1986a, 1987); and the existence of Yp–Yq homologous sequences. The comparison of locus order on the human and mouse X chromosomes suggests that inversions have also occurred on the X (Amar *et al.*, 1988; Davisson, 1987). Disruption of X–Y sequence homology can also be brought about by the importation of sequences onto the Y from other chromosomes. The proposed X–Y transpositions, various autosomal homologies detected by Y sequences (Bishop *et al.*, 1984; Affara *et al.*, 1986a), and the deposition of pseudogenes onto the Y (Heilig *et al.*, 1984; Daiger *et al.*, 1984) are indicative of such events.

Figure 8 summarizes the evolution of the X–Y homologies discussed above and indicates the likely chronology of their appearance on the sex

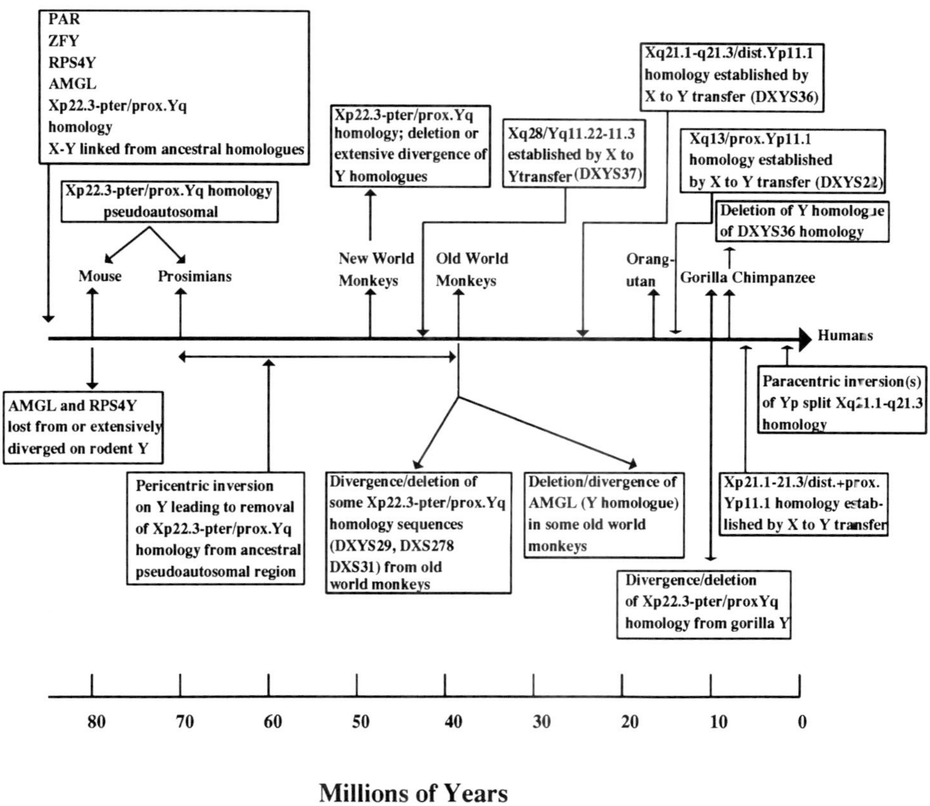

Figure 8 Proposed evolution of X–Y homologous regions in primates. PAR, pseudoautosomal region.

chromosomes. From an evolutionary perspective there are four main groups of sequences. First, the sequences in the pseudoautosomal region that date from the ancestral homologs. Second, sequences that were probably removed from the pseudoautosomal region into the differential part of the Y chromosome and are also likely to date from the ancestral homologs. Third, conserved functional genes in the differential part of the Y that date from very early times and possibly the ancestral homologs. Fourth, various categories of X–Y homology created by movements from the X to the Y during primate evolution.

V. Conclusion

By combining genetic, cytogenetic, and molecular analyses it has been possible to resolve the basic bipartite structure (pairing region/sex-specific region) of the Y chromosome into a number of sequence domains. Several regions of X–Y homology have been defined.

The pseudoautosomal region has received the most intensive study and has been shown to be strictly X–Y homologous and to measure 2.5–3.0 Mb in length. Genetic analysis with polymorphic pseudoautosomal sequences has demonstrated the existence of an obligatory crossover in male meiosis and has provided a genetic map of the pairing region. It is likely that strict X–Y homology is maintained so that the pseudoautosomal region can function to ensure pairing and segregation during gametogenesis. *MIC2, CSF2RA,* and *XE7* are the only genes to be definitively identified as pseudoautosomal, although physical analysis has revealed the presence of many CpG islands that may be associated with further loci. These remain to be characterized and it is possible that CpG islands clustered in the subtelomeric region do not mark the positions of genes, but rather play a role in the pairing function or nuclear membrane attachment.

Outside the pseudoautosomal region, a number of other X–Y homologies have been defined.

1. Conserved genes (*ZFY/ZFX, RPS4Y/RPS4X,* and *amelogenin*) demonstrating early X–Y linkage and possibly representing genes dating from the ancestral homologs to the sex chromosomes.

2. A block of sequences in Xp22.3-pter that includes the *STS, KAL,* and *GS1* genes showing homology to proximal Yq. Evolutionary studies indicate that this homology was once part of the pseudoautosomal region but that after emergence of the prosimians it was removed from the pairing region by a pericentric inversion on the Y. These sequences are now diverging on the Y as evidenced by the pseudogene status of the Y homologs of the *STS* and *KAL* genes.

3. A number of homologies occurring between Yp–Xq and Yq–Xq showing very limited divergence between the sex chromosomes. Analysis of primates shows that these homologies represent relatively recent transfers from the X to the Y and are most likely the result of duplicative transpositions during primate evolution. They are therefore not a remnant of homology between ancestral homologs that gave rise to the X and Y. Little is known of the size of these blocks of homology, their internal sequence structure, whether they contain any gene sequences or what function they fulfill.

Analysis of sequence order on the Y provides evidence for paracentric inversions on the long and short arms. Such rearrangements on the sex-specific portion of the Y may be selected for in order to suppress recombination with the X and hence preserve its sex-determining role. The introduction of recent X–Y homologies and the deposition of pseudogenes and autosomal sequences into the differential segment of the Y may also contribute in this respect.

Acknowledgments

We thank Dr. Ellen Magenis for many of the patients with dicentric isochromosomes of the Y and several of the XX males. We also thank Marie Ferguson-Smith, Adele Pelmear, and Dr. Nigel Carter for providing us with cell cultures to extract DNA for our deletion and pulsed-field gel electrophoresis studies.

References

Affara, N. A., Florentin, L., Morrison, N., Kwok, K., Mitchell, M., Cooke, A., Jamieson, D., Glasgow, L., Meredith, L., Boyd, E., and Ferguson-Smith, M. A. (1986a). Regional assignment of Y-linked DNA probes by deletion mapping and their homology with X chromosome and autosomal sequences. *Nucleic Acids Res.* **14,** 5353–5373.

Affara, N. A., Ferguson-Smith, M. A., Tolmie, J., Kwok, K., Mitchell, M., Jamieson, D., Cooke, A., and Florentin, L. (1986b). Variable transfer of Y specific sequences in XX males. *Nucleic Acids Res.* **14,** 5375–5387.

Affara, N. A., Ferguson-Smith, M. A., Magenis, R. E., Tolmie, J., Boyd, E., Cooke, A., Jamieson, D., Kwok, K., Mitchell, M., and Snadden, L. (1987). Mapping the testis determinants by an analysis of Y-specific sequences in males with apparent XX and XO karyotypes and females with XY karyotypes. *Nucleic Acids Res.* **15,** 7325–7342.

Affara, N. A., Chambers, D., O'Brien, J., Habeebu, S. S. M., Kalaitsiadaki, M., Bishop, C. E., and Ferguson-Smith, M. A. (1989). Evidence for distinguishable transcripts of the putative testis determining gene (ZFY) and mapping of homologous cDNA sequences to chromosomes X, Y and 9. *Nucleic Acids Res.* **17,** 2987–2999.

Akesson, H. O., Hagberg, B., and Wahlström, J. (1980). Y-to-X chromosome translocation observed in two generations. *Hum. Genet.* **55,** 39–42.

Alderdice, P. W., Aveling, J. V., Eales, B. A., Lewis, M. J., McAlpine, P. J., Ross, J. B., and Simms, R. J. (1983). Familial t(X;Y)(p22.3;q11) associated with short stature in 4 male and

5 female carriers and with X-linked ichthyosis and anhydrosis in 4 male carriers. *Am. J. Hum. Genet.* **35,** 124A.

Amar, L., Dandolo, L., Hanauer, A., Ryder-Cook, A., Arnaud, D., Mandel, J-L., and Avner, P. (1988). Conservation and reorganisation of loci on the mammalian X chromosome: A molecular framework for the identification of homologous subchromosomal regions in man and mouse. *Genomics* **2,** 220–230.

Arveiler, B., Vincent, A., and Mandel, J. L. (1989). Toward a physical map of the Xq28 region in man: Linking colour vision, G6PD, and coagulation factor VIII genes to an X–Y homology region. *Genomics* **4,** 460–471.

Bailey, D. M. D., Affara, N. A., and Ferguson-Smith, M. A. (1992). The X–Y homologous gene amelogenin maps to the short arms of both the X and Y chromosomes and is highly conserved in primates. *Genomics* **14,** 203–205.

Ballabio, A., Parenti, G., Carrozzo, R., Sebastio, G., Andria, G., Buckle, V., Fraser, N., Craig, I., Rocchi, M., Romeo, D., Jobsis, A. C., and Persico, M. G. (1987). Isolation and characterization of a steroid sulphatase cDNA clone: Genomic deletions in patients with X-chromosome-linked ichthyosis. *Proc. Natl. Acad. Sci. U.S.A.* **84,** 4519–4523.

Ballabio, A., Bardoni, B., Carrozzo, R., Andria, G., Bick, D., Campbell, L., Hamel, B., Ferguson-Smith, M. A., Gimelli, G., Fraccaro, M., Maraschio, P., Zuffardi, O., Guioli, S., and Camerino, G. (1989b). Contiguous gene syndromes due to deletions in the distal short arm of the human X chromosome. *Proc. Natl. Acad. Sci. U.S.A.* **86,** 10001–10005.

Ballabio, A., Carrozzo, R., Parenti, G., Gil, A., Zollo, M., Persico, M. G., Gillard, E. F., Affara, N. A., Yates, J., Ferguson-Smith, M. A., Frants, R. R., Eriksson, A. W., and Andria, G. (1989a). Molecular heterogeneity of steroid sulphatase deficiency: A multicenter study on 57 unrelated patients, at DNA and protein levels. *Genomics* **4,** 36–40.

Ballabio, A., Carrozzo, R., Gil, A., Gillard, B., Affara, N., Ferguson-Smith, M. A., Fraser, N., Craig, I., Rocchi, M., Romeo, G. and Andria, G. (1989c). Molecular characterisation of human X/Y translocations suggests their aetiology through aberrant exchange between homologous sequences on Xp and Yq. *Ann. Hum. Genet.* **53,** 9–14.

Bardoni, B., Zuffardi, O., Guiolo, S., Ballabio, A., Simi, P., Cavalli, P., Grimoldi, M. G., Fraccaro, M., and Camerino, G. (1991). A deletion map of the human Yq11 region: Implications for the evolution of the Y chromosome and tentative mapping of a locus involved in spermatogenesis. *Genomics* **11,** 443–451.

Bengtsson, B. O., and Goodfellow, P. N. (1987). The effect of recombination between the X and Y chromosomes of mammals. *Ann. Hum. Genet.* **51,** 57–64.

Bickmore, W. A., and Cooke, H. J. (1987). The evolution of homologous sequences on the human X and Y chromosomes, outside the meiotic pairing region. *Nucleic Acids Res.* **15,** 6261–6271.

Bird, A. (1986). CpG-rich islands and the function of DNA methylation. *Nature (London)* **321,** 209–213.

Bishop, C. E., Guellaen, G., Geldwerth, D., Voss, R., Fellous, M., and Weissenbach, J. (1983). Single-copy DNA sequences specific for the human Y chromosome. *Nature (London)* **303,** 831–832.

Bishop, C. E., Guellaen, G., Geldwerth, D., Fellous, M., and Weissenbach, J. (1984). Extensive sequence homologies between the Y and other human chromosomes. *J. Mol. Biol.* **173,** 403–417.

Bishop, C. E., Roberts, C., Michot, J-L., Nagamine, C., Winking, H., Guenet, J.-L., and Weith, A. (1987) The use of specific DNA probes to analyse the Sxr mutation in the mouse. *Development (Cambridge, UK)* **101,** Suppl., 167–175.

Boyd, E., Ferguson-Smith, M. A., Ferguson-Smith, M. E., Jamieson, M. E., Russel, J. E., Aitken, D. A., Tippet, P., and Sanger, R. (1981). A case of X:Y translocation which maps the Xg locus to Xp22.3-pter. *J. Med Genet.* **18,** 224.

Brown, C. J., and Willard, H. F. (1989). Noninactivation of a selectable human X-linked gene that complements a murine temperature sensitive cell cycle defect. *Am. J. Hum. Genet.* **45,** 592–598.

Brown, W. R. A. (1988). A physical map of the human pseudoautosomal region. *EMBO J.* **7,** 2377–2385.

Brown, W. R. A., and Bird, A. (1986). Long range restriction site mapping of mammalian genomic DNA. *Nature (London)* **322,** 477–481.

Buckle, V., Mondello, C., Darling, S., Craig, I. W., and Goodfellow, P. N. (1985). Homologous expressed genes in the human sex chromosome pairing region. *Nature (London)* **317,** 739–741.

Burgoyne, P. S. (1982). Genetic homology and crossing over in the X and Y chromosomes of mammals. *Hum. Genet.* **61,** 85–90.

Burgoyne, P. S. (1989). Thumbs down for the zinc finger? *Nature (London)* **342,** 860–862.

Castillo, I.-D., Cohen-Salmon, M., Blanchard, S., Lutfalla, G., and Petit, C. (1992). Structure of the X-linked Kallmann syndrome gene and its homologous pseudogene on the Y chromosome. *Nat. Genet.* **2,** 305–310.

Chandley, A. C., Goetz, P., Hargreave, T. B., Joseph, A. M., and Speed, R. M. (1984). On the nature and extent of XY pairing in meiotic prophase in man. *Cytogenet. Cell Genet.* **38,** 241–247.

Collinge, J., Boccio, A., Delisi, L., Johnstone, E., Lofthouse, R., Owen, F., Poulter, M., Risby D., Shah, T. and Crow, T. J. (1989). Evidence for a pseudoautosomal locus for schizophrenia. *Cytogenet. Cell Genet.* **51,** 978.

Cooke, H. J., and Smith, B. A. (1986). Variability at the telomeres of the human X/Y pseudoautosomal region. *Cold Spring Harbor Symp. Quant. Biol.* **51,** 213–219.

Cooke, H. J., Brown, W. R. A., and Rappold, G. A. (1984). Closely related sequences on the human X and Y chromosomes outside the pairing region. *Nature (London)* **311,** 259–261.

Cooke, H. J., Brown, W. R. A., and Rappold, D. A. (1985). Hypervariable telomeric sequences from the human sex chromosomes are pseudoautosomal. *Nature (London)* **317,** 687–692

Crow, T. J. (1988). Sex chromosomes and psychosis. The case for a pseudoautosomal locus. *Br. J. Psychiatry* **153,** 675–683.

Crow, T. J. (1990). The continuum of psychosis and its genetic origin. *Br. J. Psychiatry* **156,** 788–797.

Daiger, S. P., Wildin, R. S., and Su, T. S. (1984). Sequences on the human Y chromosome homologous to the autosomal gene for argininosuccinate synthetase. *Nature (London)* **293,** 683–684.

Darling, S. M., Banting, G. S., Pym, B., Wolfe, J., and Goodfellow, P.N. (1986) Cloning an expressed gene shared by the human sex chromosomes. *Proc. Natl. Acad. Sci. U.S.A.* **83,** 135–139.

Darlington, C. D. (1937). "Recent Advances in Cytology." Churchill, London.

Davies, K. E., Mandel, J.-L., Monaco, A. P., Nussbaum, R. L., and Willard, H. F. (1991). Report of the committee on the genetic constitution of the X chromosome. *Cytogenet. Cell Genet.* **58,** 853–966.

Davis, R. M. (1981). Localisation of male determining factors in man: A thorough review of structural anomalies of the Y chromosome. *J. Med. Genet.* **18,** 161–195.

Davisson, M. T. (1987). X-linked genetic homologies between mouse and man. *Genomics* **1,** 213–227.

Disteche, C. M., Casanova, M., Saal, H., Friedman, C., Sybert, V., Graham, J., Thuline, H., Page, D. C., and Fellous, M. (1986). Small deletions of the short arm of the Y chromosome in 46,xy females. *Proc. Natl. Acad. Sci. U.S.A.* **83,** 7841–7844.

Ellis, N. A., and Goodfellow, P. N. (1989). The mammalian pseudoautosomal region. *Trends Genet.* **5,** 406–410.

Ellis, N. A., Goodfellow, P. J., Pym, B., Smith, M., Palmer, M., Frischauf, A.-M., and Goodfellow, P. N. (1989). The pseudoautosomal boundary in man is defined by an Alu repeat sequence inserted on the Y chromosome. *Nature* (*London*) **337,** 81–84.

Ellis, N. A., Taylor, A., Bengtsson, B. O., Kidd, J., Rogers, J., and Goodfellow, P. N. (1990) Population structure of the human pseudoautosomal boundary. *Nature* (*London*) **344,** 663–665.

Ellison, J., Passage, M., Yu, L. C., Yen, P., Mohandas, T. K., and Shapiro, L. J. (1992a). Directed isolation of human genes that escape X inactivation. *Somatic Cell Mol. Genet.* **18,** 259–268.

Ellison, J., Ramos, C., Yen, P., and Shapiro, L. J. (1992b). Structure and expression of the human pseudoautosomal gene *XE7*. *Am. J. Hum. Genet.* **51,** A127.

Evans, E. P., Burtenshaw, M. D., and Cattanach, B. M. (1982). Meiotic crossing-over between the X and Y chromosomes of male mice carrying sex-reversing (Sxr) factor. *Nature* (*London*) **300,** 443–445.

Ferguson-Smith, M. A. (1965). Karyotype-phenotype correlations in gonadal dysgenesis and their bearing on the pathogenesis of malformations. *J. Med. Genet.* **2,** 142–155.

Ferguson-Smith, M. A. (1966). X–Y interchange in the aetiology of true hermaphroditism and of XX Klinefelter's syndrome. *Lancet* **2,** 475–476.

Ferguson-Smith, M. A., Sanger, R., Tippett, P., Aitken, D. A., and Boyd, E. (1982). A familial t(X;Y) translocation which assigns the Xg blood group locus to the region Xp22.3-pter. *Cytogenet. Cell Genet.* **32,** 273–274.

Ferguson-Smith, M. A., Affara, N. A., and Magenis, R. E. (1987). Ordering of Y-specific sequences by deletion mapping and analysis of X–Y interchange males and females. *Development* (*Cambridge, UK*) **101,** *Suppl.*, 41–50.

Fincham, A. G. (1991). Human developing enamel proteins exhibit a sex-linked dimorphism. *Calcif. Tissue Int.* **48,** 288–290.

Fisher, E. M. C., Beer-Romero, P., Brown, L. G., Ridley, A., McNeil, J. A., Lawrence, J. B., Willard, H. F., Bieber, F. R., and Page, D. C. (1990). Homologous ribosomal protein genes on the human X and Y chromosomes: Escape from X inactivation and possible implications for Turner syndrome. *Cell* (*Cambridge, Mass.*) **63,** 1205–1218.

Foote, S., Vollrath, D., Hilton, A., and Page, D. C. (1992). The human Y chromosome: Overlapping DNA clones spanning the euchromatic region. *Science* **258,** 60–66.

Franco, B., Guioli, S., Praglioli, A., Incerti, B., Bardoni, B., Tonlorenzi, R., Carrozzo, R., Maestrini, E., Pierretti, M., Taillon-Miller, P., Brown, C. J., Willard, H. F., Lawrence, C., Persico, M. G., Camerino, G., and Ballabio, A. (1991). A gene deleted in Kallmann's syndrome shares homology with neural cell adhesion and axonal path-finding molecules. *Nature* (*London*) **353,** 529–536.

Fraser, N., Ballabio, A., Zollo, M., Persico, G., and Craig, I. W. (1987). Identification of incomplete coding sequences for steroid sulphatase on the human Y chromosome: Evidence for an ancestral pseudoautosomal gene? *Development* (*Cambridge, UK*) **101,** Suppl., 127–132.

Geldwerth, D., Bishop, C. E., Guellaen, G., Koenig, M., Vergnaud, G., Mandel, J. L., and Weissenbach, J. (1985). Extensive DNA sequence homologies between the human Y and the long arm of the X chromosome. *EMBO J.* **4,** 1793–1743.

Gillard, E. F., Affara, N. A., Yates, J. R. W., Goudie, D. R., Lambert, J., Aitken, D. A., and Ferguson-Smith, M. A. (1987). Deletion of a DNA sequence in eight of nine families with X-linke ichthyosis (steroid sulphatase deficiency). *Nucleic Acids Res.* **15,** 3977–3985.

Goodfellow, P. J., Darling, S. M., Thomas, N. S., and Goodfellow, P. N. (1986). A pseudoautosomal gene in man. *Science* **234,** 740–743.

Goodfellow, P. J., Darling, S., Banting, G., Pym, B., Mondello, C., and Goodfellow, P. N. (1987a). Pseudoautosomal genes in man. *Development (Cambridge, UK)* **101,** Suppl., 119–125.

Goodfellow, P. J., Pritchard, C., Tippett, P., and Goodfellow, P. N. (1987b). Recombination between the X and Y chromosomes at the YG locus: Implications for the relationship between MIC2, XG and YG. *Ann. Hum. Genet.* **51,** 161–167.

Goodfellow, P. N., and Darling, S. M. (1988). Genetics of sex determination in man and mouse. *Development (Cambridge, UK)* **102,** 251–258.

Goodfellow, P. N., and Tippett, P. (1981). A human polymorphism related to Xg blood groups. *Nature (London)* **289,** 404–405.

Goodfellow, P.N., Banting, G., Sheer, D., Ropers, H. H., Caine, A., Ferguson-Smith, M. A., Povey, S., and Voss, R. (1983). Genetic evidence that a Y-linked gene in man is homologous to a gene on the X chromosome. *Nature (London)* **302,** 346–349.

Goodfellow, P. N., Pym, B., Mohandas, T., and Shapiro, L. J. (1984). The cell surface antigen locus MIC2X escapes X-inactivation. *Am. J. Hum. Genet.* **36,** 777–782.

Goodfellow, P. N., Darling, S. M., and Wolfe, J. (1985). The human Y chromosome. *J. Med. Genet.* **22,** 329–344.

Gough, N. M., Gearing, D. P., Nicola, N. A., Baker, E., Pritchard, M., Callen, D. F., and Sutherland, G. R. (1990). Localization of the human GM-CSF receptor gene to the X–Y pseudoautosomal region. *Nature (London)* **345,** 734–736.

Gubbay, J., Collignon, J., Koopman, P., Capel, B., Economou, A., Münsterberg, A., Vivian, N., Goodfellow, P. N., and Lovell- Badge, R. (1990). A gene mapping to the sex-determining region of the mouse Y chromosome is a member of a novel family of embryonically expressed genes. *Nature (London)* **346,** 245–250.

Hamvas, R. M., Zinn, A., Keer, J. T., Fisher, E. M. C., Beer-Romero, P., Brown, S. D. M., and Page, D. C. (1992). Rps4 maps near the inactivation center on the mouse X chromosome. *Genomics* **12,** 363–367.

Harbers, K., Soriano, P., Müller, U., and Jaenisch, R. (1986). High frequency of unequal recombination in the pseudoautosomal region shown by proviral insertion in transgenic mouse. *Nature (London)* **324,** 682–685.

Heilig, R., Hanauer, A., Grzeschik, K. H., Hors-Cayla, M. C., and Mandel, J. L. (1984). Actin-like sequences are present on the human X and Y chromosomes. *EMBO J.* **38,** 1803–1807.

Henke, A., Wapenaar, M., van Ommen, G.-J., Marachio, P., Camerino, G., and Rappold, G (1991). Deletions within the pseudoautosomal region help map three new markers and indicate a possible role of this region in linear growth. *Am. J. Hum. Genet.* **49,** 811–819.

Incerti, B., Guioli, S., Pragliola, A., Zanaria, E., Borsani, G., Tonlorenzi, R., Bardoni, B., Franco, B., Wheeler, D., Ballabio, A., and Camerino, G. (1992). Kallmann syndrome gene on the X and Y chromosomes: Implications for evolutionary divergence of the human sex chromosomes. *Nat. Genet.* **2,** 311–314.

Keitges, E., Rivest, M., Siniscalco, M., and Gartler, S. M. (1985). X-linkage of steroid sulphatase in the mouse is evidence for a functional Y-linked allele. *Nature (London)* **315,** 226–227.

Keitges, E., Schroderet, D. F., and Gartler, S. (1987). Linkage of the steroid sulphatase gene to the sex-reversed mutation in the mouse. *Genetics* **116,** 465–469.

Keith, D. H., Singer-Sam, J., and Ross, A. D. (1986). Active X chromosome DNA is methylated at eight CCGG sites clustered in a guanine-plus-cytosine rich island at the 5′ end of the gene for phosphoglycerate kinase. *J. Mol. Cell. Biol.* **6,** 4122–4125.

Knowlton, R. G., Nelson, C. A., Brown, V. A., Page, D. C., and Donis-Keller, H. (1989). An extremely polymorphic locus on the short arm of the human X chromosome with homology to the long arm of the Y. *Nucleic Acids Res.* **17,** 423–437.

Koenig, M., Camerino, G., Heilig, R., and Mandel, J. L. (1984). A DNA fragment from the human X chromosome short arm which detects a partially homologous sequence on the Y chromosome's long arm. *Nucleic Acids Res.* **12,** 4097–4109.

Koenig, M., Moisan, J. P., Heilig, R., and Mandel, J. L. (1985). Homologies between the X and Y chromosomes detected by DNA probes: Localisation and evolution. *Nucleic Acids Res.* **13,** 5485–5503.

Koller, P. C., and Darlington, C. D. (1934). The genetical and mechanical properties of the sex chromosomes. 1 *Rattus norvegicus. J. Genet.* **29,** 159–173.

Koopman, P., Gubbay, J., Collognon, J., and Lovell-Badge, R. (1989). Zfy gene expression patterns are not compatible with a primary role in mouse sex determination. *Nature (London)* **342,** 940–942.

Koopman, P., Münsterberg, A., Capel, B., Vivian, N., and Lovell- Badge, R. (1990). Expression of a candidate sex-determining gene during mouse testis differentiation. *Nature (London)* **348,** 450–452.

Koopman, P., Gubbay, J., Vivian, N., Goodfellow, P., and Lovell-Badge, R. (1991). Male development of chromosomally female mice transgenic for Sry. *Nature (London)* **351,** 117–121.

Kunkel, L. M., Tantravahi, U., Kurmit, D. M., Eisenhard, M., Bruns, J. P., and Latt, S. A. (1983). Identification and isolation of transcribed human X chromosome DNA sequences. *Nucleic Acids Res.* **11,** 7961–7979.

Lambson, B., Affara, N. A., Mitchell, M., and Ferguson-Smith, M. A. (1992). Evolution of DNA sequence homologies between the sex chromosomes in primate species. *Genomics* **14,** 1032–1040.

Lau, C., and Chan, K. (1989). The putative testis-determining factor and related genes are expressed as discrete-sized transcripts in adult gonadal and somatic tissues. *Am. J. Hum. Genet.* **45,** 942–952.

Lau, E. C., Mohandas, T., Shapiro, L. J., Slavkin, H. C., and Snead, M. L. (1989). Human and mouse amelogenin loci are on the sex chromosomes. *Genomics* **4,** 162–168.

Legouis, R., Hardelin, J. P., Levilliers, J., Claverie, J. M., Compain, S., Wunderle, V., Millasseau, P., Le Paslier, D., and Cohen, D. (1991). The candidate gene for the X-linked Kallmann syndrome encodes a protein related to adhesion molecules. *Cell (Cambridge, Mass.)* **67,** 423–435.

Levilliers, J., Quack, B., Weissenbach, J., and Petit, C. (1989). Exchange of the terminal portions of the X- and Y- chromosomal short arms in human XY females *Proc. Natl. Acad. Sci. U.S.A.* **86,** 2296–2300.

Ma, K., Sharkey, A., Kirsch, S., Vogt, P., Keil, R., Hargreave, T. B., McBeath, S., and Chandley, A. C. (1992). Towards the molecular localisation of the AZF locus: mapping of microdeletions in azoospermic men within 14 subintervals of interval 6 of the human Y chromosome. *Hum. Mol. Genet.* **1,** 29–33.

Magenis, R. E., Tochen, M. L., Holahan, K. P., Carey, T., Allen, L., and Brown, M. G. (1984). Turner syndrome resulting from partial deletion of Y chromosome short arm: Localisation of male determinants. *J. Pediatr.* **105,** 916–919.

Metaxotou, C., Ikkos, D., Panagiotopoulou, M., Alevizaki, M., Mavrou, A., Tsenghi, C., and Matsaniotis, N. (1983). A familial X/Y translocation in a boy with ichthyosis, hypogonadism and mental retardation. *Clin. Genet.* **24,** 380–383.

Mitchell, M. (1989). The isolation and characterisation of Y homologous and XY homologous human DNA sequences. Ph.D. Thesis, University of Glasgow, Scotland.

Müller, G., and Schempp, W. (1989). Mapping the human ZFX locus to Xp21.3 by in situ hybridization. *Hum. Genet.* **82,** 82–84.

Müller, U., Donlon, T., Schmid, M., Fitch, N., Richer, C. L., Lalande, M., and Latt, S. A.

(1986). Deletion mapping of the testis determining locus with DNA probes in XX males and in 46,XY and 46,X,dic(Y) females. *Nucleic Acids Res.* **14,** 6489–6505.

Nakahori, Y., and Nakagome, Y. (1989). Xp-Yq homologous region which encodes open reading frames corresponding to "Amelogenin." *Cytogenet. Cell Genet.* **51,** 1050 A2008.

Nakahori, Y., Takenaka, O., and Nakagome, Y. (1991a). A human X–Y homologous region encodes "amelogenin." *Genomics* **9,** 264–269.

Nakahori, Y., Tamura, T., Nagafuchi, S., Fujiede, K., Minowda, S., Fukutani, K., Fuse, H., Hayashi, K., Kuroki, Y., Fukushima, Y., Agematsu, K., Kuno, T., Kaneko, T., Yamada, S., Kitagawa, K., Nonomura, M., Fukuda, S., Kusano, M., Onigata, S., Hibi, I., and Nakagome, Y. (1991b). Molecular cloning and mapping of 10 new probes on the human Y chromosome. *Genomics* **9,** 765–769.

Newman, R., Affara, N. A., Mitchell, M., Gillard, E. F., and Ferguson-Smith, M. A. (1990). Physical mapping of deletion breakpoints around the steroid sulphatase gene in patients with X-linked ichthyosis: Evidence for tight clustering of distal and proximal breakpoints. *Proc. R. Soc. London Ser. B* **242,** 231–239.

North, M., Sargent, C., O'Brien, J., Taylor, K., Wolfe, J., Affara, N. A., and Ferguson-Smith, M. A. (1991). Comparison of ZFY and ZFX gene structure and analysis of alternative 3′untranslated regions of ZFY. *Nucleic Acids Res.* **19,** 2579–2586.

Ogata, T., Petit, C., Rappold, G., Matsuo, N., Matsumoto, T., and Goodfellow, P. N. (1992). Chromosomal localisation of a pseudoautosomal growth gene(s). *J. Med. Genet.* **29,** 624–628.

Ohno, S. (1967). "Sex Chromosomes and Sex-linked Genes." Springer-Verlag, Heidelberg.

O'Reilly, A. J., Affara, N. A., Simpson, E., Chandler, P., Goulmy, E., and Ferguson-Smith, M. A. (1992). A molecular deletion map of the Y chromosome long arm defining X and autosomal homologous regions and the localisation of the HYA locus to the proximal region of the Yq euchromatin. *Hum. Mol. Genet.* **1,** 379–385.

Page, D. C. (1986). Sex reversal: Deletion mapping the male-determining function of the human Y chromosome. *Cold Spring Harbor Symp. Quant. Biol.* **51,** 229–235.

Page, D. C., Matinville, B., Barker, D., Wyman, A., White, R., Francke, U., and Botstein, D. (1982). Single copy sequence hybridises to polymorphic and homologous loci on human X and Y chromosomes. *Proc. Natl. Acad. Sci. U.S.A.* **79,** 5352–5356.

Page, D. C., Harper, M. E., Love, J., and Botstein, D. (1983). Occurrence of a transposition from the X chromosome long arm to the Y chromosome short arm during human evolution. *Nature (London)* **311,** 119–123.

Page, D. C., Mosher, R., Simpson, E. M., Fisher, E. M. C., Mardon, G., Pollack, J., McGillivray, B., de la Chapelle, A., and Brown, L. G. (1987). The sex-determining region of the human Y chromosome encodes a finger protein. *Cell (Cambridge, Mass.)* **51,** 1091–1104.

Palmer, M., Berta, P., Sinclair, A., Pym, B., and Goodfellow, P. (1990). Comparison of ZFY and ZFX transcripts. *Proc. Natl. Acad. Sci. U.S.A.* **87,** 1681–1685.

Petit, C., Leillires, J., and Weissenbach, J. (1988). Physical mapping of the human pseudoautosomal region; comparison with genetic linkage map. *EMBO J.* **7,** 2369–2376.

Petit, C., Levilliers, J., Rouyer, F., Simmler, M-C., Nakahori, Y., Nakagome, Y., and Weissenbach, J. (1989). A physical map of the terminal part of the short arm of the X chromosome. *Cytogenet. Cell Genet.* **51,** 1059, Abstr. 2013.

Petit, C., Levillier, J., Rouyer, F., Simmler, M.-C., Herouin, E., and Weissenbach, J. (1990). Isolation of sequences from Xp22.3 and deletion mapping using sex chromosome rearrangements from human X–Y interchange sex reversals. *Genomics* **6,** 651–658.

Polani, P. E. (1982). Pairing of X and Y chromosomes, noninactivation of X-linked genes and the maleness factor. *Hum. Genet.* **60,** 207–211.

Pritchard, C. A., Goodfellow, P. J., and Goodfellow, P. N. (1987). Mapping the limits of the human pseudoautosomal region and a candidiate sequence for the male-determining gene. *Nature* (*London*) **328,** 273–275.

Race, R. R., and Sanger, R. (1975). "Blood groups in man," 6th ed. Blackwell, Oxford.

Rappold, G. A., and Lehrach, H. (1988). A long range map of the pseudoautosomal region by partial digest PFGE analysis from the telomere. *Nucleic Acids Res.* **16,** 5361–5377.

Rappold, G. A., Willson, T. A., Henke, A., and Gough, N. (1992). Arrangement and localization of the human GM-CSF receptor alpha chain gene CFS2RA within the X–Y pseudoautosomal region. *Genomics* **14,** 455–461.

Rastan, S., and Brown, S. D. M. (1990). The search for the X inactivation center. *Genet. Res.* **56,** 99–106.

Rosenfeld, R. G., Luzzati, L., Hintz, R. L., Miller, O. J., Koo, G. C., and Wachtel, S. S. (1979). Sexual and somatic determinants of the human Y chromosome: Studies in a 46,XYp-phenotypic female. *Am. J. Med. Genet.* **31,** 458–468.

Ross, M. T., Ballabio, A., and Craig, I. W. (1990). Long-range physical map around the human steroid sulphatase locus. *Genomics* **6,** 528–539.

Rouyer, F., Simmler, M.-C., Johnsson, C., Vergnaud, G., Cooke, H. J., and Weissenbach, J. (1986). A gradient of sex linkage in the pseudoautosomal region of the human sex chromosomes. *Nature* (*London*) **319,** 291–295.

Salido, E. C., Yen, P. H., Koprivnikar, K., Yu, L.-C., and Shapiro, L. (1992). The human enamel protein gene amelogenin is expressed from both the X and Y chromosomes. *Am. J. Hum. Genet.* **50,** 303–316.

Schempp, W., Weber, B., and Müller, G. (1989). Mammalian sex chromosome evolution: A convserved homologous segment on the X and Y chromosomes in primates. *Cytogenet. Cell Genet.* **50,** 201–205.

Schiebel, K., Weiss, B., Wohrle, D., and Rappold, G. (1993). A human pseudoautosomal gene, ADP/ATP translocase, escapes X-inactivation whereas a homologue on Xq is subject to X-inactivation. Nat. Genet. **3,** 82–87.

Schneider-Gadicke, A., Beer-Romero, P., Brown, L., Nussbaum, R., and Page, D. C. (1989). ZFX has a structure similar to ZFY, the putative human sex determinant, and escapes X inactivation. *Cell* (*Cambridge, Mass.*) **57,** 1247–1258.

Shapiro, L. J. (1985). Steroid sulphatase deficiency and the genetics of the short arm of the human X chromosome. *Adv. Hum. Genet.* **14,** 331–381.

Shapiro, L. J., Mohandas, T., Weiss, R., and Romeo, G. (1979). Non-inactivation of an X-chromosome locus in man. *Science* **204,** 1224–1226.

Sibley, C. G., and Alquist, J. E. (1987). DNA hybridization evidence of hominoid phylogeny: Results from an expanded data set. *J. Mol. Evol.* **26,** 99–121.

Simmler, M.-C., Rouyer, F., Vergnaud, G., Nystrom-Lahti, M., Ngo, K .Y., de la Chapelle, A., and Weissenbach, J. (1985). Pseudoautosomal DNA sequences in the pairing region of the human sex chromosomes. *Nature* (*London*) **317,** 692–697.

Sinclair, A. H., Foster, J. W., Spencer, J. A., Page, D. C., Palmer, M., Goodfellow, P. N., and Graves, J. A. M. (1988). Sequences homologous to ZFY, a candidate human sex-determining gene, are autosomal in marsupials. *Nature* (*London*) **336,** 780–783.

Sinclair, A. H., Berta, P., Palmer, M., Hawkins, J. R., Griffiths, B., Smith, M., Foster, J., Frischauf, A-M., Lovell-Badge, R., and Goodfellow, P. (1990). A gene from the human sex-determining region encodes a protein with homology to a conserved DNA-binding motif. *Nature* (*London*) **346,** 240–244.

Singh, L., and Jones, K. W. (1982). Sex reversal in the mouse (Mus musculus) is caused by a recurrent nonreciprocal crossover involving the X and an aberrant Y chromosome. *Cell* (*Cambridge, Mass.*) **28,** 205–216.

Solari, A. J. (1974). The behaviour of the XY pair in mammals. *Int. Rev. Cytol.* **38,** 273–317.

Tiepolo, L., Zuffardi, O., and Rodewald, A. (1977). Nullisomy for the distal portion of Xp in a male child with an X/Y translocation. *Hum. Genet.* **54,** 205–206.

Tiepolo, L., Zuffardi, O., Fraccaro, M., di Natale, D., Gargantini, L., Müller, C.R., and Ropers, H. H. (1980). Assignment by deletion mapping of the steroid sulphatase X-linked ichthyosis locus to Xp22.3 *Hum. Genet.* **54,** 205–206.

Tippett, P., Shaw, M. A., Green, C. A., and Daniels, G. L. (1986). The 12E7 red cell quantitative polymorphism: Control by the Y-borne locus, Yg. *Ann. Hum. Genet.* **50,** 339–347.

Verga, V., and Erickson, P. (1989). An extended long-range restriction map of the human sex-determining region on Yp, including ZFY, finds marked homology on Xp and no detectable Y sequences in an XX male. *Am. J. Hum. Genet.* **44,** 756–765.

Vergnaud, G., Page, D. C., Simmler, M.-C., Brown, L., Rouyer, F., Noel, B., Botstein, D., de la Chapelle, A., and Weissenbach, J. (1986). A deletion map of the human Y chromosome based on DNA hybridization. *Am. J. Hum. Genet.* **38,** 109–124.

Vollrath, D., Foote, S., Hilton, A., Brown, L.G.,, Beer-Romero, P., Bogan, J. S., and Page, D. C. (1992). The human Y-chromosome: A 43-interval map based on naturally occurring deletions. *Science* **258,** 52–59.

Watanabe, M., Zinn, A. R., Page, D. C., and Nishimoto, T. (1993). Functional equivalence of human X- and Y-encoded isoforms of ribosomal protein S4 consistent with a role in Turner syndrome. *Nat. Genet.* **4,** 268–271.

Weber, B., Weissenbach, J., and Schempp, W. (1987). Conservation of human-derived pseudoautosomal sequences on the sex chromosomes of the great apes. *Cytogenet. Cell Genet.* **45,** 26–29.

Weber, B., Weissenbach, J., and Schempp, W. (1989). X–Y crossing over in the chimpanzee. *Hum. Genet.* **80,** 301–303.

Weissenbach, J., and Goodfellow, P. N. (1991). Report of the committee on the genetic constitution of the Y chromosome. *Cytogenet. Cell Genet.* **58,** 967–985.

Weissenbach, J., Levilliers, J., Petit, C., Rouyer, F., and Simmler, M.-C. (1987). Normal and abnormal interchanges between the human X and Y chromosomes. *Development* (*Cambridge, UK*) **101,** Suppl. 67–74.

Wolfe, J., Erickson, R. P., Rigby, P. W. J., and Goodfellow, P. N. (1984). Cosmid clones derived from both euchromatic and heterochromatic region of the human Y chromosome. *EMBO J.* **3,** 1997–2003.

Yen, P. H., Patel, P., Chinault, A. C., Mohandas, T., and Shapiro, L. J. (1984). Differential methylation of hypoxanthine phosphoribosyltransferase genes on the active and inactive human X chromosomes. *Proc. Natl. Acad. Sci. U.S.A.* **81,** 1759–1763.

Yen, P. H., Allen, E., Marsh, B., Mohandas, T., Wang, N., Taggart, R. T., and Shapiro, L. J. (1987). Cloning and expression of steroid sulphatase cDNA and the frequent occurrence of deletions in STS deficiency: Implications for X–Y interchange. *Cell* (*Cambridge, Mass.*) **49,** 443–454.

Yen, P. H., Marsh, B., Allen, E., Tsai, S.P., Ellison, J., Connolly, L., Neiswanger, K., and Shapiro, L. J. (1988). The human X-linked steroid sulphatase gene and a Y-encoded pseudogene: Evidence for an inversion of the Y chromosome during primate evolution. *Cell* (*Cambridge, Mass.*) **55,** 1123–1135.

Yen, P. H., Ellison, J., Salido, E. C., Mohandas, T., and Shapiro, L. (1992). Isolation of a new gene from the distal short arm of the human X chromosome that escapes X-inactivation. *Hum. Mol. Genet.* **1,** 47–52.

Yi, H., Donohue, S. J., Klein, D. C., and McBride, O. W. (1993). Localization of the hydroxyindole-O-methyltransferase gene to the pseudoautosomal region: Implications for mapping of psychiatric disorders. *Hum. Mol. Genet.* **2,** 127–131.

12

XX Sex Reversal in the Human

Stephen S. Wachtel
Division of Reproductive Genetics
Department of Obstetrics and Gynecology
University of Tennessee, Memphis
Memphis, Tennessee 38163

Molecular Genetics of Sex Determination

I. Introduction

In humans, as in mammals generally, sex is determined by presence or absence of the Y chromosome: XY embryos become males, and XX embryos become females. This is the general rule but it does not hold true in every case, for depending on circumstances, a given embryo may develop as a male, female, or hermaphrodite, regardless of sex chromosome constitution. In this chapter, we discuss sex reversal in 46,XX males and 46,XX hermaphrodites. We describe the syndromes that result from development of the testis in subjects with a female sex chromosome constitution, and we consider the cytogenetic and molecular events that may cause these extraordinary conditions.

II. XX Males

A. The Syndrome

Males with the 46,XX female karyotype were first described in 1964. The patient of de la Chapelle *et al.* (1964) was a young man of grossly normal appearance with a phallus of 4 cm length and a small scrotum containing a soft left testis of about 2 cm diameter and a soft right testis of "about the size of a pea." Epididymal tissue was present on both sides and a prostate was palpated. When the patient was explored surgically, physicians found a normal internal anatomy with no evidence of female sex organs.

Histological evaluation of the gonads revealed mostly interstitial tissue containing numerous Leydig cells. There were a few small seminiferous tubules composed mostly of Sertoli cells; although some spermatogonia were present, there was no spermatogenesis. The karyotype was that of a normal female—46,XX—in all cells that were studied, including those from blood, bone marrow, skin, and testes. By contrast, cells from the father and a brother were 46,XY.

From the clinical perspective, the XX male syndrome may be classified as a variant of Klinefelter syndrome (47,XXY), although XX males are not as tall as XXY males and there is no tendency in XX males toward mental retardation (de la Chapelle, 1972). The small sterile testis is a key feature of the XX male syndrome (Fig. 1). The tubules are narrower than those in normal men and contain only Sertoli cells. Other features include thickening of the basement membranes, peritubular fibrosis, and hyalinization of some of the tubules. Leydig cell hyperplasia is common but Leydig cell aplasia has also been reported.

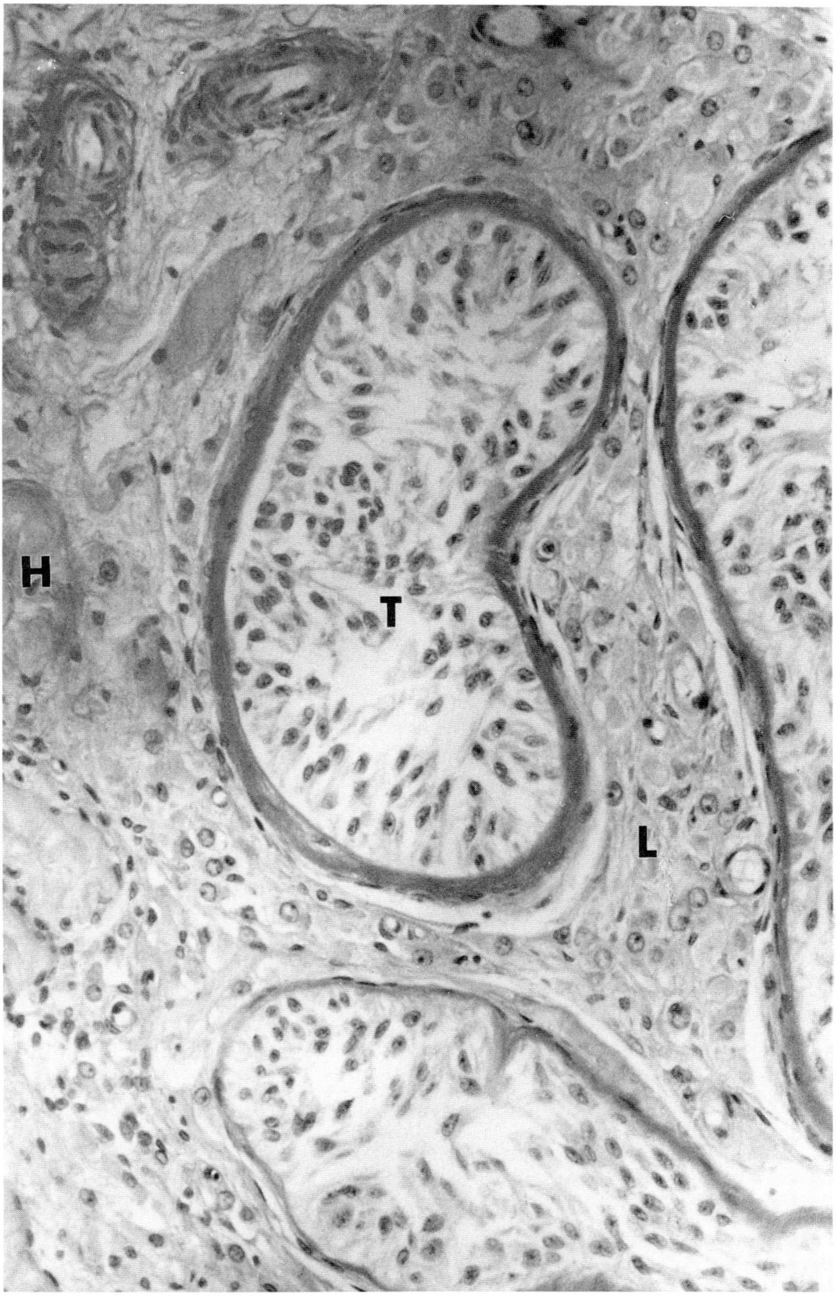

Figure 1 The XX testis. Section of testicular biopsy from a 20-year-old 46,XX male with descended testes. Some tubules are hyalinized (H). Some (T) are remarkable for thickness of the basement membrane and are lined by Sertoli cells only; germ cells are absent. Leydig cells (L) can be seen in the interstitium. This photomicrograph, the generous gift of Professor Albert de la Chapelle, appeared in Wachtel and Bard (1981) and Wachtel (1982). Reprinted by permission.

The condition occurs in about 1 in 20,000 male births. In general, affected males are normal-appearing, although body hair may exhibit a female pattern of distribution and abundance, and facial hair may be sparse. Gynecomastia is present in about one of three XX males (as in XXY males), cryptorchidism was noted in 6 of the 35 XX males surveyed by de la Chapelle (1972), and hypospadias is present in some cases. Psychosexual orientation is standard. As in XXY Klinefelter syndrome, production of androgen is deficient and there is a subnormal response to stimulation with pituitary gonadotropins.

B. Etiology

Because the testis determining gene, *TDF,* is situated on the Y chromosome, development of testis in the absence of the Y would seem to violate a fundamental principle of genetics. But that problem is circumvented easily enough by assuming conservation of Y material or *TDF* function. Several ideas have been put forward to explain how this might happen.

1. Loss of the Y Chromosome: Mosaicism

There is an X-linked blood group called Xg. In the case of XX sex reversal reported by de la Chapelle *et al.* (1964), the father was Xg(a+) and the mother was Xg(a−). A normal XX sister of the XX male was Xg(a+), indicating that she had inherited one X chromosome from her father, but the XX male himself was Xg(a−). Because Xg(a+) is a dominant trait, it was concluded that the XX male had inherited both of his X chromosomes from his mother. At the time, this was explained by postulating nondisjunction of the X pair, fertilization of an XX egg with a Y sperm, and subsequent loss of the Y in cells of an early XXY embryo (but see Section II,B,3).

Although this scheme requires two abnormal events—nondisjunction and loss of the Y chromosome—XXY cell lines have been reported in several "XX males." In the case of Dosik *et al.* (1976), for example, the 46,XX karyotype was indicated uniformly in skin fibroblasts, but a single XXY cell was detected among 201 mitogen-stimulated blood lymphocytes, raising the question whether male development in this patient could be due to an undetected population of XXY cells in the gonad (testicular cells were not evaluated).

Miro *et al.* (1978) found an XXY cell among 500 lymphocytes in one XX male and a single fluorescent Y-body among 400 oral mucosa cells in another XX male. On the basis of these data and similar data from other laboratories, it was suggested that XX/XXY mosaicism could account for 17% of all cases of XX male syndrome.

Two XX/XXY males were described by de la Chapelle *et al.*(1990). In one, cytogenetic analysis of 20 mitogen-stimulated blood cells disclosed only XX cells, but 3 years later, analysis of 100 blood cells revealed 11 with the XXY complement. In the other patient, mosaicism was immediately diagnosed: 21 XXY cells were identified among 100 blood leukocytes.

2. Y-Autosome Translocation

Koo *et al.* (1977) described a Y-22 translocation in a retarded male with the 45,X/45,X,der(22)t(22;Y) karyotype, and Verjaal *et al.* (1978) described a Y-22 translocation in a phenotypically normal male infant with the 46,X,t(Y;22)(q11;p11) karyotype. Moreover, Y-autosome translocations per se have been reported in several cases not involving sex reversal (reviewed in Wachtel, 1983; see discussion in Verjaal *et al.*, 1978). Thus it would seem reasonable to assume that at least some cases of XX sex reversal are due to Y-autosome translocation. Yet we know of no clearcut demonstration of Y-autosome translocation among human XX males.

3. Y–X Translocation

The human X and Y chromosomes are known to pair during meiosis. This occurs in the distal short arms of the X and Y due to a sharing of bases in the pseudoautosomal region (PAR), which consists of some 3000 kb of DNA in each chromosome (Ellis and Goodfellow, 1989). Homology of the PAR in the X and Y chromosomes is maintained by regular crossing-over during spermatogenesis. When the PAR is deleted from the X, as in the patient of Mohandas *et al.* (1992), X–Y pairing does not occur and crossing-over is precluded.

We have seen that Y–X crossover causes XX sex reversal in the mouse (Chapter 4, this volume). There is now considerable evidence that the same is true in the human. In fact, sex reversal in the majority of XX males is due to Y–X crossover, or more specifically, to transfer of the testis determining gene, *TDF,* from distal Yp near the PAR, to distal Xp. This could also explain the aberrant inheritance of the Xg(a) trait in XX males.

Early evidence for Y–X crossover in human XX males was provided by Madan and Walker (1974), who noted that the terminal segment of the short arm of one of the X chromosomes was longer than the terminal segment of the other. In our own study, a minute additional band was identified in the short arm of one X chromosome in another XX male (the additional band, which could not be positively identified, did not fluoresce and was no longer than the short arm of the Y) (Wachtel *et al.*, 1976). And in yet another study, Xp segments of different length were found among the X-chromosome pairs in 9 of 13 XX males (Evans *et al.*, 1979);

these differences were not found in normal XX females. A Y-specific DNA fragment of 3.5 kb was identified in 1 of 3 XX males in this study. The authors concluded that the additional material in their patients was Y-derived and that Y–X translocation could account for 70% of all cases of XX sex reversal.

4. Mutation of a Downstream Gene

Suppose that the testis-determining pathway is initiated by a gene on the Y chromosome (*TDF*) and that it proceeds through a series of steps involving other autosomal or X-linked genes:

$$\overset{TDF}{\mathrm{Y}} \rightarrow \mathrm{A} \rightarrow \mathrm{B} \rightarrow \mathrm{C} \rightarrow \mathrm{n} \rightarrow \mathrm{TESTIS}$$

Constitutive mutation at any of these downstream loci would enable testicular organogenesis, even in the absence of the Y chromosome (induction of gene A, for example, in the absence of the normally required signal from the product of *TDF*). Conversely, a deleterious mutation at any of these loci could block development of the testis, even when the Y chromosome is intact (see Tommerup *et al.*, 1993).

Gene mutation must be considered as an alternative to X–Y translocation in the etiology of XX males, because XX sex reversal sometimes occurs in familial clusters (Nicholis *et al.*, 1972; Yamada and Isurugi, 1981; also see Figure 2). Minowada *et al.* (1979) described XX sex reversal in two brothers with penoscrotal hypospadias and small intrascrotal testes. Surgical exploration, performed when the brothers were 1.3 and 2.5 years old, respectively, revealed a normal male picture internally; histology revealed "immature testes." Cytogenetic data were unremarkable and the mode of inheritance could not be ascertained.

De la Chapelle *et al.* (1978) studied normal-appearing second-cousins with the 46,XX karyotype and small, sterile testes. They were distantly related to a third XX male through a common male ancestor. Although autosomal sex-reversing genes seemed to be indicated by the pedigree, evidence of Y-specific DNA was later obtained in all three XX males by use of molecular probes (Page *et al.*, 1985).

III. XX True Hermaphrodites

A. Historical Notes

In his *Metamorphoses IV*, Ovid tells us that the first hermaphrodite was Hermaphroditus, the handsome son of Hermes and Aphrodite. When he

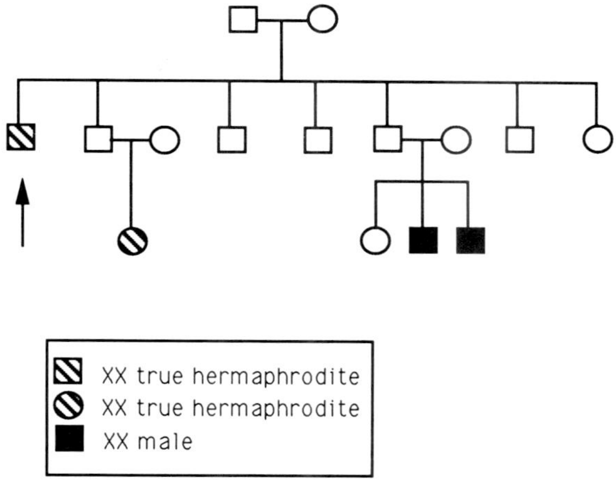

Figure 2 Familial XX sex reversal. Pedigree of family including two XX true hermaphrodites and two XX males (Skordis *et al.*, 1987). The propositus, a masculinized true hermaphrodite with scrotal ovotestes (arrow) had six normal siblings (and occurrence of genital abnormalities in other relatives was denied by his parents). A younger brother of the propositus sired a true hermaphrodite with bilateral ovaries and an appended right testis. She was reared as a female after clitoridectomy and removal of the testis. Another brother of the propositus sired two XX males, both with ambiguous external genitalia. The genealogy is consistent with occurrence of an autosomal sex-determining gene (see also Kasdan *et al.*, 1973).

was a lad of 15 years, Hermaphroditus came upon a crystal pool where he was discovered by the nymph, Salmacis. She was taken by his beauty and made advances to him "but the boy blushed rosy red, for he knew not what love is." When he threatened to run away, Salmacis withdrew and hid nearby in a thicket. Hermaphroditus thought he was alone. He removed his garments to bathe in the clear water, whereupon Salmacis could no longer contain her passion. She threw off her own clothes and leapt into the water, enwrapping him in a serpentine embrace. "Strive as you may, wicked boy," she cried, "you shall not escape me." Salmacis prayed that she might never be separated from Hermaphroditus, and the gods granted her prayer—literally. Their bodies were fused, and from two, they became one, with one face and one form, neither male nor female, but having the features of both (see account in Gordon, 1979).

Of course hermaphroditism is more than myth. It was mentioned by the Talmudic sages in the sociolegal context of the Oral Law: "An hermaphrodite (lit. *androgynos*) is in some ways like men and in others, like women; and in some ways like both men and women and in others, like neither men nor women." That designation was probably legal rather than morphologic. But there were additional paragraphs that specified the laws applying to hermaphrodites and hermaphroditism. Some of those provide hints concerning the physiology of the condition as it was recognized nearly 2000 years ago (From the *Mishnah, Bikkurim 4*):

> How is he (the hermaphrodite) like men? He is defiled by seminal issue like men; and he must dress like men; and he marries like men, but he may not be taken as a wife . . . like men, he may not be alone with women . . . and like men he must not transgress the commandment . . . 'neither shalt thou mar' (the corners of thy beard; Leviticus 19:27). How is he like women? He is defiled by red flow like women; and like women he may not be alone with men . . . if like women he has been unlawfully cohabited with, he becomes ineligible (to eat) of the priest's due (he may not marry into the lineage of the priests).
>
> How is he like both men and women? Persons incur guilt for smiting him or cursing him, as in the case of men and women; and one who kills him unwittingly is exiled, but if intentionally, is put to death . . .
>
> How is he not like men and women? . . . his valuation is not valid unlike (the case of) both men and women, and he may not be sold as a Hebrew bondsman.

From these paragraphs it seems clear that the hermaphrodite under consideration had a testis that contained Sertoli cells as well as testosterone-secreting Leydig cells, hence the references to seminal issue and beard. And it seems likely that he possessed ovarian tissue and estrogen-secreting granulosa cells, hence the reference to red flow.

It is remarkable that these criteria are used by physicians today to define hermaphroditism. According to our "modern" clinical designation, a true hermaphrodite is a person with ovarian and testicular tissue, either in separate gonads or in a single gonad called the ovotestis. The testicular part should contain seminiferous tubules and the ovarian part should contain follicles—a combination often associated with profound and tragic ambiguity of the secondary sex traits.

B. The Syndrome

The case of Saenger *et al.* (1976) is typical. The subject was a 2-year-old child with a clitoris of 4 cm length, fused labioscrotal folds and a single perineal opening. There was a right inguinal hernia but the gonads were not palpable. A vagina and uterus were present, as indicated by dye injection. Testosterone levels were elevated in blood and urine, in response to stimulation with chorionic gonadotropin in the presence of dexamethasone (which suppresses the adrenal glands). This implied presence of testicular Leydig cells. When the child was explored surgically, physicians found an underdeveloped uterus and fallopian tubes, a left ovary with epididymis, and a right ovotestis with epididymis and vas deferens. The ovotestis contained ovarian and testicular morphologies in about equal proportions: there were numerous primordial follicles with germ cells and sterile degenerative testicular tubules. Leydig cells were present in the interstitium. The karyotype was 46,XX without evidence of mosaicism. [For a comprehensive description of XX true hermaphroditism among 23 African Bantus, see van Niekerk (1974).]

In adult XX true hermaphrodites, the most striking features are ambiguous genitalia and well-developed breasts. The internal genital ducts may develop along male or female lines, or both, and the body habitus may vary from female to intersexual to male. The most common gonad in XX true hermaphroditism is the ovotestis: at least one ovotestis was found in 93 of the 116 patients (80.2%) surveyed by van Niekerk and Retief (1981), and 30 of 37 (81.1%) surveyed by Ramsay *et al.* (1988). [It might be worth noting that when alternative gonadal types are present—testis versus ovotestis, or ovotestis versus ovary, for example—the more testicular gonads tend to be situated in the right side. This has been attributed to differential growth rates in the left and right gonads of the developing embryo (see discussion by Mittwoch, 1986).]

The testes and testicular portions of ovotestes in XX true hermaphrodites resemble closely the testes in XX males. The tubules contain only Sertoli cells and there is no evidence of spermatogenesis. But the ovarian portions contain normal-appearing follicles that may be functional, and indeed, pregnancy has been reported in at least four human XX true hermaphrodites (see review of three cases in Wachtel, 1983). The patient of Narita *et al.* (1975) possessed a left ovotestis and right ovary. After removal of the ovotestis and surgical reconstruction of the external genitalia, this patient gave birth by caesarian section to a normal XY male.

C. Etiology

In XX true hermaphroditism, as in 46,XX male syndrome, we are left with the question of testicular development in the absence of the Y chromosome. As with XX male syndrome the question can be answered by assuming conservation in one form or another of Y-material or Y-function. In fact several of the etiologies discussed above for sex reversal of the XX male could also explain sex reversal of the XX hermaphrodite.

It could be argued, for example, that a residuum of Y-bearing cells in an XX/XXY mosaic embryo could generate testis versus ovary—or ovotestis, when the XXY line is confined to a small portion of the single gonad. Or it could be argued that X–Y translocation can generate testis versus ovary (or ovotestis), depending on whether the X;Y translocation chromosome or the normal X chromosome is inactivated in a developing gonad or part of a developing gonad. This is the hypothesis of Ferguson-Smith (1966), according to which variation of the relative proportion of inactivated X or X;Y chromosomes in the developing gonads can generate phenotypes ranging from male to hermaphrodite to female.

Constitutive mutation of a downstream X-linked gene could have the same effect. Given two X chromosomes, one normal (X) and one bearing the mutant gene (X^M), random lyonization could produce a variety of gonadal phenotypes—depending on the relative proportion of activated X^M chromosomes in a developing gonad.

IV. Familial Occurrence of XX Male Syndrome and XX True Hermaphroditism: The Syndromes as Alternative Faces of the Same Coin

A. Family Studies

We have seen that more than one XX male can occur in the same kindred. XX males can likewise occur in the same kindred as XX true hermaphrodites. An example is given by Kasdan *et al.* (1973). The propositus was a 2-month-old child with ambiguous external genitalia and bilateral scrotal ovotestes. Uterus and fallopian tubes were not present. The karyotype was 47,XX,mar+, indicating presence of an unidentified marker chromosome. A brother of the propositus was evaluated at the age of 3 years for hypospadias and chordee. He possessed bilateral scrotal testes; there was no evidence of ovarian tissue. The karyotype was 46,XX without the marker. A paternal uncle of the propositus was evaluated at age 31 after a 9-year marriage without children. The external genitalia were unremarkable. The testes were scrotal and small (the patient did not consent to a testicular biopsy). The karyotype was 47,XX,mar+, as in the propositus. On the basis of the genealogy, it was inferred that sex reversal in this kindred was due to a paternally transmitted autosomal gene.

More recently Skordis *et al.* (1987) described a similar family. The propositus was notable for prominent gynecomastia, small penis, penoscrotal hypospadias, and bifid scrotum containing bilateral ovotestes. The karyotype was 46,XX. An unaffected younger brother of the propositus fathered a child with ambiguous genitalia, normal uterus and tubes, proximal vagina, and bilateral ovaries with an appended right testis. As in the propositus, the karyotype was 46,XX. Another unaffected brother fathered two 46,XX males. One had a small phallus with severe chordee, hypospadias, and bifid scrotum containing hypoplastic testes, but no ovarian tissue. Similar traits were described in the other child. An autosomal sex-reversing gene was indicated by the genealogy (Fig. 2).

Neither of these cases is so striking as the family of American cocker spaniels in which an XX male with unambiguous testes was whelped by an XX true hermaphrodite with bilateral ovotestes (the pup inherited his testicles from his *mother*) (Selden *et al.*, 1978), yet these and other less dramatic observations in the human (Berger *et al.*, 1970; Palmer *et al.*, 1989; Pereira *et al.*, 1991) provide compelling evidence that XX male sex reversal and XX true hermaphroditism are closely related conditions. If two members of a family are sex-reversed by a common genetic defect, it remains to be determined why one should become a male with unambiguous testes, and the other a true hermaphrodite with gonads containing testicular and ovarian portions.

B. A Spectrum of Phenotypes

Given (*a*) the familial occurrence of sex-reversed XX males and XX true hermaphrodites, and (*b*) the persistence of uterus and tubes in some XX males, and hypospadias and marked ambiguity of the external genitalia in others, it seems clear that there is a spectrum of phenotypes in 46,XX subjects carrying Y-chromosomal loci or activated sex-determining genes downstream of the Y (Table I). At the one end is the XX female with dysgenetic or even fertile gonads, and at the other, the sterile XX male. Intermediate stages would include XX males with pronounced genital ambiguity, and XX true hermaphrodites, some of whom are virilized and some, feminized (Ramsay *et al.*, 1988; van Niekerk, 1974). Virilized hermaphrodites might appear as XX males with gynecomastia and scrotal ovotestes, whereas the more feminized hermaphrodites might actually bear offspring (see Section III,B above). In each case, the sex phenotype would be determined ultimately by the relative expression of testis-inducing *TDF* and downstream genes in the gonadal primordium. In the case of X-linked genes or genes translocated to the X, this could be influenced, in turn, by lyonization. Yet despite this continuum, there is no reason to think that cases of XX sex reversal having alternative etiologies (gene mutation versus mosaicism versus X–Y transfer) are "related conditions," as is evidently the case when XX males and XX true hermaphrodites are found in a kindred.

Table I
Phenotypes of XX Subjects

	XXM	XXM	XXTH	XXTH	XXF
Habitus	M	Eunuchoid	MI	FI	F
Genitals	M	Ambiguous	Ambiguous	Ambiguous	F
Y DNA	+	−	−(+)	−(?)	−
Gonads	Testes	Testes	OT	OT	O (dys)
Example	Petit *et al.* (1987)	Ferguson-Smith *et al.* (1990a)	Berkovitz *et al.* (1992)	Narita *et al.* (1975)	

Note. M, male; TH, true hermaphrodie; F, female; MI, masculinized intersex; FI, feminized intersex; OT, ovotestis (or ovaries and testes, etc.); O (dys), ovary (dysgenetic).

V. Molecular Genetics of XX Sex Reversal

A. Y-DNA in XX Males

As noted above, Evans *et al.* (1979) found a 3.5 kb Y-fragment in an XX male with cytogenetic evidence of X–Y interchange. A molecular demonstration of occult Y-material in XX males was provided by Guellaen *et al.* (1984). The latter study involved four sporadic cases with no evidence of mosaicism. Samples of DNA were digested with restriction enzymes, fractionated, and hybridized in Southern blots with 10 single-copy Y-specific DNA probes including 47c and 50f2. The samples from cases 1 and 4 reacted with six and five of the probes, respectively, whereas DNA from case 3 reacted with a single probe. Hybridization did not occur in case 2.

The same subset of probes was reactive in each of the Y-positive cases, signaling occurrence of overlapping chromosomal segments. The authors noted that if a single piece of the Y chromosome were present in each case, the method could be used to align Y-specific DNA fragments in a deletion map. They noted that X–Y interchange could account for presence of Y-DNA in the apparent absence of the Y chromosome, especially in patient 3, in whom aberrant inheritance of the Xg blood group had been observed.

B. Deletion Mapping of the Y Chromosome

Inasmuch as these data were consistent with X–Y interchange, and with occurrence of Y-fragments of differing sizes in the genomes of XX males, it now seemed feasible to map the Y chromosome by using panels of Y-specific restriction fragments. In the period following the report of Guellaen *et al.* (1984), a number of mapping studies were therefore undertaken (Page *et al.*, 1985; Affara *et al.*, 1986, 1987; Müller *et al.*, 1986; Müller, 1987; also see Chapter 10 this volume). In the study by Vergnaud *et al.* (1986), 23 Y-specific probes were used to screen DNA from 19 XX males, 2 XX true hermaphrodites, and 6 patients with various structural abnormalities of the Y chromosome. The results indicated that the patients carried a single contiguous segment of the Y; this enabled construction of a map containing seven intervals, with assignment of *TDF* to an interval in the short arm. Y-DNA was detected in 12 of the 19 XX males tested; the amount of Y-DNA was variable among the different subjects.

By the same method, Page *et al.* (1987) constructed a deletion map of Yp containing 13 intervals. In one of the intervals ("1A2"), they found a probe, pDP1007, that could be used to identify Y-specific DNA in every mammal tested. The amino acid sequence corresponding to pDP1007 resembled that of regulatory proteins. In light of these findings, and because 1A2 was present in the genome of an XX male who carried only 300 kb of Y-DNA, it was proposed that the 1A2 interval included *TDF* (see discussions of *ZFY* in Chapters 1 and 2, this volume).

A comprehensive deletion map of the human Y chromosome has now been published (Vollrath *et al.*, 1992). Based on a study of 96 patients with partial Y chromosomes (including 23 XX males), the map contains 132 loci, and 43 ordered intervals. Each interval is defined by chromosomal breakpoints and contains less than 800 kb DNA.

C. X–Y Interchange in XX males: Y(+) and Y(−) Classes

As noted above, Guellaen *et al.* (1984) obtained early molecular evidence for X–Y interchange in XX sex reversal. More recently, by *in situ* hybridization with probe pDP105, which identifies sequences concentrated in Yp, Andersson *et al.* (1986) detected Yp-DNA in the distal short arm of one of the X chromosomes in each of three other XX males. These findings were extended in the study by Petit *et al.* (1987), who evaluated Y-DNA in nine XX males (eight sporadic and one brother of an XX true hermaphrodite) including two of the patients of Andersson *et al.* (1986).

By use of Yp probes 47z and 50f2 (Vergnaud *et al.*, 1986), six of the nine XX males were found to be positive for Y-DNA and three, including the familial case, to be negative. Paternal Y-linked alleles were present in the six Y(+) cases (a telomeric sequence and at least one proximal locus—DXYS17 or *MIC2*—in five, and a telomeric sequence in one). On the other hand, the paternal X-pseudoautosomal alleles were deleted in each. This was the first demonstration by use of molecular methods that sex reversal in Y(+) XX males is correlated not only with gain of sequences in Yp, but also with *loss* of terminal Xp. By contrast, the entire paternal X-pseudoautosomal region was present in two of the Y(−) XX males. It is worth emphasizing that hypospadias was observed in one of the Y(−) XX males, and that the other was the brother of an XX true hermaphrodite, raising the question whether ambiguous genitalia in such cases might be related to the absence of Y-DNA.

D. Ambiguous XX Males Are Y(−)

Occurrence of alternative classes of the XX male syndrome, one Y(+) and one Y(−), was confirmed by Ferguson-Smith *et al.* (1990a). They described 10 Y(+) XX males and four Y(−) XX males, among whom the latter were notable for gynecomastia and abnormality of the external genitalia. Male sex determination of the Y(−) XX males was attributed to mutation of an X-linked testis-determining gene. In a later report, Ferguson-Smith *et al.* (1990b) noted absence of the *SRY* gene in five Y(−) XX males and pointed out that *SRY* itself is not essential for male sex determination. Another *SRY*-negative case was described by van der Auwera *et al.* (1992) among six XX males evaluated by *in situ* hybridization [in the five Y(+) cases, *SRY* was found translocated to terminal Xp]. The Y(−) patient was remarkable for gynecomastia, cryptorchid testes, and small phallus (also see Abbas *et al.*, 1990).

By the polymerase chain reaction (PCR), and by Southern blotting with 17 different Y probes, Numabe *et al.* (1992) analyzed loci including *SRY*, *ZFY*, and *PABY* in DNA from 14 XX males. Various Y sequences were identified in 8 of the XX males; none was detected in 6. Among the 6 Y(−) males, 5 exhibited hypospadias. This was not observed in any of the (Y+) XX males.

E. Most XX True Hermaphrodites Are Y(−)

Using a panel of six Y-specific probes, including 47z (DXYS5), 50f2, Fr 80-II, Fr 35-II, and the Yq-specific repeats pJA1143 (DYZ1) and pHY2.1 (DYZ2), Waibel *et al.* (1987) studied DNA from 11 XX males and 7 XX true hermaphrodites. DNA from each of the XX males was positive for at least one of the probes in Southern blots. Although hybridization did not occur with DNA from any of the XX true hermaphrodites in these trials, one of the cases was later found to be Y(+) when probed for sequences adjacent to the pseudoautosomal boundary (Jäger *et al.*, 1990).

In a study of XX true hermaphroditism among southern African blacks, Ramsay *et al.* (1988) tested DNA in 10 sporadic cases with 6 Y-specific probes (pDP132, pDP61, pDP105, pDP31, pDP97, and pY431-*Hinf*A). Hybridization was not observed in any case. Later, those authors tested pDP1007 (*ZFY*) in *Taq*I digests from several of these and other cases. Again there was no evidence of hybridization.

Four XX true hermaphrodites were studied by Damiani *et al.* (1990). Y-specific sequences including *ZFY* could not be identified in any of the four. The same is true in three cases of XX true hermaphroditism evaluated by Abbas *et al.* (1990).

Two XX true hermaphrodites were studied by Pereira *et al.* (1991) using probes for *SRY*. Both of the patients were negative for the probes. Similar data were obtained by Berkovitz *et al.* (1992) by use of a variety of Y-specific probes including *SRY*, *ZFY*, and centromeric pDP97, in five XX true hermaphrodites, including one with a terminal Xp deletion. Whereas sequences including *SRY* and *ZFY* were detected by Southern blotting or PCR in the patient with the Xp deletion, no Y-specific sequences were detected in the other four patients.

Nakagome *et al.* (1991) evaluated 25 Y-specific probes in an XX true hermaphrodite with mild gynecomastia, small penis, hypospadias, right scrotal testis, and left inguinal ovotestis. Two Y-loci were detected by PCR in this case: *SRY* and the proximal border of the pseudoautosomal region (*PABY*) (also see Palmer *et al.*, 1989).

The same two loci were detected in 3 of 30 XX true hermaphrodites evaluated by McElreavey *et al.* (1992) (in two of those cases, *SRY* was localized to terminal Xp by *in situ* hybridization; this is consistent with X–Y interchange). Thus, although Y-DNA may occasionally be found in true hermaphroditism, this seems to be rare (the methods described above do not rule out the possibility of cryptic Y-bearing cell lines in gonadal tissue, however).

VI. Conclusions

There is compelling evidence for the existence of X-linked and autosomal genes in the sex-determining pathway (review in Simpson, 1989). Thus the apparent function of the Y-situated *TDF* "master gene" is to initiate the pathway and thereby induce the expression of downstream elements. Theoretically, sex reversal could ensue when any of these genes is prevented from functioning in an XY embryo, or gratuitously induced in an XX embryo.

We have reviewed evidence that XX sex reversal can be caused by (*a*) mosaicism for Y-bearing cell lines; (*b*) transfer of Y-DNA, including *TDF*, to the distal terminus of an X chromosome; and (*c*) mutation of a preexisting autosomal gene. It seems reasonable to conclude that some cases of XX sex reversal are caused by mutation of an X-linked gene as well.

Because X–Y interchange has been documented in most XX males and a few XX true hermaphrodites, the proposal of Ferguson-Smith (1966) would seem to be vindicated: Differences between the two conditions can be due to the proportion of X;Y translocation chromosomes that are activated or lyonized in a developing gonad. But this cannot account for sex reversal in the majority of XX true hermaphrodites because they rarely possess Y-DNA, and because XX males and true hermaphrodites are sometimes found in the same family, evidently sex-reversed by autosomal genes. Whereas mosaicism, X–Y transfer, or X-linked genes readily explain how one embryo becomes a male and another a true hermaphrodite, it is less clear how an autosomal dominant can cause development of an umambiguous testis in one family member and an ovotestis in another.

Finally two kinds of XX male have been identified: One is Y(+) and one is Y(−). The latter kind is often found in familial aggregates. He is generally less virilized than are (Y+) XX males and frequently exhibits abnormalities of the external genitalia such as small phallus and hypospadias. This underscores the importance of the Y chromosome (*TDF*) in the determination of male sex and the ultimate development of the male phenotype.

References

Abbas, N. E., Toublanc, J. E., Boucekkine, C., Toublanc, M., Affara, N. A., Job, J.-C., and Fellous, M. (1990). A possible common origin of "Y-negative" human XX males and XX true hermaphrodites. *Hum.Genet.* **84,** 356–360.

Affara N. A., Florentin, L., Morrison, N., Kwok, K., Mitchell, M., Cooke, A., Jamieson, D., Glasgow, L., Meredith, L., Boyd, E., and Ferguson-Smith, M.A. (1986). Regional assignment of Y-linked DNA probes by deletion mapping and their homology with X chromosome and autosomal sequences. *Nucleic. Acids Res.* **14,** 5353–5373.

Affara, N. A., Ferguson-Smith, M. A., Magenis, R. E., Tolmie, J. L., Boyd, E., Cooke, A., Jamieson, D., Kwok, K., Mitchell, M., and Snadden, L. (1987). Mapping the testis determinants by an analysis of Y-specific sequences in males with apparent XX and XO karyotypes and females with XY karyotypes. *Nucleic Acids Res.* **15,** 7325–7342.

Andersson, M., Page, D. C., and de la Chapelle, A. (1986). Chromosome Y-specific DNA is transferred to the short arm of X chromosome in human XX males. *Science* **233,** 786–788.

Berger, R., Abonyi, D., Nodot, A., Vialatte, J., and Lejeune, J. (1970). Hermaphrodisme vrai et "Garcon XX" dans une fratrie. *Rev. Eur. Etud. Clin. Biol.* **15,** 330–333.

Berkovitz, G. D., Fechner, P. Y., Marcantonio, S. M., Bland, G., Stetten, G., Goodfellow, P. N., Smith, K. D., and Migeon, C. (1992). The role of the sex-determining region of the Y chromosome (SRY) in the etiology of 46,XX true hermaphroditism. *Hum. Genet.* **88,** 411–416.

Damiani, D., Billerbeck, A. E., Goldberg, A. C., Setian, N., Fellous, M., and Kalil, J. (1990). Investigation of the ZFY gene in XX true hermaphroditism and Swyer syndrome. *Hum. Genet.* **85,** 85–88.

de la Chapelle, A. (1972). Analytic review: Nature and origin of males with XX sex chromosomes. *Am. J. Hum. Genet.* **24,** 71–105.

de la Chapelle, A., Hortling, H., Niemi, M., and Wennström, J. (1964). XX sex chromosomes in a human male. First case. *Acta Med. Scand., Suppl.* **412,** 25–38.

de la Chapelle, A., Koo, G. C., and Wachtel, S. S. (1978). Recessive sex- determining genes in human XX male syndrome. *Cell (Cambridge, Mass.)* **15,** 837–842.

de la Chapelle, A., Hästbacka, J., Korhonen, T., and Mäenpää, J. (1990). The etiology of XX sex reversal. *Reprod. Nutr. Dev. Suppl.* **1,** 39s–49s.

Dosik, H., Wachtel, S. S., Khan, F., Spergel, G., and Koo, G. C. (1976). Evidence for the presence of Y-chromosomal genes in a phenotypic male with a 46,XX karyotype. *J. Am. Med. Assoc.* **236,** 2505–2508.

Ellis, N., and Goodfellow, P. N. (1989). The mammalian pseudoautosomal region. *Trends Genet.* **5,** 406–410.

Evans, H. J., Buckton, K. E., Spowart, G., and Carothers, A. D. (1979). Heteromorphic X chromosomes in XX males: Evidence for the involvement of X-Y interchange. *Hum. Genet.* **49,** 11–31.

Ferguson-Smith, M. A. (1966). X-Y chromosomal interchange in the aetiology of true hermaphroditism and of the XX Klinefelter's syndrome. *Lancet* **2,** 475–476.

Ferguson-Smith, M. A., Cooke, A., Affara, N. A., Boyd, E., and Tolmie, J. L. (1990a). Genotype-phenotype correlations in XX males and their bearing on current theories of sex determination. *Hum. Genet.* **84,** 198–202.

Ferguson-Smith, M. A., North, M. A., Affara, N. A. and Briggs, H.(1990b). The secret of sex. *Lancet* **336,** 809–810.

Gordon, H. (1979). Ancient ideas about sex determination. *In* "Genetic Mechanisms of Sexual Development" (H. L. Vallet and I. H. Porter, eds.), pp. 1–32. Academic Press, New York.

Guellaen, G., Casanova, M., Bishop, C., Geldwerth, D., Andre, G., Fellous, M., and Weissenbach, J. (1984). Human XX males with Y single-copy DNA fragments. *Nature (London)* **307,** 172–173.

Jäger, R. J., Ebensperger C., Fraccaro, M., and Scherer, G. (1990). A ZFY-negative 46,XX true hermaphrodite is positive for the Y pseudoautosomal boundary. *Hum. Genet.* **85,** 666–668.

Kasdan, R., Nankin, H. R., Troen, P., Wald, N., Pan, S., and Yanaihara, T. (1973). Paternal transmission of maleness in XX human beings. *N. Engl. J. Med.* **288,** 539–545.

Koo, G. C., Wachtel, S. S., Krupen-Brown, K., Mittl, L. R., Breg, W. R., Genel, M., Rosenthal, I. M., Borgaonkar, D. S., Miller, D. A., Tantravahi, R., Schreck, R. R., Erlanger, B. R., and Miller, O. J. (1977). Mapping the locus of the H-Y gene on the human Y chromosome. *Science* **198,** 940–942.

Madan, K., and Walker, S. (1974). Possible evidence for Xp+ in an XX male. *Lancet* **1,** 1223.

McElreavey, K., Rappaport, R., Vilain, E., Abbas, N., Richaud, F., Lortat- Jacob, S., Berger, R., Le Coniat, M., Boucekkine, C., Kucheria, K., Temtamy, S., Nihoul-Fekete, C., Braunner, R., and Fellous, M. (1992). A minority of 46,XX true hermaphrodites are positive for the Y-DNA sequence including SRY. *Hum. Genet.* **90,** 121–125.

Minowada, S., Kobayashi, K., Isurugi, K., Fukutani, K., Ikeuchi H., Hasegawa, T., and Yamada, K. (1979). Two XX male brothers. *Clin. Genet.* **15,** 399–405.

Miro, R., Caballin, M. R., Marsini, S., and Egozcue, J. (1978). Mosaicism in XX males. *Hum. Genet.* **45,** 103–106.

Mittwoch, U. (1986). Males, females and hermaphrodites. *Ann. Hum. Genet.* **50,** 103–121.

Mohandas, T. K., Speed, R. M., Passage, M. B., Yen, P. H., Chandley, A. C., and Shapiro, L. J. (1992). Role of the pseudoautosomal region in sex-chromosome pairing during male meiosis: Meiotic studies in a man with a deletion of distal Xp. *Am. J. Hum. Genet.* **51,** 526–533.

Müller, U. (1987). Mapping of testis-determining locus on Yp by the molecular genetic analysis of XX males and XY females. *Development (Cambridge, UK), Suppl.* **101,** 51–58.

Müller, U., Donlon, T., Schmid, T., Fitch, M., Richter, N., Lalande, C., and Latt, S. M. (1986). Deletion mapping of the testis determining locus with DNA probes in 46,XX males and 46,XY and 46,X,dic(Y) females. *Nucleic Acids Res.* **14,** 6489–6505.

Nakagome, Y., Seki, S., Fukutani, K., Nagafuchi, S., Nakahori, Y., and Tamura, T. (1991). PCR detection of distal Yp sequences in an XX true hermaphrodite. *Am. J. Med. Genet.* **41,** 112–114.

Narita, O., Manba, S., Nakanishi, T., and Ishizuka, N. (1975). Pregnancy and childbirth in a true hermaphrodite. *Obstet. Gynecol.* **45,** 593–595.

Nicholis, G. L., Hsu, L. Y., Sabetghadam, R., Kardin, M. B., Chernay, P. R., Mathur, D. P., Rose, H. G., Hirschhorn, K., and Gabrilove, J. L. (1972). Klinefelter's syndrome in identical twins with the 46,XX chromosome constitution. *Am. J. Med. Genet.* **52,** 482–491.

Numabe, H., Nagafuchi, S., Nakahori, Y., Tamura, T., Kiuchi, H., Namiki, M., Kohda, N., Fukushima, Y., Fuse, H., Kusano, M., Arai, T., Matsuzaki, Y., Fukutani, K., Isurugi, K., Kuroki, Y., Ikeuchi, T., Yoshida, M., Minowada, S., and Nakagome, Y. (1992). DNA analyses of XX and XX-hypospadiac males. *Hum. Genet.* **90,** 211–214.

Page, D. C., de la Chapelle, A., and Weissenbach, J. (1985). Chromosome Y-specific DNA in related human XX males. *Nature (London)* **315,** 224–226.

Page, D. C., Mosher, R., Simpson, E. M., Fisher, E. M. C., Mardon, G., Pollack, J., McGillivray, B., de la Chapelle, A., and Brown, L. G. (1987). The sex-determining region of the human Y chromosome encodes a finger protein. *Cell (Cambridge, Mass.)* **51,** 1091–1104.

Palmer, M. S., Sinclair, A. H., Berta, P., Ellis, N. A., Goodfellow, P. N., Abbas, N. E., and Fellous, M. (1989). Genetic evidence that ZFY is not the testis-determining factor. *Nature (London)* **342,** 937–939.

Pereira, E. T., Cabral de Almeida, J. C., Gunha, A. C. Y. R. G., Patton, M., Taylor, R., and Jeffrey, S. (1991). Use of probes for ZFY, SRY, and the Y pseudoautosomal boundary in XX males, XX true hermaphrodites, and an XY female. *J. Med. Genet.* **28,** 591–595.

Petit, C., de la Chapelle, A., Levilliers, J., Castillo, S., Noel, B., and Weissenbach, J. (1987). An abnormal X-Y interchange accounts for most but not all cases of human XX maleness. *Cell (Cambridge, Mass.)* **49,** 595–602.

Ramsay, M., Bernstein, R., Zwane, E., Page, D. C., and Jenkins, T. (1988). XX true hermaphroditism in southern African blacks: An enigma of primary sex differentiation. *Am. J. Hum. Genet.* **42,** 4–13.

Saenger, P., Levine, L. S., Wachtel, S. S., Korth-Schutz, S., Doberne, Y., Koo, G. C., Lavengood, R. X., German, J. L., and New, M. I. (1976). Presence of H-Y antigen and testis in 46,XX true hermaphroditism. Evidence for Y-chromosomal function. *J.Clin. Endocrinol. Metab.* **43,** 1234–1239.

Selden, J. R., Wachtel, S. S., Koo, G. C., Haskins, M. E., and Patterson, D. F. (1978). Genetic basis of XX male syndrome and XX true hermaphroditism: Evidence in the dog. *Science* **201,** 644–646.

Simpson, J. L. (1989). Genetic heterogeneity in XY sex reversal: Potential pitfalls in isolating the testis-determining factor (TDF). *In* "Evolutionary Mechanisms in Sex Determination" (S. S. Wachtel, ed.), pp. 265–277. CRC Press, Boca Raton, FL.

Skordis, N., Stetka, D. G., MacGillivray, M. H., and Greenfield, S. P. (1987). Familial 46,XX males coexisting with familial 46,XX true hermaphrodites in same pedigree. *J. Pediatr.* **110,** 244–248.

Tommerup, N., Schempp, W., Meinecke, P., Pedersen, S., Bolund, L., Brandt, C., Goodpasture, C., Guldberg, P., Held, K. R., Reinwein, H., Saugstad, O. D., Scherer, G., Skjeldal, O., Toder, R., Westvik, J., van der Hagen, C. B., and Wolf, U. (1993). Assignment of an autosomal sex reversal locus (*SRA1*) and campomelic dysplasia (*CMPD1*) to 17q24.3–q25.1. *Nat. Genet.* **4,** 170–174.

van der Auwera, B., Van Roy, N., De Paepe, A., Hawkins, J. R., Liebaers, I., Castedo, S., Dumon, J., and Speleman, F. (1992). Molecular cytogenetic analysis of XX males using Y-specific DNA sequences, including SRY. *Hum. Genet.* **89,** 23–28.

van Niekerk, W. A. (1974). "True Hermaphroditism. Clinical, Morphologic and Cytogenetic Aspects." Harper & Row, Hagerstown, MD.

van Niekerk, W. A., and Retief, A. E. (1981). The gonads of human true hermaphrodites. *Hum. Genet.* **58,** 117–122.

Vergnaud, G., Page, D. C., Simmler, M.-C., Brown, L., Rouyer, F., Noel, B., Botstein, D., de la Chapelle, A., and Weissenbach, J. (1986). A deletion map of the human Y chromosome based on DNA hybridization. *Am. J. Hum. Genet.* **38,** 109–124.

Verjaal, M., Treffers, P. E., Nagai, Y., and Leschot, N. J. (1978). Prenatal diagnosis of a de novo Y/22 translocation. *J. Med.Genet.* **15,** 475–486.

Vollrath, D., Foote, S., Hilton, A., Brown, L. G., Beer-Romano, P., Bogan, J. S., and Page, D. C. (1992). The human Y-chromosome: A 43-interval map based on naturally occurring deletions. *Science* **258,** 52–59.

Wachtel, S. S. (1983). "H-Y Antigen and the Biology of Sex Determination." Grune & Stratton, New York.

Wachtel, S. S., and Bard, J. (1981). The XX testis. *In* "The Intersex Child" (Z. Laron, ed.), pp. 116-132. Karger, Basel.

Wachtel, S. S., Koo, G. C., Breg, W. R., Thaler, H. T., Dillard, G. M., Rosenthal, I. M., Dosik, H., Gerald, P. S., Saenger, P., New, M., Lieber, E., and Miller, O. J. (1976). Serologic detection of a Y- linked gene in XX males and XX true hermaphrodites. *N. Engl. J. Med.* **295,** 750–754.

Waibel, F., Scherer, G., Fraccaro, M., Hustinx, T. W. J., Weissenbach, J., Wieland, J., Mayerová, M., Back, E., and Wolf, U. (1987). Absence of Y-specific DNA sequences in human 46,XX true hermaphrodites and in 45,X mixed gonadal dysgenesis. *Hum.Genet.* **76,** 332–336.

Yamada, K., and Isurugi, K. (1981). H-Y antigen studies in thirty patients with abnormal gonadal differentiation: Correlations among sex chromosome complement, H-Y antigen and gonadal type. *Jpn. J. Hum. Genet.* **26,** 227–235.

13

XY Sex Reversal in the Human

Stephen S. Wachtel and Joe Leigh Simpson
Division of Reproductive Genetics
Department of Obstetrics and Gynecology
University of Tennessee, Memphis
Memphis, Tennessee 38163

Molecular Genetics of Sex Determination

I. Introduction

Normal differentiation of the human ovary requires the presence of two X chromosomes. In 45,X embryos—those lacking a second sex chromosome—the ovary appears to develop normally for about the first trimester of gestation, after which there is a progressive loss of germ cells and follicles, and an increase in the amount of connective tissue (Carr *et al.*, 1968). Although a few germ cells and follicles may persist into adulthood, the 45,X gonad is usually represented at or around the time of birth by a white fibrous streak, called a streak gonad. In affected adults, the streak gonad is about 2–3 cm long and 0.5 cm thick, and contains whorls of fibrous tissue resembling the tissue found in the normal ovary. The rete ovarii may be present and a few hilar cells, but streak gonads are, in general, endocrinologically inert.

According to Ohno (1967), the mammalian X and Y chromosomes originated from an ancestral pair of homologous autosomes, and the Y lost most of its genetic material to become a specialized testis inducer while the X remained more or less invariant (see discussion in Wolf *et al.*, 1992). Traditionally, it has been assumed that the genes necessary for normal ovarian development were among those lost from the Y. However when the Y fails to induce a testis, the gonadal primordium initiates ovarian development, so it can be deduced that some of the genes necessary for ovarian development are retained. Still, there is a progressive loss of germ cells and follicles, and a corresponding increase in the appearance of connective tissue, as in the 45,X gonad. The result is a 46,XY dysgenetic gonad, a fibrous streak devoid of germ cells and follicles, that closely resembles the 45,X streak gonad.

In 46,XY females, as in 45,X females, germ cells and follicles occasionally persist beyond the perinatal period. For example, germ cells and occasional follicles were observed in the "underdeveloped ovaries" of a 3-month-old 46,XY female studied by Cussen and MacMahon (1979). When the child was reexamined $3\frac{1}{2}$ years later, the gonads were characterized as small, firm and "flattened," and follicles were absent. As another example, Russell *et al.* (1982) described follicular structures in the gonads of a 17-year-old female with 46,XY karyotype and amenorrhea of 7 months duration. This patient had experienced irregular menses starting at age 14, but when the gonads were removed at age 17, histologic examination revealed only degenerative follicles and a few hilus cells (Fig. 1).

The condition that we have described is called XY gonadal dysgenesis, or sometimes "pure gonadal dysgenesis" because the somatic signs of Turner syndrome—short stature, webbing of the neck, cubitus valgus—

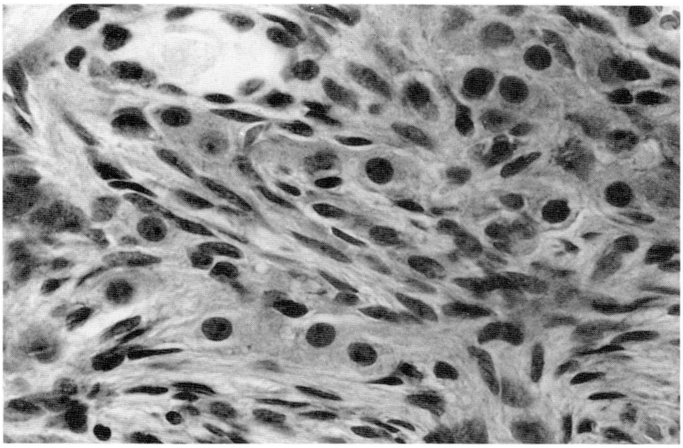

Figure 1 46,XY gonadal dysgenesis. Section of left gonad exhibiting ovarian stroma and degenerative follicle from 17-year-old female evaluated for secondary amenorrhea. From Russell *et al.* (1982). Reprinted by permission.

are seldom encountered in XY females. In this chapter we review: (*a*) evidence for X-linked, autosomal, and Y-linked genes in the etiology of XY gonadal dysgenesis; (*b*) the molecular genetics of the condition, with special reference to *SRY*, identified recently as the human testis-determining gene (Chapter 2); (*c*) the relationship between serological H-Y antigen and the occurrence of germ cell tumors in XY gonadal dysgenesis; and (*d*) related syndromes in which dysgenetic XY gonads are found in association with severe congenital abnormalities.

II. The Syndrome

In 46,XY gonadal dysgenesis, also called Swyer syndrome (Swyer, 1955), the chief signs are delayed puberty and primary amenorrhea. The habitus is somewhat eunuchoid but appearance is usually feminine. Height ranges from average to tall. Although the gonads are usually characterized as dysgenetic ovaries, some have been described as dysgenetic testes. Most often the gonads resemble those in 45,X females with Turner syndrome. A striking feature of XY gonadal dysgenesis that distinguishes it from 45,X Turner syndrome is the high incidence of gonadal tumors—especially gonadoblastoma and dysgerminoma (Fig. 2). In our earlier survey (Wachtel, 1983), gonadal tumors were known to be present in 17

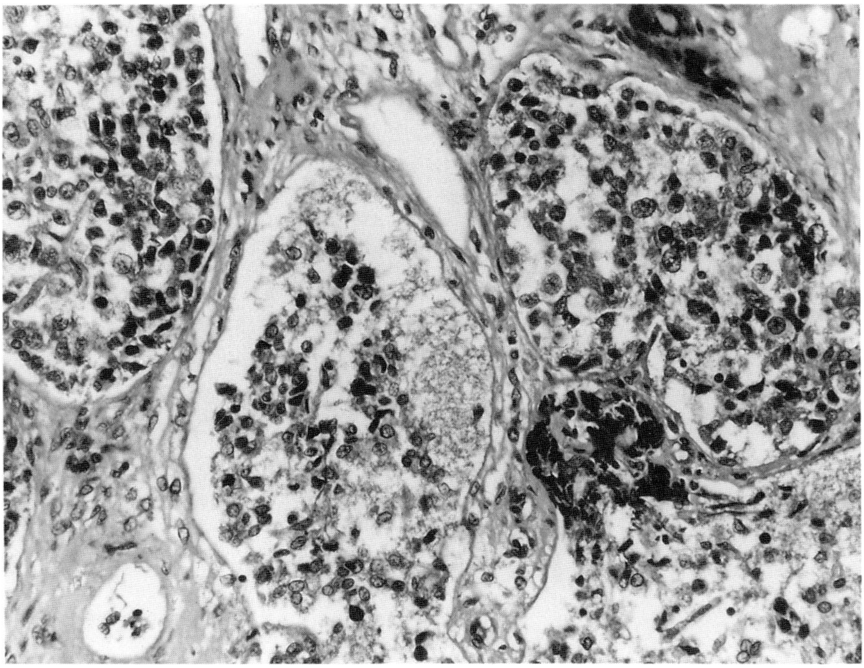

Figure 2 Section of gonadoblastoma-dysgerminoma from the patient in Villanueva *et al.* (1984). Note tubular structures containing nests of germ cells and Sertoli-like cells (see Scully 1970). Published in Warner *et al.* (1985). Reprinted by permission.

of 46 (37%) patients with the condition (see also Simpson, 1976, and discussion below).

Secretion of gonadotropins (LH, FSH) is elevated and there is a reduction in circulating testosterone and estradiol. Although breast development is generally minimal, the breasts are well developed in some cases. In untreated cases, breast development often occurs together with estrogen-producing gonadal tumors (Simpson *et al.*, 1982); but no strict correlation could be found between oncogenesis and breast development by Warner *et al.* (1985). Irrespective of this, axillary and pubic hair are usually sparse. The external genitalia are female but the clitoris may be enlarged in some cases. The labia minora may likewise be poorly developed in untreated cases, and the vagina may be narrow or infantile. Without administration of sex steroids, the Müllerian derivatives are also underdeveloped. With hormonal treatment, however, secondary sexual development is essentially normal, including uterine size. Thus, the uterus of a patient with XY gonadal dysgenesis can sometimes sustain a pregnancy. This is accomplished by *in vitro* fertilization of donor ova with donor (husband's) sperm and transfer of the resulting embryo to the uterus of the patient, or alternatively, by direct transfer of an intact embryo from the uterus of a healthy donor (Cornet *et al.*, 1990).

III. Genetic Basis of XY Gonadal Dysgenesis

A. X-Linked Genes

The most obvious explanation for development of the XY streak gonad is deletion or mutation within the Y-linked testis-determining gene. Indeed, we have evidence of lesions in the Y in some cases of XY gonadal dysgenesis (see below), but other cases can be attributed directly to failure of X-linked genes. Familial occurrence was described in 18 kindreds surveyed by Simpson *et al.* (1981); X-linked recessive inheritance has now been indicated in three of those, and in another kindred studied more recently (Sternberg *et al.*, 1968; Espiner *et al.* 1970; German *et al.* 1978; and Mann *et al.* 1983.

The first kindred (Sternberg *et al.*, 1968) included three cases of XY gonadal dysgenesis: the 13-year-old proband, a maternal cousin, and a maternal aunt. When the gonads of the proposita were removed, histological examination disclosed a focus of poorly defined tubules containing "primitive Sertoli (or granulosa) cells." This was classified as a sex cord tumor.

The second kindred (Espiner *et al.*, 1970) included a 21-year-old proposita with secondary amenorrhea, two sisters with primary amenorrhea, and two nieces, i.e., three sibships in two generations. Streak gonads were present in each of the patients; there was no evidence of neoplasia.

The third kindred (German *et al.*, 1978; Simpson *et al.*, 1976) included two affected sisters, a maternal cousin, and a maternal aunt. Both sisters had streak gonads; the cousin was diagnosed with bilateral gonadoblastoma. Histology was not available in the maternal aunt, who had died some 50 years earlier after removal of an abdominal tumor.

The fourth family with apparent X-linked XY gonadal dysgenesis was described by Müller (1983) and Mann *et al.* (1983). This included three affected sisters, an affected maternal second cousin, and a maternal aunt. Gonadal tumors were present in each of the five subjects. One of the siblings and the second cousin and maternal aunt were typed positive for the serological H-Y antigen (H-Ys^{+}); the other siblings were not studied serologically. The maternal grandmother of the affected siblings had a

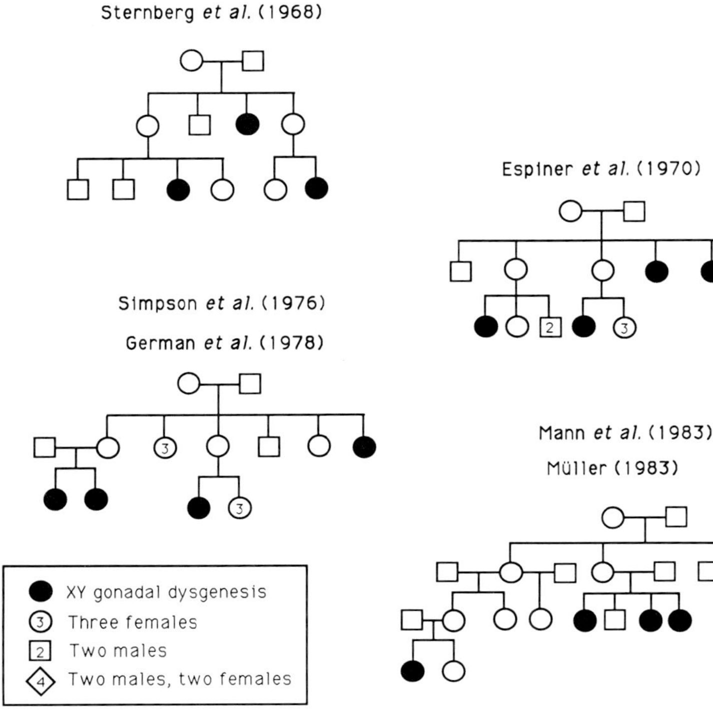

Figure 3 Abbreviated pedigrees. 46,XY gonadal dysgenesis in familial aggregates consistent with X-linked inheritance. An extended pedigree of the family of Müller is shown in Mann *et al.* (1983).

sister who was infertile and presumed to be an affected XY female. The maternal great grandmother had three infertile sisters, each presumed to be an affected XY female (Mann *et al.*, 1983).

Abbreviated pedigrees of each of these families are shown in Fig. 3. The genealogies favor existence of a gene or group of genes on the X chromosome, the mutation of which can block organogenesis of the normal testis. Physical rearrangements of the X have been described in certain patients with XY gonadal dysgenesis and other somatic lesions (Bernstein *et al.*, 1980; Stern *et al.*, 1990; P. Arn, personal communication; and see discussion below). However, a male-limited autosomal dominant gene is not ruled out.

According to Ohno's Law, a gene X-linked in one mammalian species is likely to be X-linked in another (Ohno, 1967; also see Chapter 6, this volume). So it might be worth describing here the unusual mode of sex determination found in the wood lemming, *Myopus schisticolor*, a small mouselike rodent inhabiting the mossy woodlands of Scandinavia. This animal is remarkable for a skewed sex ratio of 1 : 4 (male : female). There are two kinds of female in this species—one producing male and female offspring, and the other producing only female. The two kinds of female are indistinguishable phenotypically, but the former has a 32,XX karyotype, and the latter, a 32,XY karyotype. The female Y chromosome is the same as that of the male. In fact it is the X that is rearranged. G-banding has revealed two classes of X chromosome in this species (Fig. 4): one (denoted X^*) is found in XY females (X^*Y); the other (denoted X) is found in XY males (Herbst *et al.*, 1978). This indicates occurrence of an X-linked gene, the mutation (rearrangement) of which blocks testicular differentiation and thereby leads to development of the XY ovary, which, in contrast to the situation in humans, is usually fertile in rodents. A

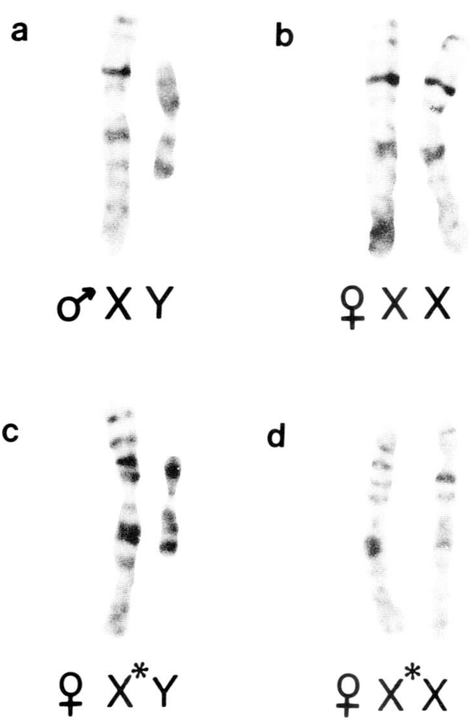

Figure 4 Sex chromosomes of the Scandinavian wood lemming, *Myopus schisticolor*. G-banding patterns in: (a) normal male; (b) normal female; (c) X^*Y female, which bears only female progeny (due to elimination of the Y and "nondisjunctional" doubling of the X early in the life of the fetal ovary); (d) carrier X*X female. From Herbst *et al.* (1978). Reprinted by permission.

similar though pathological condition has been described in the human; see Section VI below.

B. Autosomal Genes

According to the attractive hypothesis of Wolf *et al.* (1980), the primary genes for testis determination are not on the sex chromosomes, but on autosomes. These genes are said to be repressed by X-linked genes that escape inactivation, and derepressed by a gene on the Y (Adinolfi *et al.*, 1982). It follows that XY gonadal dysgenesis could be caused by failure of any of these genes, and in some cases, manifested as an autosomal recessive or male-limited autosomal dominant trait. After excluding families in which X-linked inheritance seemed likely, and another family in which the mutant gene seemed to be inherited from a male, Simpson *et al.* (1981) analyzed segregation ratios in 24 kindreds. Among those were 12 families with a single case of XY gonadal dysgenesis, 6 families with two cases and 6 with three cases. The proportion of affected relatives was not different from the proportion that would be expected if the condition were inherited as a male-limited autosomal recessive trait.

Autosomal etiology for some cases of XY gonadal dysgenesis would seem likely because most cases are sporadic (and see Section VII,C). Moreover, as we shall see in the next section, mutations are not often identified within the Y-linked testis-determining gene (*SRY*).

C. Y-Linked Genes

Loss of the short arm of the Y chromosome has been correlated with gonadal dysgenesis in several cases, including, for example, phenotypic females with 46,X,i(Yq), denoting isochromosome of the long arm of the Y, and absence of the short arm (Yp) (see Wachtel, 1983). Another example is the patient of Rosenfeld *et al.* (1979), a female with a deletion in the Y short arm [46,X,del(Yp)]. At birth the proband exhibited the external genitalia of a normal female, but also marked lymphedema of the hands and feet and increased nuchal skin folds. When the patient was explored surgically at 3 months of age, normal uterus and tubes were found. But evaluation of the gonads revealed fibrous stroma; follicles were absent. At 9 years of age, the patient manifested various Turner stigmata: webbing of the neck, low posterior hairline, cubitus valgus, pigmented nevi, short stature (25th percentile). Reevaluation of the karyotype disclosed a subtle deletion within Yp; the deletion was not present in the Y chromosome of the father (see next section).

This study provided evidence for (*a*) a Yp-located testis-determining gene, the loss or mutation of which could cause 46,XY gonadal dysgenesis; and (*b*) a gene or group of genes in Yp that act in concert with homologous X-linked ("anti-Turner") genes, to block development of the Turner stigmata in normal males (see Rosenfeld *et al.*, 1979). In fact, Fisher *et al.* (1990) later isolated two genes, one on the human X chromosome and one on the Y, that appeared to encode isoforms of a ribosomal protein (differing at 19 of 263 amino acids). Because the X-linked gene was not inactivated, and because the Y-linked gene was mapped to a segment implicated in Turner syndrome (1A1B), it was proposed that absence of either gene could cause appearance of the Turner stigmata.

IV. *SRY* and the The Molecular Genetics of XY Gonadal Dysgenesis

Elsewhere in this volume, we discuss formally the molecular biology of the human Y chromosome, with special reference to *SRY* (sex determining region Y), the prime candidate for the testis determining gene. In brief, *SRY* is located in the smallest region of the Y known to be sex-determining, and microinsertion of the murine homolog (called *Sry*) can cause testicular differentiation in transgenic XX male mice.

The *SRY* gene is identified by probe pY53.3, which consists of a 2.1 kb *Hind*III sequence; a 0.9-kb subclone of pY53.3 is delineated by two *Hinc*II restriction sites (Sinclair *et al.*, 1990). This flanks an open reading frame containing a conserved amino acid motif (Fig. 5). Of the 80 amino acids contained within this motif in the human, 64 were present in the rabbit (80% homology); only 54% homology was observed outside the conserved motif. Given the degree of conservation within that motif (a strong homology has been noted with proteins of the yeast *Schizosaccharomyces pombe*), it seemed reasonable to assume a critical function for that part of the SRY protein in man. If *SRY* is the testis-determining gene, it would

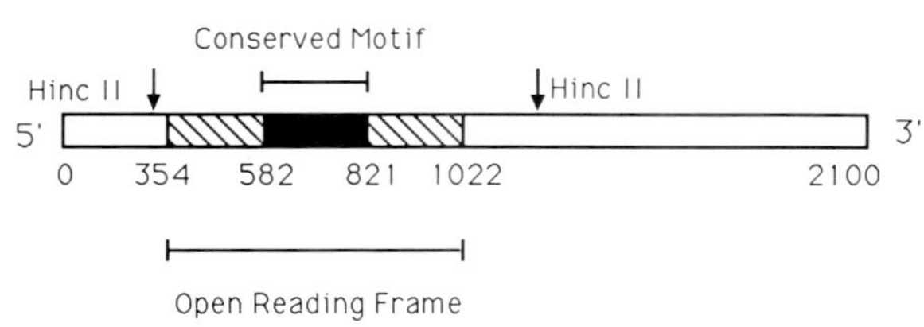

Figure 5 Graphic representation of 2.1-kb probe pY53.3 that identifies *SRY*. Numbers represent base pairs from the 5′ end. See text. Modified after Sinclair *et al.* (1990), by permission.

follow that mutations likely to block its function—base changes or deletions within the conserved motif, for example—could upset normal development of the testis. In other words, some women with XY gonadal dysgenesis would be expected to have mutations within this portion of *SRY*.

For that reason the nucleotide sequence of the conserved motif was compared in normal males and affected XY females. In an early study by Berta *et al.* (1990), a segment of 609 bp including the conserved motif was sequenced in 50 normal males and in 11 females with XY gonadal dysgenesis. No mutations were found in the male controls or in 9 of the affected females. A single base change (G to C, predicting a change from valine to leucine in the SRY protein) was detected in one of the XY females, but the same change was present in her father, thus indicating a benign polymorphism presumably unrelated to failure of testicular differentiation in the affected female (but see Vilain *et al.*, 1992). However, a *de novo* mutation (G to A, predicting a change from methionine to isoleucine) was found in another XY female; this mutation was not present in her father. Both mutations were mapped near the 5′ boundary of the conserved motif. In another study (Jäger *et al.*, 1990), a 422-bp sequence including the conserved motif was sequenced in 12 females with XY gonadal dysgenesis, and a 4-base deletion (AGAG), found in one. This region of the father's Y chromosome was intact. No other variants were identified.

Table I
SRY Mutations in Patients with 46,XY Gonadal Dysgenesis

Study	No. Cases	Mutations detected	Percentage mutations
Berta *et al.* (1990)	11	2[a]	18
Jäger *et al.* (1990)	12[b]	1	8
Behzadian *et al.* (1991)	5	0	0
Jäger *et al.* (1991,1992)	3	1[c]	33
Hawkins *et al.* (1992a)	10	2	20
Beer-Romero *et al.* (1992)	9	1	11
Hawkins *et al.* (1992b)	5	3[a]	60
Pivnick *et al.* (1992)	8	0	0
Totals	63	10 (7)[d]	15.9 (11.1)

[a] One of the mutations was also present in the father of the patient.
[b] Two patients in this study were siblings; no other family members were affected.
[c] This mutation was also present in the father, two brothers, and an uncle of the patient.
[d] The number and percentage of *de novo* mutations are given in parentheses.
For additional data, see Affara *et al.* (1993).

The sequencing studies so far reported are cataloged in Table I. No mutations were found by Behzadian *et al.* (1991) in 5 cases of XY gonadal dysgenesis, or by Pivnick *et al.* (1992) in 8 cases. A mutation was identified by Jäger *et al.* (1991,1992), but the same change was present in the father, two brothers, and an uncle of the patient. Beer-Romano *et al.* (1992) found a single *de novo* mutation among 9 cases of XY gonadal dysgenesis. Two studies were performed by Hawkins *et al.* (1992a,b). In the first, they identified 2 *de novo* mutations among 10 cases; in the second, they identified 2 *de novo* mutations including a single-base deletion, and one mutation also present in the patient's father. It was suggested (Hawkins *et al.*, 1992b) that the increased frequency of mutations in the second study was due to the histological confirmation of "complete" gonadal dysgenesis in the patients under evaluation. The authors noted that gonadal dysgenesis had not been confirmed histologically in most reports involving sequencing of the *SRY* gene. The high frequency of erroneous diagnoses in cases referred for sequencing was also noted by Pivnick *et al.*, (1992).

Yet, at the time of writing, only 7 *de novo* mutations have been discovered within the *SRY* gene among 63 patients with XY gonadal dysgenesis. Most of the mutations were single-base changes, one was a single base deletion, and one was a 4-base deletion. Three other mutations were found not only in the affected women, but also in their fathers. In addition to these observations, two SRY mutations (one silent) were found in the ovotestis of a 46,XY true hermaphrodite (Braun *et al.*, 1993).

V. Interaction between *SRY* and Downstream Genes

Of interest is the sequencing of *SRY* in a family studied by Vilain *et al.* (1992). In this family, gonadal histology *was* performed. As a result, 46,XY gonadal dysgenesis was confirmed in three females: the propositus, her younger sister and a paternal aunt. The same single-base substitution was found in the conserved domain of the *SRY* gene not only in the three affected females, but also in two of the normal males, including the father of the propositus [the propositus and her father are described previously in Berta *et al.* (1990)]. To explain the alternative phenotypes in subjects with a common *SRY* variant, the authors suggested these possibilities: (*a*) expression of SRY product at near threshold levels allowing SRY activity to be influenced by environmental factors; (*b*) fortuitous polymorphism in five subjects among whom three carry mutant downstream genes; and (*c*) interaction of *SRY* with independently segregating alleles of another gene. [In the family described by Jäger *et al.*(1992), a T-to-C transition resulted in substitution of a serine for phenylalanine at amino acid position 109. The substitution was found in an XY female, and several of her male relatives, as noted above, but it was not detected among 176 male controls.]

Taken together, the foregoing studies seem to indicate that although base changes or deletions within the Y-situated *SRY* gene can cause failure of testicular differentiation and thus XY gonadal dysgenesis, mutations in "downstream" X-linked or autosomal genes have the same effect. It should be emphasized that the sequencing studies cited here were focused mainly within the conserved motif of the *SRY* gene (which evidently encodes a DNA binding protein), and that mutations in other parts of the *SRY* gene might alter the testis-determining protein in such a way to render it nonfunctional. Of course this possibility cannot be ruled out until the entire gene is sequenced in each of the patients with XY gonadal dysgenesis.

VI. Germ Cell Tumors in XY Gonadal Dysgenesis

Women with XY gonadal dysgenesis have an increased risk of developing germ cell tumors (gonadoblastoma, dysgerminoma) (see Simpson, 1976). There are indications that this may be related to the expression or nonexpression of the H-Ys ("male") antigen. Because of its phylogenetic stability, the serological H-Y antigen (H-Ys), was once considered a prime candidate for the mammalian testis-determining gene (the more likely candidate is *SRY;* see Chapter 2). Thus a number of laboratories set out to evaluate presence or absence of H-Ys antigen in subjects in whom gonadal sex did not conform to chromosomal sex or body sex—in XX males and XX true hermaphrodites, for example, and in women with XY gonadal dysgenesis. If H-Ys were the product of the testis-determining gene, it was to be expected that the antigen would be present in subjects with testes and absent in subjects lacking a testis, despite karyotype or body type.

This expectation was borne out in XX males and XX true hermaphrodites (Wachtel, 1983), but the results seemed contradictory in women with XY gonadal dysgenesis, in whom alternative H-Ys$^+$ and H-Ys$^-$ phenotypes were identified. The apparent contradiction could be resolved by assuming that H-Ys antigen and at least one other factor—a receptor, for example—were required for testicular differentiation. If development of the embryonic testis were secondary to reaction of an inducer and its receptor, failure of the inducer (as in H-Ys$^-$ women) would be tantamount to failure of the receptor (as in H-Ys$^+$ women); the presumptive testis-inducer would be inoperative in both cases.

When Warner *et al.* (1985) surveyed 55 women with XY gonadal dysgenesis, 41 were H-Ys$^+$ and 14 were H-Ys$^-$. Gonadal tumors were found almost exclusively in the H-Ys$^+$ class. Overall, gonadal tumors were present in 25 of the 55 cases surveyed (45%); of the 25 women with tumors, 24 were H-Ys$^+$. We also know of 5 more patients with XY gonadal dysgenesis: three were typed H-Ys$^-$ and two, H-Ys$^+$; gonadal tumors were not described in any of the patients in this group (Wachtel *et al.*, 1986; Jaswaney *et al.*, 1989; and our unpublished results). The precise relationship between H-Ys, gonadal tumors, and heritability remains to be clarified, but the alternative H-Ys phenotypes in XY gonadal dysgenesis imply alternative etiologies in this syndrome and thereby support the genealogic data reviewed above.

VII. Related Syndromes

A. Genito-Palato-Cardiac (Gardner–Silengo–Wachtel) Syndrome

1. The Clinical Condition

Various reports indicate the existence of a novel syndrome characterized by genital, palatal, and cardiac anomalies. Greenberg *et al.* (1987) described two affected siblings and recommended that the condition be called the Gardner–Silengo–Wachtel syndrome in consideration of those who had contributed most to its delineation. Gardner *et al.* (1970) had described a female with cleft palate, micrognathia, kyphoscoliosis, and equinovarus. The external genitalia were female and the müllerian structures were developed, despite the presence of testicular gonads. Three years later, Silengo *et al.* (1974) described a 46,XY female with low-set ears, cleft palate, micrognathia, and several other congenital defects. Uterus and fallopian tubes were present; the gonads resembled dysgenetic testes with Sertoli cells.

Wolman *et al.* (1980) described a 46,XY female infant with abnormal facies, cleft palate, thoracic dysplasia, micrognathia, macroglossia, and webbed neck. When the child died at 5 months of age, necropsy revealed normal uterus and tubes, and two small ovarian gonads containing ovarian stroma and structures resembling primitive seminiferous cords or Pflüger's cords. The 46,XY siblings described by Greenberg *et al.* (1987) exhibited similar anomalies, but one had female external genitalia, müllerian derivatives, and ovarian gonads, whereas the other exhibited first-degree penile hypospadias and empty scrotal sac with pelvic testes; uterus and tubes were absent in the latter. Those cases (Simpson, 1989; Wachtel, 1983) are consistent with occurrence of a discrete class of 46,XY sex reversal involving multiple congenital abnormalities that result in early demise. This could be due to multiple effects of a single gene or, alternatively, to a chromosomal abnormality involving multiple loci. Patients with XY sex reversal and a rearrangement of the X chromosome have in fact been described.

2. XY Sex Reversal and Xp Duplication

Bernstein *et al.* (1980) discovered an anomalous additional band in the short arm of the X chromosome in a profoundly retarded girl with multiple congenital abnormalities that included ventricular septal defect, cleft palate, asymmetric skull and facies, asymmetric chest with widely spaced nipples, prognathic jaw, low-set ears, and clinodactyly of the fifth digit. Initially the karyotype was designated as 46,Xp+Y, the additional band representing a duplication within the short arm [46,dup(X)(p21 → pter)Y]. The Y chromosome appeared to be normal. The proband died at 5 years of age. Postmortem studies revealed hypoplastic uterus and fallopian tubes. Ovaries could not be identified macroscopically, but microscopic examination of the uterine adnexae disclosed a small area of ovarian stroma containing scattered degenerative follicles. There was no sign of testicular morphology. The external genitalia were those of a normal female.

In this case the abnormal X chromosome (Xp+) was inherited from the mother of the proband, who in turn inherited it from the grandmother. The first child of the mother died shortly after birth and was not studied. The proband was the second child (Fig. 6). The third child was a normal female carrying the abnormal X chromosome (46,XXp+). When amniocentesis revealed the 46,Xp+Y karyotype in fetal cells during a fourth pregnancy, the parents elected to terminate the pregnancy at 20 weeks. The fetus was a female exhibiting multiple abnormalities similar to those of the proband. External and internal genitalia resembled those of a normal female fetus of 20 weeks' gestation. The gonads were fetal ovaries containing numerous follicles and germ cells (Fig. 7); there was no evidence of testicular tissue.

Other cases with X duplications have been reported. Stern *et al.* (1990) described a sex-reversed female with facial dysmorphism, hypertelorism, cleft palate, and hyperthyroidism. The karyotype was given as 46,Y,der dup(X)(p21 → p22.3). Again, the abnormal X was transmitted by the mother. The external genitalia were female; labia minora were hypoplastic.

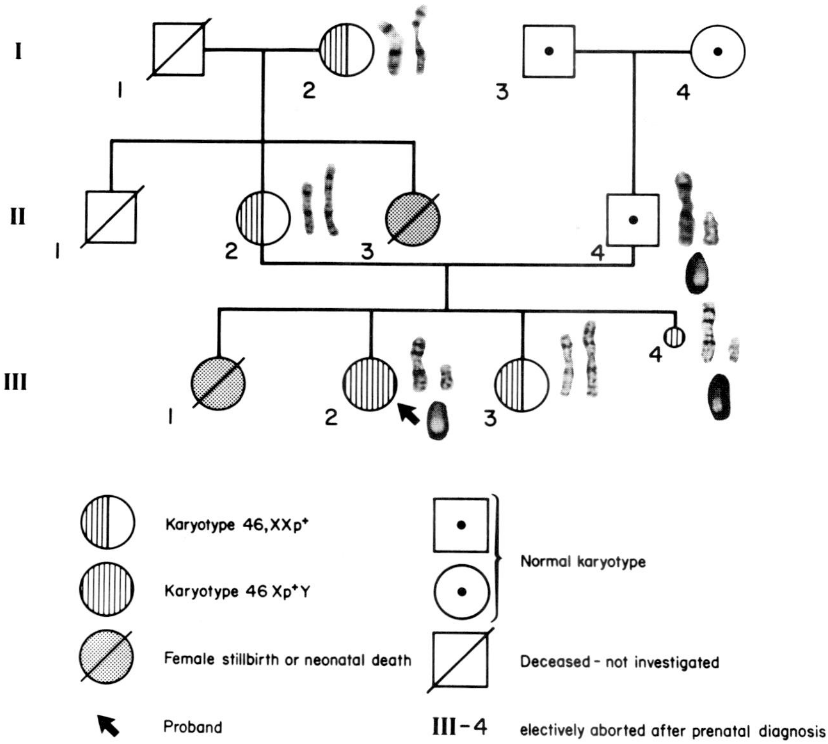

Figure 6 Inheritance of dup(Xp) chromosome (referred to as Xp+) in three generations. The abnormal band was found in the short arm of the X chromosome in the maternal grandmother (I-2), mother (II-2) and normal female sibling (III-3) of the proband, and in the proband (III-2; arrow) and her fetal sibling (III-4). The father's chromosomes (II-4) were unremarkable. G-banding with quinacrine-stained inserts. An example of the Genito-palato-cardiac syndrome (see Fig. 5 and text). From Bernstein *et al.* (1980). Reprinted by permission.

Two other females with a duplicated Xp chromosome and a Y showed multiple somatic abnormalities (Scherer *et al.*, 1989). The patients were evidently unrelated. Although the gonads were not studied, it seems reasonable to assume that they were ovaries or streaks. Endocrinologic evaluation revealed primary hypogonadism in one case, and there was no masculinization in either patient. The abnormal X chromosome resembled the one described by Bernstein *et al.* (1980): the duplicated region of the X included Xp21.2 → Xp22.2 (which contains the *ZFX* gene). Another case (Arn *et al.*, 1992) involved a "female" infant with ambiguous external genitalia and an XY complement with duplicated Xp21.

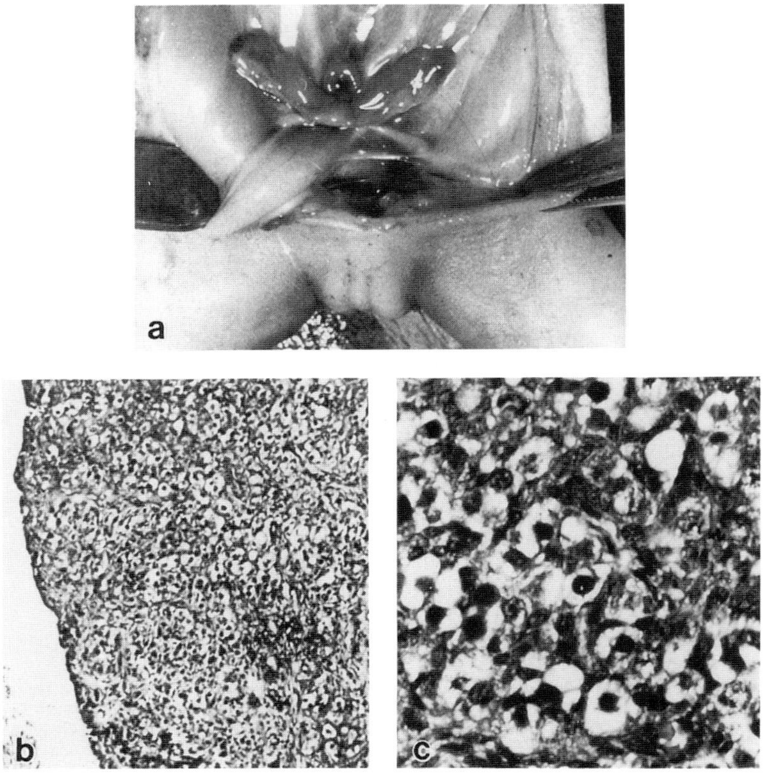

Figure 7 Female anatomy of 46,Xp+Y fetus with Genito-palato-cardiac syndrome. (a) Gross anatomy of internal and external genitalia. (b) Section of fetal ovary (original magnification ×90). (c) Section of fetal ovary showing numerous follicles (original magnification ×360). From Bernstein *et al.* (1980). Reprinted by permission.

In a related study, Ogata *et al.* (1992) described a sex-reversed Japanese child with dysmorphic features including frontal bossing, antimongoloid slant, low set ears and cleft palate. The child died at 3 years of age after chronic systemic lupus erythematosis (SLE)-like symptoms and cachexia. Necropsy revealed normal müllerian derivatives and no evidence of wolffian structures. The ovaries were represented by streak gonads containing ovarian stroma with germ cells and gonadoblastoma. Cytogenetic analysis revealed a 46,X,Yp+ karyotype. Evidently the Yp+ chromosome had been derived by transfer to Yp11.3 of an Xp fragment consisting of Xp21 → Xp22.3. As a result of the transfer, terminal Y-material had been lost (*DXYS15* → telomere), and the X-specific fragment consisting of (*DXS84* → *PABX*) and the proximal pseudoautosomal region

(*MIC2* → *DXYS17*) had been duplicated. The *SRY* gene was intact and there was no deletion in the Y-specific region.

Each of the foregoing cases involves female development and multiple congenital abnormalities in children remarkable for a duplication in Xp21. This is consistent with occurrence of an Xp-situated gene in the testis-determining pathway, downstream of *SRY*.

B. Smith–Lemli–Opitz Syndrome Type II

The Smith–Lemli–Opitz syndrome was first reported in 1964 (reviewed in Simpson, 1989). Affected individuals show characteristic facial dysmorphia, syndactyly between the second and third toes, and profound mental retardation. Males and females both can be affected. Genital abnormalities do not occur in females but hypospadias and cryptorchidism are common in males. Curry *et al.* (1987) defined a related but distinct condition that is now called Smith–Lemli–Opitz syndrome Type II. In a survey of 19 cases, they found facial characteristics similar to those in Smith–Lemli–Opitz syndrome (Type I). However, there were, in addition, more severe congenital abnormalities: cleft palate, microglossia, cardiac defects. These resembled but are considered distinct from those in the genito-palato-cardiac syndrome (Greenberg *et al.*, 1987). Severe genital ambiguity (sex reversal) was noted in some of the affected 46,XY patients, unlike the situation in Smith–Lemli–Opitz Type I. However, the gonads of the 46,XY patients were always testicular, unlike the situation in the genito-palato-cardiac syndrome. For a comprehensive review of these and other, related, conditions, see Simpson (1989).

C. Campomelic Dysplasia

This condition is notable for bowing of the long bones, small scapulae, brachydactyly, clinodactyly, "bell-shaped" chest, micrognathia and other skeletal anomalies, small mouth, cleft palate, and low-set dysplastic ears. Affected infants usually die within a few weeks of birth, after experiencing feeding and respiratory difficulties, and apneic episodes. The syndrome is due to a gene, *CDMP1*, on chromosome 17 (Tommerup *et al.*, 1993), and is often accompanied by XY gonadal dysgenesis: some two-thirds of patients with 46,XY are sex reversed (Table II). For example, a patient described in Bricarelli *et al.* (1981)—a female with campomelic dysplasia and 46,XY karyotype—had normal external genitalia, but when the child expired, necropsy disclosed normal uterus and tubes, and normal-appearing gonads (Case 1). Histologic evaluation revealed ovarian stroma, rete tubules, and a few scattered germ cells; follicles were scarce. A brother of the patient likewise exhibited signs of campomelic dysplasia, but he was unambiguously male with 46,XY karyotype and normal testes. Another 46,XY female with this syndrome was described in Bricarelli *et al.* (1981). In this patient (Case 5), external and internal genitalia appeared normal, and histologic evaluation of the gonads revealed ovarian stroma with medullary cords and "tubules similar to those of a rete ovarii of a newborn girl." Primary follicles were numerous, some containing degenerative oocytes.

Another sex-reversed child with campomelic dysplasia was described by Puck *et al.* (1981). This stillborn 46,XY female had a slightly enlarged clitoris, and uterus and fallopian tubes normal for a female of the same age, but the gonads contained undifferentiated stroma and primitive cords resembling "dysgenetic testicular tissue." The parents were consanguineous (see also Tommerup *et al.*, 1993).

D. Gonadal Dysgenesis in Families with Male Pseudohermaphroditism

XY gonadal dysgenesis has been reported in familial aggregates with male pseudohermaphroditism (testicular development in subjects with a female or intersexual body type). An early example was described by Barr *et al.* (1967). The proband was a hirsute female with 46,XY karyotype, enlarged clitoris, small uterus, and bilateral testes. A sister was a 46,XY female with small uterus and streak ovaries (and a nephew exhibited hypospadias). A similar family was described by Chemke *et al.* (1970). The proband was a 46,XY male pseudohermaphrodite with ambiguous external genitalia, uterus and fallopian tubes, small left testis with epididymis and vas deferens, and degenerative right testis connected to the right fallopian tube. Two sisters were diagnosed with 46,XY gonadal dysgenesis. One sister had streak gonads with ovarian stroma and scattered tubular elements, and the other had a right streak gonad and left dysgerminoma.

These conditions could be due to a mutant testis inducer with variable expression. However, it may be worth noting that pigmented nevi and cubitus valgus were present in one of the females studied by Chemke *et al.* (1970) and multiple nevi and clinodactyly V in the other, raising the question of mosaicism for 45,X cell lines in that family. Differences in the phenotype of the brother and sisters could then be due to alternative patterns of mosaicism.

Table II
Syndromes with Gonadal Dysgenesis

Syndrome	Karyotype	Habitus	Somatic phenotype	Inheritance[a]	Gonads
Turner	45,X[b]	Short female	Multiple anomalies	Loss of X, Y genes	Streaks
Swyer	46,XY	Eunuchoid	Variable; see text	A, X, Y	Streaks, tumors
Gardner–Silengo–Wachtel	46,XY; 46,Xp+Y; 46,XYp+	Female	Multiple anomalies	A (?), X	Dysgenetic ovaries[c]
Campomelic dysplasia[d]	46,XY	Female	Dwarfism, multiple anomalies	A	Dysgenetic ovaries
Wood lemming female[e]	32,XY	Female	Normal	X	Fertile ovaries

[a] A, X, Y; autosomal, X-linked, Y-linked.
[b] Turner syndrome has been described in association with many other karyotypes, including 45,X/46,X,i(Xq), 46,XYp− etc.
[c] This is the usual gonadal type; testicular-like architecture was described in the patient of Wolman *et al.* (1980). [Gonadal dysgenesis is found in association with other karyotypes (46,XX, for instance), and other syndromes (see Simpson, 1989).]
[d] Gonadal dysgenesis in this syndrome is due to the *SRA1* sex reversal gene, linked closely to *CMPD1* on chromosome 17q (Tommerup *et al.*, 1993).
[e] This normal condition is included here for comparison; the question whether there are fertile XY women remains to be answered.

VII. Conclusions

Taken together, the data that we have reviewed (Table II) indicate the following: (*a*) at least three genes are involved in testicular differentiation; (*b*) these genes are located on the Y-chromosome, the X chromosome, and an autosome(s), respectively; (*c*) male development is set in motion by the product of the Y-linked (*SRY*) gene; (*d*) the SRY product is a DNA-binding protein that influences the function of a "downstream" locus or group of loci.

Sporadic XY gonadal dysgenesis can result from mutation in any of the above three genes. Familial cases can result from inheritance of autosomal or X-linked mutants. Particular mutations within the *SRY* conserved motif may effect sex reversal in some XY family members but not others; this could depend on genetic background, i.e., on the interaction of SRY and downstream gene products.

XY gonadal dysgenesis can be stratified into two groups: one in which the H-Ys antigen is expressed and one in which it is not expressed. The risk of of gonadal tumors seems far greater in patients in whom the antigen is expressed.

References

Adinolfi, M., Polani, P., and Zenthon, J. (1982). Genetic control of H-Y synthesis: A hypothesis. *Hum. Genet.* **61,** 1–2.

Affara, N. A., Chalmers, I. J., and Ferguson-Smith, M. A. (1993). Analysis of the SRY gene in 22 sex-reversed XY females identifies four new point mutations in the conserved DNA binding domain. *Hum. Molec. Genet.* **2,** 785–789.

Arn, P., Wachtel, G., de Mello, M., and Wachtel, S. S. (1992). Duplication within the X-chromosomal region containing *ZFX* causes XY sex reversal. *Annu. Meet., Soc. Gynecol. Invest.,* San Antonio, *Sci. Program Abstr.* Abstr. No. 152, p. 184.

Barr, M. L., Carr, D. H., Plunkett, E. R., Soltan, H. C., and Wiens, R. G. (1967). Male pseudohermaphroditism and pure gonadal dysgenesis in sisters. *Am. J. Obstet. Gynecol.* **99,** 1047–1055.

Beer-Romano, P., Reindollar, R. H., Fusaris, K., Gray, M. R., and Page, D. C. (1992). Mutations identified by rapid screening of the SRY locus in 46,XY gonadal dysgenesis and true hermaphroditepatients with denaturing gradient gel electrophoresis (DGGE). *Annu. Meet. Soc. Gynecol. Invest., San Antonio Sci. Program Abstr.,* Abstr. No. 146, p. 181.

Behzadian, M. A., Tho, S. P. T., and McDonough, P. G. (1991). The presence of the testicular determining sequence, SRY, in 46,XY females with gonadal dysgenesis (Swyer syndrome). *Am. J. Obstet. Gynecol.* **165,** 1887–1890.

Bernstein, R., Jenkins, T., Dawson, T., Wagner, J., Dewald, G., Koo, G. C., and Wachtel, S. S. (1980). Female phenotype and multiple abnormalities in siblings with a Y-chromosome and partial X- chromosomal duplication: H-Y antigen and Xg blood group findings. *J. Med. Genet.* **17,** 291–300.

Berta, P., Hawkins, J. R., Sinclair, A. H., Taylor, A., Griffiths, B. L., Goodfellow, P. N., and

Fellous, M. (1990). Genetic evidence equating *SRY* and the testis-determining factor. *Nature* (*London*) **348,** 448–450.

Braun, A., Kammerer, S., Cleve, H., Löhrs, U., Schwarz, H.-P., and Kuhnle, U. (1993). True hermaphroditism in a 46,XY individual, caused by a postzygotic somatic point mutation in the male gonadal sex-determining locus (SRY): molecular genetics and histological findings in a sporadic case. *Am. J. Hum. Genet.* **52,** 578–585.

Bricarelli, F. D., Fraccaro, M., Lindsten, J., Müller, U., Baggio, P., Carbone, L. D. L., Hjerpe, A., Lindgren, F., Mayerová, A., Ringertz, H., Ritzen, E. M., Rovetta, D. C., Sicchero, C., and Wolf, U. (1981). Sex-reversed XY females with campomelic dysplasia are HY negative. *Hum. Genet.* **57,** 15–22.

Carr, D. H., Haggar, R. A., and Hart, A. G. (1968). Germ cell in theovaries of XO female infants. *Am. J. Clin. Pathol.* **49,** 521–526.

Chemke, J., Carmichael, R., Stewart, J. M., Geer R. H., and Robinson, A. (1970). Familial gonadal dysgenesis. *J. Med. Genet.* **7,** 105–111.

Cornet, D., Alvarez, S., Antoine, J. M., Tibi, Ch., Mandelbaum, J., Plachot, M., and Salat-Baroux, J. (1990). Pregnancies following ovum donation in gonadal dysgensis. *Hum. Reprod.* **5,** 291–293.

Curry, C. J. R., Carey, J. C., Holland, J. S., Chopra, D., Fineman, R., Golabi, M., Sherman, S., Pagon, R. A., Allanson, J., Shulman, S., Barr, M., McGravey, V., Dabiri, C., Schimke, N., Ives, E., and Hall, B. D. (1987). Smith-Lemli-Opitz Syndrome Type II: Multiple congenital anomalies with male pseudohermaphroditism and frequent early lethality. *Am. J. Med. Genet.* **26,** 45–57.

Cussen, L. J., and MacMahon, R. A. (1979). Germ cells and ova in dysgenetic gonads of a 46,XY female dizygotic twin. *Am. J. Dis. Child.* **133,** 373–375.

Espiner, E. A., Veale, A. M. O., Sands, V. E., and Fitzgerald, P. H. (1970). Familial syndrome of streak gonads and normal male karyotype in five phenotypic females. *N. Engl. J. Med* **283,** 6–1.

Fisher, E. M. C., Beer-Romero, P., Brown, L. G., Ridley, A., McNiel, J. A., Lawrence, J. B., Willard, H. F., Bieber, F. R., and Page, D. C. (1990). Homologous ribosomal protein genes on the human X and Y chromosomes: Escape from X inactivation and implications for Turner syndrome. *Cell* (*Cambridge, Mass.*) **63,** 1205–1218.

Gardner, L. I., Assemany, S. R., and Neu, R. I. (1970). A 46,XY female: Anti-androgenic effect of oral contraceptive? *Lancet* **2,** 667–668.

German, J., Simpson, J. L., Chaganti, R. S. K., Summitt, R. L., Reid, L. B., and Merkatz, I. R. (1978). Genetically determined sex-reversal in 46,XY humans. *Science* **202,** 53–56.

Greenberg, F., Greesick, M. V., Carpenter, R. J., Law, S. W., Hoffman, L. P., and Ledbetter, D. H. (1987). The Gardner-Silengo-Wachtel Syndrome: Male pseudohermaphroditism with micrognathia, cleft palate, and conotruncal cardiac defect. *Am. J. Hum. Genet.* **26,** 59–64.

Hawkins, J. R., Taylor, A., Berta, P., Levilliers, J., van der Auwera, B., and Goodfellow, P. N. (1992a). Mutational analysis of SRY: Nonsense and missense mutations in XY sex reversal. *Hum.Genet.* **88,** 471–474.

Hawkins, J. R., Taylor, A., Goodfellow, P. N., Migeon, C. J., Smith, K. D., and Berkovitz, G. D. (1992b). Evidence for increased prevalence of *SRY* mutations in XY females with complete rather than partial gonadal dysgenesis. *Am. J. Hum. Genet.* **51,** 979–984.

Herbst, E. W., Fredga, K., Frank, F., Winking, H., and Gropp, A. (1978). Cytological identification of two X-chromosome types in the wood lemming (*Myopus schisticolor*). *Chromosoma* **69,** 185–191.

Jäger, R. J., Anvret, M., Hall, K., and Scherer, G. (1990). A human XY female with a frame shift mutation in the candidate testis-determining gene SRY. *Nature* (*London*) **348,** 452–454.

Jäger, R. J., Pfeiffer, R. A., and Scherer, G. (1991). A familial amino acid substitution can lead to conditional sex inversion. *Am. J. Hum. Genet.* **49,** Suppl., 219.

Jäger, R. J., Harley, V. R., Pfeiffer, R. A., Goodfellow, P. N., and Scherer, G. (1992). A familial mutation in the testis-determining gene SRY shared by both sexes. *Hum. Genet.* **90,** 350–355.

Jaswaney, V., Braithwaite, S., Demas, S., Elias, S., Simpson, J. L., Müller, U., and Wachtel, S. (1989). Novel methods in the study of gonadal dysgenesis and associated neoplasms. *Am. J. Hum. Genet.* **45,** Suppl., A25.

Mann, J. L., Corkery, J. J., Fisher, H. J. W., Cameron, A. H., Mayerová, A., Wolf, U., Kennaugh, A. A., and Woolley, V. (1983). The X-linked recessive form of XY gonadal dysgenesis with a high incidence of gonadal germ cell tumors: Clinical and genetic studies. *J. Med. Genet.* **20,** 264–270.

Müller, U. (1983). Genetics of H-Y antigen and its gonad-specific receptor. *Bibl. Anat.* **24,** 111–118.

Ogata, T., Hawkins, J. R., Taylor, A., Matsuo, N., Hata, J., and Goodfellow, P. N. (1992). Sex reversal in a child with a 46,X,Yp+ karyotype: Support for the existence of a gene(s) located in distal Xp, involved in testis formation. *J. Med. Genet.* **29,** 226–230.

Ohno, S. (1967). "Sex Chromosomes and Sex-linked Genes." Springer-Verlag, New York.

Pivnick, E. K., Wachtel, S., Woods, D., Simpson, J. L., and Bishop, C. E. (1992). Mutations in the conserved domain of SRY are uncommon in XY gonadal dysgenesis. *Hum. Genet.* **90,** 308–310.

Puck, S. M., Haseltine, F., and Franke, U. (1981). Absence of H-Y antigen in an XY female with campomelic dysplasia. *Hum. Genet.* **57,** 23–27.

Rosenfeld, R. G., Luzzatti, L., Hintz, R. L., Miller, O. J., Koo, G. C., and Wachtel, S. S. (1979). Sexual and somatic determinants of the human Y chromosome. Studies in a 46,XYp− phenotypic female. *Am. J. Hum. Genet.* **31,** 458–468.

Russell, M. H., Wachtel, S. S., Davis, B. W., Cahill, L. T., Groos, E., Nidblack, G. D., and Burr, I. M. (1982). Ovarian development in 46,XY gonadal dysgenesis. *Hum. Genet.* **60,** 196–199.

Scherer, G., Schempp, W., Baccichetti, C., Lenzini, E., Bricarelli, F. D., Carbone, L. D., and Wolf, U. (1989). Duplication of an Xp segment that includes the ZFX locus causes sex inversion in man. *Hum. Genet.* **81,** 291–294.

Scully, R. E. (1970). Gonadoblastoma. A review of 74 cases. *Cancer* (*Philadelphia*) **25,** 1340–1356.

Silengo, M., Kaufman, R. L., and Kissane, J. (1974). A 46,XY infant with uterus, dysgenetic gonads and multiple anomalies. *Humangenetik* **25,** 65–68.

Simpson, J. L. (1976). "Disorders of Sexual Differentiation: Etiology and Clinical Delineation." Academic Press, New York.

Simpson, J. L. (1989). Genetic heterogeneity in XY sex reversal: Potential pitfalls in isolating the testis-determining factor (TDF). *In:* "Evolutionary Mechanisms in Sex Determination" (S. S. Wacthel, ed.), pp. 264–277. CRC Press, Boca Raton, FL.

Simpson, J. L., Summitt, R. L., German, J., and Merkatz, I. R. (1976). Etiology of XY gonadal dysgenesis. *Gynecol. Invest.* **7,** 37.

Simpson, J. L., Blagowidow, N., and Martin, A. O. (1981). XY gonadal dysgenesis: Genetic heterogeneity based upon clinical observations, H-Y antigen status and segregation analysis. *Hum.Genet.* **58,** 91–97.

Simpson, J. L., Chaganti, R. S. K., Mouradian, J., and German, J. (1982). Chronic renal disease, myotonic dystrophy, and gonadoblastoma in XY gonadal dysgenesis. *J. Med. Genet.* **19,** 73–76.

Sinclair, A. H., Berta, P., Palmer, M. S., Hawkins, J. R., Griffiths, B. L., Smith, M. J., Foster, J. W., Frischauf, A.-M. Lovell-Badge, R., and Goodfellow, P. N. (1990). A gene from the human sex-determining region encodes a protein with homology to a conserved DNA-binding motif. *Nature (London)* **346,** 240–244.

Stern, H. J., Garrity, A. M., Saal, H. M., Wangsa, D., and Disteche, C. M. (1990). Duplication of Xp21 and sex reversal: Insight into the mechanism of sex determination. *Am. J. Hum. Genet.* **47,** Suppl., A41.

Sternberg, W. H., Barclay, D. L., and Kloepfer, H. W. (1968). Familial XY gonadal dysgenesis. *N. Engl. J. Med.* **278,** 695–700.

Swyer, G. I. M. (1955). Male pseudohermaphroditism: A hitherto undescribed form. *Br. Med. J.* **2,** 709–712.

Tommerup, N., Schempp, W., Meinecke, P., Pedersen, S., Bolund, L., Brandt, C., Goodpasture, C., Guldberg, P., Held, K. R., Reinwein, H., Saugstad, O. D., Scherer, G., Skjeldal, O., Toder, R., Westvik, J., van der Hagen, C. B., and Wolf, U. (1993). Assignment of an autosomal sex reversal locus (*SRA1*) and campomelic dysplasia (*CMPD1*) to 17q24.3–q25.1. *Nat. Genet.* **4,** 170–174.

Vilain, E., McElreavey, K., Jaubert, F., Raymond, J.-P., Riuchaud, F., and Fellous, M. (1992). Familial case with sequence variant in the testis-determining region associated with two sex phenotypes. *Am. J. Hum. Genet.* **50,** 1008–1011.

Villanueva, A. L., Benirschke, K., Campbell, J., Wachtel, S. S., and Rebar, R. W. (1984). Complete development of secondary sex characteristics in a case of 46,XY pure gonadal dysgenesis. *Obstet. Gynecol.* **64,** Suppl., 68S–72S.

Wachtel, S. S. (1983). "H-Y Antigen and the Biology of Sex Determination." Grune & Stratton, New York.

Wachtel, S. S., Brunner, M., and Lehn-Jensen, H. (1986). H-Y antigen, sex determination and gender control. *In* "Perinatal Genetics: Diagnosis and Treatment" (I. H.. Porter N. H. Hatcher, and A. M. Willey, eds.), pp. 181–198. Academic Press, Orlando, FL.

Warner, B. A., Monsaert, R. P., Stumpf, P. G., Kulin, H. E., and Wachtel, S. S. (1985). 46,XY gonadal dysgenesis: Is oncogenesis related to H-Y phenotype or breast development? *Hum. Genet.* **69,** 79–85.

Wolf, U., Fraccaro, M., Mayerová, A., Hecht, T., Maraschio, P., and Hameister, H. (1980). A gene controlling H-Y antigen on the X chromosome. *Hum. Genet.* **54,** 149–154.

Wolf, U., Schempp, W., and Scherer, G. (1992). Molecular biology of the human Y chromosome. *Rev. Physiol. Biochem. Pharmacol.* **121,** 147–213.

Wolman, S. R., McMorrow, L. E., Roy, S., Koo, G. C., Wachtel, S. S., and David, R. (1980). Abberant testicular differentiation in 46,XY gonadal dysgenesis: Morphology, endocrinology, serology. *Hum.Genet.* **55,** 1–5.

14

Phenotypic Correlations of X-Chromosome Loss

E. Kirk Neely and Ron G. Rosenfeld
Department of Pediatrics
Stanford University School of Medicine
Stanford, California 94305

Molecular Genetics of Sex Determination

I. Historical Introduction

The phenotypic appearance attributable to X-chromosome monosomy or its variants was first described by Otto Ullrich in 1930 in a case report in the German literature (refer to Ullrich, 1949). Professor Henry Turner of the University of Oklahoma subsequently characterized seven patients with "infantilism, congenital webbed neck, and cubitus valgus" (Turner, 1938). Although not mentioned in the title, each of these patients also exhibited extreme short stature. Patient No. 7, for instance, was a prepubertal 16-year-old measuring 48.5 in., in comparison with her twin sister, who was 63 in. tall and sexually mature. Turner compared the "osseous and sexual retardation" with that of hypopituitarism and accordingly administered injections of various pituitary extracts, presumably of bovine origin, to three of the girls for several months. Growth rate and breast appearance were unaltered by treatment with pituitary extract (he did note more abundant pubic hair), but breast development and menses were subsequently established in one girl by injection of an estrogen-containing preparation.

Rudimentary ovaries as the cause of sexual "infantilism" were identified by Wilkins and Fleischmann (1944), and the terms gonadal dysgenesis and streak gonads are used interchangeably. The recognizable infantile phenotype of lymphedema and neck folds was described by Ullrich (1949) and is sometimes referred to as the Bonnevie–Ullrich syndrome. Elucidation of the chromosomal basis of Turner syndrome began with the discovery that the patients were sex chromatin negative (Polani *et al.*, 1954); it was later postulated that the phenomenon might be due to the absence of one X chromosome. Shortly after new culture techniques allowed the correct number of human chromosomes to be elucidated, the syndrome was definitively correlated with X-chromosome monosomy by Ford *et al.* (1959) in a prepubertal 14-year-old with short stature and characteristic features (also noted to be "backward in school"). Other manifestations of the syndrome were described and termed collectively "Turner stigmata" (Haddad and Wilkins, 1959). The development of routine cytogenetic analysis in the early 1960s rapidly led to elucidation of the diverse chromosomal complements associated with Turner stigmata; the relative proportions of karyotypes found in several large series over the past two decades are presented in Table I.

II. Major Features of Turner Syndrome

From a clinical perspective, recognition of X-chromosome loss in an individual patient depends on the phenotype. Due to the routine avail-

Table I
Percentages of Major Sex Chromosome Subtypes in Selected Large Series of Patients with Turner Syndrome[a]

Karyotype	Palmer and Reichman (1976)	Hall *et al.* (1982)	Park *et al.* (1983)	Held *et al.* (1992)
45,X	58.2	55.0	61.2	20.7
46,X,iso(Xq)	7.3	5.5	6.0	5.7
45,X/46,XX	8.2	13.4	11.2	17.2
45,X/46,X,iso(Xq)	11.8	4.7	7.8	12.6
45,X/46,X,ringX	5.5	3.9	1.7	3.5
45,X/46,XY	5.4	3.1	2.6	2.3
45,X/46,X,del(Xp)	0.9	0.8	2.6	2.3
45,X/46,X,del(Xq)	0	1.6	1.7	3.5
45,X/47,XXX or 45,X/46,XX/47,XXX	1.8	0.8	0.9	6.9
45,X/46,X+marker	0	0.8	0	18.4

[a] Note the decreased frequency of 45,X and increased mosaicism, particularly in secondary cell lines containing small marker chromosomes, in the series of Held *et al.* (1992).

ability of prenatal diagnosis, an increasing proportion of girls with Turner syndrome (TS) is now identified by prenatal karyotyping; many are carried to term, although several centers report that a majority of families are electing to terminate these pregancies. Prenatal diagnosis currently accounts for less than 10% of children with the diagnosis of TS, since the primary indication for amniocentesis (or chorionic villous sampling) is the higher incidence of trisomy with advancing maternal age, whereas the mean maternal age associated with X-chromosome loss is much younger (Warburton *et al.*, 1980). A greater number of patients with TS, approximately one-third, are identified at birth by pterygium coli (webbed neck) or lymphedema of the extremities; occasionally TS is diagnosed in infancy during an evaluation of cardiac manifestations. Since each of these physical findings is more common in patients with X monosomy than in patients with mosaicism or structural abnormality of the X, a disproportionate number of infants diagnosed with TS will exhibit the 45,X karyotype. TS is additionally suspected in many girls in mid-childhood, when growth failure results in a height noticeably below the normal range. Other patients will not be recognized until pubertal failure or short stature in the teenage years leads to medical evaluation. Finally, the diagnosis may be made on referral for primary or secondary amenorrhea, infertility or recurrent miscarriage. Patients also are occasionally detected by any of the diverse array of less common, but functionally noticeable, features such as bony abnormalities, learning problems, or hearing loss.

Since it is not possible to determine the proportion of patients with X-chromosome loss, particularly mosaics, whose manifestations are mild enough to be overlooked clinically, ascertainment bias may be seen in attempts to relate phenotype with karyotype, especially in the older literature. Although evaluation of patients diagnosed prenatally may soon reveal a broader population of subjects with X-chromosome loss and minimal phenotypic findings, it seems likely that most patients with 45,X karyotype are identified by some combination of congenital lymphedema, short stature, or delayed puberty/primary amenorrhea. In the following sections we describe the clinical pathology and consequences of some of these major manifestations of TS. Other anomalies in Turner syndrome are listed in Table II and are described in greater detail elsewhere (Lippe, 1990; Rosenfeld and Grumbach, 1990).

A. Embryonic Lethality and Lymphedema

Turner syndrome occurs with an incidence of approximately 1 in 2500 live female births (Hook and Warburton, 1983). However, it is estimated that up to 99% of Turner conceptuses are spontaneously aborted, and, conversely, that 5–10% of spontaneously aborted fetuses have 45,X karyotypes. Thus, it is probable that more than 1% of conceptions involve X-chromosome loss. Because the peak of fetal deaths occurs at 11–13 weeks of gestational age, it is evident that most 45,X fetuses will be spontaneously aborted prior to the gestational age at which amniocentesis is usually performed. Amniocentesis, therefore, is late enough in gestation that the frequency of karyotypes approximates that seen postnatally. We anticipate that larger series from chorionic villous sampling will not only discern a percentage of 45,X pregnancies greater than is seen in amniocentesis, but may also vary in the percentages of mosaicism found.

Several investigators have contended that all fetuses with the 45,X karyotype are spontaneously aborted and, conversely, that all liveborn individuals with TS are X-chromosome mosaics secondary to early mitotic errors (Hecht and Macfarlane, 1969). This suggestion is based on discrepancies between fetal tissue and liveborn individuals in the incidence of mosaicism. The proportion of individuals with Turner phenotype and 45,X karyotype is cited as approximately 50–60% in most studies (Table I). In contrast, fetuses with X-chromosome loss are overwhelmingly 45,X. In their examination of spontaneous abortions, Hook and Warburton (1983) reported that fetal demise involved only 45,X (92% of abortuses) or 45,X/46,XX karyotypes (8%), but not other struc-

Table II
Features of Turner Syndrome

Feature	Incidence (%)
Skeletal growth disturbances	
Short stature	100
Short neck	40
Abnormal upper-to-lower-segment ratio	97
Cubitus valgus	47
Short metacarpals	37
Madelung's deformity	8
Scoliosis	13
Genu valgum	35
Micrognathia	60
High-arched palate	38
Lymphatic obstruction	
Webbed neck	25
Low posterior hairline	42
Rotated ears	Common
Edema of hands, feet	22
Nail dysplasia	13
Characteristic dermatographics	35
Germ cell defects	
Gonadal failure	96
Infertility	>99
Miscellaneous defects	
Strabismus	18
Ptosis	11
Multiple pigmented nevi	26
Cardiovascular anomalies	55
Hypertension	7
Renal and renovascular anomalies	39
Associated disorders	
Hashimoto's thyroiditis	34
Hypothyroidism	10
Alopecia	2
Vitiligo	2
Gastrointestinal disorders	3
Carbohydrate intolerance	40

Note. Adapted from Lippe (1990).

tural abnormalities or mosaicisms. In more recent studies, Hassold *et al.* (1992) documented only about 1% mosaicism in fetuses by cytogenetic techniques and a 15% frequency by DNA hybridization studies (including 4/29 with Y-chromosomal material detected in one series). He estimated that mosaicism with 45,X is, at minimum, three to five times more common in liveborns than in spontaneous abortions, a figure in agreement with Hook and Warburton (1983), and implying a significant feto-protective effect of mosaicism. However, the same data make it clear that mosaicism is not sufficient to ensure survival.

By employing cytogenetic and molecular techniques (Southern blotting and PCR), Held *et al.* (1992) documented a higher level of mosaicism in lymphocytes and fibroblasts of TS patients than is usually reported, largely by the detection of marker chromosomes in smaller cell populations. Mosaicism was nevertheless undetectable in 21% of patients studied. Evaluating multiple tissues from 10 reported 45,X patients, Burns *et al.* (1979) could demonstrate mosaicism in only 2. However, advocates of occult mosaicism argue that mosaicism may be present in multiple tissues without detectability in blood cells or fibroblasts. Extrapolating from evidence of selection against mosaicism *in vitro*, investigators have further suggested that mosaicism during fetal development may have been eliminated by the time of birth. In particular, mosaicism in the trophoblast or in placental tissue (confined placental mosaicism) may be all that is required to avoid lethality. Thus, advocates of obligatory mosaicism in survivors to birth cite spatial and temporal explanations of why mosaicism may be undetectable. Despite the attractiveness of occult mosaicism as an explanation for the survival of infants with TS, the theory is unconfirmed and does not provide a definitive solution to this puzzling phenomenon.

Several groups have described the morphologic findings commonly associated with fetal demise. Canki *et al.* (1988) documented in detail the pathology of 45,X abortuses, classifying them into four groups. Approximately half consisted of a sac with a dilated umbilical cord but little embryonic tissue, with findings suggesting a predisposition to fetal abdominal wall rupture. Another 25% were macerated embryos at developmental age 6–7 weeks, despite a gestational age of 12–13 weeks. Five percent of abortuses were larger second trimester fetuses with cystic hygromas, lymphedema, ascites, and frequent coarctation of the aorta. The remainder (20%) consisted merely of tissue fragments. The authors concluded that fetal demise may be attributable to a generalized edema related to vascular abnormalities or to abnormal fluid balance. Lymphatic vessel hypoplasia and hypoalbuminemia have been noted by other investigators (Singh and Carr, 1966; Shepard *et al.*, 1986); thus, intrauterine edema may be the common determining factor in both fetal lethality and the multiple organ and bone abnormalities seen in patients with TS. If this were true, the severity of fetal edema, as modulated by other developmental factors, would be the key to lethality and to expression of the "complete" array of tissue anomalies. Lymphedema as the primary determinant of syndrome "severity" is supported by the strong correlation in incidence of coarctation of the aorta and neck webbing (Clarke, 1984), but such a correlation is not found between most other features of TS.

B. Organ Anomalies

Proper management of TS requires immediate assessment of potentially life-threatening cardiac and renal anomalies. Echocardiography should be performed as soon as the diagnosis of TS is confirmed in order to detect cardiac malformations, the two most common being bicuspid aortic valve and coarctation of the aorta. In studies of second trimester fetal pathology, a high frequency of heart vessel defects (juxtaductal coarctation, diminution of the ascending aorta, dilated pulmonary arteries) has been noted in association with edema, nuchal cystic hygromas, and lymphatic aberrations at the base of the blood vessels, implying a causal effect of lymphedema (Lacro *et al.*, 1988). However, cardiac anatomy is normal in many fetuses with severe edema as well, and the etiology of cardiac defects remains speculative.

Coarctation of the aorta was the first cardiac anomaly noted in TS and remains the most significant clinically. In large patient series, coarctation of the aorta is reported in approximately 20% of patients (range 15–68%) and is more frequent in association with a 45,X karyotype. The advent of high resolution echocardiography has revealed other anomalies with potentially greater frequency in TS. Miller *et al.* (1983) documented isolated bicuspid aortic valve in 34% of patients, in addition to bicuspid aortic valve coexisting with coarctation. Similarly, echocardiography has revealed isolated aortic root dilation in approximately 10% of patients. Allen *et al.* (1986) noted a significantly increased aortic root diameter in 45,X subjects compared with control subjects or with patients with X mosaicism or structural abnormality, who manifested an intermediate aortic root diameter. Each of these anomalies places patients at risk for catastrophic aortic dissection and aneurysmal rupture (Lin *et al.*, 1986), in addition to the risk of bacterial endocarditis and the development of aortic stenosis. Surgical intervention is utilized for significant coarctation, and medical management is directed at routine antibacterial prophylaxis and control of hypertension.

Renal abnormalities were documented by Lippe *et al.* (1988) in 33% of 141 patients with various karyotypes. The most frequent abnormalities were double collecting systems (8% of all patients studied), horseshoe kidney (7%), and rotational abnormalities (6%). Abnormalities requiring urologic referral (and/or surgery) included ureteropelvic and ureterovesicle junction obstruction (6% together) and absent kidney (3%). The incidence of renal dysmorphology in girls with 45,X karyotype was 45%, compared with 18% in all other karyotypes combined. None of the ten 46,X,i(Xq) patients in Lippe's study exhibited renal abnormalities, but in other series 45,X and 46,X,i(Xq) patients had similar percentages of renal findings. Since dysmorphic renal structures can be adequately visualized by either ultrasonography or intravenous pyelography, a single screening renal ultrasound is now recommended for all patients at the time of diagnosis. Renal and cardiac anomalies occur independently.

C. Short Stature

Ranke *et al.* (1983) described growth in the absence of hormonal therapy in 150 German girls with Turner syndrome. These data, as well as more recent growth standards of Naeraa and Nielsen (1990), provide a basis for evaluating therapeutic regimens designed to augment growth. Lyon *et al.* (1985) combined Ranke's data with data from three other studies of growth in untreated girls with Turner syndrome and constructed a standard growth curve. Standard plots permit an accurate projection of adult height from the current height standard deviation scores, and demonstrate a correlation coefficient of 0.95 between first measured height at 3–12 years of age and final adult height (Rosenfeld and Grumbach, 1990; Lyon *et al.*, 1985).

These and other excellent descriptions of growth resulting from X-chromosome loss have identified developmental phases in the evolution of short stature. Intrauterine growth is frequently retarded, with a mean reduction of 540 g in birth weight and of 2.8 cm in birth length. Growth velocity in the first 3 years of life is relatively normal, but a more marked and progressive decline occurs relative to the height velocity of normal children during the remainder of childhood (Fig. 1). The 5th percentile of the normal female curve and the 95th percentile of the TS curve diverge at about 9 years of age. Growth failure becomes even more clinically apparent in adolescence, due to the absence of a pubertal growth spurt, and the height nadir relative to normal female

Figure 1 Mean height and height velocity in untreated European girls with Turner syndrome from four European studies, plotted on a growth curve for normal females. Note the steady decline in height velocity beginning in early childhood and the absence of a pubertal growth spurt. From Lyon *et al.* (1985).

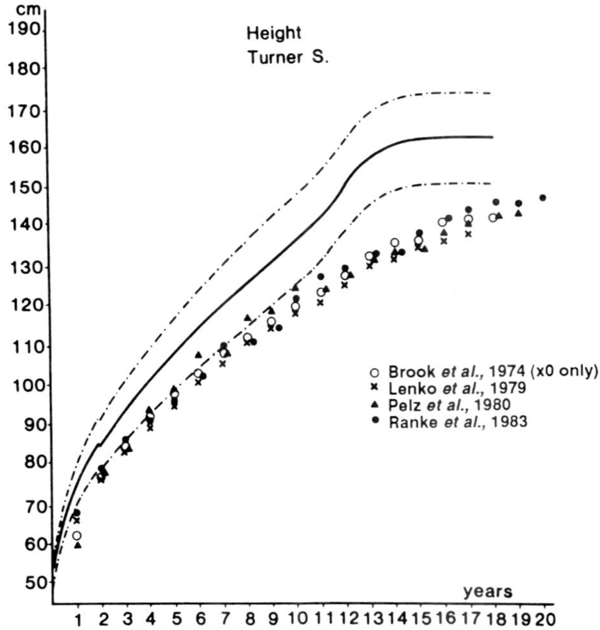

Height
Turner S.
cm
190
180
170
160
150
140
130
120
110
100
90
80
70
60
50
years
1 2 3 4 5 6 7 8 9 10 11 12 13 14 15 16 17 18 19 20
○ Brook et al., 1974 (x0 only)
× Lenko et al., 1979
▲ Pelz et al., 1980
● Ranke et al., 1983

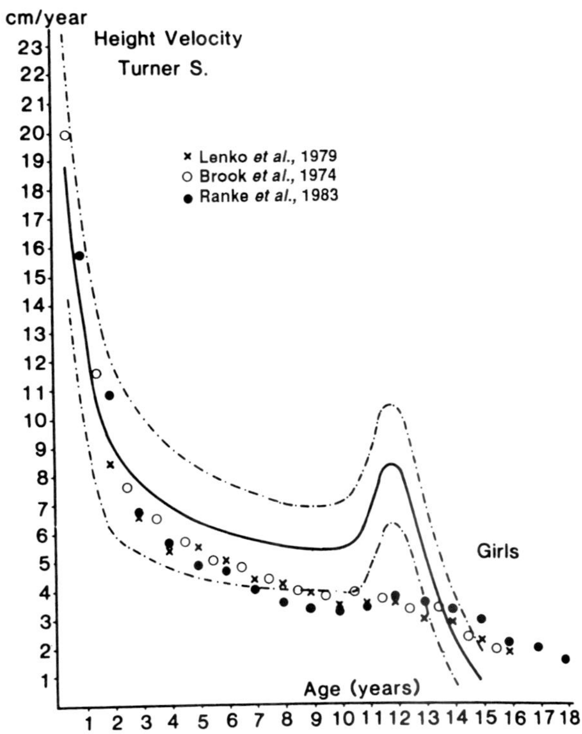

cm/year
Height Velocity
Turner S.
23
22
21
20
19
18
17
16
15
14
13
12
11
10
9
8
7
6
5
4
3
2
1
× Lenko et al., 1979
○ Brook et al., 1974
● Ranke et al., 1983
Girls
Age (years)
1 2 3 4 5 6 7 8 9 10 11 12 13 14 15 16 17 18

growth curves is typically found at 14 years of age. In association with failed puberty, girls with TS exhibit delayed epiphyseal fusion, with typical closure at 18–20 years; ironically, this continued slow growth during late adolescence allows for growth therapy over a prolonged period.

Interpretation of final height data is complicated by many factors, including karyotype, parental heights and earlier hormonal treatment with estrogens and androgens. The effect of spontaneous puberty on adult height has not been conclusively described (Massa *et al.*, 1990). Furthermore, adult heights may be underestimated due to ascertainment bias, since TS is less likely to be diagnosed in the more phenotypically normal patients. In the study by Lyon *et al.* (1985), the mean final height was 142.9 ± 7.3 cm, a figure close to the height reported for the combined European–American experience over the last 20 years (Rosenfeld and Grumbach, 1990). However, several investigators have documented taller adult heights in patient series, including adult heights of approximately 147 cm in two European studies and in that by Sybert from the United States (1984), although she notes a slightly greater frequency of mosaicism in her series. Resolution of the discrepancies regarding untreated final heights is important for assessing the benefits of growth therapies, because studies with human growth hormone have not utilized long-term control groups (Ranke and Rosenfeld, 1991; Rosenfeld *et al.*, 1992).

The growth failure associated with X-chromosome loss potentially results from the interaction of skeletal dysplasia, growth hormone secretory dysfunction and estrogen deficiency, but essentially the cause remains unclear. There are no conclusive data concerning the histologic appearance of bone and cartilage in children with TS, and increased bone resorption and defective osteoblast renewal and differentiation have been reported (Rosenfeld and Grumbach, 1990). Consistent with the latter possibility are observations that tibial cartilage from TS patients has narrower growth zones than normal bone has. Pathological bone development theoretically represents an intrinsic defect in ossification, due to loss of critical genes on the X chromosome, or a secondary effect of intrauterine edema, in a manner comparable to the development of cardiac anomalies.

Evidence of disordered growth hormone (GH) secretion *later* in childhood and in adolescence is convincing. Although most patients with Turner syndrome are not GH deficient as classically defined by provocative GH testing (i.e., they have maximal serum GH levels >7–10 ng/ml), some investigators have documented relatively low peak stimulated GH levels or actual GH deficiency in a minority of patients.

TS adolescents, but not younger children, are reported to have reduced mean 24-h GH levels, peak amplitudes, and peak frequencies when compared with age-matched controls (Ross *et al.*, 1985; Massarano *et al.*, 1989b). Similarly, serum levels of insulin-like growth factor I (IGF-I), which mediates most actions of GH, fall in the lower range of normal in adolescence, but not in childhood (Cuttler *et al.*, 1985). This failure of the normal pubertal rise in GH and IGF-I secretion in Turner adolescents is plausibly related to a primary hypothalamic–pituitary abnormality or to estrogen deficiency; the latter concept is supported by an increase in GH concentrations in response to estrogen therapy (Mauras *et al.*, 1990). However, growth failure usually predates the reduction in spontaneous GH secretion seen in TS, making doubtful the primacy of GH secretory dysfunction in the etiology of short stature; preadolescent growth failure probably evolves secondary to an intrinsic bone defect.

D. Bone Abnormalities

The most common bone abnormalities are listed in Table II. Cubitus valgus and Madelung's wrist deformity are easily seen on physical examination, and short fourth metacarpal is often documented by X-ray of the wrist and hand obtained for assessment of skeletal maturation. These abnormalities appear to be seen equally in 45,X monosomy, structural abnormality, or mosaicism. Less frequent skeletal abnormalities include an obtuse cranial base angle, thinning of parietal bones, pectus excavatum, premature fusion of the ossification centers of the sternum, "drumstick" appearance of the distal phalanges, fusion of the tarsal bones, decreased bitrochanteric breadth, and a male appearance of the pelvic inlet. Scoliosis can occur. Although most groups have documented short-leggedness and abnormal body segment ratios in TS, and many patients exhibit a short, square body habitus (the appearance of widely spaced nipples may be attributable to this phenomenon), some investigators have reported a normal ratio of trunk to leg length.

Deficits in bone mineral have also been widely reported in TS, and several lines of evidence suggest that a relative osteopenia, like the short stature of TS, may be a product of an underlying skeletal dysplasia interacting with hormonal deficiency. Radiographic evidence of a coarse trabecular pattern of vertebrae and carpal bones, reduced metacarpal cortical thickness, and failure to shift to endosteal bone growth during adolescence (Rosenfeld and Grumbach, 1990; Barr, 1974; Fransancho *et al.*, 1970; Preger *et al.*, 1968) has been cited as evidence of osteopenia beginning in childhood and continuing into adulthood. Bone mineral status is of particular concern because ovarian failure probably contributes to insufficient acquisition and maintenance of bone mass.

Early reports of osteopenia in adults with TS employed computerized tomographic scans, but more recently single and dual photon absorptiometry, followed by dual-energy X-ray absorptiometry, have been utilized to report bone mineral content (BMC = g) and density (BMD = g/cm^2), the latter measure compensating for surface area of bones. In adults with TS, initial limited studies documented a markedly reduced BMC of the calcaneus or radius in some patients, with less reduction in those on estrogen replacement (Smith *et al.*, 1982; Naeraa *et al.*, 1991). The evidence for osteopenia in children and adolescents with TS is not convincing, and an increased incidence of fractures has never been documented. Furthermore, the reported decreases in mean radial BMC in children with TS have little significance, because BMC is highly dependent on body and bone size. Studies measuring BMD have given contradictory results (Kirkland *et al.*, 1990; Ross *et al.*, 1991) and have not effectively delineated when relative demineralization begins. Our group recently found that adolescents with TS receiving GH therapy (as virtually all our patients do) exhibit normal bone mineral properties (Neely *et al.*, 1993). Thus, it is not clear when estrogen replacement is required for bone accretion or maintenance, nor whether bone mass as currently measured has marked functional significance in TS.

E. Cognitive Function

Although early studies reported an increased risk of mental retardation in patients with X monosomy, later studies have reported an increased incidence of retardation (as opposed to other cognitive impairment) only in patients with ring X karyotypes (Fryns *et al.*, 1990), and even this may be doubted. Modern investigations have generally revealed only mild and specific impairments in intellectual functioning, with frequently documented reductions in performance IQ relative to a normal verbal IQ (Bender *et al.*, 1984; Rovet and Netley, 1982). The incidence of this type of cognitive impairment appears to be minimal in 45,X/46,XX mosaicism. In a large study of children and adults, Garron (1977) found specific difficulties on tasks assessing spatial and numerical abilities; these were unrelated to karyotype or to physical stigmata.

Money (1973) coined the term "space-form blindness" to describe the characteristic cognitive deficits in Turner patients and postulated a bilateral impairment of parietal lobe function. McCauley *et al.* (1987) also argued that right hemispheric dysfunction links cognitive and psychosocial (affective) problems. Although a reduction in performance IQ has been confirmed in many studies, the evidence for right-hemisphere localization as the causal link has been criticized. Investigations using EEG, event-related potentials, and positron emission tomography have failed to locate any lateralizing abnormalities in TS. As a result, investigators more recently have accounted for the heterogeneous cognitive deficits by postulating a global problem involving attention and information processing, as well as visual memory, perhaps attributable to early alterations in brain function (lymphedema?) instead of lateralization. From a practical perspective, few studies have examined actual academic outcome in large numbers of girls (Pennington *et al.*, 1982). In our experience, few girls are referred for testing or special education, possibly because patients with TS are not disruptive in the classroom setting. Nonetheless, it is crucial that patients with TS receive psychological and educational evaluation early in their school careers. Hearing evaluation, due to the high incidence of conductive and sensorineural hearing loss, is highly recommended as an adjunct measure to improve social and academic skills of girls with TS.

F. Autoimmunity

The frequency of thyroid autoimmunity is increased in TS patients of all karyotypes (Papendieck *et al.*, 1987). Estimates of prevalence in TS of autoantibodies specific to thyroid antigens range from 25 to 60% (Lippe, 1990; Hall and Gilchrist, 1990), in contrast with prevalence for thyroid autoimmunity of 1 to 2% in the general population. The incidence of frank hypothyroidism is lower than the incidence of serum antibodies, but develops nonetheless in 10-20% of TS patients. Although the prevalence of other autoimmune diseases, such as rheumatoid arthritis and inflammatory bowel disease, is slightly increased in the TS population, there does not appear to be a comparably increased incidence of polyglandular autoimmunity in antithyroid antibody-positive TS patients.

Whether the increase in specific thyroid autoimmunity in Turner syndrome is familial is controversial, but parents of patients have a higher than expected incidence of thyroid autoimmunity. This has led to the speculation that familial thyroid autoimmunity might have a protective effect against the usual lethality of X chromosome monosomy *in utero* (Schatz *et al.*, 1990). It also has recently been determined that verbal and performance IQ are diminished in a subset of Turner patients with positive serum thyroid antibodies, in comparison with autoimmune-negative TS patients and normal short age-matched females. Neither the mechanism relating thyroid autoimmunity with cognitive impairment in Turner syndrome nor the genetic and pathophysiologic basis for an increased risk of autoimmunity in TS in general are understood at this time.

G. Gonadal Dysgenesis

Gonadal dysgenesis, along with short stature, is the hallmark of X-chromosome loss. In more than 90% of 45,X girls, the ovaries are long, pale streaks consisting of normal fibrous stroma but devoid of primordial follicles. The gonadal ridges of abortuses contain primordial germ cells that develop normally until late first trimester, at which time oocyte formation and folliculogenesis fail to occur normally and connective tissue proliferates instead (Grumbach and Conte, 1992). Surviving oocytes undergo premature atresia within a time frame that may be completed before birth or may continue into adolescence or adulthood. The phenotypic expression of declining numbers of functioning ovarian follicles varies from absent pubertal development to puberty stalling in mid-development, failure to progress to menses, and secondary amenorrhea. Ovarian failure is reflected in reduced feedback on the hypothalamic–pituitary axis and consequent elevations in serum FSH and LH. Markedly elevated gonadotropin levels are detectable from birth to approximately 4 years of age, but then fall to high normal levels before rising again to very high levels at the normal age of puberty (Conte *et al.*, 1975).

Massarano *et al.* (1989a) reported the ultrasound appearance of ovaries in 104 girls of varying ages with TS. The ovaries of 70 subjects were classified as streak gonads; the ovarian volumes of the other 34 subjects generally ranged in the lower range of normal. In contrast, uterine lengths fell in the middle of the normal range. Non-streak ovaries were most easily detected in the 0- to 4-year and 10+-year age groups, paralleling the gonadotropin stimulation pattern mentioned above. Girls with ring X and Xdel(p) karyotypes demonstrated a higher incidence of both non-streak ovaries and spontaneous breast development, supporting the argument that loss of the short arm alone is associated with a higher prevalence of retained ovarian function, as discussed below.

III. Phenotype–Karyotype Correlations

The availability of cytogenetic analysis in the 1960s provided the opportunity to compare phenotypic findings associated with the 45,X karyotype with those seen with X-chromosome structural abnormalities and mosaicisms. Following analysis of 30 patients, Ferguson-Smith (1965) reviewed 307 published cases in order to document phenotype–karyotype correlations. He hypothesized that 45,X subjects demonstrated the "complete Turner syndrome" and, more precisely, that this was attributable to monosomy of loci on the short arm. In the review of Ferguson-Smith (1965), patients with 45,X karyotype exhibited a higher incidence of neck webbing, congenital lymphedema, and cardiac malformations than did patients with any of the X-structural abnormalities or mosaicisms. Spontaneous pubertal development and menses, which occurred in 8% of patients with X monosomy, were more likely in girls with 45,X/46,XX mosaicism (21%) and deletions of the short arm (25%). Short stature was a universal feature in 45,X and 46,X,i(Xq) or mosaics; in contrast, normal stature occurred in 20% of 45,X/46,XX, 50% of 45,X/47,XXX, and 63% of long arm deletions. These results led Ferguson-Smith (1965) to conclude that his short arm postulate was confirmed: (*a*) patients with 45,X monosomy exhibited the highest incidence of the major medical features (but not minor features such as pigmented nevi), (*b*) loss of the short arm of an X chromosome [that is, i(Xq) or del(Xp)] resulted in equally severe manifestations, but (*c*) loss of the long arm [i(Xp) or del(Xq)] was often associated with normal stature. He concluded that large deletions of any part of the X are usually associated with streak gonads, and he hypothesized that monosomy of certain presumptive Y-homologous loci on the short arm, which normally escaped X-inactivation, resulted in manifestations equivalent to complete X monosomy.

These observations have held up well. Several patient series over the last 25 years have corroborated the greater prevalence of lymphedema, coarctation of the aorta, and neck webbing (or low hairline) associated with a 45,X karyotype, in contrast with mosaicism or with heterozygosity for X-structural abnormalities. Various studies have confirmed the higher incidence of spontaneous puberty in certain mosaics (approximately 20% in 45,X/46,XX as opposed to 10% in 45,X). In their patients, Palmer and Reichman (1976) noted a particularly low incidence of neck webbing and coarctation of the aorta associated with isochromosome of the long arm or ring formation. In a subsequent review of the literature on phenotypic features of long arm isochromosomes, Otto *et al.* (1981) concluded that all manifestations seen with 45,X karyotype, other than short stature and gonadal dysgenesis (which may be biased as the mode of ascertainment), occurred in reduced frequency in 45,46,X,i(Xq) or i(Xq) mosaicisms. There is considerable diversity, however, in reports regarding the incidence of other anomalies. Furthermore, many findings, such as the multitude of bony abnormalities, appear to occur equally in X monosomy and mosaicisms. Due to the great diversity in phenotypic expression, it is clinically prudent to consider that phenotypic characteristics of karyotypes overlap. Every endocrinologist can relate anecdotes of patients with Turner syndrome discerned by minor features alone.

There are several unresolved difficulties in interpreting phenotype–karyotype relationships in Turner syndrome. These include the problem of undetected mosaicism, the unpredictability of tissue karyotypes assumed on the basis of blood cell karyotype, and the incomplete nature of X inactivation. Such background problems have exacerbated any possible resolution of the fundamental paradoxes of phenotype–karyotype correlations. A particularly obvious X-chromosome quandary is the basic equivalence of phenotypes resulting from loss of distinctly separate regions (i.e., the short or long arm). Even more problematic is the variability of the X monosomy phenotype (assuming the absence of occult mosaicism). Why does a small percentage of patients with identical karyotypes exhibit certain renal and cardiac anomalies, and another group of affected individuals escape ovarian failure? And what aspect of X-chromosome monosomy selects the survivors to birth? There are several speculative answers to these questions, including random or partial X-inactivation, position effect, genomic imprinting, and, simply, the involvement of multiple related X-chromosome gene products. Correlation of phenotype with karyotype in large series of patients has not resolved these types of questions. Molecular analysis holds promise, but up to now has merely highlighted the questions.

IV. Evidence from X-Chromosome Deletions and Translocations

Development of newer banding techniques in the 1970s and DNA probes in the 1980s allowed phenotypic correlations with a more detailed localization of X-chromosome breakpoints. On the basis of some of this information, Simpson (1979) presented classic reviews of phenotype–karyotype correlations in the 1970s. Therman and Susman

(1990) more recently reviewed and summarized the literature on phenotypes of nonmosaic adults with long or short arm X deletions (Table III). Although most of the published cases are terminal deletions, molecular techniques have detected several interstitial deletions as well. However, interpretation of data from deletions and translocations is troubled by several of the methodologic concerns in phenotype–karyotype correlations in general, such as inclusion of data prior to improvement in banding technology, occult mosaicism, and undetected multiple breakpoints.

In agreement with the early conclusions of Ferguson-Smith (1965), virtually any feature of TS can be seen with major deletion of either Xp or Xq. Nevertheless, a general summary can be advanced that short arm deletions more frequently result in short stature, and long arm deletions more commonly lead to ovarian dysgenesis. Ovarian dysgenesis, including primary and secondary amenorrhea, resulted from 93 and 65% of Xq- and Xp- karyotypes, respectively. In contrast, short stature was reported in 43 and 88% of the Xq- and Xp- cases, respectively. Turner stigmata (features other than gonadal dysgenesis and growth failure) are frequently associated with short arm deletion, but the prevalence of Turner stigmata is, in fact, markedly reduced compared with a popula-

Table III
Phenotypic Features of Adults with X Monosomy (45,X), Short Arm Deletion (46,XXp-), or Long Arm Deletion (46,XXq-)[a]

Feature	45,X n = 332	46,XXp- n = 52	46,XXq- n = 67
Short stature	100%	88%	43%
Gonadal dysgenesis	91	65	93
Cubitus valgus	77	25	16
Short neck	77	38	21
"Shield chest"	74	35	13
Low hairline	72	19	9
Pigmented nevi	64	27	19
Delayed bone age	64	17	10
Nail anomaly	57	8	7
Short metacarpal	55	29	12
Renal anomaly	44	8	6
Webbed neck	42	2	1
Hypertension	37	8	7
Cardiac anomaly	23	2	0
Thyroid disease	18	6	3

[a] Adapted from Therman and Susman (1990).

tion of girls with complete X monosomy. In particular, in the review of short and long arm deletions by Therman and Susman (1990), lymphedema was rarely seen, and cardiovascular anomaly was reported in only one patient.

In the phenotype–karyotype summary of Simpson (1979), 23 short arm deletions were listed, and other instructive cases have been described since then. In a series of five Xp- cases reported by Kalousek *et al.* (1979), all patients exhibited short stature, but Turner stigmata were notably variable: pigmented nevi and short fourth metacarpal were present in a case of 46,X,delX(pter-p21) but absent in a patient with a larger deletion at X(pter-p11.2). All of these patients with short arm deletions were menarchal, although the adolescent with a breakpoint at p11.2 exhibited secondary amenorrhea. Similarly, there was unpredictable variability in 11 patients with short arm deletions reported by Fryns *et al.* (1981), with 4 females of normal stature, including a mother and daughter (discovered fortuitously) with no features of Turner syndrome despite the presence of deletion at pter-p11.6. Phenotypic characteristics of the seven 46,X,delX(pter-p11.2) karyotypes in the series ranged from no Turner stigmata to "typical Turner," and from normal gonadal function to primary amenorrhea. In contrast, in a family with a balanced X;1 translocation with the breakpoint at Xp11, all affected females manifested a more complete syndrome of short stature, primary amenorrhea, and stigmata (Leichtman *et al.*, 1978). Location of a breakpoint on the short arm correlates poorly with the resulting phenotype.

These and other series of short arm deletions have led to consensus, however, that short stature is associated with deletion of the distal tip of Xp. Fraccaro *et al.* (1977) described short stature as the only component of TS present in several fertile females with delX(pter-p21). In two cases of familial transmission of Xp deletions with breakpoints at p22.32, both male and female carriers were short (Curry *et al.*, 1984). An interstitial deletion at p11.4-p22.3 resulted in short stature, although an inherited interstitial deletion in a mother and two daughters at p11-p22 did not (Herva *et al.*, 1979). Thus, evidence exists for an unidentified locus or loci controlling stature at Xp22, although its absence on the second X is not invariably correlated with short stature. On the other hand, localization of a critical region of the proximal short arm involved in normal ovarian function and identification of the anti-Turner loci on the short arm controlling somatic stigmata have been more confounding.

Although only some short arm deletions result in gonadal dysgenesis, the majority of long arm deletions and translocations are associated with some degree of ovarian dysfunction. The "critical region" hypothesis states that Xq13-q24 in its entirety is required for normal gonadal function (Skibsted *et al.*, 1984). However, interstitial deletions at Xq13::Xq26 and Xq13::Xq21.3 have been present in patients with apparently normal ovarian function (de la Chapelle *et al.*, 1975). Additionally, deletions with breakpoints from Xq25 to the distal tip are associated with gonadal dysgenesis, albeit of reduced severity, such as late secondary amenorrhea or early menopause (Krauss *et al.*, 1987). Essentially the entire long arm, therefore, is involved in ovarian development and maintenance. Not surprisingly, patients with identical breakpoints in this region may have very different manifestations of ovarian failure, and deletion breakpoints with apparently identical degrees of ovarian failure are scattered along the length of the long arm.

V. X-Chromosome Loss and the Y Chromosome

It is currently hypothesized that the phenotype resulting from X-chromosome loss is attributable to halved expression of gene loci that are not X-inactivated and are present on the X and Y. Manifestation of Turner syndrome features from Y-chromosome deletions might therefore aid in localization of "Turner genes." Reports of XY females ("pure" gonadal dysgenesis) are most often not attributable to detectable cytogenetic abnormality of the Y chromosome and are likely due to very small deletions or point mutations in the sex-determining region. Such individuals do not usually exhibit TS characteristics. In contrast, XY females with documented Y short arm deletions, as demonstrated by banding or by DNA probes, do manifest features of TS. XYp- females reported by Rosenfeld *et al.* (1979) and Magenis *et al.* (1984) exhibited lymphedema and redundant neck folds, but not organ anomalies or short stature. Gonadoblastoma was detected in the latter report and in a patient reported by Disteche *et al.* (1986). In two other XY females evaluated by Levilliers *et al.* (1989), congenital lymphedema and nuchal skin folds were seen in association with exchange of paternal X(pter-p22.3) for the homologous Y region, i.e., normal exchange of the pseudoautosomal region but extending proximally to p22.3. These studies and another recent case (Blagowidow *et al.*, 1989) suggest that lymphedema (as well as failure of male sex determination) can result from loss of the distal short arm of the Y, specifically interval 1 or 2. The study by Levilliers *et al.* (1989) describing lymphedema without apparent loss of chromosomal material implies an asymmetry between loci on X and Y, bringing into question the assumption of functionally identical X and Y copies of putative Turner loci.

X-chromosomal monosomy occurs in mosaicism with 46,XY cells in approximately 5% of patients with Turner syndrome in large series, although only a minority of 45,X/46,XY (or 45,X/47,XXY) patients manifest TS features. The associated phenotype, ranging from female to male, is more appropriately considered in discussions of pseudohermaphroditism, since it involves ambiguous genitalia and mixed gonadal dysgenesis (streak ovary with contralateral dysgenetic testicular elements, resulting in asymmetry of Wolffian and Müllerian structures). In an early survey of the 45,X/46,XY literature, Hsu *et al.* (1974) concluded that 25% of cases were phenotypic females with some features of Turner syndrome, 60% presented with ambiguous genitalia and mixed gonadal dysgenesis, and the remaining 15% were phenotypic males with evident "undermasculinization." The phenotypic females, who occasionally exhibit mild degrees of virilization such as clitoromegaly, ordinarily have bilateral streak gonads; some of these patients escape diagnosis until evaluation for primary amenorrhea. Infants with the more masculinized phenotype are evaluated for small phallus, hypospadias, or undescended gonads; most of these individuals have Müllerian structures as well. Approximately two-thirds of 45,X/46,XY individuals diagnosed at birth are raised as females.

Analysis of 45,X/46,XY individuals discovered prenatally has revolutionized our understanding of the phenotypic spectrum (Wheeler *et al.*, 1988; Hsu, 1989; Chang *et al.,* 1990). Recent surveys indicate that 85–95% of the fetuses or liveborns with this karyotype are, in fact, normal phenotypic males, rendering blatant the earlier ascertainment bias. In Hsu's (1989) survey of *prenatally* diagnosed cases, of which two-thirds could be confirmed by postnatal karyotype, only 10% exhibited mixed gonadal dysgenesis. In a recent literature review of 92 cases, Chang *et al.* (1990) reported only 4 exceptions to normal male genitalia. Furthermore, no Turner characteristics were reported other than a single case of cystic hygroma. Although the liveborn individuals with 45,X/46,XY karyotype have not been carefully (or longitudinally) scrutinized for features of TS, these data imply that characteristics of TS are much less common in 45,X/46,XY than in other 45,X mosaicisms (the alternative implication is that 45,X/46,XX is equally benign and markedly underreported as well). No correlations of phenotype with percentages of 45,X versus 46,XY cells have been documented.

The high incidence of gonadal malignancy in mixed gonadal dysgenesis receives considerable clinical attention (and is discussed in Chapter 13). In a series of 30 cases of gonadoblastoma, 10 were associated with 45,X/46,XY mosaicism (Scully, 1970), and Simpson (1979) and others have estimated the risk of developing gonadoblastoma (or dysgerminoma) with this karyotype at 15–20%. Although this figure may be exaggerated in light of recognition of the generally normal phenotype in 45,X/46,XY mosaicism, detection of Y material in TS appears desirable. Numerous investigators, therefore, have searched for Y-chromosomal material, previously undetectable by cytogenetics, by combinations of Southern blotting, *in situ* hybridization, or PCR with Y-specific primers (Johnson *et al.*, 1991; Lindgren *et al.*, 1992; Medlej *et al.*, 1992). These techniques have been particularly helpful in determining the Y origin of marker chromosomes or small rings and can be recommended when either is present. Molecular analysis of cells from series of patients with TS has revealed an incidence of previously undetected Y material (in mosaicism) ranging from 0 to 15%. In a recent series of 40 cases evaluated by PCR with Y primers, a single Y sequence was detected (Medlej *et al.*, 1992).

Utility of these techniques as an adjunct to routine karyotyping is not yet clear, in part because the risk of gonadoblastoma with small fragments of Y chromosome is unclear. Nevertheless, barring contradictory information, routine probing for Y material will probably be adopted in the near future. Future molecular studies may additionally identify the gene normally involved in suppression of gonadoblastoma. Such characterization may allow us to determine which TS patients with Y chromosomal material are at risk for malignancy and would benefit from prophylactic gonadectomy.

VI. Recent Molecular Approaches to the Phenotype–Genotype Problem

A. Parental Origin of the X Chromosome

Assignment of parental origin of the remaining X chromosome has been undertaken for several reasons, not only to look for clues to the etiology of X monosomy, but also to evaluate the possibility that parental origin might explain the vast phenotypic variation associated with the 45,X karyotype. The recent discovery of genomic imprinting has highlighted the theoretical basis by which parental origin could lead to the phenotypic diversity seen in Turner syndrome. Through assessment of Xg blood groups, Sanger *et al.* (1971) initially determined that loss of the paternal X occurred in 77% of cases. With the availability of X-chromosomal DNA probes, restriction fragment length polymorphism (RFLP) analysis was applied to this task first by Hassold *et al.* (1985) and subsequently by several other groups. The results have verified the finding of Sanger *et al.* (1971) that the X in 45,X monosomy is maternal (X^m) approximately 80% of the time, although the ratio of X^m to X^p in mosaicism and i(Xq) karyotypes may be closer to parity.

Although this information has not markedly altered notions regarding the developmental error involved in monosomy X, it has allowed for partial clarification of the phenotypic effects of parental origin of the X. The combined data from Hassold *et al.* (1988) and Jacobs *et al.* (1990), reveal no shift in the prevalence in fetal demise of either a maternally or a paternally imprinted X. Parental origin of the X therefore seems an unlikely factor in this most striking phenotypic determination. Furthermore, two studies performed to date have failed to discern other clinical differences in 45,X subjects on the basis of parental origin of the X chromosome. Although the data of Mathur *et al.* (1991) suggested an absence of renal abnormalities with X^p, and Lorda-Sanchez *et al.* (1992) reported no cardiac anomalies with X^p, the studies combined revealed no distinctions between X^m and X^p groups. Larger compilations of X^p phenotypes will be required for a definitive answer regarding differences in clinical phenotypes. Analysis of all determinations of parental origin suggests that the reduced maternal age in 45,X cases is confined to mothers of X^p individuals (Lorda-Sanchez *et al.*, 1992).

B. The Search for Specific Turner Loci

The usual explanation of the Turner phenotype invokes haploinsufficiency of a gene or genes on the X chromosome in which gene dosage is tightly regulated. According to this model, the locus must have X and Y copies and must escape X-inactivation, and transcription would be widespread during development. Cytogenetic evidence suggests that three regions of the X chromosome are not inactivated: pericentromeric regions on the long and short arm, and the distal portion of the short arm, including the pseudoautosomal region. Several candidate genes that escape X-inactivation have been identified in recent years. Several genes in the pseudoautosomal region, such as *STS, Xg, MIC2,* and the gene for the GM-CSF receptor, are known to be transcribed from both X chromosomes. Genes within interval 1 and 2 of the short arm (proximal to the pseudoautosomal region) known to escape inactivation include *ZFX, RPS4X,* and *KALIG-1* (Kallman syndrome gene on Xp). *ZFX,* which has motifs suggestive of a transcription factor, maps to distal Xp. *RPS4X* and its homolog *RPS4Y* are ubiquitously transcribed and encode isoforms (19 amino acids are distinct) of a component of the small ribosomal subunit (Fisher *et al.*, 1990). Whereas *RPS4Y* maps to the distal Yp proximal to the sex-determining region, *RPS4X* and the mouse homolog *Rps4* map tantalizingly close to the respective X inactivation centers (Xq13.1 in the human, just proximal to XIC). The relevance of *ZFX* and *RPS4X* to Turner syndrome is suggested by the contrasting findings in humans and in mice (prenatal lethality and phenotypic abnormality are not seen in XO mice). Ashworth *et al.* (1991) have demonstrated that *Zfx* and *Rps4* undergo normal X-inactivation in mice, circumstantially nominating these loci as the controlling elements in fetal lethality and the postnatal manifestations of TS in humans.

Although satisfying the hypothetical criteria established for Turner loci, initial evidence contradicts each of these genes as a putative Turner suppressor. Page *et al.* (1990) observed in an X,t(Y;22) female lacking *ZFY* that the consequent monosomy for this locus did not result in features of TS. As for *RPS4X,* Just *et al.* (1992) demonstrated normal transcription rates in fibroblasts of Turner patients with deletions of pter-p11.23 and in fibroblasts from a patient with isochromosome of the long arm. Thus, whereas *RPS4X* and *ZFX* are not X-inactivated, the evidence that gene dosage is halved by loss of one copy of the gene has not been confirmed. Presumably other noninactivated X-chromosome genes will be identified, and the search for putative Turner genes will continue.

VII. Conclusion

The phenotype associated with X-chromosome monosomy in humans can be summarized by (*a*) overwhelming fetal lethality, possibly associated with severe lymphedema, (*b*) a high incidence of short stature and gonadal dysgenesis, and (*c*) diverse dysmorphic and dysfunctional features of variable expression. Early phenotype–karyotype correlations from large series of patients revealed that X-chromosome loss with mosaicism results in a reduced frequency of major manifestations in comparison with X monosomy. Molecular assessment for the presence of mosaicism in spontaneous abortions has suggested a strong, but not absolute, fetoprotective effect of mosaicism. Loss of the short arm, determined cytogenetically, has been associated with short stature, whereas loss of the long arm almost universally results in ovarian dysgenesis of variable degrees. However, the overlap in features of long and short arm deletions remains perplexing, and the genes responsible for Turner syndrome have been difficult to localize.

Novel banding methods and DNA techniques, while adding details to phenotype mapping of the X, have reinforced the marked phenotypic diversity associated with similar translocation or deletion breakpoints. As is clear from phenotypic correlations derived by mapping of breakpoints, no single locus can satisfy a simple version of the paradigm established for a putative Turner gene. The apparent failure to localize genotypically the phenotypic manifestations of X-chromosomal loss implies other levels of control of X-chromosome expression. One potential control mechanism is parental imprinting of the X, but RFLP analysis has not revealed any phenotypic correlations of parental origin. Deletion, translocation, inversion, or other structural alteration of the X chromosome may prevent normal sex chromosome pairing in meiosis, thus leading to infertility, *or* may alter mitosis, with the potential for protean developmental disarray. Even subtle alteration of the X chromosome may influence the process of X-inactivation, leading to functional monosomy of one or a few loci, even though the locus may not itself have been physically eliminated (position effect). Since both X chromosomes appear to be active in the oocyte, gonadal dysgenesis could result from monosomy of loci that are required during gonadal development, but are X inactivated in subsequent embryonic stages. Elucidation of X-chromosome inactivation, and the effects of structural abnormalities upon this process, should contribute to our understanding of the etiology of the Turner phenotype.

References

Allen, D. B., Hendrich, A., and Levy, J. M. (1986). Aortic dilatation in Turner syndrome. *J. Pediatr.* **109,** 302–305.

Ashworth, A., Rastan, S., Lovell-Badge, R., and Kay, G. (1991). X-chromosome inactivation may explain the difference in viability of XO humans and mice. *Nature (London)* **351,** 406–408.

Barr, D. G. O. (1974). Bone deficiency in Turner's syndrome measured by metacarpal dimensions. *Arch. Dis. Child.* **49,** 821–822.

Bender, B., Puck, M., Salbenblatt, J., and Robinson, A. (1984). Cognitive development of unselected girls with complete and partial X monosomy. *Pediatrics* **73,** 175–182.

Blagowidow, N., Page, D. C., Huff, D., and Mennuti, M. T. (1989). Ullrich-Turner syndrome in an XY female fetus with deletion of the sex determining portion of the Y chromosome. *Am. J. Med. Genet.* **34,** 159–162.

Burns, J. L., Hall, J. G., Powers, E., Callis, J. B., and Hoehn, H. (1979). No evidence for chromosomal mosaicism in multiple tissues of 10 patients with 45,XO Turner syndrome. *Clin. Genet.* **15,** 22–28.

Canki, N., Warburton, D., and Byrne, J. (1988). Morphological characteristics of monosomy X in spontaneous abortions. *Ann. Genet.* **31,** 4–13.

Chang, H. J., Clark, R. D., and Bachman, H. (1990). The phenotype of 45,X/46,XY mosaicism: An analysis of 92 prenatally diagnosed cases. *Am. J. Hum. Genet.* **46,** 156–168.

Clarke, E. B. (1984). Web neck and congenital heart defects: A pathogenic association in 45XO Turner syndrome? *Teratology* **29,** 355–361.

Conte, F. H., Grumbach, M. M., and Kaplan, S. L. (1975). A diphasic pattern of gonadotropin secretion in patients with the syndrome of gonadal dysgenesis. *J. Clin. Endocrinol. Metab.* **40,** 670–676.

Curry, C. J. R., Magenis, R. E., Brown, M. *et al.* (1984). Inherited chondrodysplasia punctata due to a deletion of the terminal short arm of an X chromosome. *N. Engl. J. Med.* **311,** 1010–1015.

Cuttler, L., Van Vliet, G., Conte, F. A., Kaplan, S. L., and Grumbach, M. M. (1985). Somatomedin-C levels in children and adolescents with gonadal dysgenesis: Differences from age-matched normal females and effect of chronic estrogen replacement. *J. Clin. Endocrinol. Metab.* **60,** 1087–1092.

de la Chapelle, A., Schroder, J., Haahtela, T., and Aro, P. (1975). Deletion mapping of the human X chromosome. *Hereditas* **80,** 113–120.

Disteche, C. M., Casanova, M., Saal, H. *et al.* (1986). Small deletions of the short arm of the Y chromosome in 46,XY females. *Proc. Natl. Acad. Sci. U.S.A.* **83,** 7841–7844.

Ferguson-Smith, M. A. (1965). Karyotype-phenotype correlations in gonadal dysgenesis and their bearing on the pathogenesis of malformations. *J. Med. Genet.* **2,** 142–145.

Fisher, E. M. C., Beer-Romero, P., Brown, L. G. *et al.* (1990). Homologous ribosomal protein genes on the human X and Y chromosomes: Escape from X inactivation and possible implications for Turner syndrome. *Cell (Cambridge, Mass.)* **63,** 1205–1218.

Ford, C. E., Miller, O. J., Polani, P. E., de Almeida, J. C., and Briggs, J. H. (1959). A sex chromosome anomaly in a case of gonadal dysgenesis (Turner's syndrome). *Lancet* **1,** 711–713.

Fraccaro, M., Maraschio, P., Pasquali, F., and Scappaticci, S. (1977). Women heterozygous for deficiency of the (p21-pter)region of the X chromosome are fertile. *Hum. Genet.* **39,** 283–292.

Fransancho, A. R., Garn, S. M., and Ascoli, W. (1970). Subperiosteal and endosteal bone apposition during adolescence. *Hum. Biol.* **42,** 639–662.

Fryns, J. P., Petit, P., and Van den Berghe, H. (1981). The various phenotypes in Xp deletion; observations in eleven patients. *Hum. Genet.* **57,** 385–387.

Fryns, J. P., Kleczkowska, A., and Van den Berghe, H. (1990). High incidence of mental retardation in Turner syndrome patients with ring chromosome X formation. *Genet. Couns.* **38,** 161–165.

Garron, D. A. (1977). Intelligence among persons with Turner's syndrome. *Behav. Genet* **7,** 105–127.

Grumbach, M. M., and Conte, F. A. (1992). Disorders of sex differentiation. *In* "Williams Textbook of Endocrinology" (J. D. Wilson and D. W. Foster, eds.), pp. 853–952. Saunders, Philadelphia.

Haddad, H. M., and Wilkins, L. (1959). Congenital anomalies associated with gonadal aplasia. *Pediatrics* **23,** 885.

Hall, J. G., and Gilchrist, D. M. (1990). Turner syndrome and its variants. *Pediatr. Clin. North Am.* **37,** 1421–1440.

Hall, J. G., Sybert, V. P., Williamson, R. A., Fisher, N. L., and Reed, S. D. (1982). Turner's syndrome. *West. J. Med.* **137,** 32–44.

Hassold, T., Kumlin, E., Takaesu, N., Leppert, M. (1985). Determination of the parenteral origin of sex chromosome monosomy using restriction fragment length polymorphisms. *Am. J. Hum. Genet.* **37,** 965–972.

Hassold, T., Benham, F., and Leppert, M. (1988). Cytogenetic and molecular analysis of sex chromosome monosomy. *Am. J. Hum. Genet.* **42,** 534–541.

Hassold, T., Pettay, D., Robinson, A., and Uchida, I. (1992). Molecular studies of parental origin and mosaicism in 45,X conceptuses. *Hum. Genet.* **89,** 647–652.

Hecht, F., and Macfarlane, J. P. (1969). Mosaicism in Turner's syndrome reflects the lethality of XO. *Lancet* **2,** 1197–1198.

Held, K. R., Kerber, S., Kaminsky, E. *et al.* (1992). Mosaicism in 45,X Turner syndrome: Does survival in early pregnancy depend on the presence of two sex chromosomes? *Hum. Genet.* **88,** 288–294.

Herva, R., Kaluzewski, B., and de la Chapelle, A. (1979). Inherited interstitial del(Xp) with minimal clinical consequences: With a note on the location of genes controlling phenotypic features. *Am. J. Med. Genet.* **3,** 43–58.

Hook, E. B., and Warburton, D. (1983). The distribution of chromosomal genotypes associated with Turner's syndrome: Livebirth prevalence rates and evidence for diminished fetal mortality and severity in genotypes associated with structural X abnormalities or mosaicism. *Hum. Genet.* **62,** 24–27.

Hsu, L. Y. F. (1989). Prenatal diagnosis of 45,X/46,XY mosaicism—a review and update. *Prenatal Diagn.* **9,** 31–48.

Hsu, L. Y. F., Kim, H. J., Paciuc, S., Steinfeld, L., and Hirshhorn, K. (1974). Nonfluorescent and non-heterochromatic Y chromosome in 45,X/46,XY mosaicism. *Ann. Genet.* **17,** 5–9.

Jacobs, P. A., Betts, P. R., Cockwell, A. E., Crolla, J. A., Mackenzie, M. J., Robinson, D. O., Youings, S. A. (1990). A cytogenetic and molecular reappraisal of a series of patients with Turner's syndrome. *Ann. Hum. Genet.* **54,** 209–223.

Johnson, V. P., McDonough, P. G., Cheung, S. W., and Sun, L. (1991). Sex chromosome marker: Clinical significance and DNA characterization. *Am. J. Med. Genet.* **39,** 97–101.

Just, W., Geerkens, C., Held, K. R., and Vogel, W. (1992). Expression of RPS4X in fibroblasts from patients with structural aberrations of the X chromosome. *Hum. Genet.* **89,** 240–242.

Kalousek, D., Schiffrin, A., Berguer, A.-M. *et al.* (1979). Partial short arm deletions of the X chromosome and spontaneous pubertal development in girls with short stature. *J. Pediatr.* **94,** 891–894.

Kirkland, R. T., Lin, T.-H., LeBlanc, A. D., Kirkland, J. L., and Evans, H. J. (1990). Effects of hormonal therapy on bone mineral density in Turner Syndrome. *In* "Turner Syndrome" (R. G. Rosenfeld and M. M. Grumbach, eds.), pp. 319–325. Dekker, New York.

Krauss, C. M., Turksoy, R. N., Atkins, L., McLaughlin, C., Brown, L. G., and Page, D. C. (1987). Familial premature ovarian failure due to an interstitial deletion of the long arm of the X chromosome. *N. Engl. J. Med.* **317,** 125–131.

Lacro, R. V., Jones, K. L., and Benirschke, K. (1988). Coarctation of the aorta in Turner syndrome: A pathologic study of fetuses with nuchal cystic hygromas, hydrops fetalis, and female genitalia. *Pediatrics* **81,** 445–451.

Leichtman, D. A., Schmickel, R. D., Gelehrter, T. D., Judd, W. J., Woodbury, M. C., and Meilinger, K. L. (1978). Familial Turner syndrome. *Ann. Intern. Med.* **89,** 473–476.

Levilliers, J., Quack, B., Weissenbach, J., and Petit, J. (1989). Exchange of terminal portions of X and Y chromosomal short arms in human XY females. *Proc. Natl. Acad. Sci. U.S.A.* **86,** 2296–3000.

Lin, A. E., Lippe, B. M., Geffner, M. E. *et al.* (1986). Aortic dilation, dissection, and rupture in patients with Turner syndrome. *J. Pediatr.* **109,** 820–822.

Lindgren, V., Chen, C., Bryke, C. R., Lichter, P., Page, D. C., and Yang-Feng, T. L. (1992). Cytogenetic and molecular characterization of marker chromosomes in patients with mosaic 45,X karyotypes. *Hum. Genet.* **88,** 393–398.

Lippe, B. M. (1990). Primary ovarian failure. *In* "Clinical Pediatric Endocrinology" (S. A. Kaplan, ed.), pp. 325–366. Saunders, Philadelphia.

Lippe, B. M., Geffner, M. E., Dietrich, R. B., Boechat, M. I., and Kangarloo, H. (1988). Renal malformations in patients with Turner syndrome: Imaging in 141 patients. *Pediatrics* **83,** 852–856.

Lorda-Sanchez, I., Binkert, F., Maechler, M., and Schinzel, A. (1992). Molecular study of 45,X conceptuses: Correlation with clinical findings. *Am. J. Med. Genet.* **42,** 487–490.

Lyon, A. J., Preece, M. A., and Grant, D. B. (1985). Growth curve for girls with Turner syndrome. *Arch. Dis. Child.* **60,** 932–935.

Magenis, R. E., Tochen, M. L., Holahan, K. P., Carey, T., Allen, L., and Brown, M. G. (1984). Turner syndrome resulting from partial deletion of Y chromosome short arm: Localization of male determinants. *J. Pediatr.* **105,** 916–919.

Massa, G., Vanderschueren-Lodeweyckx, M., and Malvaux, P. (1990). Linear growth in patients with Turner syndrome: Influence of spontaneous puberty and parental height. *Eur. J. Pediatr.* **149,** 246–250.

Massarano, A. A., Adams, J. A., Preece, M. A., and Brook, C. G. D. (1989a). Ovarian ultrasound appearances in Turner Syndrome. *J. Pediatr.* **114,** 568–573.

Massarano, A. A., Brook, C. G. D., Hindmarsh, P. C. *et al.* (1989b). Growth hormone secretion in Turner's syndrome and influence of oxandrolone and ethinyl estradiol. *Arch. Dis. Child.* **64,** 587–592.

Mathur, A., Stekol, L., Schatz, D., Maclaren, N. K., Scott, M. L., and Lippe, B. (1991). The parental origin of the single X chromosome in Turner syndrome: Lack of correlation with parental age or clinical phenotype. *Am. J. Hum. Genet.* **48,** 682–686.

Mauras, N., Rogol, A. D., and Veldhuis, J. D. (1990). Increased hGH production rate after low-dose estrogen therapy in prepubertal girls with Turner's syndrome. *Pediatr. Res.* **28,** 626–630.

McCauley, E., Kay, T., Ito, J., and Treder, R. (1987). The Turner Syndrome: Cognitive deficits, affective discrimination, and behavior problems. *Child Dev.* **58,** 464–473.

Medlej, R., Lobaccaro, J. M., Berta, P. *et al.* (1992). Screening for Y-derived sex determining gene SRY in 40 patients with Turner syndrome. *J. Clin. Endocrinol. Metab.* **75,** 1289–1292.

Miller, M. J., Geffner, M. E., Lippe, B. M. *et al.* (1983). Echocardiography reveals a high incidence of bicuspid aortic valve in Turner syndrome. *J. Pediatr.* **102,** 47–50.

Money, J. (1973). Turner's syndrome and parietal lobe functions. *Cortex* **9,** 385–393.

Naeraa, R. W., and Nielsen, J. (1990). Standards for growth and final height in Turner's syndrome. *Acta Paediatr. Scand.* **79,** 182–190.

Naeraa, R. W., Brixen, K., Hansen, R. M., Hasling, C., Mosekilde, L., Andresen, J.-H., Charles, P., and Nielsen, J. (1991). Skeletal size and bone mineral content in Turner's syndrome: Relation to karyotype, estrogen treatment, physical fitness, and bone turnover. *Calcif. Tissue Int.* **49,** 77–83.

Neely, E. K., Marcus, R., Rosenfeld, R. G., and Bachrach, L. K. (1993). Bone mineral apparent density is normal in adolescents with Turner syndrome. *J. Clin. Endocrinol. Metab.* (in press).

Otto, P. G., Vianna-Morgante, A. M., Otto, P. A., Wajntal, A. (1981). The Turner phenotype and the different types of human X isochromosome. *Hum. Genet.* **57,** 159–164.

Page, D. C., Fisher, E. M. C., McGillivray, B., and Brown, L. G. (1990). Additional deletion in sex-determining region of human Y chromosome resolves paradox of X,t(Y;22) female. *Nature (London)* **346,** 279–281.

Palmer, C. G., and Reichman, A. (1976). Chromosomal and clinical findings in 110 females with Turner syndrome. *Hum. Genet.* **35,** 35–42.

Papendieck, L. G., Iorcansky, S., Coco, R., Rivarola, M. A., and Bergada, C. (1987). High incidence of thyroid disturbances in 49 children with Turner syndrome. *J. Pediatr.* **111,** 258–261.

Park, E., Bailey, J.D., and Cowell, C. A. (1983). Growth and maturation of patients with Turner's syndrome. *Pediatr. Res.* **17,** 1–7.

Pennington, B. F., Bender, B., Puck, M., Salbenblatt, J., and Robinson, A. (1982). Learning disabilities in children with sex chromosome anomalies. *Child Dev.* **53,** 1182–1192.

Polani, P. E., Hunter, W. F., and Lennox, B. (1954). Chromosomal sex in Turner's syndrome with coarctation of the aorta. *Lancet* **2,** 120.

Preger, L., Steinbach, H. L., Moskowitz, P., Scully, A. L., and Goldberg, M. B. (1968). Roentgenographic abnormalities in phenotypic females with gonadal dysgenesis. *Am. J. Radiol.* **106,** 899–910.

Ranke, M. B., and Rosenfeld, R. G., eds. (1991). "Turner Syndrome: Growth Promoting Therapies." Excerpta Medica, Amsterdam.

Ranke, M. B., Pfluger, H., Rosendahl, W. *et al.* (1983). Turner syndrome: Spontaneous growth in 150 cases and review of the literature. *Eur. J. Pediatr.* **141,** 81–88.

Rosenfeld, R. G., and Grumbach, M. M., eds. (1990). "Turner Syndrome." Dekker, New York.

Rosenfeld, R. G., Luzzatti, L., Hintz, R. L., Miller, O. J., Koo, G. C., and Wachtel, S. S. (1979). Sexual and somatic determinants of the human Y chromosome: Studies in a 46,XYp- phenotypic female. *Am. J. Hum. Genet.* **31,** 458–468.

Rosenfeld, R. G., Frane, J., Attie, K. M. *et al.* (1992). Six year results of a randomized prospective trial of human growth hormone and oxandrolone in Turner syndrome. *J. Pediatr.* **121,** 49–55.

Ross, J. L., Long, L. M., Loriaux, D. L., and Cutler, G. B., Jr. (1985). Growth hormone secretory dynamics in Turner syndrome. *J. Pediatr.* **106,** 202–206.

Ross, J. L., Long, L. M., Feuillan, P., Cassorla, F., and Cutler, G. B. (1991). Normal bone density of the wrist and spine and increased wrist fractures in girls with Turner's syndrome. *J. Clin. Endocrinol. Metab.* **73,** 355–359.

Rovet, J., and Netley, C. (1982). Processing deficits in Turner's syndrome. *Dev. Psychol.* **18,** 77–94.

Sanger, R., Tippett, P., and Gavin, J. (1971). Xg groups and sex abnormalities in people of Northern European ancestry. *J. Med. Genet.* **8,** 417–426.

Schatz, R., Maclaren, N. K., and Lippe, B. M. (1990). Autoimmunity in Turner syndrome. *In* "Turner Syndrome" (R. G. Rosenfeld and M. M. Grumbach, eds.), pp. 205–215. Dekker, New York.

Scully, R. E. (1970). Gonadoblastoma: A review of 74 cases. *Cancer (Philadelphia)* **25,** 1340–1356.

Shepard, T. H., Wener, M. H., Myrhe, S. A., and Hickok, D. E. (1986). Lowered plasma albumin concentrations in fetal Turner syndrome. *J. Pediatr.* **108,** 114–116.

Simpson, J. L. (1979). Gonadal dysgenesis and sex chromosome abnormalities: Phenotypic-karyotypic correlations. *In* "Genetic Mechanisms of Sexual Development" (H. L. Vallet and I. H. Porter, eds.), pp. 365–405. Academic Press, New York.

Singh, R. P., and Carr, D. H. (1966). The anatomy and histology of XO embryos and fetuses. *Anat. Rec.* **155,** 369–381.

Skibsted, L., Westh, H., and Niehbur, E. (1984). X long arm deletions: A review of non-mosaic cases studied with banding techniques. *Hum. Genet.* **67,** 1–5.

Smith, M. A., Wilson, J., and Price, W. H. (1982). Bone demineralisation in patients with Turner's syndrome. *J. Med. Genet.* **19,** 100–103.

Sybert, V. P. (1984). Adult height in Turner syndrome with and without androgen therapy. *J. Pediatr.* **104,** 365–369.

Therman, E., and Susman, B. (1990). The similarity of phenotypic effects caused by Xp and Xq deletions in the human female: A hypothesis. *Hum. Genet.* **85,** 175–183.

Turner, H. H. (1938). A syndrome of infantilism, congenital webbed neck, and cubitus valgus. *Endocrinology (Baltimore)* **23,** 566–574.

Ullrich, P. (1949). Turners syndrome and status Bonnevie-Ullrich. *Am. J. Hum. Genet.* **1,** 179–202.

Warburton, D., Kline, J., Stein, Z., and Susser, M. (1980). Monosomy X: A chromosomal anomaly associated with young maternal age. *Lancet* **1,** 167.

Watanabe, M., Zinn, A. R., Page, D. C., and Nishimoto, T. (1993). Functional equivalence of human X- and Y-encoded isoforms of ribosomal protein S4 consistent with a role in Turner syndrome. *Nat. Genet.* **4,** 268–271.

Wheeler, M., Peakman, D., Robinson, A., Henry, G. (1988). 45X/46XY mosaicism: Contrast of prenatal and postnatal diagnosis. *Am. J. Med. Genet.* **29,** 565–571.

Wilkins, L., and Fleischmann, W. (1944). Ovarian agenesis: Pathology, association with clinical symptoms, and their bearing on the theories of sex differentiation. *J. Clin. Endocrinol. Metab.* **4,** 357–368.

15

Molecular Genetics of Androgen Insensitivity Syndromes in Humans

Leonard Pinsky, Mark Trifiro, Lenore K. Beitel, and Morris Kaufman

Lady Davis Institute for Medical Research
Sir Mortimer B. Davis–Jewish General Hospital
McGill University
Montreal, Quebec, Canada H3T 1E2

Molecular Genetics of Sex Determination

I. Introduction

This chapter has four purposes: (*a*) to introduce the androgen-response apparatus that underlies normal androgen sensitivity; (*b*) to consider briefly the history and frequency of androgen insensitivity syndromes in humans, and certain of their nosologic, dymorphologic, pathophysiologic, and differential diagnostic aspects; (*c*) to discuss the properties of the androgen receptor gene and its product in the context of its membership in the superfamily of genes whose products are ligand-activated, DNA-binding, transcription-regulating proteins; and (*d*) to illustrate the basic insights and clinical utility already provided by the 80 or so androgen receptor gene mutations that we and others have found among unrelated subjects with various clinical–biochemical phenotypes of androgen insensitivity.

II. The Androgen-Response Apparatus

Androgens affect differentiation, morphogenesis, maturation, and terminal function of their target cells, tissues, and organs, by promoting the expression of some genes and suppressing that of others. Most of this regulation appears to occur at the level of transcription (Berger and Watson, 1989; Rundlett *et al.*, 1990).

The androgen receptor (AR) is an intracellular protein, predominantly or exclusively nuclear (Ruizeveld de Winter *et al.*, 1991), that transduces the signal embodied in an androgen to a gene whose expression is androgen-regulated. The AR acts by binding an androgenic ligand noncovalently to form an androgen–receptor (A–R) complex (Fig. 1). The initial A–R complex is heteropolymeric and has relatively low affinity for androgen or DNA. After "transformation," a homodimeric AR has a greater affinity for androgen, and the A–R complex is able to bind to androgen-response elements (AREs). These are specific sequences of nucleotides that are typically located upstream (5′) of the principal promoter elements of an androgen-regulated gene. AREs act like enhancers: They promote transcription initiation. Negative AREs (nAREs) act like silencers: They suppress transcription initiation. A–R complexes may also suppress gene transcription by binding to the positive REs of other transcriptional regulatory proteins (Levine and Manley, 1989). The rate of transcription of an androgen-regulated gene, at one time, probably reflects the vector of multiple positive and negative transcriptional regulatory proteins (not all DNA-binding), which may interact with each other (Rennie *et al.*, 1993), and with components of

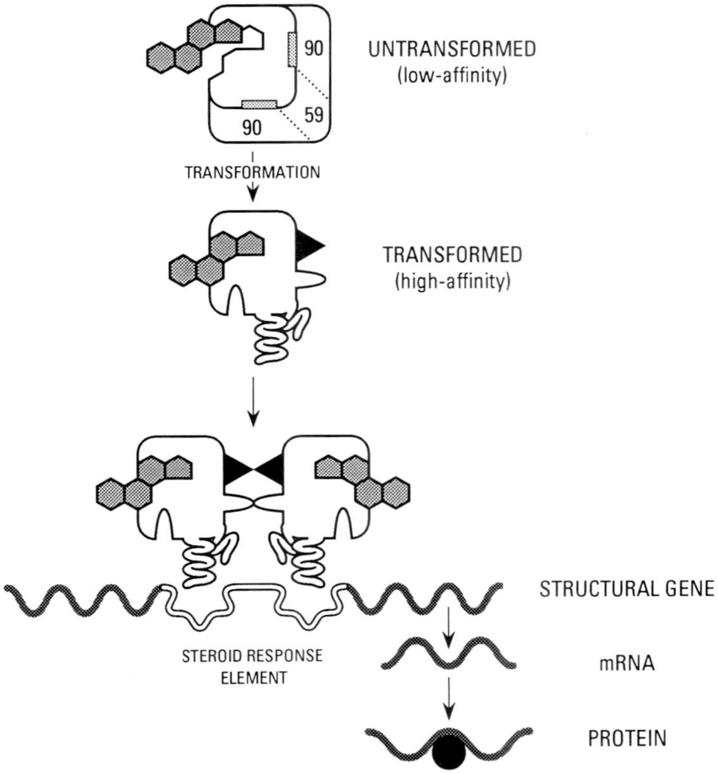

Figure 1 A structure–function model of the androgen receptor according to the modal features of the steroid receptor family. The untransformed version is composed of an inner, low-affinity, androgen-binding molecule whose potential DNA-binding and homodimerization sites are physically and/or functionally masked by two heteromeric molecules of Heat Shock Protein (hsp)-90, and one of 59 kilodaltons (Rehberger *et al.*, 1992). The transformed AR binds androgen with higher affinity, and has realized its potential for DNA-binding (a recognition α-helix pointing toward one-half of the steroid response element), and homodimerization (a solid triangle, representing a dimerization element in the ABD; an open projection, representing a dimerization element of the C-terminal zinc finger). Ultimately, each member of a dimerized A–R complex binds to one-half of an imperfectly inverted hexanucleotide repeat that constitutes an androgen-response element. This binding activates (or represses) transcription of a nearby ("androgen-regulated") gene depending on the vectorial effect of other, interactive transcriptional regulatory cofactors that will vary with target gene, time, and/or cell type.

the RNA polymerase II transcription–initiation complex (Johnson and McKnight, 1989). In fact, many androgen-regulated genes are co-regulated by other hormones (Berger and Watson, 1989).

III. Androgen Insensitivity

A. Terminology–Nosology

"Androgen insensitivity syndrome" (AIS) refers to an aggregate of developmental defects attributable to a single constitutional defect in the androgen-response apparatus. Conventionally, the term applies to an XY person who has deficient or defective activity of the AR due to mutation at its X-linked locus. The term AIS is often preceded by complete, incomplete, partial, or mild (minimal); these indicate severity and distribution (generality or focality) of the syndrome in a single subject, or in several members of one family. In the complete form (CAIS) an XY female presents with inguinal testes in infancy, or with primary amenorrhea and absent or sparse sexual hair at puberty. In one of its mild forms (MAIS), an otherwise normal adult male may have a small but nonhypospadic penis and scrotum, and may be infertile because of oligo-azoospermia. In another mild form, fertility is preserved despite gynecomastia or skeletal undermusculature. Between the extremes (partial, PAIS) there is congenital external genital undermasculinization (masculinization refers to genital morphogenesis; virilization refers to pubertal growth and maturation of male genital and somatic characteristics) (varying from posteriorly fused labia majora with or without clitoromegaly to different degrees of penoscrotal hypospadias) that may or may not be coordinate with signs of pubertal extragenital androgen insensitivity.

Usually, the clinical phenotype of androgen insensitivity varies little among affected members of one family, especially in those at the extremes of the phenotypic spectrum. Expressivity is more variable among families with PAIS: In certain ones, the phenotype can vary from nearly normal male to nearly normal female. Some of this variation must represent "background" genetic differences that render androgen sensitivity a continuous trait in the normal population. However, in the families with markedly variable expressivity, it is likely that the mutant AR itself has a threshold effect on final androgen sensitivity.

As defined above, AIS excludes nearly all androgen-sufficient males who are undermasculinized and/or undervirilized as part of a syndrome whose other features are not explainable by androgen insensitivity.

Table I
Classification of Constitutional Androgen Insensitivity

Feature	Type
Severity	Complete: female external genitalia
	Partial: any external genital ambivalence
	Minimal: male external genitalia
Anatomic distribution	Universal: all targets affected
	Multifocal: more than one affected
	Isolated: only one affected
Chronology	Apparent at birth; unchanging with time
	Apparent at birth; changing with time
	Adult (pubertal) onset
Androgen selectivity	Resistance to T alone; T and DHT
	Resistance to natural and synthetic androgen
	Resistance to selected natural or synthetic androgen
Inheritance	X-linked recessive
	Autosomal recessive, male-limited

Most such subjects will have a primary fault of histogenesis or morphogenesis that has nullified the normal androgen-response system. A few will have a contiguous gene syndrome involving the AR locus.

Normal androgen responsiveness depends on the presence of sufficient amounts of the correct androgen in target cells, and ends with appropriate changes in the expression of androgen-regulated genes. In other words, multiple pre- and post- (or para-) receptor events condition the ultimate response of target cells to androgen. A prereceptor cause of androgen insensitivity is reduced activity of 5α-reductase (type 2) (Andersson *et al.*, 1991), one of the enzymes responsible for converting testosterone (T) to 5α-dihydrotestosterone (DHT). This autosomal recessive disorder interferes with those portions of male sexual development that depend on DHT (e.g., the prostatic and external genital primordia), but spares those that are responsive to testosterone (T) (e.g., the Wolffian ducts). In other words, 5α-reductase deficiency is an androgen-selective PAIS (see Chapter 16, this volume).

A clinical classificatory scheme that can encompass all forms of androgen insensitivity is given in Table I.

B. History

There are biblical references to families whose women were infertile putatively because of CAIS (Goodman and Plato, 1982). Pettersson and

Bonnier (1937) are credited for the first full clinical description of such a family. Keenan *et al.* (1974) were the first to find deficient androgen-binding activity in CAIS. In the same year, two groups (Walsh *et al.*, 1974; Imperato-McGinley *et al.*, 1974) first described 5α-reductase deficiency.

C. Frequency

There are no reliable estimates of the incidence of AIS. The one usually quoted for CAIS, 1 : 50,000, is based on the finding of testes in inguinal hernias of infant girls, and is obviously an underestimate. In our experience, PAIS is at least as common as CAIS but is more difficult to diagnose.

The incidence of MAIS is even less certain than that of PAIS for two reasons: First, its clinical expression blends with the lower limit of normal male sexual development; second, the laboratory diagnosis of subtle AR dysfunction is difficult.

D. Differential Diagnosis

The various AI syndromes must be distinguished from clinically comparable phenotypes that reflect specific defects in testis formation or testicular biosynthesis of androgen. The latter two categories are treated in Chapter 16. Here we address certain pathophysiologic, dysmorphologic, familial, and genital skin fibroblast features that are relevant to the diagnosis of AR-dependent AIS.

In CAIS, resistance of the gonadotropin-releasing hormone (GnRH)–luteinizing hormone (LH) axis to the feedback inhibitory effects of T classically results in LH and T production that is above those of normal males. However, single determinations of circulating LH or T may not be elevated, even when mean levels on repeated samples are. Interestingly, the blood level of androstenedione may be elevated when that of T is not (Boyar *et al.*, 1978). In PAIS and especially in MAIS, the GnRH–LH axis may be spared so that the stimulus to increased synthesis of LH and T is absent.

PAIS is often proffered as a diagnosis of exclusion in two situations: When an infant male suspect, less than 3 months of age, has normal basal levels of T, and when an prepubertal male suspect has a normal T response to the challenge of human chorionic gonadotrophin or Gn–RH stimulation. It should be noted that a diagnosis of PAIS made by exclusion overlooks the possibility of transient prenatal androgen insufficiency during the critical period of genital morphogenesis.

In all classes of AR-dependent AIS, some degree of pubertal feminization results from testicular synthesis of estrogen and peripheral aromatization of androgen to estrogen. The extent of feminization varies directly with the degree of AI conferred by a particular AR mutation.

The constitutive pattern of external genital development is female. Hence, in CAIS the external genitalia are unambiguously female, although the labia minora may be hypoplastic, indicating an androgen effect on their normal development. In PAIS various degrees of residual androgen sensitivity cause various degrees of labioscrotal fusion and phallic growth; this yields various degrees of hypospadias. In MAIS, fusion of the scrotal folds is complete, although the midline raphe may be prominent. The phallus is clearly penile, but may be short with first-degree hypospadias.

Although Müllerian duct regression is considered to be androgen-independent, subjects with CAIS may have Müllerian duct rudiments, and a vagina that is not short clinically. Likewise, although Wolffian duct differentiation is T-dependent, subjects with CAIS may have rudimentary Wolffian duct derivatives.

X-linkage and variable expressivity make the maternal family history a cornerstone in the differential diagnosis of AIS. Aside from 46,XY affected individuals whose sexual phenotype may vary appreciably, 46,XX female carriers may have delayed, sparse, or asymmetrical pubic and/or axillary hair, or delayed menarche. Often the family history is negative because about one-third of all subjects represent novel mutations in their mother's egg, and another one-third are the only affected individuals conceived by their heterozygous mothers.

The biochemical phenotype of AIS is defined initially by the specific androgen-binding activity in genital skin fibroblasts. (As a group, normal genital skin fibroblast lines have two- to three-fold more androgen-binding activity than their nongenital counterparts; some of the latter have activities near the lower limit of the assay's sensitivity). The activity may be negligible (R^-, R-neg, ≤ 5 fmol/mg protein with DHT as ligand), or quantitatively normal (15–40 fmol/mg protein; R-positive; R^+) but qualitatively abnormal. Usually the latter is expressed by an increased rate of dissociation of the A–R complexes. Often, this is accompanied by an increased apparent equilibrium dissociation constant (K_d). Other intracellular indices of a qualitative AR defect are: (*a*) thermolability of A–R complex formation or retention; (*b*) failure of up-regulation of the androgen-binding activity during prolonged incubation of the cells with a synthetic, nonmetabolizable androgenic ligand; or (*c*) failure of A–R complexes to localize in nuclei. Extracellular indices include resistance of A–R complexes to the stabilizing effect of molybdate during prolonged ultracentrifugation through sucrose gradients, and lability of DHT-R complexes under conditions that transform normal ones to the DNA-binding state.

Occasionally the activity is easily measurable but clearly less than normal (R^{def}; R-def). Such a finding is usually accompanied by one or more of the qualitative abnormalities listed above. Uncommonly the activity appears to be normal quantitatively and qualitatively. In this situation a defect in the androgen receptor's DNA-binding domain (DBD) is likely.

IV. Structure–Function Organization of the Human Androgen-Receptor Gene and Its Product

The human AR gene (*AR*) is located in the q11-q12 region of the X chromosome (Brown *et al.*, 1989). It is 80–90 kb long, but most of it is intronic or in flanking untranscribed regions (Kuiper *et al.*, 1989; Lubahn *et al.*, 1988). The 5′ UTR, encoded by exon 1, is ~1.1 kb in length and contains one major and one minor site of transcription initiation in a 13-bp region. The promoter lacks TATA and CCAAT-like elements, but has GC-rich sequences including a putative Sp-1 binding motif at −50 to −60, and an adenine/guanine-rich region at −185 to −230 (Tilley *et al.*, 1990; Faber *et al.*, 1991). The 3′ UTR, encoded by exon 8, is 6.8 kb long. There are two $poly(A)^+$ signals. Two major mRNA species are reported: the larger, ~10 or 11 kb; the smaller, ~7 or 8 kb. They differ in length of the 3′ UTR because of alternative splicing (Faber *et al.*, 1991).

The *AR* gene is composed of eight exons among which 2730-57 nucleotides (nt) are translated to yield a protein of 910–919 aa (Fig. 2), calculated to be ~98–99 kDa (Brinkmann *et al.*, 1989; Chang *et al.*, 1988; Lubahn *et al.*, 1988; Tilley *et al.*, 1989). Its intron–exon organization and the assembly of its exons place it in the superfamily of genes whose products—including the receptors for thyroxine, vitamin D and retinoic acid—are ligand-activated, DNA-binding, transcription-regulating proteins. The AR is a prototypic member of the subfamily of steroid receptor (SR) proteins that includes the receptors for glucocorticoid (GR), progesterone (PR), and mineralocorticoid (MR). The family members illustrate beautifully the modular mode of molecular evolution (Amero *et al.*, 1992). However, the functional autonomy of the modules is incomplete.

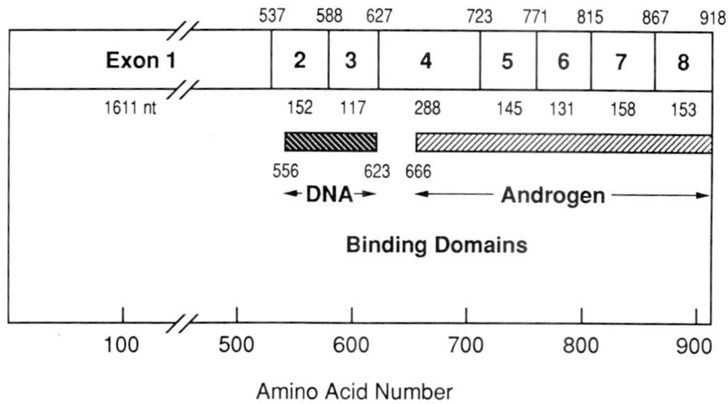

Figure 2 The *hAR* gene and major binding domains of the hAR. The numerical coordinates are those of Chang *et al.* (1988).

The N-terminal domain (~529–539 aa) has a transcriptional modulatory function and is encoded by exon 1 (Faber *et al.*, 1989). Its aa sequence is poorly conserved (16–25%) relative to the sequences of other members of the AR subfamily. It has a relatively high concentration of acidic residues between positions 100 and 325. Simental *et al.* (1991) localized its functionally important region to aa 140–337. Jenster *et al* (1991) identified a more proximal region (aa 51–211). There are several homopolymeric sequences: The longest are a 15- to 31-mer of polyglutamine starting at codon 189, and a 16- to 27-mer of polyglycine starting at codon 449 (Fig. 3). Polymorphic variation in the size of the

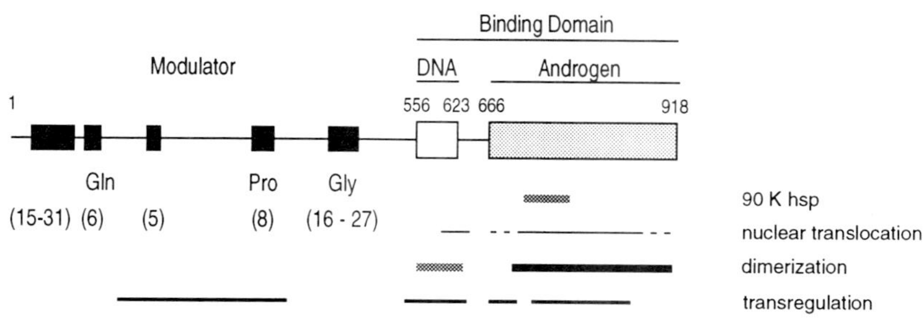

Figure 3 Proven or putative functional domains and subdomains (regions) of the 918-amino acid-long hAR using the numerical coordinates of Chang *et al.* (1988). The solid boxes indicate the major poly-amino acid repeats of the N-terminal domain, and the subjacent numbers in parentheses equal the polymorphic or fixed sizes of the repeats. The conventional limits of the major binding domains are identified by the open and hatched boxes. The real or potential distribution of various other functions of the hAR are indicated by the various inferior bars.

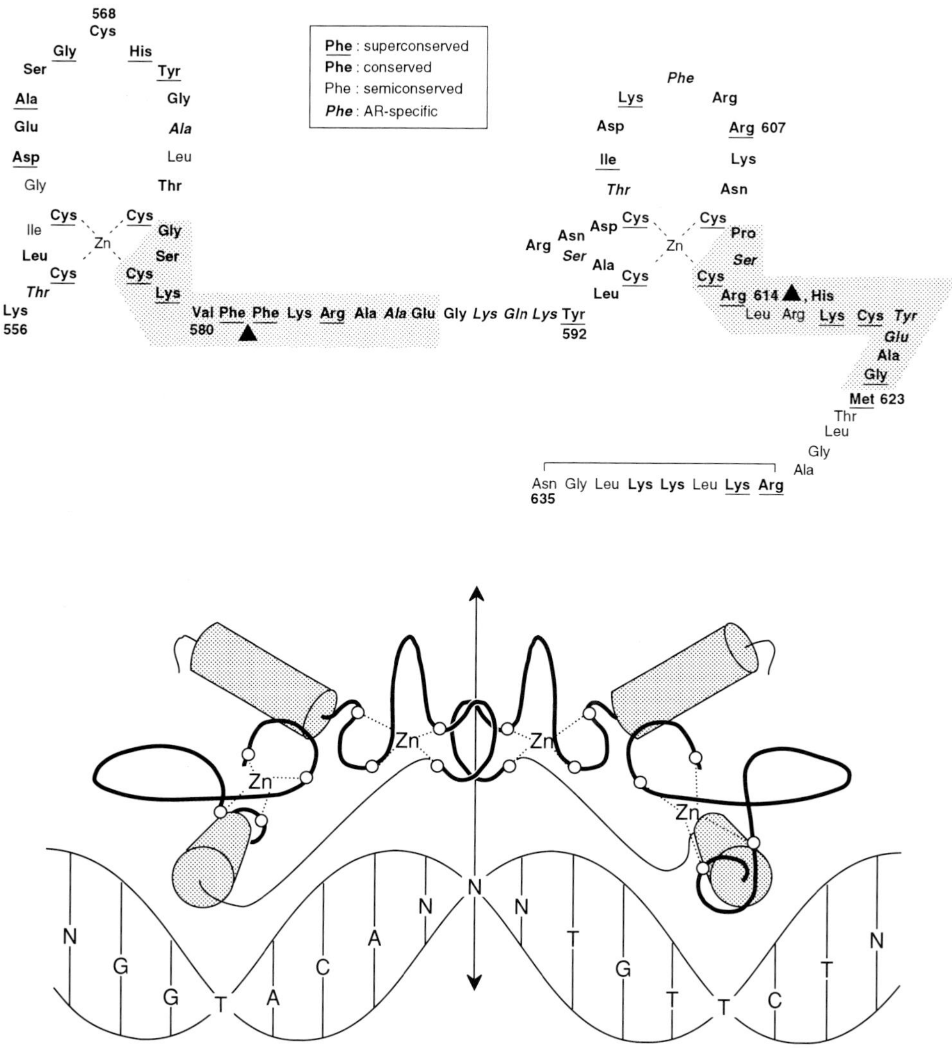

Figure 4 The DNA-binding domain of the AR. (Top) The underlined bolded letters denote residues that are invariant in the AR subfamily and are strongly conserved in the ER, and in the receptors for thyroxine, vitamin D, and retinoic acid. The bolded letters identify residues that are invariant in the AR subfamily. The regular letters identify residues that are conservatively substituted among members of the AR subfamily. The italicized bolded letters with asterisks refer to residues that are unique to the AR. Certain residues are numbered for landmark purposes. The N-terminal hatched sequence would represent the recognition α-helix. The inverted triangle beneath the consecutive Phe's at

homopolymeric tracts accounts for most of the reported length variation among the human androgen receptor (hAR) cDNA clones. The AR migrates as a doublet on SDS–PAGE because about half of the molecules in a sample are phosphorylated (Kuiper *et al.,* 1992). The AR deleted of aa 51–211 is phosphorylation deficient and migrates as a singlet. Since protein kinase agonists can mimic steroid action in some steroid-free target cells, phosphorylation may influence AR activity (Beck *et al.,* 1992).

The central domain, encoded by exons 2 and 3, contains 66 amino acids. They share 75–80% aa sequence homology with the hPR, hGR, and hMR, the hAR subfamily, and about 55% with the hER. Cysteine, lysine, and arginine are prominent among the conserved residues (Fig. 4). Two-dimensional nuclear magnetic resonance spectroscopy (Härd *et al.,* 1990) and crystallography (Luisi *et al.,* 1991) studies on the GR indicate that this domain features two α-helices, each flanked N-terminally by a finger that is rooted in a tetrahedral arrangement of 4 cysteines coordinated by a zinc ion (Fig. 4). This configuration, presumably in its dimerized form (Fawell *et al.,* 1990; Guiochan-Mantel *et al.,* 1989), is responsible for sequence-specific binding of A–R complexes to AREs. Specifically, it is the N-terminal α-helix of each receptor monomer that "fits" into successive major grooves on one surface of the DNA, and thereby interacts with the imperfect inverted (palindromic) hexanucleotide repeat that composes each half-site of a steroid response element (SRE) for a SR of the AR subfamily (Figs. 4 and 6). Such a "fit" had been predicted after particular amino acid substitutions of 3 different residues in this region [corresponding to Gly-576 (amino acids are numbered according to Chang *et al.* (1988) as shown in Fig. 5), Ser-577, and Val-580 in the hAR] changed the recognition specificity of a GR from a GRE to an ERE, and vice versa (Luisi *et al.,* 1991).

In addition to dimerization, the DBD of the AR probably contributes to a third function of the intact receptor: transcriptional regulation. This is suggested by point mutagenesis in the DBD of the hGR (Hollenberg and Evans, 1988) and rGR (Schena *et al.,* 1989) that permits them

---

positions 581, 582 denotes an in-frame, 3-bp deletion that eliminates one of the Phe's in a family with CAIS. Arg 614 is substituted by His in an unrelated CAIS family and it is eliminated by an in-frame, 3-bp deletion in a third CAIS family. The bracket that overlies residues 628 to 635 identifies a nuclear localization signal (NLS-1). (Bottom) A model of the dimeric DNA-binding domain of the GR showing the arrangement of α-helices (cylinders) and zinc fingers in relation to a GRE. The AR would be expected to have similar features; for instance, each of the N-terminal (recognition) α-helices would extend from Gly 576 to Glu 587 or Gly 588. Likewise, each of the C-terminal α-helices would extend from Pro 611 to Gly 622. From Muller and Renkawitz (1991), with permission.

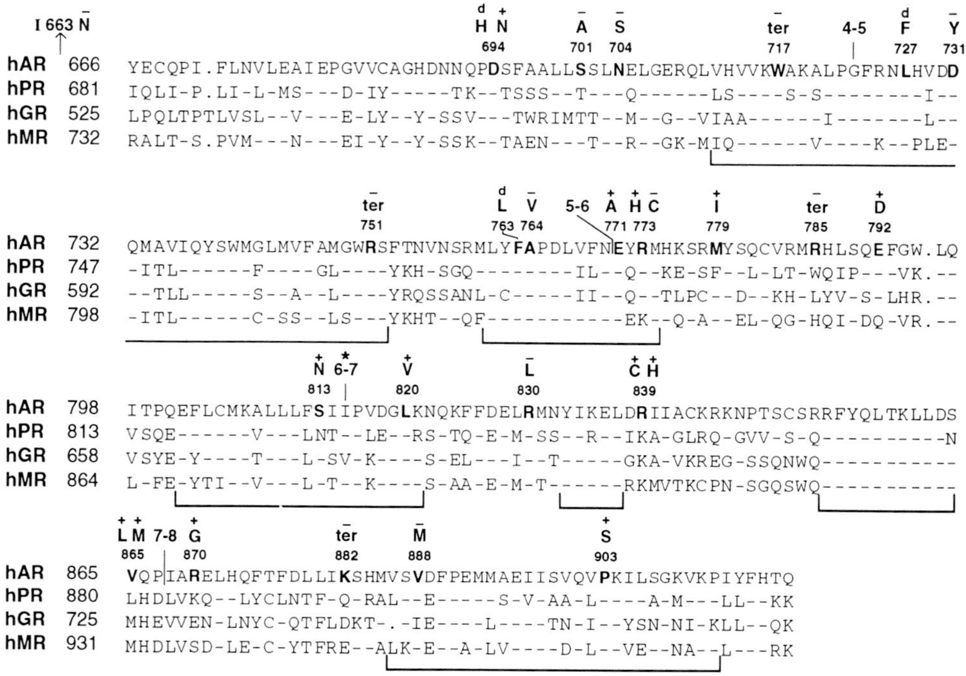

Figure 5 The deduced amino acid sequences of the steroid-binding domains of the human androgen receptor (hAR) and the members of its subfamily according to Chang *et al.* (1988): PR, progesterone receptor; GR, glucocorticoid receptor; MR, mineralocorticoid receptor. Dashes represent aa identity. Dots equal gaps to allow for maximal match both within the subfamily and with the estrogen receptor, a member of a separate subfamily. The vertical lines between adjacent residues in the AR sequence, and the sets of hyphenated numbers above them define exon junctions. Bold brackets below the hMR sequence indicate regions of high homology among members of the subfamily. "Homology" is defined as identity or conservative substitution of the aa at the corresponding position in the hAR. Bold letters and numbers in and above the hAR sequence identify the sites and predicted aa substitutions due to single nucleotide alterations in subjects/families with androgen insensitivity (AI). ter, premature termination codon; +, R-positive; −, R-negative, d, R-deficient. The asterisk (*) above the exon 6–7 junction denotes an A → T transition at the first position following the exon 6–intron 6 donor splice site.

to bind the GRE of MMTV, but not to transactivate a GRE-linked reporter gene.

Not far from the C-terminal Zn finger (Fig. 4), beginning at position 628, there is an 8 aa sequence, rich in lysine and arginine, and highly homologous with the nuclear localizing signal of the SV40 large T antigen (Guiochon-Mantel *et al.*, 1989). A similar sequence is conserved in

```
5' - A G T A C G T G A T G T T C T - 3'

5' - G G T A C A A A A T G T T C T - 3'

5' - T G T A C A G G A T G T T C T - 3'

     A
5' - G G T A C A n n n T G T T C T - 3'
     T
```

Figure 6 The imperfect inverted (palindromic) hexanucleotide repeats (half-sites), separated by three nucleotide spacers, that act as GREs/PREs and AREs.

other SRs. It probably acts constitutively to achieve partial nuclear translocation of unliganded AR, and collaborates with a second nuclear translocation signal that appears in the androgen-binding domain after it has bound androgen (Simental *et al.*, 1991; Jenster *et al.*, 1991).

The C-terminal domain of the AR, ~250 aa, encoded by exons 4–8, is primarily responsible for androgen binding. It is relatively hydrophobic, and its 50–55% aa homology with each of the three other members of its subfamily is distributed unevenly (Fig. 5). In some regions most of the aa positions are invariant, three-fourths preserved, or substituted conservatively. Evolutionary conservation of structure predicts functional importance: This is borne out in the discussion below of the numerous point mutations discovered in this domain among patients with androgen insensitivity.

Nothing is known about the shape of the hAR molecule before it binds androgen, or about the stereochemistry of the hormone-binding pocket in the androgen-binding domain (ABD) that allows it to recognize and bind androgens preferentially. The analysis of point mutations that impair androgen-binding activity or impair androgen-binding specificity will help to elucidate some of these unknowns.

Partly on inference by homology with other members of the AR subfamily, the ABD is likely to have three outstanding structure–function features. The first is that its entire length must be intact for normal binding to occur. A C-terminal truncation that removes 12 aa eliminates androgen-binding activity completely (Jenster *et al.*, 1991). This implies that aa throughout the ABD participate directly or indirectly in formation of the androgen-binding pocket. The second is that an unliganded ABD exerts a dominant inhibitory influence on all other functions of the intact AR molecule. The third feature is that an androgen-bound ABD acquires several subsidiary functions that complement those in other parts of the SR molecule. These functions are dimerization, transactivation, and nuclear translocation (Fig. 3).

The dominant inhibitory influence of an ABD is relieved physiologically when it is occupied by an agonistic ligand, or experimentally when it is largely or completely deleted (Rundlett *et al.*, 1990; Jenster *et al.*, 1991). It is possible that ligand binding disengages the AR from a heteropolymeric complex of other molecules (hsp-90, hsp-70, p59) that conspire to keep the A–R complex in an untransformed state (Fig. 1). However, that cannot be the only function of a ligand, since the binding of an androgen antagonist can promote the binding of an A–R complex to an ARE, and yet not transactivate an androgen-regulated gene (Kemppainen *et al.*, 1992). This is the rationale for inferring one (or more) agonist-dependent transactivation region(s) in a SBD (Webster *et al.*, 1988).

One candidate for such a region has very high aa sequence homology throughout the superfamily and has been implicated in the binding of hsp-90. In the AR it extends from about Val 714 to Met 741 (Fig. 5).

Discovery of the dimerization function in the steroid-occupied SBD was stimulated by the recognition of several possible protein dimerization elements (Fawell *et al.*, 1990; Lees *et al.*, 1990) in a portion of the mER that has high homology among members of the superfamily. One of these elements, a hydrophobic heptad repeat, begins at Leu 858 in the hAR (Fig. 5) and includes Val 865, Leu 872, and Leu 879. Point mutagenesis in the mER showed that substitution of Arg at the position homologous to Leu 858 of the hAR abolishes ER dimerization and specific ER–ERE binding, but not estrogen binding. The same results were produced by substitution of Ala at the position homologous to Arg 854 in the hAR. In contrast, substitution of Arg at the mER position homologous to Val 865 in the hAR abolished estrogen binding, dimerization, and ERE binding. The same region has been analyzed in the context of α-helix/β-turn units, four of which are responsible for dimerization of the progesterone-binding protein uteroglobin. The ultraconserved stretch in the subfamily extending from Tyr833 to Leu837 in the AR becomes significant in regard to this motif.

V. Genotype–Phenotype Correlation of Androgen Receptor Gene Mutation in Humans

Below we discuss all proven or putative pathogenic *hAR* gene alterations found in our laboratory, and selected ones (compiled until November 6,

1992) in the *AR* gene mutation database maintained by Dr. Mark Patterson at Cambridge University.

We have discovered 32 *AR* alterations among 38 families with various clinical forms of androgen resistance (Pinsky *et al.*, 1992) (Figs. 4 and 5). Twenty-eight of them are single-nucleotide alterations affecting the ABD, 8 as alternatives at four separate codons; 3 affect the DBD; one is a complete gene deletion (Trifiro *et al.*, 1991). About one-third of the point mutations have occurred at CpG dinucleotides. Twenty-three of 28 in the first category result in missense codons with a variety of consequences for receptor and clinical phenotype; four result in nonsense codons, predicting premature termination of translation; and one alters a consensus donor splice site. The pathogenicity of 14 missense mutations in the ABD has been proven experimentally by: (*a*) recreating the mutation in an expression plasmid containing the AR cDNA; (*b*) transfecting the plasmid into AR-free host cells; and (*c*) demonstrating that the mutant AR is defective as an androgen-binding molecule and/or as a transcriptional activator of the growth hormone (GH) reporter gene in a cotransfected androgen-responsive plasmid containing all the GREs/AREs in the long terminal repeat of the mouse mammary tumor virus (MMTV). In 8 of the remaining 10 missense alterations, the presumption of pathogenicity is based on substitution of dissimilar aa at positions and/or in regions that are highly conserved in the AR subfamily of steroid receptors.

A. DNA-Binding Domain Mutations in the hAR

Each of the three DBD mutations we have found caused CAIS. Deletion of Phe 581 (or 582) disrupts the DNA-binding, N-terminal α-helix. Arg 614 occupies the position homologous to Arg 496 in the rGR; Luisi *et al.* (1991) have identified the latter as making phosphate contacts with a GRE. Hence, it is not surprising that perturbation of Arg 614 causes CAIS. The equivalent to Arg614His in the AR has been found in the GR of a mouse lymphoma cell line resistant to glucocorticoid cytolysis (Danielsen *et al.*, 1986).

Other missense mutations causing CAIS have been found at Cys 558, Cys 575 and Arg 616 (Zoppi *et al.*, 1992); those causing PAIS have been found at Gly 567 (Chang *et al.*, 1991), Ala 595 (Klocker *et al.*, 1992), Arg 607 (Wooster *et al.*, 1992), and Leu 615 (de Bellis *et al.*, 1991). A particularly informative subject with PAIS has Arg616Pro and Ser596Gly (Zoppi *et al.*, 1992). Since Arg616Pro alone produces CAIS, the concomitance of Ser596Gly is ameliorative. More such genotype–phenotype correlations will help to elucidate the structure–function properties of the androgen receptor's DBD.

A family with CAIS whose genital skin fibroblasts (GSF) have supernormal androgen-binding activity has an AR gene deleted of the exon 3 sequence that encodes the C-terminal Zn finger (Quigley *et al.*, 1992a). The A–R complexes in the GSF of the subject with two missense substitutions in the DBD of her AR (Zoppi *et al.*, 1992) have an increased rate of dissociation. Androgen-binding activity is low in the GSF of subjects with two of the three DBD mutations we have identified. Thus, DBD alterations may affect androgen-binding activity, indicating less than complete autonomy among the three major structure–function domains of the AR.

B. Androgen-Binding Domain Mutations in the hAR

Twenty-seven of our 28 mutations in the ABD are exonic and span positions 663–903 (Fig. 5). This indicates that widely dispersed residues can affect the androgen-binding pocket.

Trp717ter is the most interesting nonsense mutant we have identified in the ABD (Sai *et al.*, 1990). It is equal or similar to ones generated by Rundlett *et al.* (1990) in the rAR, and by Marcelli *et al.* (1990) and Jenster *et al.* (1991) in the hAR. All three are constitutively active in host cells transiently cotransfected with an ARE-linked reporter gene; in contrast, all three siblings with the Trp717ter mutation have R^- CAIS. Thus, the transient overexpression–cotransfection assay may not be a faithful indicator of a mutant androgen receptor's expression in a human being. In this instance, the most likely cause of the disparity is a decreased level of AR mRNA and protein. The association of nonsense codons with reduced mRNA is recognized frequently but is poorly understood (Mashima *et al.*, 1992).

Eleven of the 28 ABD mutations have created or ablated restriction sites that allow one to genotype the mutant locus easily by use of PCR-fragment pattern. Eleven have occurred at invariant positions in the AR subfamily, and 7, at positions that are conserved in three-fourths of the members or are conservatively substituted (Arg or Lys; Val or Ile). Ile663Asn, Ser701Ala, Val865Met, Val865Leu, and Arg870Gly have occurred at unconserved positions. Four of the 23 missense mutations have occurred in pairs of unrelated families with concordant clinical and biochemical phenotypes (Asp694Asn, Arg773Cys, Arg773His, Arg839His).

Unrelated families with concordant phenotypes have been observed by others for Gly749Asp (Batch *et al.*, 1992), Arg751Gln (Brown *et al.*,

1992), Leu761Phe (Batch *et al.*, 1992; Brown *et al.*, 1992), Phe763Leu (Pinsky *et al.*, 1992; Marcelli *et al.*, 1992), Arg773Cys, Arg773His (Prior *et al.*, 1992; Brown *et al.*, 1990; Marcelli *et al.*, 1991; Batch *et al.*, 1992), Arg839His (de Bellis *et al.*, 1991), Val865Met, and Val865Leu (Kazemi-Esfarjani *et al.*, 1993; Brown *et al.*, 1990; Saunders *et al.*, 1992).

The Asp694His mutation causes CAIS, and the GSF have deficient androgen-binding activity that dissociates at least as fast as that with the Asp694Asn mutation (Ris-Stalpers *et al.*, 1991).

The region from Leu 761 to Met 774 is highly conserved; as expected, mutations have been found at seven of its sites. CAIS is caused by Leu761Phe (Batch *et al.*, 1992; Brown *et al.*, 1992), Phe763Leu (Pinsky *et al.*, 1992; Marcelli *et al.*, 1992), Ala764Val (Pinsky *et al.*, 1992), Ala764Thr (Batch *et al.*, 1992), Pro765Ser (Marcelli *et al.*, 1992), Arg773Cys, and Arg773His (Prior *et al.*, 1992). In contrast, Tyr762Cys (McPhaul *et al.*, 1991) and Glu771A1a (Pinsky *et al.*, 1992) cause PAIS. GSF with either of the last two mutant ARs have increased rates of A–R complex dissociation despite normal equilibrium dissociation constants (Grino *et al.*, 1988; Kaufman *et al.*, 1990).

The Tyr762Cys AR has only 12 glutamines in its exon 1 polyglutamine tract (McPhaul *et al.*, 1991). It causes profound hypospadias but permits considerable virilization at puberty (Grino *et al.*, 1989). Expression of AR cDNAs containing either mutation, or both, revealed that both are necessary to produce the qualitative AR abnormalities observed in GSF. Thus, structural abnormalities of the N-terminal domain can affect ABD function. This exemplifies incomplete functional autonomy among the three major structural modules of the AR.

Overexpression of the AR in cells transiently transfected with mutant AR cDNAs permits qualitative observations on mutant ARs with low androgen-binding activities in GSF. By this means, the Arg773Cys AR thermolability within 30 min at 37°C in COS cells was found to be fully reversible within 1 or 2 h at 22°C (Vasiliou *et al.*, 1992). The Ala764Val mutant AR behaves similarly.

Two unrelated families have an androgen-selective mutant AR due to Ser813Asn phenotype. In one family, sons of three sisters had MAIS (Pinsky *et al.*, 1984); in the other, a single subject had PAIS (Pinsky *et al.*, 1985). This mutant AR forms complexes with DHT and mibolerone (MB) that dissociate faster than normal, whereas complexes with T or methyltrienolone dissociate normally. Ser 813 is in a highly conserved region of the AR subfamily (Fig. 5). The position is filled by Ser in the GR and MR, and by Asn in the PR; but the mutant AR does not bind progesterone. The LNCaP cell line has a Thr876Ala mutant AR (Veldscholte *et al.*, 1990). It can bind androgen, but also estradiol, progesterone, antiandrogen, and antiestrogen; all of them are agonistic. Human

ER mutants with altered estrogen and antiestrogen discrimination have been reported by Pakdel and Katzenellenbogen (1992). Ligand-sensitive mutations hold important clues to the stereochemistry of steroid-specific binding by SRs, and to the contribution of steroids to the combinatorial attributes of S–R complexes acting as trans-regulatory signals.

The Leu820Val alteration is provocative. Leu, Val, and Ile are apparently functional alternatives at many positions in the SBD of the AR subfamily. But, position 820 is invariantly Leu in all the members (Fig. 5) and is close to Ser 813. This region seems intolerant of even subtle substitution. The Leu820Val AR causes severe perineoscrotal hypospadias, yet it dissociates from androgens only twice as fast as normal. We suspect that the region has another function (transactivation?).

We have observed two unrelated families (four affected members) with Arg839His, and one family (five affected members in two generations) with Arg839Cys. Patients with Arg839His had ambiguous external genitalia of a predominantly female character and were reared as females. Those with Arg839Cys had a scrotum with various degrees of hypospadias and were reared as males.

Position 854 in the hAR is conserved as Arg in all the SRs. Arg854 in the hAR corresponds to Arg507 in the mER, a residue critical for mER dimerization (White *et al.*, 1991). Arg854Cys in the hAR has caused CAIS in two families (de Bellis *et al.*, 1991; Tincello *et al.*, 1992). In contrast, Arg854His has caused PAIS in two families (Chang *et al.*, 1991; Batch *et al.*, 1992). We need to know whether impaired dimerization is a pathogenetic factor for either Arg 854 mutation. The Cys substitution is clinically more severe than the His alternative at codon 854; the reverse is true at codon 839. Such comparisons will help to define the structure–function relations of the human androgen receptor's ABD.

Val865Met and Val865Leu are a provocative pair. The former causes CAIS, the latter, PAIS. The homologous position is occupied by Leu in the PR and by Met in the GR and MR; yet, each substitution is pathogenic. In GSFs the Val865Met mutant has a severe, generalized, androgen-affinity defect; the Val865Leu mutant has a less severe, androgen-selective affinity defect that is coordinate with its less severe clinical phenotype (Kazemi-Esfarjani *et al.*, 1993). Finally, hAR Val 865 is homologous to mER Ile 518, a hydrophobic residue in a 22 aa segment required for mER dimerization (Lees *et al.*, 1991).

In COS-1 cells, Val865Leu A–R complexes had near-normal dissociation rates at 37°C; Val865Met A–R complexes had clearly increased rates. Thus, the milder, androgen-selective dissociative defect of the Val865Leu mutant, defined in GSF, disappeared in the COS-1 cells (Kazemi-Esfarjani *et al.*, 1993). Apparently, parareceptor "factors" affect the dissociative behavior of mutant A–R complexes; they may also contribute to the target-selective phenotypes observed in PAIS. As expected, the Val865Met receptor appeared to be less active than the Val865Leu receptor in transactivating the androgen-responsive MMTV–GH reporter plasmid in COS-1 cells incubated with MB for 72 h. However, when GH production was plotted against mean MB-binding activity during the 72-h incubation, the two mutants transactivated normally. This happened because the normal AR maintained its MB-binding activity in COS-1 cells, whereas the Val865Met AR lost much more activity than the Val865Leu AR did. Importantly, these losses of MB-binding activity were not accompanied by proportionate losses of immunoreactivity. This implies that prolonged incubation with MB caused both mutant ARs to adopt a non-binding state. The same interpretation has been applied to the behavior of the Cys447Ala mutation in the hER (Reese and Katzenellenbogen, 1992). The nonbinding state of SRs has been conceptualized as part of an intracellular process of SR recycling. Thus, the Val865Leu and Val865Met ARs may suffer from abnormal recycling, at least in some cells. We did not measure dimerization of either mutant AR directly. However, they transactivated normally per unit MB-binding activity, suggesting that neither mutant AR has impaired dimerization. This suggestion fits with the fact that neither Val865Met nor Val865Leu alters the hydrophobic character of the position. Yet, their different degrees of androgen-affinity defect indicate the structural subtlety that underlies the steroid-binding specificity of the AR subfamily.

C. Splice Site Mutations

The one intronic mutation among our 28 ABD mutants is an A → T transition in the first position of the consensus sequence at the exon 6–intron 6 donor splice site. The subject has R^- CAIS. Ris-Stalpers *et al.* (1990) found a similar mutation at the exon 4–intron 4 junction. It activates a cryptic splice donor, causing aa 682–722 to be deleted.

D. Mutation of the N-Terminal Modulatory Domain

Expansion of the poly-Gln tract (tandem CAG repeat) in exon 1 from about 15–31 to 40–50 or more glutamines is the cause of spinal and bulbar muscular atrophy (SBMA) (La Spada *et al.*, 1991). This adult-onset neurodegenerative disorder (Kennedy syndrome) is often associated with MAIS (gynecomastia; oligo-azoospermia). SBMA has never been reported in an androgen-resistant subject with any other AR mutation. It rarely, if ever, affects females (Harding *et al.*, 1982). Hence, the SBMA mutation may act by a recessive, sex-limited gain-of-function. A loss-of-function is possible, but only if the normal AR prevents SBMA independently of its androgen- and DNA-binding abilities, and if the two known adults with complete AR coding sequence deletions (Trifiro *et al.*, 1991; Quigley *et al.*, 1992b) eventually develop SBMA.

Androgen is known to support growth, differentiation, and regeneration of motor neurones in mammals (Araki *et al.*, 1991). How an expanded poly-Gln tract renders an AR injurious to certain male motor neurons is an open question with implications for various disorders in which the sex ratio departs significantly from unity. The candidates include some autoimmune diseases, some behavioral abnormalities, and some neoplasms.

VI. Conclusions

Androgen insensitivity due to deficient or defective activity of the hAR is far more prevalent than is insensitivity to all other steroid hormones and related ligands combined. The phenotypes caused by mutation at the X-linked AR locus range from nearly normal male to nearly normal female. Molecular–genetic analysis of such mutations is useful for diagnosis, prognosis, and therapy, for elucidating the stereochemistry of steroid-specific binding by different steroid receptors, and for learning how steroid–receptor complexes act to influence the rate of transcription of genes under their control.

Acknowledgments

We are indebted to Albert O. Brinkmann, Jan Trapman, and their colleagues at Erasmus University, Rotterdam, and to Shutsung Liao and his colleagues at the University of Chicago for helping us to initiate molecular–genetic analysis of hAR mutations. This chapter was written during tenure of grants from the Medical Research Council of Canada, the Fonds de la Recherche en Santé du Québec.

References

Amero, A. A., Kretsinger, R. H., Moncrief, N. D., Yamamoto, K. R., and Pearson, W. R. (1992). The origin of nuclear receptor proteins: A single precursor distinct from other transcription factors. *Mol. Endocrinol.* **6,** 3–7.

Andersson, S., Berman, D. M., Jenkins, E. P., and Russell, D. W. (1991). Deletion of steroid 5α-reductase 2 gene in male pseudohermaphroditism. *Nature* (*London*) **354,** 159–161.

Araki, I., Harada, Y., and Kuno, M. (1991). Target-dependent hormonal control of neuron size in the rat spinal nucleus of the bulbocavernosus. *J. Neurosci.* **11,** 3025–3033.

Batch, J. A., Williams, D. M., Davies, H. R., Brown, B. D., Evans, B. A. J., Hughes, I. A., and Patterson, M. N. (1992). Androgen receptor gene mutations identified by SSCP in fourteen subjects with androgen insensitivity syndrome. *Hum. Mol. Genet.* **1,** 497–503.

Beck, C. A., Weigel, N. L., and Edwards, D. P. (1992). Effects of hormone and cellular modulators of protein phosphorylation on transcriptional activity, DNA binding, and phosphorylation of human progesterone receptors. *Mol. Endocrinol.* **6,** 607–620.

Berger, F. G., and Watson, G. (1989). Androgen-regulated gene expression. *Annu. Rev. Physiol.* **51,** 51–65.

Boyar, R. M., Moore, R. J., Rosner, W., Aiman, J., Chipman, J., Madden, J. D., Marks, J. F., and Griffin, J. E. (1978). Studies of gonadotropin-gonadal dynamics in patients with androgen insensitivity. *J. Clin. Endocrinol. Metab.* **47,** 1116–1122.

Brinkmann, A. O., Faber, P. W., van Rooij, H. C. J., Kuiper, G. G. J. M., Ris, C., Klaassen, P., van der Korput, J. A. G. M., Voorhorst, M. M., van Laar, J. H., Mulder, E., and Trapman, J. (1989). The human androgen receptor: Domain structure, genomic organization and regulation of expression. *J. Steroid Biochem.* **34,** 307–310.

Brown, C. J., Goss, S. J., Lubahn, D. B., Joseph, D. R., Wilson, E. M., French, F. S., and Willard, H. F. (1989). Androgen receptor locus on the human X chromosome: Regional localization to Xq11-12 and description of a DNA polymorphism. *Am. J. Hum. Genet.* **44,** 264–269.

Brown, T. R., Lubahn, D. B., Wilson, E. M., French, F. S., Migeon, C. J., and Corden, J. L. (1990). Functional characterization of naturally occurring mutant androgen receptors from subjects with complete androgen insensitivity. *Mol. Endocrinol.* **4,** 1759–1772.

Brown, T. R., Newmark, J., and Ghirri, P. (1992). Six naturally occurring mutations in the 144-base pairs of exon E of the human androgen receptor gene. *Program, Endocr. Soc., 74th Annu. Meet.,* San Antonio, TX, Abstr. 1506, p. 428.

Chang, C., Kokontis, J., and Liao, S. (1988). Structural analysis of complementary DNA and amino acid sequences of human and rat androgen receptors. *Proc. Natl. Acad. Sci. U.S.A.* **85,** 7211–7215.

Chang, Y. T., Migeon, C. J., and Brown, T. R. (1991). Human androgen insensitivity syndrome due to androgen receptor gene point mutations in subjects with normal androgen receptor levels but impaired biological activity. *Program, Endocr. Soc., 73rd Annu. Meet.,* Washington, DC, Abstr. 28, p. 37.

Danielsen, M., Northrop, J. P., and Ringold, G. M. (1986). The mouse glucocorticoid receptor: Mapping of functional domains by cloning, sequencing and expression of wild-type and mutant receptor proteins. *EMBO J.* **5,** 2513–2522.

de Bellis, A., Quigley, C. A., Cariello, N. F., Ho, K. C., and French, F. S. (1991). DNA and steroid binding domain mutations of the androgen receptor cause partial or complete androgen insensitivity. *Program, Endocr. Soc., 73rd Annu. Meet.,* Washington, DC, Abstr. 1141, p. 316.

Faber, P. W., Kuiper, G. G. J. M., van Rooji, H. C. J., van der Korput, J. A. G. M., Brinkmann, A. O., and Trapman, J. (1989). The N-terminal domain of the human androgen receptor is encoded by one, large exon. *Mol. Cell. Endocrinol.* **61,** 257–262.

Faber, P. W., van Rooji, H. C. J., van der Korput, H. A. G. M., Baarends, W. M., Brinkmann, A. O., Grootegoed, J. A., and Trapman, J. (1991). Characterization of the human androgen receptor transcription unit. *J. Biol. Chem.* **266,** 10743–10749.

Fawell, S. E., Lees, J. A., White, R., and Parker, M. G. (1990). Characterization and colocalization of steroid binding and dimerization activities in the mouse estrogen receptor. *Cell (Cambridge, Mass.)* **60,** 953–962.

Goodman, R. M., and Plato, C. (1982). A talmudic reference to a family with probable testicular feminization syndrome. *Koroth* **8,** 40–47.

Grino, P. B., Griffin, J. E., Cushard, W. G., Jr., and Wilson, J. D. (1988). A mutation of the androgen receptor associated with partial androgen resistance, familial gynecomastia, and fertility. *J. Clin. Endocrinol. Metab.* **66,** 754–761.

Grino, P. B., Isidro-Gutierrez, R. F., Griffin, J. E., and Wilson, J. D. (1989). Androgen resistance associated with a qualitative abnormality of the androgen receptor and responsive to high dose androgen therapy. *J. Clin. Endocrinol. Metab.* **68,** 578–584.

Guiochon-Mantel, A., Loosfelt, H., Lescop, P., Sar, S., Atger, M., Perrot-Applanat, M., and Milgrom, E. (1989). Mechanisms of nuclear localization of the progesterone receptor: Evidence for interaction between monomers. *Cell (Cambridge, Mass.)* **57,** 1147–1154.

Härd, T., Kellenbach, E., Boelens, R., Maler, B. A., Dahlman, K., Freedman, L. P., Carlstedt-Duke, J., Yamamoto, K. R., Gustafsson, J.-A., and Kaptein, R. (1990). Solution structure of the glucocorticoid receptor DNA-binding domain. *Science* **249,** 157–160.

Harding, A. E., Thomas, P. K., Baraitser, M., Bradbury, P. G., Morgan-Hughes, J. A., and Ponsford, J. R. (1982). X-linked recessive bulbospinal neuronopathy: A report of ten cases. *J. Neurol., Neurosurg. Psychiat.* **45,** 1012–1019.

Hollenberg, S. M., and Evans, R. M. (1988). Multiple and cooperative trans-activation domains of the human glucocorticoid receptor. *Cell (Cambridge, Mass.)* **55,** 899–906.

Imperato-McGinley, J., Guerrero, L., Gautier, T., and Peterson, R. E. (1974). Steroid 5α-reductase deficiency in man: an inherited form of male pseudohermaphroditism. *Science* **186,** 1213–1215.

Jenster, G., van der Korput, H. A. G. M., van Vroonhoven, C., van der Kwast, T. H., Trapman, J., and Brinkmann, A. O. (1991). Domains of the human androgen receptor involved in steroid binding, transcriptional activation, and subcellular localization. *Mol. Endocrinol.* **5,** 1396–1404.

Johnson, P. F., and McKnight, S. L. (1989). Eukaryotic transcriptional regulatory proteins. *Annu. Rev. Biochem.* **58,** 799–839.

Kaufman, M., Pinsky, L., Gottlieb, B., Schweitzer, M., Brezezinski, A., von Westarp, C., and Ginsberg, J. (1990). Androgen receptor defects in patients with minimal and partial androgen resistance classified according to a model of androgen-receptor complex energy states. *Hormone Res.* **33,** 87–94.

Kazemi-Esfarjani, P., Beitel, L. K., Trifiro, M., Kaufman, M., Rennie, P., Sheppard, P., Matusik, R., and Pinsky, L. (1993). Substitution of Valine-865 by methionine or leucine in the human androgen receptor causes complete or partial androgen insensitivity, respectively with distinct androgen receptor phenotypes. *Mol. Endocrinol.* **7,** 37–46.

Keenan, B. S., Meyer, W. J., III, Hadjian, A. J., Jones, H. W., and Migeon, C. J. (1974). Syndrome of androgen insensitivity in man: Absence of 5α-dihydrotestosterone binding protein in skin fibroblasts. *J. Clin. Endocrinol. Metab.* **38,** 1143–1146.

Kemppainen, J. A., Lane, M. V., Sar, M., and Wilson, E. M. (1992). Androgen receptor phosphorylation, turnover, nuclear transport, and transcriptional activation. *J. Biol. Chem.* **267,** 968–974.

Klocker, H., Kaspar, F., Eberle, J., Uberreiter, S., Radmayr, C., and Bartsch, G. (1992). Point mutation in the DNA binding domain of the androgen receptor in two families with Reifenstein syndrome. *Am. J. Hum. Genet.* **50,** 1318–1327.

Kuiper, G. G. J. M., Faber, P. W., van Rooji, H. C. J., van der Korput, J. A. G. M.,

Ris-Staplers, C., Klaassen, P., Trapman, J., and Brinkmann, A. O. (1989). Structural organization of the human androgen receptor gene. *J. Mol. Endocrinol.* **2,** R1–R4.

Kuiper, G. G. J. M., de Ruiter, P. E., and Brinkmann, A. O. (1992). Androgen receptor heterogeneity in LNCaP cells is caused by a hormone independent phosphorylation step. *J. Steroid Biochem. Mol. Biol.* **41,** 697–700.

La Spada, A. R., Wilson, E. M., Lubahn, D. B., Harding, A. E., and Fischbeck, K. H. (1991). Androgen receptor gene mutations in X-linked spinal and bulbar muscular atrophy. *Nature (London)* **352,** 77–79.

Lees, J. A., Fawell, S. E., White, R., and Parker, M. G. (1990). A 22-amino acid peptide restores DNA-binding activity to dimerization-defective mutants of the estrogen receptor. *Mol. Cell. Biol.* **10,** 5529–5531.

Levine, M., and Manley, J. L. (1989). Transcriptional repression of eukaryotic promoters. *Cell (Cambridge, Mass.)* **59,** 405–408.

Lubahn, D. B., Joseph, D. R., Sar, M., Tan, J., Higgs, H. N., Larson, R. E., French, F. S., and Wilson, E. M. (1988). The human androgen receptor: Complementary deoxyribonucleic acid cloning, sequence analysis and gene expression in prostate. *Mol. Endocrinol.* **2,** 1265–1275.

Luisi, B. F., Xu, W. X., Otwinowski, Z., Freedman, L. P., Yamamoto, K. R., and Sigler, P. B. (1991). Crystallographic analysis of the interaction of the glucocorticoid receptor with DNA. *Nature (London)* **352,** 497–505.

Marcelli, M., Tilley, W. D., Wilson, C. M., Griffin, J. E., Wilson, J. D., and McPhaul, M. J. (1990). Definition of the human androgen receptor gene structure permits the identification of mutations that cause androgen resistance: Premature termination of the receptor protein at amino acid residue 588 causes complete androgen resistance. *Mol. Endocrinol.* **4,** 1105–1116.

Marcelli, M., Tilley, W. D., Zoppi, S., Griffin, J. E., Wilson, J. D., and McPhaul, M. J. (1991). Androgen resistance associated with a mutation of the androgen receptor at amino acid 772 (Arg→Cys) results from a combination of decreased messenger ribonucleic acid levels and impairment of receptor function. *J. Clin. Endocrinol. Metab.* **73,** 318–325.

Marcelli, M., Zoppi, S., Wilson, C. M., Griffin, J. E., Wilson, J. D., and McPhaul, M. J. (1992). Amino acid substitutions in a small segment of exon 5 of the human androgen receptor (AR) gene cause complete testicular feminization (CTF) by different mechanisms. *Program, Endocr. Soc., 74th Annu. Meet.*, San Antonio, TX, Abstr. 224, p. 107.

Mashima, Y., Murakami, A., Weleber, R. G., Kennaway, N. G., Clarke, L., Shiono, T., and Inana, G. (1992). Nonsense-codon mutations of the ornithine aminotransferase gene with decreased levels of mutant mRNA in gyrate atrophy. *Am. J. Hum. Genet.* **51,** 81–91.

McPhaul, M. J., Marcelli, M., Tilley, W. D., Griffin, J. E., Isidro-Gutierrez, F., and Wilson, J. D. (1991). Molecular basis of androgen resistance in a family with a qualitative abnormality of the androgen receptor and responsive to high-dose androgen therapy. *J. Clin. Invest.* **87,** 1413–1421.

Muller, M., and Renkawitz, R. (1991). The glucocorticoid receptor. *Biochim. Biophys. Acta* **1088,** 171–182.

Pakdel, F., and Katzenellenbogen, B. S. (1992). Human estrogen receptor mutants with altered estrogen and antiestrogen ligand discrimination. *J. Biol. Chem.* **267,** 3429–3437.

Pettersson, G., and Bonnier, G. (1937). Inherited sex-mosaic in man. *Hereditas* **23,** 49–69.

Pinsky, L., Kaufman, M., Killinger, D. W., Burko, B., Shatz, D., and Volpe, R. (1984). Human minimal androgen insensitivity with normal dihydrotestosterone-binding capacity in cultured genital skin fibroblasts: Evidence for an androgen-selective qualitative abnormality of the receptor. *Am. J. Hum. Genet.* **36,** 965–978.

Pinsky, L., Kaufman, M., and Chudley, A. E. (1985). Reduced affinity of the androgen receptor for 5α-dihydrotestosterone but not methyltrienolone in a form of partial androgen resistance. *J. Clin. Invest.* **75,** 1291–1296.

Pinsky, L., Trifiro, M., Kaufman, M., Beitel, L. K., Mhatre, A., Kazemi-Esfarjani, P., Sabbaghian, N., Lumbroso, R., Alvarado, C., Vasiliou, M., and Gottlieb, B. (1992). Androgen resistance due to mutation of the androgen receptor. *Clin. Invest. Med.* **15,** 456–472.

Prior, L., Bordet, S., Trifiro, M. A., Mhatre, A., Kaufman, M., Pinsky, L., Wrogemann, K., Belsham, D. D., Pereira, F., Greenberg, C., Trapman, J., Brinkmann, A. O., Chang, C., and Liao, S. (1992). Replacement of arginine 773 by cysteine or histidine in the human androgen receptor causes complete androgen insensitivity with different receptor phenotypes. *Am. J. Hum. Genet.* **51,** 143–155.

Quigley, C. A., Simental, J. A., Evans, B. A., Lubahn, D. B., Hughes, I. A., and French, F. S. (1992a). Androgen insensitivity due to deletion of the second zinc finger of the androgen receptor. *Mol. Endocrinol.* **6,** 1103–1112.

Quigley, C. A., Friedman, K. J., Johnson, A., Lafreniere, R. G., Silverman, L. M., Lubahn, D. B., Brown, T. R., Wilson, E. M., Willard, H. F., and French, F. S. (1992b). Complete deletion of the androgen receptor gene: Definition of the null phenotype of the androgen insensitivity syndrome and determination of carrier status. *J. Clin. Endocrinol. Metab.* **74,** 927–933.

Reese, J. C., and Katzenellenbogen, B. S. (1992). Characterization of a temperature-sensitive mutation in the hormone binding domain of the human estrogen receptor. *J. Biol. Chem.* **267,** 9868–9873.

Rehberger, P., Rexin, M., and Gehring, U. (1992). Heterotetrameric structure of the human progesterone receptor. *Proc. Natl. Acad. Sci. U.S.A.* **89,** 8001–8005.

Rennie, P. S., Bruchovsky, N., Leco, K. J., Sheppard, P. C., McQueen, S. A., Cheng, H., Snoek, R., Hamel, A., Bock, M. E., MacDonald, B. S., Nickel, B. E., Chang, C., Liao, S., Cattini, P. A., and Matusik, R. J. (1993). Characterization of two *cis*-acting DNA elements involved in the androgen regulation of the probasin gene. *Mol. Endocrinol.* **7,** 23–36.

Ris-Stalpers, C., Kuiper, G. G. J. M., Faber, P. W., Schweikert, H. U., van Rooji, H. C. J., Zegers, N. D., Hodgins, M. B., Degenhart, H. J., Trapman, J., and Brinkmann, A. O. (1990). Aberrant splicing of androgen receptor mRNA results in synthesis of a nonfunctional receptor protein in a patient with androgen insensitivity. *Proc. Natl. Acad. Sci. U.S.A.* **87,** 7866–7870.

Ris-Stalpers, C., Trifiro, M., Kuiper, G. G. J. M., Jenster, G., Romalo, G., Sai, T., van Rooji, H. C. J., Kaufman, M., Rosenfield, R. L., Liao, S., Schweikert, H. U., Trapman, J., Pinsky, L., and Brinkmann, A. O. (1991). Substitution of aspartic acid 686 by histidine or asparagine in the human androgen receptor leads to a functionally inactive protein with altered hormone-binding characteristics. *Mol. Endocrinol.* **5,** 1562–1569.

Ruizeveld de Winter, J. A., Trapman, J., Vermey, M., Mulder, E., Zegers, N. D., and van der Kwast, T. H. (1991). Androgen receptor expression in human tissues: An immunohistochemical study. *J. Histochem. Cytochem.* **39,** 927–936.

Rundlett, S. E., Wu, X. P., and Miesfeld, R. L. (1990). Functional characterizations of the androgen receptor confirm that the molecular basis of androgen action is transcriptional regulation. *Mol. Endocrinol.* **4,** 708–714.

Sai, T., Seino, S., Chang, C., Trifiro, M., Pinsky, L., Mhatre, A., Kaufman, M., Lambert, B., Trapman, J., Brinkmann, A. O., Rosenfield, R. L., and Liao, S. (1990). An exonic point mutation of the androgen receptor gene in a family with complete androgen insensitivity. *Am. J. Hum. Genet.* **46,** 1095–1100.

Saunders, P. T. K., Padayachi, T., Tincello, D. G., Shalet, S. M., and Wu, F. C. W. (1992). Point mutations detected in the androgen receptor gene of three men with partial androgen insensitivity syndrome. *Clin. Endocrinol. (Oxford)* **37,** 214–220.

Schena, M., Freedman, L. P., and Yamamoto, K. R. (1989). Mutations in the glucocorticoid receptor zinc finger region that distinguish interdigitated DNA binding and transcriptional enhancement activities. *Genes Dev.* **3,** 1590–1601.

Simental, J. A., Sar, M., Lane, M. V., French, F. S., and Wilson, E. M. (1991). Transcriptional activation and nuclear targeting signals of the human androgen receptor. *J. Biol. Chem.* **266,** 510–518.

Tilley, W. D., Marcelli, M., Wilson, J. D., and McPhaul, M. J. (1989). Characterization and expression of a cDNA encoding the human androgen receptor. *Proc. Natl. Acad. Sci. U.S.A.* **86,** 327–331.

Tilley, W. D., Marcelli, M., and McPhaul, M. J. (1990). Expression of the human androgen receptor gene utilizes a common promoter in diverse human tissues and cell lines. *J. Biol. Chem.* **265,** 13776–13781.

Tincello, D. G., Hargreave, T. B., Wu, F. C., Padayachi, T., and Saunders, P. T. (1992). Mutations in the androgen receptor gene of patients with androgen insensitivity. *J. Endocrinol.* **131,** 86 (Abstr. 87).

Trifiro, M., Gottlieb, B., Pinsky, L., Kaufman, M., Prior, L., Belsham, D. D., Wrogemann, K., Brown, C. J., Willard, H. F., Trapman, J., Brinkmann, A. O., Chang, C., Liao, S., Sergovich, F., and Jung, J. (1991). The 56/58 kDa androgen-binding protein in male genital skin fibroblasts with a deleted androgen receptor gene. *Mol. Cell. Endocrinol.* **75,** 37–47.

Vasiliou, M., Beitel, L., Mhatre, A., Kazemi-Esfarjani, P., Trifiro, M., Kaufman, M., and Pinsky, L. (1992). Androgen insensitivity due to amino acid substitutions at 4 of 14 positions in an ultra-conserved region of the androgen-binding domain of the human androgen receptor. *Clin. Invest. Med.* **15,** Suppl., A40.

Veldscholte, J., Ris-Stalpers, C., Kuiper, G. G. J. M., Jenster, G., Berrevoets, C., Claassen, E., van Rooij, H. C. J., Trapman, J., Brinkmann, A. O., and Mulder, E. (1990). A mutation in the ligand binding domain of the androgen receptor of human LNCaP cells affects steroid binding characteristics and response to anti-androgens. *Biochem. Biophys. Res. Commun.* **173,** 534–540.

Walsh, P. C., Madden, J. D., Harrod, M. J., Goldstein, J. L., MacDonald, P. C., and Wilson, J. D. (1974). Familial incomplete male pseudohermaphroditism, type 2. Decreased dihydrotestosterone formation in pseudovaginal perineoscrotal hypospadias. *N. Engl. J. Med.* **291,** 944–949.

Webster, N. J. G., Green, S., Jin, J. R., and Chambon, P. (1988). The hormone-binding domains of the estrogen and glucocorticoid receptors contain an inducible transcription activation function. *Cell (Cambridge, Mass.)* **54,** 199–207.

White, R., Fawell, S. E., and Parker, M. G. (1991). Analysis of oestrogen receptor dimerisation using chimeric proteins. *J. Steroid Biochem. Mol. Biol.* **40,** 333–341.

Wooster, R., Mangion, J., Eeles, R., Smith, S., Dowsett, M., Averill, D., Barrett-Lee, P., Easton, D. F., Ponder, B. A. J., and Stratton, M. R. (1992). A germline mutation in the androgen receptor gene in two brothers with breast cancer and Reifenstein syndrome. *Nat. Genet.* **2,** 132–134.

Zoppi, S., Marcelli, M., Deslypere, J.-P., Griffin, J. E., Wilson, J. D., and McPhaul, M. J. (1992). Amino acid substitutions in the DNA-binding domain of the human androgen receptor are a frequent cause of receptor-binding positive androgen resistance. *Mol. Endocrinol.* **6,** 409–415.

16

Male Pseudohermaphroditism

Milo Zachmann
Department of Pediatrics
University of Zurich
Kinderspital
CH-8032 Zurich
Switzerland

Molecular Genetics of Sex Determination

I. Introduction

Male pseudohermaphroditism (MPH) traditionally refers to female or intersexual genital development in individuals with histologically normal or almost normal testes and with a 46,XY karyotype (Aarskog, 1970; Prader, 1974; Park *et al.*, 1975; Imperato-McGinley and Peterson, 1976; Simpson, 1976, 1978; Savage *et al.*, 1978; Migeon, 1980; Berkovitz *et al.*, 1984). It comprises a large group of disorders (Table I), whose pathogenesis has been clarified in a number of defects, but remains unclear in several syndromes. If peripheral cells fail to respond to androgens secreted normally by the fetal testes, female or intersexual external genital development occurs. This is MPH due to the various types of androgen insensitivity, which will not be dealt with here. If the conversion of testosterone (T) to dihydrotestosterone (DHT) is disturbed by reduced 5α-reductase activity, the external genitalia will also be intersexual or female, but male internal genitalia and later male puberty will develop. Besides these "peripheral" causes of MPH, there

Table I
Causes of Male Pseudohermaphroditism

Defects of testosterone biosynthesis
20,22-Desmolase deficiency
3β-Hydroxysteroid dehydrogenase deficiency
17α-Hydroxylase deficiency
17,20-Desmolase deficiency
Combined deficiency of cytochrome P450 C17α
17-Ketosteroid reductase deficiency
Defect of testosterone metabolism
5α-Reductase deficiency
Testicular dysgenesis
Pure gonadal dysgenesis
Swyer's syndrome
Asymmetrical mixed gonadal dysgenesis (XO/XY)
Syndrome of rudimentary testes
Leydig cell hypoplasia or aplasia
LH receptor defect
LH-deficiency (?)
MPH in various syndromes
Drash syndrome
Smith–Lemli–Opitz syndrome
Syndromes with cleft palate
MPH with campomelic dwarfism
Other syndromes

Note. Not included: androgen insensitivity, older terms like "hereditary perineal hypospadias" or Reifenstein's syndrome, for reasons discussed in the text.

are a number of hereditary enzyme defects, where the biosynthesis of T is not possible or reduced. MPH may also occur when Leydig cells are not present or not functioning properly.

Apart from these disorders, the mechanisms of which are relatively clear, there are others, whose pathogenesis is not understood. There may be hypospadias without a detectable endocrine cause (Gearhart *et al.*, 1988), or a sex chromosome anomaly. Persistence of Müllerian structures has been termed "internal" MPH. This condition also will not be discussed here. MPH may be associated with other malformations as part of a dysmorphic syndrome, or occur in tumor-associated disease. In such cases, T production, conversion, and receptor sensitivity are usually normal.

In the following sections, I emphasize defects of T biosynthesis, 5α-reductase deficiency, and some of the more important syndromes associated with MPH. Clinical signs and classic biochemical findings in the various disorders are first presented. Then, some relevant newer studies regarding more fundamental aspects and molecular biology are discussed.

II. Defects of Testosterone Biosynthesis

The classic enzyme defects were described from the 1950s to the 1970s. Two causing MPH were observed first in Zurich: Prader and Gurtner (1955) and Prader and Siebenmann (1957) described lipoid adrenal hyperplasia (or as it is now called, deficiency of the cytochrome P450 SCC) before specific steroid assays were available, by clinical observation and autopsy studies. Later, we described 17,20-desmolase (also called 17,20-lyase or cytochrome P450 C17α deficiency) (Zachmann *et al.*, 1972, 1982). These two defects demonstrate well how concepts on regulation, specificity, and function of steroid converting enzymes have changed.

All defects were recognized indirectly. The existence of an enzyme deficiency was postulated and named solely on the basis of precursor "pile-up" and product deficiency. This method was with few exceptions successful, and all defects have been confirmed by more sophisticated techniques. Among these defects, some also lead to reduced cortisol and/or aldosterone production, causing special types of congenital adrenal hyperplasia (CAH; lipoid adrenal hyperplasia, deficiency of 3β-hydroxysteroid dehydrogenase, and partly 17α-hydroxylase). In others, which are more difficult to recognize in prepubertal children, the pathways to cortisol and aldosterone are functioning normally, and only sex hormone production is affected (deficiencies of 17,20-desmolase and 17-ketosteroid reductase). Since 46,XX individuals with all defects have been described, the inheritance of all is autosomal recessive. Because of the more severe symptomatology, most cases have, however, been detected and described in 46,XY individuals. In 46,XX individuals, the clinical symptomatology is less impressive: With the exception of the types associated with impaired cortisol secretion and salt-wasting, the signs are limited to the absence of development of estrogen-dependent secondary sex characteristics and menarche at a pubertal age.

A. Lipoid Adrenal Hyperplasia

In this condition (Prader syndrome, 20,22-desmolase-, cytochrome P450 SCC deficiency), the defect precludes conversion of cholesterol to pregnenolone (Prader and Gurtner, 1955; Prader and Siebenmann, 1957). With the exception of some insignificant cholesterol metabolites, no steroids can be produced (Degenhart, 1984). Newborns of both sexes suffer from adrenal insufficiency and salt-wasting. In 46,XY newborns, the external genitalia are female or intersexual. Autopsy in early cases revealed large adrenals packed with lipoid material. Since no or only small quantities of steroids are produced, no diagnostically useful metabolites are found in plasma or urine. Also, serum or urinary cholesterol is not appreciably increased, because the quantity of cholesterol, which may pile up as a precursor, is far too small in comparison with the normally large cholesterol pool. For this reason, differential diagnosis between 46,XY or 46,XX individuals with the condition and girls with congenital adrenal hypoplasia is clinically difficult or impossible if there are female external genitalia. Examination of the adrenals with computer-assisted tomography or magnetic resonance imaging, however, reveals large adrenals in the former (Ogata *et al.*, 1988). Hauffa *et al.* (1985) and Müller *et al.* (1991) reported long-term studies in typical patients. It was first thought that salt-wasting was more severe in these patients than, e.g., in patients with 21-hydroxylase deficiency, because most early patients died at a very young age. This is, however, probably not the case, and the reported high incidence of deaths was due rather to the later diagnosis because of absence of intersexual genitalia.

B. Congenital Adrenal Hyperplasia Due to 3β-Hydroxysteroid Dehydrogenase Deficiency

This type of CAH (original references cited in Bongiovanni, 1984) differs from the more frequent types (21- and 11β-hydroxylase deficiency) in that there is no or only a slight overproduction of androgens. In this rare defect (Grendrel *et al.,* 1979), steroid biosynthesis is blocked at the stage of the Δ5-compounds (pregnenolone, 17αOH-pregnenolone, and DHEA, dehydroepiandrosterone). DHEA and DHEAS are the only androgens that may be produced in excess, and classically, all increased steroids are Δ5-compounds (Whorwood *et al.,* 1990). Besides increased levels of pregnenolone and 17αOH-pregnenolone in plasma, the reduced metabolites pregnenediol and pregnenetriol are found in urine. In young patients, in whom fetal adrenal tissue is active, 16α-hydroxylated compounds (16αOH-DHEA, 16αOH-pregnenolone, urinary metabolite 3α,16α,20α-pregnenetriol) are found. Since 21- and 11β-hydroxylase activities are intact, 17αOH-pregnenolone can be further hydroxylated in position 21 and 11β, resulting in the appearance of unusual, normally not occurring 21-hydroxylated compounds. Because DHEA is a weak androgen, no or only mild virilization is observed. 46,XY patients have incompletely masculinized external genitalia with varying degrees of MPH. The genitalia in females show slight hypertrophy of the clitoris or are normal. In most cases, there is salt-wasting, which may, however, be compensated (Mendonca *et al.,* 1987; Pang *et al.,* 1983).

Since the defect is also located in the testes and ovaries, no or insufficient male secondary sex characteristics develop at a pubertal age in boys, and spontaneous pubertal development and menarche do not occur in girls, even with adequate gluco- and mineralocorticoid treatment. Thus, in addition to the usual replacement, these patients mostly require sex hormones at a pubertal bone age (Zachmann *et al.,* 1979). Adolescents with some spontaneous pubertal development have, however, been described (Martin *et al.,* 1980; Rosenfield *et al.,* 1980; Nahoul *et al.,* 1989).

Some patients show peculiarities, which make them different from classic cases (Zachmann *et al.,* 1970; Cravioto *et al.,* 1986). Surprisingly, one of our patients excreted large quantities of pregnanetriol, normally a metabolite of 17OH-progesterone (17OHP). We could demonstrate that there was a mixture of 3α- and 3β-, rather than the usual 3α-pregnanetriol in this patient. It thus must have been formed by non-specific dehydrogenation in the liver or peripheral tissues. This was confirmed in other cases (Bongiovanni *et al.,* 1971). The same girl was studied at a pubertal age (Zachmann *et al.,* 1979); we found that stimulation with hMG revealed the defect in the ovaries, because DHEA, which normally does not respond to hMG, did increase, but not estrogens. We also studied an affected newborn girl from the same village, who was a relative of the first patient (Zachmann *et al.,* 1988). Although she had the same genetic defect, the newborn was metabolically different and excreted several unusual steroids: 16α-hydroxylated compounds were predominant, whereas pregnenetriol was only slightly increased. This was due to the fact that this young patient still had active fetal adrenal tissue with efficient 16α-hydroxylating activity. The phenomenon may cause diagnostic confusion when there is no known case in the family (Wolthers *et al.*, 1987).

Mild 3β-hydroxysteroid dehydrogenase deficiency has been thought to cause apparent "acquired virilization" in adult females (Bongiovanni, 1981), premature pubarche (Nishi and Tezuka, 1992), and hypogonadism and infertility in adult males (Frank Raue *et al.*, 1989). In adult women with hirsutism, however, this defect appears to be a very rare cause, and the incidence seems to have been overestimated.

C. 17α-Hydroxylase Deficiency

Deficiency of 17α-hydroxylase blocks the biosynthesis of cortisol, T, and estrogens (Biglieri *et al.*, 1966; Mantero and Scaroni, 1984; Blau *et al.*, 1987; Rohmer *et al.*, 1990; Biglieri and Kater, 1991). It mostly results in female or intersexual external genitalia in 46,XY newborns. Patients with almost complete labioscrotal fusion, with no development of a utriculovaginal pouch, or with micropenis have been observed, however (New, 1970; Bricaire *et al.*, 1972; Jones *et al.*, 1982). In 46,XX individuals, the external and internal genitalia are female. Pubertal development is absent or insufficient in both sexes. There is no adrenal hyperplasia or salt-wasting, and the lack of cortisol is compensated by accumulation of corticosterone. During surgery or severe stress, adrenal insufficiency may occur, however (Abad *et al.*, 1980). Besides, the production and excretion of all steroids not hydroxylated in position 17 (progesterone and its urinary metabolite pregnanediol, 11-deoxycorticosterone (DOC) and its metabolite THDOC, tetrahydro-11-deoxycorticosterone) are increased. DOC is a weak mineralocorticoid, which prevents salt-wasting. If present in large quantities, it may cause hypertension and hypokalemic alkalosis (Tourniaire *et al.*, 1976). Aldosterone is generally low, because the renin/angiotensin system is suppressed by DOC, although it may occasionally be high (Shima *et al.*, 1991). Usually, hypertension is not present in children, but develops after years without treatment. Then, however, it may become malignant and lead to cerebral hemorrhage (Yanase *et al.*, 1982). In cases with malignant hypertension, the stronger mineralocorticoids 18OH-DOC and/or 18OH-corticosterone may be pathogenetically important.

A 46,XY phenotypic female has been described, whose corticosterone/deoxycorticosterone ratio was low, suggesting an associated deficiency of 11β-hydroxylase (D'Alberton *et al.*, 1989). It is thus possible that cases exist with multiple hydroxylase deficiencies, as has been observed in other conditions (Peterson *et al.*, 1985a). The association with 17,20-desmolase deficiency is discussed below.

D. 17,20-Desmolase Deficiency

Deficiency of 17,20-desmolase (or 17,20-lyase) prevents transformation of 17αOH-pregnenolone to DHEA and of 17OHP to androstenedione (Zachmann *et al.*, 1972). The result is a lack of androgens and estrogens, whereas the biosynthesis of glucocorticoids and aldosterone are not disturbed. Affected males have female or intersexual external genitalia; pubertal development is absent or insufficient. There are also reports on affected 46,XX individuals (Larrea *et al.*, 1983). Incubation studies in testicular tissue of the original patients confirmed the defect. The two original patients could be reevaluated at a pubertal age (Zachmann *et al.*, 1982). They had some spontaneous male puberty and similar steroid results as in early childhood (Zachmann *et al.*, 1972). In addition, we found a new, unrelated patient, who was different in some aspects. This 46,XY patient had been reared as a girl, had female external genitalia, and exhibited complete absence of pubertal development. T and androstenedione were extremely low and did not respond to stimulation. One would therefore have thought that the defect was more complete than in the other patients with ambiguous genitalia. This was true for the delta4 pathway, but on the delta5 pathway, it was surprising to find normal DHEA responding to ACTH. A similar case has been reported (Goebelsmann *et al.*, 1976), whereas most of the other cases had ambiguous genitalia and low DHEA (Forest *et al.*, 1980; Campo *et al.*, 1981; Kaufman *et al.*, 1983). The features of other reported cases up to that date (Forest *et al.*, 1980, 1985; De Peretti *et al.*, 1984; Larrea *et al.*, 1983) have been summarized (Zachmann and Prader, 1984).

Since then, more cases, partially with combined defects (see below), have been reported (Kagimoto *et al.*, 1988; Yanase *et al.*, 1988, 1989; Winter *et al.*, 1989; Jones *et al.*, 1992). A peculiar child with the 18q deletion syndrome and ambiguous genitalia has been described whose endocrine results suggested the defect (Chasalow *et al.*, 1986). To determine whether 17,20-desmolase deficiency was common in del(18q) syndrome, six other patients were examined. All DHEAS levels were lower than in controls. The authors thought that a structural or regulatory gene for 17,20-desmolase might be located on the long arm of chromosome 18. This could not be confirmed in later studies (Matteson *et al.*, 1986).

We concluded from the results in our patients and those reported by others that the difference between the patients with intersexual genitalia and low DHEA, and those with completely female genitalia and normal DHEA, could not be merely gradual, but more fundamental, and characterized two clinical and biochemical types (Zachmann *et al.*, 1982): In type 1, both the delta4 and delta5 pathways seemed to be affected, but the defect was not complete, causing MPH with ambiguous external genitalia; in type 2, only the delta4 pathway appeared to be affected. There, the defect was more complete than in the first group, leading to female external genitalia. However, cleavage on the delta5 pathway was possible, and DHEA could be formed. In view of recent findings (see below), this difference is difficult to understand.

E. Combined 17α-Hydroxylase/17,20-Desmolase Deficiency

As will be discussed, steroid 17α-hydroxylation and 17,20-desmolase activity have been found to be regulated by a single cytochrome P450 C17α. It is therefore not surprising that cases with a combined 17α-hydroxylase/17,20-desmolase deficiency have been observed. On the contrary, it is unclear why cases with exclusive 17α-hydroxylase or 17,20-desmolase deficiency exist.

Female pseudohermaphroditism was observed in two 46,XX subjects with high corticosterone levels (Roger *et al.*, 1982). In retrospect, these must probably be regarded as the first observed cases with a combined defect.

A 16-year-old boy with normal external genitalia, but slow and incomplete pubertal development, has been reported (Bosson *et al.*, 1988). His data also suggested a combined 17α-hydroxylase and 17,20-desmolase deficiency. Other cases with a combined defect have since been reported (Yanase *et al.*, 1988, 1989; Winter *et al.*, 1989; Winter, 1991; Augarten *et al.*, 1992).

F. 17-Ketosteroid Reductase Deficiency

In this defect (also called 17β-hydroxysteroid dehydrogenase or 17-oxidoreductase deficiency; Zurbrügg and Neher, 1966; Saez *et al.*, 1971, 1972), the conversion of androstenedione to T (Horton and Tait, 1966) and of DHEA to androstene-3β,17β-diol is reduced or not possible. Plasma androstenedione, DHEA, and estrone (mostly originating from androstenedione in the adrenals) are increased, whereas T is low

(Akesode *et al.*, 1977; Virdis *et al.*, 1978; Harkness *et al.*, 1980; Leinonen *et al.*, 1983; Virdis and Saenger, 1984; Balducci *et al.*, 1985; Caufriez, 1986). Estradiol-17β is variable, but, since it is normally predominantly formed from T, usually low. It is probable that some of the patients reported in the older literature, who were diagnosed as having Reifenstein's syndrome (Medina *et al.*, 1989), actually had 17-ketosteroid reductase deficiency.

Individuals with 46,XY demonstrate MPH of various degrees. Although some patients have been reared as girls (Millan *et al.*, 1983), the intersexual external genitalia tend to be more masculinized than in other defects (Prader, 1954), and most patients are reared as boys. Because of the androgen action of androstenedione and androstene-3β,17β-diol, pubertal development also is generally male. Gynecomastia frequently develops, however, due to the effect of the increased estrone. Whether gynecomastia develops (Givens *et al.*, 1974), is more a consequence of the androgen/estrogen ratio than of the absolute quantities of circulating estrogens (Zachmann *et al.*, 1986).

Most patients were diagnosed at or after puberty, when significant virilization, but not fertility, occurred (Dumic *et al.*, 1985), and information on findings in prepubertal patients is scant. Two prepubertal siblings with a marked defect have been reported. Androstenedione was only slightly elevated, but did not allow the diagnosis. Steroid measurements after hCG showed subnormal peak T and elevated androstenedione (Arnhold *et al.*, 1988). Other studies confirm that the symptoms are aggravated in puberty (Wilson *et al.*, 1987). This is in agreement with our own experience in two cases, where androstenedione was not increased before puberty (Gregory *et al.*, 1992). The diagnosis can, however, be made in infancy with ACTH- and hCG-stimulation tests (von Schnakenburg *et al.*, 1980). *In vitro* studies in testis tissues from normal infants showed that active 17-ketosteroid reductase is normally present in immature human testes (Oshima *et al.*, 1980).

In another prepubertal case, a 46,XY phenotypically female infant showed normal androgens shortly after birth, but did not have the physiologic T rise at ~3 months of age (Forest *et al.*, 1973). Stimulation with hCG resulted in high androstenedione/T and DHEA/androstene-3β,17β-diol ratios (Wit *et al.*, 1988). Possibly a partial defect may also be present in phenotypically male patients with micropenis and bilateral cryptorchidism (Ulloa Aguirre *et al.*, 1985). Extensive studies on familial MPH due to the defect were carried out in Arabs from the Gaza area (Rösler and Kohn, 1983; Gross *et al.*, 1986; Eckstein *et al.*, 1989). Prenatal diagnosis of the defect has been possible by endocrine studies in amniotic fluid at midgestation (Nivelon *et al.*, 1985, 1987).

In a family with nine children, three with MPH due to 17-ketosteroid reductase deficiency and four with congenital hypothyroidism were observed. Congenital hypothyroidism affected one of the patients with MPH. Neither the locus for the 17-ketosteroid reductase nor that for congenital hypothyroidism was linked to the major histocompatibility complex (Lanes *et al.*, 1983). Coexistence of 17-ketosteroid reductase deficiency with 5α-reductase deficiency has been found within a Turkish kindred (Akgün *et al.*, 1986; Imperato-McGinley *et al.*, 1987).

III. Defect of Testosterone Metabolism (5α-Reductase Deficiency)

Even with normal biosynthesis of and normal receptor sensitivity to T, male sex differentiation may be incomplete and MPH may occur because of insufficient conversion of T to DHT in various tissues; this is due to defective 5α-reductase activity. The defect was studied in a large Dominican kindred (Imperato-McGinley *et al.*, 1974; Peterson *et al.*, 1977), but cases from other countries have been reported (Savage *et al.*, 1980; Okon *et al.*, 1980; Mauvais-Jarvis *et al.*, 1981; Price *et al.*, 1984; Deslypere *et al.*, 1985; Lo Iudice *et al.*, 1986; Greene *et al.*, 1987; Herdt and Davidson, 1988; Ivarsson *et al.*, 1988; Imperato-McGinley *et al.*, 1991). The study in the Dominican Republic revealed that 5α-reductase deficiency occurs there with a frequency of 13% of all cases of MPH (Imperato-McGinley *et al.*, 1986b).

In this condition, chromosomal males have female or intersexual external genitalia with male internal genitalia. The testes are usually located in the inguinal canal. The patients are frequently reared as females, but pubertal development is male, and the previously small phallus grows considerably. Sexuality develops in a male way, but sometimes remains female (Cantu *et al.*, 1980). The voice changes, and muscles develop markedly. Body hair is of the female type, there is no gynecomastia or acne, and the prostate remains small. The testes often descend, and normal spermiogenesis is possible. Cases without ambiguous genitalia, but with micropenis have been observed (Carpenter *et al.*, 1990; Ng *et al.*, 1990). In postpubertal patients, plasma T may be slightly increased, but DHT is absent or low. The increased T/DHT ratio is reflected by an increased urinary etiocholanolone/androsterone ratio or 5β/5α ratio. The ring A reduced urinary 5β- and 5α-metabolites of T—androstenedione, 11βOH-androstenedione, cortisol, and corticosterone—were measured by gas chromatography. Assays of the four pairs of urinary 5β- and 5α-steroid metabolites revealed decreased conversion of the parent steroids to 5α-reduced urinary metabolites with respect not only to etiocholanolone and androsterone, but also to the other steroids (Imperato-McGinley *et al.*, 1985; Peterson *et al.*, 1985b; Odame *et al.*, 1992). Besides these ratios, plasma levels of the two isomers resulting from the 5α-reduction of T provide a method for diagnosing the condition. With this technique, an intermediate abnormality was found in other family members consistent with the heterozygous state (Corrall *et al.*, 1984).

The development of DHT-dependent signs (male differentiation of the external genitalia, male pattern of body hair, acne, and growth of prostate) is insufficient, whereas that of the T-dependent signs (growth of the penis, change of voice, spermiogenesis, sexuality, muscle gain) is normal.

Another, presumably genetically different, type shows the same physical findings with a normal plasma T/DHT ratio. The 5α-reductase deficiency in these patients is present in the genital area only (Walsh *et al.*, 1974). For this type, the term "MPH of type 2" has been suggested. Results from studies of 5α-reductase activity in hair roots also support the concept that 5α-reductase deficiency is genetically heterogenous (Schmidt and Schweikert, 1986). Because of the physiologically low basal plasma T and unmeasurable DHT in infancy after about 6 months of age (Forest *et al.*, 1973), the diagnosis is impossible from basal samples between the age of about 6 months and the onset of puberty. To assess 5α-reductase activity in children with MPH in this group, we measured plasma T and DHT following stimulation of endogenous production with hCG (Zachmann, 1972) or after administration of exogenous T, and compared the response to these two tests (Greene *et al.*, 1987). Also, in other reports, 5α-reductase deficiency could be confirmed by hCG stimulation in prepubertal children (Saenger *et al.*, 1978; Saenger, 1984; Stanhope *et al.*, 1986). Since cortisol metabolites in urine are higher in infancy and childhood than are the individual 17-ketosteroids, their 5β/5α ratio also can be used for the diagnosis in this age group (Imperato-McGinley *et al.*, 1986a).

In our patients with MPH not due to 5α-reductase deficiency, who had been stimulated with hCG, and in patients studied by Savage *et al.* (1980), the same age-related changes of the T and DHT levels and of the T/DHT ratios were observed as in normal children (Pang *et al.*, 1979). This suggests that 5α-reductase activity is diminished with increasing age in patients with MPH due to other causes. Secondary 5α-reductase deficiency may also occur in androgen insensitivity (Jukier *et al.*, 1984; Wilson *et al.*, 1988). Concentrations of 3α-androstenediol glucuronide in plasma and/or random urine samples were also found to be useful for the diagnosis (Paulson *et al.*, 1986; Ertel *et al.*, 1989).

Recently, it has been found that patients with the polycystic ovary syndrome have increased 5α-reductase activity in liver and skin at the expense of androgen excess and that this may be the underlying abnormality in the syndrome (Stewart *et al.*, 1990).

IV. Testicular Dysgenesis and Leydig Cell Aplasia or Hypoplasia

Several types of testicular agenesis and dysgenesis may cause MPH and/or hypergonadotropic hypogonadism.

Pure gonadal dysgenesis involves individuals with functionless rudimentary streak gonads as in Turner syndrome (Chemke *et al.*, 1970; Simpson, 1976; Mann *et al.*, 1983). However, neither short stature nor dysmorphic features are present. The karyotype is XX or XY. In XY individuals, this is called Swyer syndrome (Swyer, 1955; Amarose *et al.*, 1977; Dorus *et al.*, 1977). The inheritance in XX individuals is autosomal recessive. Swyer syndrome is transmitted through healthy women, suggesting X-chromosomal recessive inheritance. Malignant gonadoblastomas or dysgerminomas are found in 20–30% of XY patients (Simpson, 1976). For this reason, streaks should be removed. Cases with special chromosomal findings have been reported (Sorgo *et al.*, 1991; Maciel Guerra *et al.*, 1991). The term asymmetrical mixed gonadal dysgenesis is applied to a combination of intersexual genitalia with a testis on one side and a rudimentary gonad (streak) on the other (Donahoe *et al.*, 1979; Kofmann *et al.*, 1981). Short stature and dysmorphic features similar to those in girls with Turner syndrome are present. 45,X/46,XY mosaicism is found in the majority of cases. Development of malignant tumors (gonadoblastoma or seminoma) or carcinoma *in situ* is common on the side with the streak or the testis (Aarskog, 1970; Nagel *et al.*, 1984; Müller *et al.*, 1985).

In the syndrome of rudimentary testes, testicular function may be sufficient for male differentiation (Bergada *et al.*, 1962; Najjar *et al.*, 1974). The patients thus may not present with MPH, but their testes are extremely small and show immature tubules and Sertoli cells with few germ cells. There also exists an intermediate form with sexual ambiguity (Wolman *et al.*, 1980; Messinis and Nillius, 1981; Dunkel *et al.*, 1984).

In addition, there are hereditary syndromes causing MPH with aplasia (Berthezène *et al.*, 1976; Martinez Mora *et al.*, 1991) or hypoplasia of the Leydig cells (Brown *et al.*, 1978; Lee *et al.*, 1982; Rogers *et al.*, 1982; Eil *et al.*, 1984; Arnhold *et al.*, 1985; Toledo *et al.*, 1985; Saldanha *et al.*, 1987; Toledo, 1992). The Leydig cells may also be resistant to the action of gonadotropins secondary to an LH receptor defect (Ochoa *et al.*, 1981; David *et al.*, 1984), or there may be an anomaly in their differentiation (Schwartz *et al.*, 1981). Müllerian structures are absent.

Whether MPH can be caused by prenatal LH deficiency in addition to hypogenitalism and cryptorchidism, has not yet been proven. The observation in one patient, however, suggests that this might be possible (Prader *et al.*, 1976).

V. MPH in Various Syndromes

A. Drash Syndrome

This syndrome (Drash *et al.*, 1970) consists of the triad of MPH, progressive renal insufficiency, and Wilms tumor. It appears early in life. The first sign is genital ambiguity. Proteinuria, hematuria, and hypertension appear very soon. Renal biopsy may reveal a variety of glomerular and interstitial changes. The nephropathy progresses rapidly, is unresponsive to steroids, and eventually requires dialysis. A Wilms tumor may appear as a mass on ultrasound or may not be recognized until nephrectomy or even autopsy. Young children with gonadal dysgenesis and nephropathy are at high risk for nephroblastoma and gonadal tumors, and prophylactic nephrectomy and gonadectomy should be carried out early (Jensen *et al.*, 1989; Heppe *et al.*, 1991). Early diagnosis is essential and with dialysis or renal transplantation the prognosis might improve in the future (Goldman *et al.*, 1981; Eddy and Mauer, 1985).

B. Smith–Lemli–Opitz Syndrome

This syndrome of microcephaly, growth and mental retardation, unusual facial appearance, syndactyly of toes 2 and 3, and genital abnormalities was originally not thought to be associated with MPH. Major structural malformations and early death have been uncommon in the reports. More recently, patients with major abnormalities, MPH, and early lethality have been reported (Belmont *et al.*, 1987; Lachman *et al.*, 1991; Pankau *et al.*, 1992).

C. Syndromes Associated with Cleft Palate

Two syndromes have been described in which MPH is associated with cleft palate: The Gardner–Silengo–Wachtel or genito-palato-cardiac syndrome includes MPH, micrognathia, cleft palate, and a conotruncal cardiac defect (Wachtel, 1983; Greenberg *et al.*, 1987). Another similar autosomal recessive syndrome has been reported, which includes a peculiar face, deafness, cleft palate, MPH, and growth and psychomotor retardation (Ieshima *et al.*, 1986).

D. MPH with Campomelic Dwarfism

Infants with campomelic dwarfism have been presented with a male karyotype, but female phenotype (Hovmöller *et al.*, 1977). In the gonads, there were ovarian stroma containing tubules similar to the rete ovarii found in normal newborn girls, and undifferentiated germ cells.

VI. Recent Developments in Molecular Biology Related to MPH

Theories concerning the action and genetic regulation of steroid converting enzymes have had to be revised during the last years, because fundamental research in biochemistry and molecular biology has made much progress. This concerns mainly the oxygenating enzymes cytochromes P450 and 3β-hydroxysteroid dehydrogenase. It is not possible in this short review to discuss all the important recent work on steroid hydroxylating enzymes and the regulation of gene expression. Excellent review articles on this topic are available, however (Waterman and Simpson, 1989; Waterman *et al.*, 1990). In the context of MPH, two hydroxylases are of particular importance, the cytochrome P450 SCC, which is located in the inner mitochondrial membrane, and the cytochrome P450 C17α, which resides in the endoplasmic reticulum. The P450 gene superfamily can be classified into two major groups: those that metabolize exogenous substrates and those that metabolize endogenous substrates. Those of interest here belong to the latter class, which also includes enzymes required for cholesterol biosynthesis, bile acid formation, fatty acid metabolism, and vitamin D hydroxylation. Both cytochromes are encoded by single-copy genes in all species investigated to date. The human and bovine P450 C17α genes have been structurally characterized, as has been the human P450 SCC gene. They represent two different gene families, yet common features associated with cAMP-dependent transcription of these genes are known (for detailed references, see Waterman *et al.*, 1990).

Concerning the cytochrome P450 C17α (17α-hydroxylase and 17,20-desmolase), it has been shown that a single cytochrome P450 is regulating steroid hydroxylation in position 17 and the side-chain cleavage of 17αOH-pregnenolone and 17OHP (Hall, 1986, 1989).

Studies had been performed by Nakajin *et al.* (1984) to test the hypothesis that cytochrome P450 from testicular microsomes consists of a single protein with the two enzymatic activities. Three lines of evidence supporting the hypothesis were obtained. (*a*) The enzyme appeared to be homogeneous by immunochemical criteria with anti-P450 IgG; (*b*) optima for pH and temperature were the same for both enzymatic activities; and (*c*) a variety of inhibitors affected both activities to the same extent. Estradiol-17β was found to inhibit both activities (Onoda and Hall, 1981). It became evident from the work mentioned and other work that steroid 17α-hydroxylase and 17,20-desmolase activities reside within the same polypeptide chain (cytochrome P450 C17α), and consequently human deficiencies can be characterized by defects in each one of these activities as in the classic reports, or by simultaneous defects in both activities, as in the reported combined cases. Why there are such differences in clinical presentation in the presence of one enzyme is at present still mostly unclear. Concerning the genes regulating the cytochrome P450 C17α, the following studies have been important among others. Matteson *et al.* (1986) identified several human P450 C17α cDNA clones in a human adrenal cDNA library. A short clone containing the 3′-terminal 650 bases of the full-length sequence was used to examine Southern blots of DNA from normal persons and from a panel of mouse–human somatic cell hybrid lines. The pattern of hybridization of this cDNA to normal human DNA cut with eight restriction endonucleases suggested that the human genome has two P450 C17α genes. The pattern of hybridization to the somatic cell hybrid cell lines, each containing a limited, known number of human chromosomes, indicated the human adrenal P450 C17α gene lies on chromosome 10. Cloning of human adrenal and testis cDNAs indicated that the same gene is expressed in both these tissues (Chung *et al.*, 1987). Cloning and sequencing of the human gene for P450 C17α were achieved (Picado-Leonard and Miller, 1987), and similarity with the gene for P450 C21 (21-hydroxylase) was observed. More and more, it became evident that the nature of the regulation of steroid hydroxylase gene expression is multifactorial, involving cAMP-dependent and -independent mechanisms (Waterman and Simpson, 1989; Waterman *et al.*, 1990).

Voutilainen and Miller (1986) studied the ontogeny and tissue specificity of expression of the genes for P450 C17α in the human fetus. RNA from fetal tissues was probed with homologous human P450 C17α cloned in their laboratory. At 20–21 weeks gestation, P450 C17α mRNA was most abundant in the adrenal, followed by testis and ovary, but was undetectable in the placenta. None was detected in other tissues. Testis samples from 13–26 weeks gestation were studied, and it was found that P450 C17α mRNA was most abundant at 14–16 weeks and diminished by 20–26 weeks. These mRNA data correlated well with previously reported changes in gonadal steroidogenesis with gestational age.

Molecular techniques have also been applied to the study of patients with 17α-hydroxylase/17,20-desmolase deficiency. The exonic structure of the human P450 C17α gene and the sequences at the exon–intron boundaries and at the site of initiation of transcription have been determined (Kagimoto *et al.*, 1988). The single gene in the human genome that encodes this protein is the sole member of a unique gene family (P450XVII) within the P450 supergene family. A protocol for exonic sequencing of the P450 C17α gene has been established by these authors, which allows structural analysis of the gene from patients with 17α-hydroxylase and/or 17,20-desmolase deficiency. This procedure was applied to the mutant gene from an individual with combined 17α-hydroxylase/17,20-desmolase deficiencies. A 4-base duplication was found in exon 8 producing a protein with an altered C-terminal amino acid sequence resulting in loss of both enzymatic activities.

Yanase *et al.* (1988) determined the exonic sequence of the mutant P450 C17α gene from one Japanese patient with combined 17α-hydroxylase/17,20-desmolase deficiencies. A stop codon (TGA), resulting from a single point mutation, was found at the position of amino acid 17 in exon 1 of the P450 C17α gene. The stop codon in the N-terminal region of this gene led to the absence of a functional P450 C17α protein in the adrenals and gonads.

The cytochrome P450 C17α gene from an individual with partial combined 17α-hydroxylase/17,20-desmolase deficiency could be characterized structurally and the homozygous mutation was found to be a deletion of the phenylalanine codon (TTC) at either amino acid position 53 or 54 in exon 1 (Yanase *et al.*, 1989). Reconstruction of this mutation into a human P450 C17α cDNA followed by expression in COS-1 cells led to production of the same amount of immunodetectable P450 C17α protein, as is found with expression of the normal human P450 C17α cDNA. However, 17α-hydroxylase activity of this mutant protein in intact cells was less than that observed on expression of the wild-type enzyme, and 17,20-desmolase activity of the mutant was also less than that observed with the normal enzyme. Apparently the deletion of a single phenylalanine in the N-terminal region of P450 C17α alters its folding in such a way that both enzymatic activities are markedly decreased. A compound heterozygosity was found in one of our patients with female external genitalia (Yanase *et al.*, 1992) and in another patient with ambiguous genitalia (Ahlgren *et al.*, 1992). Canadian Mennonites and individuals from Friesland have recently been shown to share the same molecular basis of the defect (Imai *et al.*, 1992).

Winter *et al.* (1989) studied six phenotypic female children with combined 17α-hydroxylase/17,20-desmolase deficiency. The diagnosis was confirmed by extensive basal and ACTH-stimulated steroid studies. Despite virtual absence of both enzymatic activities of cytochrome P450 C17α, Northern blot analysis demonstrated abundant amounts of RNA that hybridized to a cDNA specific for this P450 enzyme. Immunoblot analysis of polyacrylamide gel electrophoresis-resolved testicular microsomes showed an apparently normal content of an immunoreactive protein with a molecular weight similar to that of authentic P450 C17α. The results of these studies suggest that the patients also have a point mutation in the gene for P450 C17α. It has to be assumed that the mutant gene is transcribed, but gives rise to a protein defective in normal 17α-hydroxylase and 17,20-desmolase activities.

Concerning the cytochrome P450 SCC (20,22-desmolase), the concept on side-chain cleavage of cholesterol and the pathophysiology of lipoid adrenal hyperplasia also had to be changed. Earlier, three different enzymes had been postulated for the conversion. Now, it is known that this cleavage is regulated by a single cytochrome P450 SCC. It is the rate-limiting, hormonally regulated enzyme that cleaves the cholesterol side chain. Because the C atoms 20 and 22 reside at a distance of 1.5 Å only, they can be oxygenated at the same time, and the cholesterol molecule does not have to be moved on the cytochrome. In the labile situation of two neighboring hydroxy groups, the side chain is removed, so to speak, by itself.

As for molecular studies on this enzyme, translation of bovine adrenocortical mRNA and immunoprecipitation with rabbit anti-bovine P450 SCC indicated that P450 SCC mRNA represents 1% of the total. DNA complementary to bovine adrenocortical mRNA was cloned, and a clone containing sequences complementary to P450 SCC mRNA was identified by hybrid-selected translation (Matteson *et al.*, 1984). Four independent clones containing a part of the P450 SCC cholesterol desmolase gene were isolated from human genomic libraries using bovine P450 SCC cDNA as a probe (Morohashi *et al.*, 1987). These clones covered the entire P450 SCC gene except part of the first intron. The gene was found to be at least 20 kb long and to be split into 9 exons by 8 introns. Sequence analysis revealed that the 9 separated exons code for 521 amino acids, showing 72% homology with bovine P450 SCC. An unusual exon–intron junctional sequence that begins with GC was found in the sixth intron. A putative extension peptide consisting of 39 amino acids was found in the sequence of human P450 SCC by comparison with the bovine counterpart. Two conserved regions were found in the extension peptide of these two forms of P450 SCC, suggesting a functional role of the portions in the mitochondrial localization and processing of P450 SCC precursor. The mature form of human P450 SCC was found to have only one cysteine residue, located in the center of the HR2 region. Voutilainen and Miller (1986) studied the ontogeny and tissue specificity of expression of the genes for P450 SCC in the human fetus. RNA from fetal tissues was probed with homologous human P450 SCC cloned in their laboratory. At 20–21 weeks gestation, P450 SCC mRNA was most abundant in the adrenals, followed by testis, placenta, and ovary. None was detected in other tissues. When testis samples from 13 to 26 weeks gestation were studied, P450 mRNA was most abundant at 14–16 weeks and was diminished by 20–26 weeks.

With respect to the 3β-hydroxysteroid dehydrogenase deficiency, the nucleotide sequence of two highly homologous genes encoding 3β-hydroxysteroid dehydrogenase in three patients from two pedigrees have been described (Rhéaume *et al.*, 1992). The first two families are from Eastern Switzerland and include our only two related cases with the defect (Zachmann *et al.*, 1970, 1979, 1988). The second family is American without relationship to our family. Nonsense and frameshift mutations were both found in the type II 3β-HSD gene, which is the predominant one expressed in the adrenals and gonads, but no mutation was found in the type I 3β-HSD gene, which is mainly expressed in the placenta and peripheral tissues.

A comprehensive review on the purification, properties, and action of 17β-hydroxysteroid dehydrogenase (17-ketosteroid reductase, or 17β-hydroxysteroid-oxidoreductase) has been given by Inano and Tamaoki (1986). The gene regulating this enzyme has not been characterized or mapped.

cDNA clones encoding a human 5α-reductase have been isolated (Jenkins *et al.*, 1992). Those authors subsequently reported that the gene encoding this enzyme is normal in subjects with 5α-reductase deficiency. They showed that in contrast to the major 5α-reductase in the prostate and cultured skin fibroblasts, the cDNA-encoded enzyme exhibits a neutral to basic pH optimum and is much less sensitive to inhibition by finasteride. These results provide genetic, biochemical, and pharmacological support for the existence of at least two isozymes in man.

VII. Importance for the Clinician and Conclusions

Although recent studies in biochemistry and molecular biology have greatly changed and amplified our understanding of how steroids are normally produced and converted, how their gene expression is regulated, and how enzyme defects are brought about, some discrepancies between basic research results and clinical experience have to be pointed out, for which no easy explanations can be given (Zachmann, 1989): The deficiencies of 17α-hydroxylase and 17,20-desmolase are combined in some patients, but in the majority of cases are two distinct conditions. Our original patients (Zachmann *et al.*, 1972) did not show reduced steroid 17α-hydroxylation. In one patient, the steroid pattern changed with age suggesting conversion from "pure" 17,20-desmolase to combined deficiency (Zachmann *et al.*, 1992). It is difficult to understand how this is possible when one gene and enzyme have been shown to regulate both reactions.

With respect to the side-chain cleavage of cholesterol, there also is such an inequity. As mentioned, it has been clearly demonstrated that one cytochrome regulates 20- and 22-hydroxylation. Yet, earlier, still valid studies have shown that in lipoid adrenal hyperplasia, pregnenolone can be formed under certain conditions, when 20αOH-cholesterol, but not when cholesterol, is added to an incubation system (Degenhart *et al.*, 1972). In view of these discrepancies, it must be assumed that there are additional, yet undiscovered, genetically determined modifying factors that regulate the action and substrate specificity of a cytochrome.

References

Aarskog, D. (1970). Clinical and cytogenetic studies in hypospadias. *Acta Paediatr. Scand.*, Suppl. **203,** 1–62.

Abad, L., Parrilla, J. J., Marcos, J., Gimeno, F., and Lopez Bernal, A. (1980). Male pseudohermaphroditism with 17-hydroxylase deficiency. A case report. *Br. J. Obstet. Gynaecol.* **87,** 1162–1165.

Ahlgren, R., Yanase, T., Simpson, E. R., Winter, J. S., and Waterman, M. R. (1992). Compound heterozygous mutations (Arg 239----stop, Pro 342----Thr) in the CYP17 (P45017α) alpha) gene lead to ambiguous external genitalia in a male patient with partial combined 17α-hydroxylase/17,20-lyase deficiency. *J. Clin. Endocrinol. Metab.* **74,** 667–672.

Akesode, F. A., Meyer, W. J., III, and Migeon, C. J. (1977). Male pseudohermaphroditism with gynaecomastia due to testicular 17-ketosteroid reductase deficiency. *Clin. Endocrinol. (Oxford)* **7,** 443–452.

Akgün, S., Ertel, N. H., Imperato-McGinley, J., Sayli, B. S., and Shackleton, C. H. (1986). Familial male pseudohermaphroditism due to 5α-reductase deficiency in a Turkish village. *Am. J. Med.* **81,** 267–274.

Amarose, A. P., Kyranzis, A. A., Dorus, E., and Azizi, F. (1977). Clinical, pathologic, and genetic findings in a case of 46, XY pure gonadal dysgenesis (Swyer's Syndrome). I. Dysgerminoma and gonadoblastoma. *Am. J. Obstet. Gynecol.* **127,** 824–828.

Arnhold, I. J., Mendoncca, B. B., Bloise, W., and Toledo, S. P. (1985). Male pseudohermaphroditism resulting from Leydig cell hypoplasia (letter). *J. Pediatr.* **106,** 1057–1060.

Arnhold, I. J., Mendoncca, B. B., Diaz, J. A., Nogueira, C., Batista, M. C., Madureira, G., Oliveira, D., Nicolau, W., and Bloise, W. (1988). Prepubertal male pseudohermaphroditism due to 17-ketosteroid reductase deficiency: Diagnostic value of a hCG test and lack of HLA association. *J. Endocrinol. Invest.* **11,** 319–322.

Augarten, A., Pariente, C., Gazit, E., Chayen, R., Goldfarb, H., and Sack, J. (1992). Ambiguous genitalia due to partial activity of cytochromes P450 c17 and P450c21. *J. Steroid Biochem. Mol. Biol.* **41,** 37–41.

Balducci, R., Toscano, V., Wright, F., Bozzolan, F., Di Piero, G., Maroder, M., Panei, P., Sciarra, F., and Boscherini, B. (1985). Familial male pseudohermaphroditism with gynaecomastia due to 17β-hydroxysteroid dehydrogenase deficiency. A report of 3 cases. *Clin. Endocrinol. (Oxford)* **23,** 439–444.

Belmont, J. W., Hawkins, E., Hejtmancik, J. F., and Greenberg, F. (1987). Two cases of severe lethal Smith-Lemli-Opitz syndrome. *Am. J. Med. Genet.* **26,** 65–67.

Bergada, C., Cleveland, W. W., Jones, H. W., Jr., and Wilkins, L. (1962). Variants of embryonic testicular dysgenesis: Bilateral anorchia and the syndrome of rudimentary testes. *Acta Endocrinol. (Copenhagen)* **40,** 521–536.

Berkovitz, G. D., Lee, P. A., Brown, T. R., and Migeon, C. J. (1984). Etiologic valuation of male pseudohermaphroditism in infancy and childhood. *Am. J. Dis. Child.* **138,** 755–759.

Berthezène, F., Forest, M. G., Grimaud, J. A., Claustrat, B., and Mornex, R. (1976). Leydig-cell agenesis. *N. Engl. J. Med.* **295,** 969–972.

Biglieri, E. G., and Kater, C. E. (1991). 17α-hydroxylation deficiency. *Endocrinol. Metab. Clin. North Am.* **20,** 257–268.

Biglieri, E. G., Herron, M. A., and Brust, N. (1966). 17-Hydroxylase deficiency in man. *J. Clin. Invest.* **45,** 1946–1954.

Blau, N., Zachmann, M., Kempken, B., Staudenmann, W., Möhr, E., and Curtius, H. C. (1987). Identification of new steroids in patients with 17α-hydroxylase deficiency by capillary gas chromatography/mass spectrometry. *Biomed. Environ. Mass Spectrom.* **14,** 633–637.

Bongiovanni, A. M. (1981). Acquired adrenal hyperplasia: With special reference to 3β-hydroxysteroid dehydrogenase. *Fertil. Steril.* **35,** 599–608.

Bongiovanni, A. M. (1984). Congenital adrenal hyperplasia due to 3β-hydroxysteroid dehydrogenase deficiency. *In* "Adrenal Diseases in Childhood" (M. I. New and L. S. Levine, eds.), Vol. 13, pp. 72–82. Karger, New York.

Bongiovanni, A. M., Eberlein, W. R., and Moshang, T. (1971). Urinary excretion of pregnane triol and pregnenetriol in two forms of congenital adrenal hyperplasia. *J. Clin. Invest.* **50,** 2751–2754.

Bosson, D., Wolter, R., Toppet, M., Franckson, J. R., and de Peretti, E. (1988). Partial 17, 20-desmolase and 17α-hydroxylase deficiencies in a 16-year-old boy. *J. Endocrinol. Invest.* **11,** 527–533.

Bricaire, H., Luton, J. P., Laudat, P., Legrand, J. C., Turpin, G., Corvol, P., and Lemmer, M. (1972). A new male pseudohermaphroditism associated with hypertension due to a block of 17-hydroxylation. *J. Clin. Endocrinol. Metab.* **35,** 67–72.

Brown, D. M., Markland, C., and Dehner, L. P. (1978). Leydig cell hypoplasia: A cause of male pseudohermaphroditism. *J. Clin. Endocrinol. Metab.* **46,** 1–7.

Campo, S., Moteagudo, C., Nicolau, G., Pellizzari, E., and Belgorosky, A. (1981). Testicular function in prepubertal male pseudohermaphroditism. *Clin. Endocrinol.* (*Oxford*) **14,** 11–22.

Cantu, J. M., Corona Rivera, E., Diaz, M., Medina, C., and Esquinca, E. (1980). Postpubertal female psychosexual orientation in incomplete male pseudohermaphroditism type 2 (5α-reductase deficiency). *Acta Endocrinol.* (*Copenhagen*) **94,** 273–279.

Carpenter, T. O., Imperato-McGinley, J., Boulware, S. D., Weiss, R. M., Shackleton, C. H., Griffin, J. E., and Wilson, J. D. (1990). Variable expression of 5α-reductase deficiency: Presentation with male phenotype in a child of Greek origin. *J. Clin. Endocrinol. Metab.* **71,** 318–322.

Caufriez, A. (1986). Male pseudohermaphroditism due to 17-ketoreductase deficiency: Report of a case without gynecomastia and without vaginal pouch. *Am. J. Obstet. Gynecol.* **154,** 148–149.

Chasalow, F. I., Blethen, S. L., Knight, S. M., and Taysi, K. (1986). 18q deletion syndrome in a child with steroid 17,20-lyase deficiency. *Steroids* **47,** 421–429.

Chemke, J., Carmichael, R., Stewart, J. M., Geer, R. H., and Robinson, A. (1970). Familial XY gonadal dysgenesis. *J. Med. Genet.* **7,** 105–111.

Chung, B. C., Picado-Leonard, J., Haniu, M., Bienkowski, M., Hall, P. F., Shively, J. E., and Miller, W. L. (1987). Cytochrome P450C17 (steroid 17α-hydroxylase/17,20-lyase) Cloning of human adrenal and testis cDNAs indicates the same gene is expressed in both tissues. *Proc. Natl. Acad. Sci. U.S.A.* **84,** 407–411.

Corrall, R. J., Wakelin, K., O'Hare, J. P., O'Brien, I. A., and Ishmail, A. A. (1984). 5α-reductase deficiency: Diagnosis via abnormal plasma levels of reduced testosterone derivatives. *Acta Endocrinol.* (*Copenhagen*) **107,** 538–543.

Cravioto, M. D., Ulloa Aguirre, A., Bermudez, J. A., Herrera, J., Lisker, R., Mendez, J. P., and Pérez Palacios, G. (1986). A new inherited variant of the 3β-hydroxysteroid dehydrogenase-isomerase deficiency syndrome: Evidence for the existence of two isoenzymes. *J. Clin. Endocrinol. Metab.* **63,** 360–367.

D'Alberton, A., Reschini, E., Motta, T., and Catania, A. (1989). Male pseudohermaphroditism due to 17-hydroxylase deficiency. *J. Endocrinol. Invest.* **12,** 193–196.

David, R., Yoon, D. J., Landin, L., Lew, L., Sklar, C., Schinella, R., and Golimbu, M. (1984). A syndrome of gonadotropin resistance possibly due to a luteinizing hormone receptor defect. *J. Clin. Endocrinol. Metab.* **59,** 156–160.

Degenhart, H. J. (1984). Prader's syndrome (congenital lipoid adrenal hyperplasia). *In* "Adrenal Diseases in Childhood" (M. I. New and L. S. Levine, eds.), Vol. 13, pp. 125–144. Karger, New York.

Degenhart, H. J., Visser, H. K. A., Boon, H., and O'Doherty, N. J. (1972). Evidence for deficient 20a-cholesterol hydroxylase activity in adrenal tissue of a patient with lipoid adrenal hyperplasia. *Acta Endocrinol.* (*Copenhagen*) **71,** 512–518.

De Peretti, E., Pradon, M., and Forest, M. G. (1984). 17,20-desmolase deficiency in a female newborn, paradoxically virilized in utero. *J. Steroid Biochem.* **20,** 455–458.

Deslypere, J. P., Coucke, W., Robbe, N., and Vermeulen, A. (1985). 5α-reductase deficiency: An infrequent cause of male pseudohermaphroditism. *Acta Clin. Belg.* **40,** 240–246.

Donahoe, P. K., Crawford, J. D., and Hendren, W. H. (1979). Mixed gonadal dysgenesis, pathogenesis, and management. *J. Pediatr. Surg.* **14,** 287–300.

Dorus, E., Amarose, A. P., Koo, G. G., and Wachtel, S. S. (1977). Clinical, pathologic, and genetic findings in a case of 46,XY pure gonadal dysgenesis (Swyer's syndrome). II. Presence of H-Y antigen. *Am. J. Obstet. Gynecol.* **127,** 829–831.

Drash, A., Shermann, F., Hartmann, W. H., and Blizzard, R. M. (1970). A syndrome of pseudohermaphroditism, Wilms'tumor, hypertension, and degenerative renal disease. *J. Pediatr.* **76,** 585–593.

Dumic, M., Plavsic, V., Fattorini, I., and Ille, J. (1985). Absent spermatogenesis despite early bilateral orchidopexy in 17-ketoreductase deficiency. *Horm. Res.* **22,** 100–106.

Dunkel, L., Perheetupa, J., Tapanainen, J., and Vihko, R. (1984). Hypergonadotropic hypogonadism in newborn males with primary testicular failure. *Acta Paediatr. Scand.* **73,** 740–744.

Eckstein, B., Cohen, S., Farkas, A., and Rösler, A. (1989). The nature of the defect in familial male pseudohermaphroditism in Arabs of Gaza. *J. Clin. Endocrinol. Metab.* **68,** 477–485.

Eddy, A. A., and Mauer, S. M. (1985). Pseudohermaphroditism, glomerulopathy, and Wilms tumor (Drash syndrome): Frequency in endstage renal failure. *J. Pediatr.* **106,** 584–587.

Eil, C., Austin, R. M., Sesterhenn, I., Dunn, J. F., Cutler, G. B., and Johnsonbaugh, R. E. (1984). Leydig cell hypoplasia causing male pseudohermaphroditism: Diagnosis 13 years after prepubertal castration. *J. Clin. Endocrinol. Metab.* **58,** 441–448.

Ertel, N. H., Akgün, S., Samojlik, E., Kirschner, M. A., and Imperato-McGinley, J. (1989). Decreased 3β-androstanediol glucuronide levels in plasma and random urines in male pseudohermaphroditism caused by 5α-reductase deficiency. *Metab. Clin. Exp.* **38,** 817–821.

Forest, M. G., Cathiard, A. M., and Bertrand, J. A. (1973). Evidence of testicular activity in early infancy. *J. Clin. Endocrinol. Metab.* **37,** 148–151.

Forest, M. G., Lecornu, M., and de Peretti, E. (1980). Familial male pseudohermaphroditism due to 17-20-desmolase deficiency. I. In vivo endocrine studies. *J. Clin. Endocrinol. Metab.* **50,** 826–833.

Forest, M. G., de Peretti, E., Monneret, M. P., Lecoq, A., and Bertrand, J. (1985). Kinetics of the testicular steroidogenic response to stimulation by placental gonadotropin hormone. VI. Enzyme deficiencies of testicular biosynthesis. *Arch. Fr. Pediatr.* **42,** 187–193.

Frank Raue, K., Raue, F., Korth-Schütz, S., Vecsei, P., and Ziegler, R. (1989). Clinical features and diagnosis of mild 3-β-hydroxysteroid dehydrogenase deficiency in men. *Dtsch. Med. Wochenschr.* **114,** 331–334.

Gearhart, J. P., Linhard, H. R., Berkovitz, G. D., Jeffs, R. D., and Brown, T. R. (1988). Androgen receptor levels and 5α-reductase activities in preputial skin and chordee tissue of boys with isolated hypospadias. *J. Urol.* **140,** 1243–1246.

Gendrel, D., Chaussain, J. L., Roger, M., and Job, J. C. (1979). L'hyperplasie surrénale congénitale par bloc de la 3β-hydroxystéroide déshydrogénase. *Arch. Fr. Pédiatr.* **36,** 647–655.

Givens, J. R., Wiser, W. L., Summitt, R. L., Kerber, I. J., Andersen, R. N., Pittaway, D. E., and Fish, S. A. (1974). Familial male pseudohermaphroditism without gynecomastia due to deficient testicular 17-ketosteroid reductase activity. *N. Engl. J. Med.* **291,** 938–944.

Goebelsmann, U., Zachmann, M., Davajan, V., Israel, R., Mestman, J. H., and Mishell, D. R. (1976). Male pseudohermaphroditism consistent with 17,20-desmolase deficiency. *Gynecol. Invest.* **7,** 138–156.

Goldman, S. M., Garfinkel, D. J., Oh, K. S., Dorst, J. P., and Russell, H. (1981). The Drash syndrome: Male pseudohermaphroditism, nephritis, and Wilms tumor. *Radiology* **141,** 87–91.

Greenberg, F., Gresik, M. V., Carpenter, R. J., Law, S. W., Hoffman, L. P., and Ledbetter, D. H. (1987). The Gardner-Silengo-Wachtel or genito-palato-cardiac syndrome: Male pseudohermaphroditism with micrognathia, cleft palate, and conotruncal cardiac defect. *Am. J. Med. Genet.* **26,** 59–64.

Greene, S. A., Zachmann, M., Manella, B., Hesse, V., Hoepffner, W., Willgerodt, H., and Prader, A. (1987). Comparison of two tests to recognize or exclude 5α-reductase deficiency in prepubertal children. *Acta Endocrinol. (Copenhagen)* **114,** 113–117.

Gregory, J. W., Aynsley-Green, A., Hughes, I. A., Werder, E. A., and Zachmann, M. (1992). 17-Ketoreductase deficiency presenting in infancy. *Horm. Res.* **37,** Suppl. 4, 31.

Gross, D. J., Landau, H., Kohn, G., Farkas, A., Elrayyes, E., el Shawwa, R., Lasch, E. E., and Rösler, A. (1986). Male pseudohermaphroditism due to 17β-hydroxysteroid dehydrogenase deficiency: Gender reassignment in early infancy. *Acta Endocrinol. (Copenhagen)* **112,** 238–246.

Hall, P. F. (1986). Cytochromes P450 and the regulation of steroid synthesis. *Steroids* **48,** 131–196.

Hall, P. F. (1989). Cytochromes P-450 in steroidogenesis: Are these enzymes more specific than those of drug metabolism? *Clin. Exp. Pharmacol. Physiol.* **16,** 485–489.

Harkness, R. A., Thistlethwaite, D., Darling, J. A., and Skakkeback, N. E. (1980). Neutral 17β-hydroxysteroid oxidoreductase deficiency in testes causing male pseudohermaphroditism in an infant. *J. Inherited Metab. Dis.* **2,** 51–54.

Hauffa, B. P., Miller, W. L., Grumbach, M. M., Conte, F. A., and Kaplan, S. L. (1985). Congenital adrenal hyperplasia due to deficient cholesterol side-chain cleavage activity (20,22-desmolase) in a patient treated for 18 years. *Clin. Endocrinol. (Oxford)* **23,** 481–493.

Heppe, R. K., Koyle, M. A., and Beckwith, J. B. (1991). Nephrogenic rests in Wilms tumor patients with the Drash syndrome. *J. Urol.* **145,** 1225–1228.

Herdt, G. H., and Davidson, J. (1988). The Sambia "turnim-man": Sociocultural and clinical aspects of gender formation in male pseudohermaphrodites with 5α-reductase deficiency in Papua New Guinea. *Arch. Sex Behav.* **17,** 33–56.

Horton, T., and Tait, J. F. (1966). Androstenedione production and interconversion rates measured in peripheral blood and studies on the possible sites of conversion to testosterone. *J. Clin. Invest.* **45,** 301–313.

Hovmöller, M. L., Osuna, A., Eklöf, O., Fredga, K., Hjerpe, A., Lindsten, J., Ritzén, M., Stanescu, V., and Svenningsen, N. (1977). Camptomelic dwarfism. A genetically determined mesenchymal disorder combined with sex reversal. *Hereditas* **86,** 51.

Ieshima, A., Koeda, T., and Inagaki, M. (1986). Peculiar face, deafness, cleft palate, male pseudohermaphroditism, and growth and psychomotor retardation; a new autosomal recessive syndrome? *Clin. Genet.* **30,** 136–141.

Imai, T., Yanase, T., Waterman, M. R., Simpson, E. R., and Pratt, J. J. (1992). Canadian Mennonites and individuals residing in the Friesland region of The Netherlands share the same molecular basis of 17α-hydroxylase deficiency. *Hum. Genet.* **89,** 95–96.

Imperato-McGinley, J., and Peterson, R. E. (1976). Male pseudohermaphroditism: The complexities of male phenotypic development. *Am. J. Med.* **61,** 251–272.

Imperato-McGinley, J., Guerrero, L., Gautier, T., and Peterson, R. E. (1974). Steroid 5α-reductase deficiency in man, an inherited form of male pseudohermaphroditism. *Science* **186,** 1213–1215.

Imperato-McGinley, J., Peterson, R. E., and Gautier, T. (1985). Decreased urinary C19 and C21 steroid 5α-metabolites in parents of male pseudohermaphrodites with 5α-reductase deficiency: Detection of carriers. *Clin. Endocrinol. (Oxford)* **60,** 553–558.

Imperato-McGinley, J., Gautier, T., Pichardo, M., and Shackleton, C. H. (1986a). The diagnosis of 5α-reductase deficiency in infancy. *J. Clin. Endocrinol. Metab.* **63,** 1313–1318.

Imperato-McGinley, J., Gautier, T., Peterson, R. E., and Shackleton, C. H. (1986b). The prevalence of 5α-reductase deficiency in children with ambiguous genitalia in the Dominican Republic. *J. Urol.* **136,** 867–873.

Imperato-McGinley, J., Akgün, S., Ertel, N. H., Sayli, B., and Shackleton, C. H. (1987). The coexistence of male pseudohermaphrodites with 17-ketosteroid reductase deficiency and 5α-reductase deficiency within a Turkish kindred. *Clin. Endocrinol. (Oxford)* **27,** 135–143.

Imperato-McGinley, J., Miller, M., Wilson, J. D., and Peterson, R. E. (1991). A cluster of male pseudohermaphrodites with 5α-reductase deficiency in Papua New Guinea. *Clin. Endocrinol. (Oxford)* **34,** 293–298.

Inano, H., and Tamaoki, B. (1986). Testicular 17β-hydroxysteroid dehydrogenase: Molecular properties and reaction mechanism. *Steroids* **48,** 1–26.

Ivarsson, S. A., Nielsen, M. D., and Lindberg, T. (1988). Male pseudohermaphroditism due to 5α-reductase deficiency in a Swedish family. *Eur. J. Pediatr.* **147,** 532–535.

Jenkins, E. P., Andersson, S., Imperato-McGinley, J., and Wilson, J. D. (1992). Genetic and pharmacological evidence for more than one human steroid 5α-reductase. *J. Clin. Invest.* **89,** 293–300.

Jensen, J. C., Ehrlich, R. M., Hanna, M. K., Fine, R. N., and Grunberger, I. (1989). A report of 4 patients with the Drash syndrome and a review of the literature. *J. Urol.* **141,** 1174–1176.

Jones, H. W., Jr., Lee, P. A., Rock, J. A., Archer, D. F., and Migeon, C. J. (1982). A genetic male patient with 17α-hydroxylase deficiency. *Obstet. Gynecol.* **59,** 254–259.

Jones, K. L., Freidenberg, G. R., Buchta, R., and Derenoncourt, A. (1992). Male pseudohermaphroditism resulting from 17α-monooxygenase (P-450C17) deficiency in two unrelated Guamanians. *Am. J. Dis. Child.* **146,** 592–595.

Jukier, L., Kaufmann, M., Pinsky, L., and Peterson, R. (1984). Partial androgen resistance associated with secondary 5α-reductase deficiency: Identification of a novel qualitative androgen receptor defect and clinical implications. *J. Clin. Endocrinol. Metab.* **59,** 679–688.

Kagimoto, M., Winter, J. S., Kagimoto, K., Simpson, E. R., and Waterman, M. R. (1988). Structural characterization of normal and mutant human steroid 17α-hydroxylase genes: Molecular basis of one example of combined 17α-hydroxylase/17,20 lyase deficiency. *Mol. Endocrinol.* **2,** 564–570.

Kaufman, F. R., Costin, G., Goebelsmann, U., Stanczyk, F. Z., and Zachmann, M. (1983). Male pseudohermaphroditism due to 17,20-desmolase deficiency. *J. Clin. Endocrinol. Metab.* **57,** 32–36.

Kofmann, S., Pérez-Palacios, G., Medina, M., Escobar, N., Garcia, M., Ruz, L., Mutchinick, O., and Lisker, R. (1981). Clinical and endocrine spectrum in patients with the 45,X/46,XY karyotype. *Hum. Genet.* **58,** 373–376.

Lachman, M. F., Wright, Y., Whiteman, D. A., Herson, V., and Greenstein, R. M. (1991).

Brief clinical report: A 46,XY phenotypic female with Smith-Lemli-Opitz syndrome. *Clin. Genet.* **39,** 136–141.

Lanes, R., Brown, T. R., de Bustos, E. G., Valverde, B., and Pieretti, R. B. (1983). Sibship with 17-ketosteroid reductase (17-KSR) deficiency and hypothyroidism. Lack of linkage of histocompatibility leukocyte antigen and 17-KSR loci. *J. Clin. Endocrinol. Metab.* **57,** 190–196.

Larrea, F., Lisker, R., Banuelos, R., Bermudez, J. A., and Herrera, J. (1983). Hypergonadotrophic hypogonadism in an XX female subject due to 17,20 steroid desmolase deficiency. *Acta Endocrinol.* (*Copenhagen*) **103,** 400–405.

Lee, P. A., Rock, J. A., Brown, T. R., Fichman, K. M., Migeon, C. J., and Jones, H. W., Jr. (1982). Leydig cell hypofunction resulting in male pseudohermaphroditism. *Fertil. Steril.* **37,** 675–679.

Leinonen, P., Dunkel, L., Perheentupa, J., and Vihko, R. (1983). Male pseudohermaphroditism due to deficiency of testicular 17-ketosteroid reductase. *Acta Paediatr. Scand.* **72,** 211–214.

Lo Iudice, G., Esposito, V., Federico, P., and D'Alessandro, B. (1986). Male pseudohermaphroditism due to 5α-reductase deficiency. Clinical and endocrinological data in a 17-year-old patient. *Minerva Endocrinol.* **11,** 7–13.

Maciel Guerra, A. T., Farah, S. B., Garmes, H. M., and Pinto Junior, W. (1991). True agonadism: Report of a case analyzed with Y-specific DNA probes. *Am. J. Med. Genet.* **41,** 444–445.

Mann, J. R., Corkery, J. J., Fisher, H. J. W., Cameron, A. H., Mayerova, A., Wolf U., Kennaughs, A. A., and Woolley, V. (1983). The X-linked recessive form of XY gonadal dysgenesis with a high incidence of gonadal germ cell tumors: Clinical and genetic studies. *J. Med. Genet.* **20,** 264–270.

Mantero, F., and Scaroni, C. (1984). Enzymatic defects of steroidogenesis: 17-hydroxylase deficiency. *In* "Adrenal Diseases in Childhood" (M. I. New and L. S. Levine, eds.), Vol. 13, pp. 83–94. Karger, New York.

Martin, F., Perheentupa, J., and Adlercreutz, H. (1980). Plasma and urinary androgens and oestrogens in a pubertal boy with 3β-hydroxysteroid dehydrogenase deficiency. *J. Steroid Biochem.* **13,** 197–201.

Martinez Mora, J., Saez, J. M., Toran, N., Isnard, R., Perez Iribarne, M. M., Egozcue, J., and Audi, L. (1991). Male pseudohermaphroditism due to Leydig cell agenesia and absence of testicular LH receptors. *Clin. Endocrinol.* (*Oxford*) **34,** 485–491.

Matteson, K. J., Chung, B. C., and Miller, W. L. (1984). Molecular cloning of DNA complementary to bovine adrenal P450scc mRNA. *Biochem. Biophys. Res. Commun.* **120,** 264–270.

Matteson, K. J., Picado Leonard, J., Chung, B. C., Mohandas, T. K., and Miller, W. L. (1986). Assignment of the gene for adrenal P450c17 (steroid 17α-hydroxylase/17,20 lyase) to human chromosome 10. *J. Clin. Endocrinol. Metab.* **63,** 789–791.

Mauvais-Jarvis, P., Kuttenn, F., Mowszowicz, I., and Wright, F. (1981). Different aspects of 5α-reductase deficiency in male pseudohermaphroditism and hypothyroidism. *Clin. Endocrinol.* (*Oxford*) **14,** 459–469.

Medina, M., Castorena, G., Herrera, J., Bermudez, J. A., and Zarate, A. (1989). Modulation of luteinizing hormone secretion by estrogens in patients with Reifenstein's syndrome. *Fertil. Steril.* **52,** 239–242.

Mendonca, B. B., Bloise, W., Arnhold, I. J., Batista, M. C., Toledo, S. P., Drummond, M. C., Nicolau, W., and Mattar, E. (1987). Male pseudohermaphroditism due to nonsalt-losing 3β-hydroxysteroid dehydrogenase deficiency: Gender role change and absence of gynecomastia at puberty. *J. Steroid Biochem.* **28,** 669–675.

Messinis, I. E., and Nillius, S. J. (1981). 46,XY male pseudohermaphroditism due to early foetal testicular dysgenesis. *Acta Endocrinol.* (*Copenhagen*) **98,** 308–311.

Migeon, C. J. (1980). Male pseudohermaphroditism. *Ann. Endocrinol.* **41,** 311–343.

Millan, M., Audi, L., Martinez Mora, J., Martinez de Osaba, M. J., Viguera, J., Esmatjes, E., Peig, M., and Vilardell, E. (1983). 17-ketosteroid reductase deficiency in an adult patient without gynaecomastia but with female psychosexual orientation. *Acta Endocrinol.* (*Copenhagen*) **102,** 633–640.

Morohashi, K., Sogawa, K., Omura, T., and Fujii Kuriyama, Y. (1987). Gene structure of human cytochrome P-450(SCC), cholesterol desmolase. *J. Biochem.* (*Tokyo*) **101,** 879–887.

Müller, J., Skakkebaek, N. E., Ritzén, M., Plöen, L., and Pedersen, K. E. (1985). Carcinoma in situ of the testis in children with 45,X/45,XY gonadal dysgenesis. *J. Pediatr.* **106,** 431–436.

Müller, J., Torsson, A., Damkjaer Nielsen, M., Petersen, K. E., Christoffersen, J., and Skakkebaek, N. E. (1991). Gonadal development and growth in 46,XX and 46,XY individuals with P450scc deficiency (congenital lipoid adrenal hyperplasia). *Horm. Res.* **36,** 203–208.

Nagel, T., Camaro, M., Tagatz, G., Okagaki, T., Cervenka, J., Cahill, L., Wachtel, S. S., King, R., and Twiggs, L. B. (1984). Gonadal tumors in patients with gonadal dysgenesis and sex chromosomal rings and fragments. *Am. J. Obstet. Gynecol.* **150,** 76–82.

Nahoul, K., Perrin, C., Leymarie, P., and Job, J. C. (1989). Elevated levels of plasma 4-ene steroids in a case of congenital deficiency of 3β-hydroxysteroid dehydrogenase. *Ann. Endocrinol.* **50,** 58–63.

Najjar, S. S., Takla, R. J., and Nassar, V. H. (1974). The syndrome of rudimentary testes: Occurrence in five siblings. *J. Pediatr.* **84,** 119–122.

Nakajin, S., Shinoda, M., Haniu, M., Shively, J. E., and Hall, P. F. (1984). C21 steroid side chain cleavage enzyme from porcine adrenal microsomes. Purification and characterization of the 17α-hydroxylase/C17,20-lyase cytochrome P-450. *J. Biol. Chem.* **259,** 3971–3976.

New, M. I. (1970). Male pseudohermaphroditism due to 17α-hydroxylase deficiency. *J. Clin. Invest.* **49,** 1930–1941.

Ng, W. K., Taylor, N. F., Hughes, I. A., Taylor, J., Ransley, P. G., and Grant, D. B. (1990). 5α-reductase deficiency without hypospadias. *Arch. Dis. Child.* **65,** 1166–1167.

Nishi, Y., and Tezuka, T. (1992). Mild adrenal 3β-hydroxysteroid dehydrogenase deficiency in children with accelerated growth, premature pubarche and/or hirsutism. *Eur. J. Pediatr.* **151,** 19–23.

Nivelon, J. L., Forest, M. G., Nivelon Chevallier, A., Turc, C., Dauvergne, M., and Tenenbaum, D. (1985). Prenatal diagnosis in a familial form of male pseudohermaphroditism due to 17-keto reductase deficiency. *J. Genet. Hum.* **33,** 179–186.

Nivelon, J. L., Forest, M. G., Nivelon Chevallier, A., and Sidaner, I. (1987). Second prenatal diagnosis in a familial form of male pseudohermaphroditism caused by 17 ketoreductase deficiency: Prediction confirmed by a normal third male infant. *J. Genet. Hum.* **35,** 159–163.

Ochoa, S., Larraza, O., and Perez, A. E. (1981). Inherited male pseudohermaphroditism due to gonadotrophin unresponsiveness. *Acta Endocrinol.* (*Copenhagen*) **98,** 148–155.

Odame, I., Donaldson, M. D., Wallace, A. M., Cochran, W., and Smith, P. J. (1992). Early diagnosis and management of 5α-reductase deficiency. *Arch. Dis. Child.* **67,** 720–723.

Ogata, T., Ishikawa, K., Kohda, E., and Matsuo, N. (1988). Computed tomography in the early detection of congenital lipoid adrenal hyperplasia. *Pediatr. Radiol.* **18,** 360–361.

Okon, E., Livni, N., Rösler, A., Yorkoni, S., Segal, S., Kohn, G., and Schenker, J. G.

(1980). Male pseudohermaphroditism due to 5α-reductase deficiency. Ultrastructure of the gonads. *Arch. Pathol. Lab. Med.* **104,** 363–367.

Onoda, M., and Hall, P. F. (1981). Inhibition of testicular microsomal cytochrome P-450 (17α-hydroxylase/C-17,20-lyase) by estrogens. *Endocrinology* (*Baltimore*) **109,** 763–767.

Oshima, H., Ishiwata, D., Hatakeyama, S., and Yokokawa, M. (1980). In vitro steroidogenesis in testes of three infants, two with ambiguous external genitalia and one with true precocious puberty: Evidence for the presence of active 17β-hydroxysteroid oxidoreductase in immature human testes. *Endocrinol. Jpn.* **27,** 59–64.

Pang, S., Levine, L. S., Chow, D., Sagiani, F., Saenger, P., and New, M. I. (1979). Dihydrotestosterone and its relation to testosterone in infancy and childhood. *J. Clin. Endocrinol. Metab.* **48,** 821–826.

Pang, S., Levine, L. S., Stoner, E., Opitz, J. M., Pollack, M. S., and Dupont, B. (1983). Non-salt-losing congenital adrenal hyperplasia due to 3β-hydroxysteroid dehydrogenase deficiency with normal glomerulosa function. *J. Clin. Endocrinol. Metab.* **56,** 808–818.

Pankau, R., Partsch, C. J., Funda, J., and Sippell, W. G. (1992). Hypothalamic-pituitary-gonadal function in two infants with Smith-Lemli-Opitz syndrome. *Am. J. Med. Genet.* **43,** 513–516.

Park, I. J., Aimakhu, V. E., and Jones, H. W. (1975). An etiologic and pathogenetic classification of male hermaphroditism. *Am. J. Obstet. Gynecol.* **123,** 505–518.

Paulson, R. J., Serafini, P. C., Catalino, J. A., and Lobo, R. A. (1986). Measurements of 3α,17β androstanediol glucuronide in serum and urine and the correlation with skin 5α-reductase activity. *Fertil. Steril.* **46,** 222–226.

Peterson, R. E., Imperato-McGinley, J., Gautier, T., and Sturla, E. (1977). Male pseudohermaphroditism due to steroid 5α-reductase deficiency. *Am. J. Med.* **62,** 170–191.

Peterson, R. E., Imperato-McGinley, J., Gautier, T., and Shackleton, C. H. (1985a). Male pseudohermaphroditism due to multiple defects in steroid-biosynthetic microsomal mixed-function oxidases. A new variant of congenital adrenal hyperplasia. *N. Engl. J. Med.* **313,** 1182–1191.

Peterson, R. E., Imperato-McGinley, J., Gautier, T., and Shackleton, C. H. (1985b). Urinary steroid metabolites in subjects with male pseudohermaphroditism due to 5α-reductase deficiency. *Clin. Endocrinol.* (*Oxford*) **23,** 43–53.

Picado-Leonard, J., and Miller, W. L. (1987). Cloning and sequence of the human gene for P450 C17 (steroid 17α-hydroxylase/17,20-lyase)—Similarity with the gene for P450 C21. *J. Mol. Biol.* **6,** 439–448.

Prader, A. (1954). Der Genitalbefund beim Pseudo-Hermaphroditismus des kongenitalen adrenogenitalen Syndroms. *Helv. Paediatr. Acta* **9,** 231–248.

Prader, A. (1974). Male pseudohermaphroditism. *Helv. Paediatr. Acta* **29,** Suppl. 34, 79–86.

Prader, A., and Gurtner, H. P. (1955). Das Syndrom des Pseudohermaphroditismus masculinus bei kongenitaler Nebennierenrinden-Hyperplasie ohne Androgenüberproduktion (adrenaler Pseudohermaphroditismus masculinus). *Helv. Paediatr. Acta* **10,** 397–412.

Prader, A., and Siebenmann, R. E. (1957). Kongenitale Lipoidhyperplasie der Nebennierenrinde. *Helv. Paediatr. Acta* **12,** 569–595.

Prader, A., Illig, R., and Zachmann, M. (1976). Prenatal LH-deficiency as possible cause of male pseudohermaphroditism, hypospadias, hypogenitalism and cryptorchidism. *Pediatr. Res.* **10,** 893.

Price, P., Wass, J. A., Griffin, J. E., Leshin, M., Savage, M. O., and Large, D. M. (1984).

High dose androgen therapy in male pseudohermaphroditism due to 5α-reductase deficiency and disorders of the androgen receptor. *J. Clin. Invest.* **74,** 1496–1508.

Rhéaume, E., Simard, J., Morel, Y., Mebarki, F., Zachmann, M., Forest, M. G., New, M., and Labrie, F. (1992). Congenital adrenal hyperplasia due to point mutations in the type II 3β-hydroxysteroid dehydrogenase gene. *Nat. Genet.* **1,** 239–245.

Roger, M., Merceron, R. E., Girard, F., Canlorbe, P., Dehennin, L., Konopka, P., Seneze, J., and Toublanc, J. E. (1982). Dexamethasone-suppressible hypercorticosteronism in two 46,XX subjects with ambiguous genitalia and ovarian cysts. Partial defect of 17α-hydroxylase or 17-20-desmolase. *Horm. Res.* **16,** 23–31.

Rogers, R. M., Garcia, A., van den Berg, L., Petrik, P. K., Snihurowych, W. M., and Crockford, P. M. (1982). Leydig cell hypogenesis: A rare cause of male pseudohermaphroditism and a pathological model for the understanding of normal sexual differentiation. *J. Urol.* **128,** 1325–1329.

Rohmer, V., Barbot, N., Bertrand, P., Nahoul, K., Bigorgne, J. C., and Forest, M. G. (1990). A case of male pseudohermaphroditism due to 17α-hydroxylase deficiency and hormonal profiles in the nuclear family. *J. Clin. Endocrinol. Metab.* **71,** 523–529.

Rosenfield, R. L., Rich, B. H., Wolfsdorf, J. I., Cassorla, F., Parks, S., Bongiovanni, A. M., Wu, C. H., and Shackleton, C. H. (1980). Pubertal presentation of congenital delta5-3β-hydroxysteroid dehydrogenase deficiency. *J. Clin. Endocrinol. Metab.* **51,** 345–353.

Rösler, A., and Kohn, G. (1983). Male pseudohermaphroditism due to 17β-hydroxysteroid dehydrogenase deficiency: Studies on the natural history of the defect and effect of androgens on gender role. *J. Steroid Biochem.* **19,** 663–674.

Saenger, P. (1984). Abnormal sexual differentiation. *J. Pediatr.* **104,** 1–17.

Saenger, P., Goldman, A. S., and Levine, L. S. (1978). Prepubertal diagnosis of steroid 5α-reductase deficiency. *J. Clin. Endocrinol. Metab.* **46,** 627–634.

Saez, J. M., De Peretti, E., Morera, A. M., David, M., and Bertrand, J. (1971). Familial male pseudohermaphroditism with gynecomastia due to a testicular 17-ketosteroid reductase defect. *J. Clin. Endocrinol. Metab.* **32,** 604–610.

Saez, J. M., Morera, A. M., De Peretti, E., and Bertrand, J. (1972). Further in vivo studies in male pseudohermaphroditism with gynecomastia due to a testicular 17-ketosteroid reductase defect. *J. Clin. Endocrinol. Metab.* **34,** 598–600.

Saldanha, P. H., Arnhold, I. J., Mendoncca, B. B., Bloise, W., and Toledo, S. P. (1987). A clinico-genetic investigation of Leydig cell hypoplasia. *Am. J. Med. Genet.* **26,** 337–344.

Savage, M. O., Chaussain, J. L., Evain, D., Roger, M., Canlorbe, P., and Job, J. C. (1978). Endocrine studies in male pseudohermaphroditism in childhood and adolescence. *Clin. Endocrinol. (Oxford)* **8,** 219–231.

Savage, M. O., Preece, M. A., Jeffcoate, S. L., Ransley, P. G., Rumsby, G., Mansfield, M. D., and Williams, D. I. (1980). Familial male pseudohermaphroditism due to deficiency of 5α-reductase. *Clin. Endocrinol. (Oxford)* **12,** 397–406.

Schmidt, J. A., and Schweikert, H. U. (1986). Testosterone and epitestosterone metabolism of single hairs in 5 patients with 5α-reductase-deficiency. *Acta Endocrinol. (Copenhagen)* **113,** 588–592.

Schwartz, M., Imperato-McGinley, J., Peterson, R. E., Cooper, G., Morris, P. L., McGillivray, M., and Hense, T. (1981). Male pseudohermaphroditism secondary to an abnormality in Leydig cell differentiation. *J. Clin. Endocrinol. Metab.* **53,** 123–127.

Shima, H., Kawanaka, H., Yabumoto, Y., Okamoto, E., and Ikoma, F. (1991). A case of 17α-hydroxylase deficiency with chromosomal karyotype 46, XY and high plasma aldosterone concentration. *Int. Urol. Nephrol.* **23,** 611–618.

Simpson, J. L. (1976). "Disorders of Sexual Differentiation: Etiology and Clinical Delineation." Academic Press, New York.

Simpson, J. L. (1978). Male pseudohermaphroditism: genetics and clinical delineation. *Hum. Genet.* **44,** 1–49.

Sorgo, W., Gortner, L., Bartmann, P., Streb, H. P., Oberhoffer, R., Teller, W. M., Zachmann, M., Heymer, B., Graf, M., Lattermann, U., Jäger, R. J., and Ebensperger, C. (1991). Gonadal agenesis in a 46,XY female with multiple malformations and positive SRY testing. *Horm. Res.* **35,** 124–131.

Stanhope, R., Milroy, E. J., and Brook, C. G. D. (1986). Steroid 5α-reductase deficiency and posterior urethral valves. *Horm. Res.* **23,** 213–214.

Stewart, P. M., Beastall, G. H., Shackleton, C. H., and Edwards, C. R. W. (1990). 5α-reductase activity in polycystic ovary syndrome. *Lancet* **335,** 431–433.

Swyer, G. I. M. (1955). Male pseudohermaphroditism: A hitherto undescribed form. *Br. Med. J.* **2,** 709–712.

Toledo, S. P. (1992). Leydig cell hypoplasia leading to two different phenotypes: Male pseudohermaphroditism and primary hypogonadism not associated with this (letter). *Clin. Endocrinol. (Oxford)* **36,** 521–522.

Toledo, S. P., Arnhold, I. J., Luthold, W., Russo, E. M., and Saldanha, P. H. (1985). Leydig cell hypoplasia determining familial hypergonadotropic hypogonadism. *Prog. Clin. Biol. Res.* **200,** 311–314.

Tourniaire, J., Audi-Parera, L., Loras, B., Blum, J., Castelnovo, P., and Forest, M. G. (1976). Male pseudohermaphroditism with hypertension due to a 17α-hydroxylase deficiency. *Clin. Endocrinol. (Oxford)* **5,** 53–61.

Ulloa Aguirre, A., Bassol, S., Poo, J., Mendez, J. P., Mutchinick, O., Robles, C., and Pérez-Palacios, G. (1985). Endocrine and biochemical studies in a 46,XY phenotypically male infant with 17-ketosteroid reductase deficiency. *J. Clin. Endocrinol. Metab.* **60,** 639–643.

Virdis, R., and Saenger, P. (1984). 17β-hydroxysteroid dehydrogenase deficiency. *In* "Adrenal Diseases in Childhood" (M. I. New and L. S. Levine, eds.), Vol. 13, pp. 110–124. Karger, New York.

Virdis, R., Saenger, P., and Senior, B. (1978). Endocrine studies in a pubertal male pseudohermaphrodite with 17-ketosteroid reductase deficiency. *Acta Endocrinol. (Copenhagen)* **87,** 212–224.

von Schnakenburg, K., Bidlingmaier, F., Engelhardt, D., Butenandt, O., Unterburger, P., and Knorr, D. (1980). 17-ketosteroid reductase deficiency—plasma steroids and incubation studies with testicular tissue. *Acta Endocrinol. (Copenhagen)* **94,** 397–403.

Voutilainen, R., and Miller, W. L. (1986). Developmental expression of genes for the steroidogenic enzymes P450scc (20,22-desmolase), P450c17 (17α-hydroxylase/17,20-lyase) and P450c21 (21-hydroxylase) in the human fetus. *J. Clin. Endocrinol. Metab.* **63,** 1145–1150.

Wachtel, S. S. (1983). "H-Y antigen and the Biology of Sex Determination." Grune & Stratton, New York.

Walsh, P. C., Madden, J. D., Harrod, M. J., Goldstein, J. L., McDonald, P. C., and Wilson, J. D. (1974). Familial incomplete male pseudohermaphroditism, Type II, *N. Engl. J. Med.* **291,** 944–949.

Waterman, M. R., and Simpson, E. R. (1989). Regulation of steroid hydroxylase gene expression is multifactorial in nature. *Recent Prog. Horm. Res.* **45,** 533–563.

Waterman, M. R., Lund, J., Ahlgren, R., Wu, D., and Simpson, E. R. (1990). Regulation of steroid hydroxylase gene expression and steroid hormone biosynthesis. *Biochem. Soc. Trans.* **18,** 26–28.

Whorwood, C. B., Montalto, J., Sandars, S. R., and Connelly, J. F. (1990). Urinary 5-enesteroid excretion in non-classical congenital adrenal hyperplasia due to 3β-

hydroxysteroid dehydrogenase deficiency (NC-3BHSD). *J. Steroid Biochem.* **35,** 735–739.

Wilson, S. C., Hodgins, M. B., and Scott, J. S. (1987). Incomplete masculinization due to a deficiency of 17β-hydroxysteroid dehydrogenase: Comparison of prepubertal and peripubertal siblings. *Clin. Endocrinol. (Oxford)* **26,** 459–469.

Wilson, S. C., Oakey, R. E., and Scott, J. S. (1988). Steroid metabolism in testes of patients with incomplete masculinization due to androgen insensitivity or 17β-hydroxysteroid dehydrogenase deficiency and normally differentiated males. *J. Steroid Biochem.* **29,** 649–655.

Winter, J. S. (1991). Clinical, biochemical and molecular aspects of 17-hydroxylase deficiency. *Endocr. Res.* **17,** 53–62.

Winter, J. S., Couch, R. M., Muller, J., Perry, Y. S., Ferreira, P., Baydala, L., and Shackleton, C. H. (1989). Combined 17-hydroxylase and 17,20 desmolase deficiencies: Evidence for synthesis of a defective cytochrome P450c17. *J. Clin. Endocrinol. Metab.* **68,** 309–316.

Wit, J. M., van Hooff, C. O., Thijssen, J. H., and Van den Brande, J. L. (1988). In vivo and in vitro studies in a 46,XY phenotypically female infant with 17-ketosteroid reductase deficiency. *Horm. Metab. Res.* **20,** 367–374.

Wolman, S. R., McMorrow, L. E., Roy, S., Koo, G. C., Wachtel, S. S., and David, R. (1980). Aberrant testicular differentiation in 46,XY gonadal dysgenesis: Morphology, endocrinology, serology. *Hum. Genet.* **55,** 321–325.

Wolthers, B. G., de Vries, I. J., Volmer, M., and Nagel, G. T. (1987). Detection of 3β-hydroxysteroid dehydrogenase deficiency in a newborn by means of urinary steroid analysis. *Clin. Chim. Acta* **169,** 109–116.

Yanase, E., Nagai, T., Kato, M., and Morimoto, S. (1982). A case of male pseudohermaphroditism due to 17α-hydroxylase deficiency. *Jpn. J. Med.* **21,** 128–134.

Yanase, T., Kagimoto, M., Matsui, N., Simpson, E. R., and Waterman, M. R. (1988). Combined 17α-hydroxylase/17,20-lyase deficiency due to a stop codon in the N-terminal region of 17α-hydroxylasecytochrome P-450. *Mol. Cell. Endocrinol.* **59,** 249–253.

Yanase, T., Kagimoto, M., Suzuki, S., Hashiba, K., Simpson, E. R., and Waterman, M. R. (1989). Deletion of a phenylalanine in the N-terminal region of human cytochrome P-450(17α) results in partial combined 17α-hydroxylase/17,20-lyase deficiency. *J. Biol. Chem.* **264,** 18076–18082.

Yanase, T., Waterman, M. R., Zachmann, M., Winter, J. S. D., Simpson, E. R., and Kagimoto, M. (1992). Molecular basis of apparent isolated 17,20-lyase deficiency: Compound heterozygous mutations in the C-terminal region (Arg496−+Cys, Gln461−+Stop) actually cause combined 17α-hydroxylase/17,20-lyase deficiency. *Biochim. Biophys. Acta* **1139,** 275–279.

Zachmann, M. (1972). The evaluation of testicular endocrine function before and in puberty.. *Acta Endocrinol. (Copenhagen)* **70,** Suppl. 164, 5–94.

Zachmann, M. (1989). Clinical steroid research—is it still worthwhile? *Horm. Res.* **31,** 106–114.

Zachmann, M., and Prader, A. (1984). 17,20-Desmolase deficiency. *In* "Adrenal Diseases in Childhood" (M. I. New and L. S. Levine, eds.), Vol. 13, pp. 95–109, Karger, New York.

Zachmann, M., Völlmin, J. A., Mürset, G., Curtus, H. C., and Prader, A. (1970). Unusual type of congenital adrenal hyperplasia probably due to deficiency of 3β-hydroxysteroid dehydrogenase. Case report of a surviving girl and steroid studies. *J. Clin. Endocrinol. Metab.* **30,** 719–726.

Zachmann, M., Völlmin, J. A., Hamilton, W., and Prader, A. (1972). Steroid 17,20-desmolase deficiency. A new cause of male pseudohermaphroditism. *Clin. Endocrinol. (Oxford)* **1,** 369–385.

Zachmann, M., Forest, M. G., and De Peretti, E. (1979). 3β-hydroxysteroid dehydrogenase deficiency. Follow-up study in a girl with pubertal bone age. *Horm. Res.* **11,** 292–302.

Zachmann, M., Werder, E., and Prader, A. (1982). Two types of male pseudohermaphroditism due to 17,20-desmolase deficiency. *J. Clin. Endocrinol. Metab.* **55,** 487–490.

Zachmann, M., Eiholzer, U., Muritano, M., Werder, E., and Manella, B. (1986). Treatment of pubertal gynaecomastia with testolactone. *Acta Endocrinol.* (*Copenhagen*) **112,** Suppl. 279, 218–226.

Zachmann, M., Kempken, B., Anner, I., and Pezzoli, V. (1988). Age-related differences in the excretion of delta5 steroids in two related girls with 3β-hydroxysteroid dehydrogenase deficiency. *Pediatr. Res.* **24,** 543 (abstr.).

Zachmann, M., Kempken, B., Manella, B., and Navarro, E. (1992). Conversion from pure 17,20-desmolase-to combined 17,20-desmolase/17α-hydroxylase deficiency with age. *Acta Endocrinol.* (*Copenhagen*) **127,** 97–99.

Zurbrügg, R., and Neher, R. (1966). Steroidchemische Untersuchungen bei testikulärer Feminisierung. *Ann. Paediatr.* **206,** 213.

17

Molecular Genetics of Steroid 21-Hydroxylase Deficiency

Maria I. New and Christopher Crawford
Department of Pediatrics
Cornell University Medical College
New York, New York 10021

Molecular Genetics of Sex Determination

Congenital adrenal hyperplasia (CAH) is the name given to clinical disorders of adrenal steroidogenesis. They are all transmitted as autosomal recessive traits. The most common enzymatic error, steroid 21-hydroxylase deficiency, increases sex steroid levels in fetal life, causing phenotypic intersex in females.

I. Cortisol Biosynthesis: Enzymes and Defects

Biochemically (steroid) 21-hydroxylation is a component step in the biosynthesis of the potent corticosteroid hormones cortisol and aldosterone. One microsomal-system cytochrome P450 enzyme, cytochrome $P450_{C21}$ or P450XXI, is expressed specifically in the adrenal cortex and, accepting both 17-deoxy- and 17α-hydroxysteroids as substrates, is the form responsible for normal adrenal cortical production of these steroids. Molecular studies isolated specific transcripts from bovine and then human adrenal cortex (White *et al.*, 1984b,c). Duplication of the murine genes was found (White *et al.*, 1984a), and the local arrangement in the human genome was established (Carroll *et al.*, 1985; White *et al.*, 1985). Messenger RNA (mRNA) for apparently the same cytochrome P450 was identified in mouse liver (Amor *et al.*,1985). Extraadrenal 21-hydroxylating activity was deduced by metabolic conversion (Winkel *et al.*, 1980a,b; Speiser *et al.*, 1991) and identified *in vitro* in liver (Müller-Eberhard *et al.*, 1985), kidney, and other peripheral tissues (Casey *et al.*, 1985), and the structural–catalytic distinctness of some of these was established (Murray, 1992). Drug-metabolizing microsomal cytochrome P450 activities were noted in adrenal cortex (Otto *et al.*, 1991; Matsuo *et al.*, 1992).

The presence of a glucocorticoid hormone(s) in blood plasma is essential for the balanced maintenance of many metabolic functions. The cortisol levels circulating in the blood are regulated by the negative feedback system between the adrenal gland and the anterior pituitary gland, in the system called the hypothalamic–pituitary–adrenal (HPA) axis.

A. Normal Adrenal Steroid Synthesis

Five enzymatic steps govern the biosynthesis of cortisol from cholesterol in the adrenal cortex. Inadequate cortisol biosynthesis may be caused by some degree of blockage of any one of the five enzymatic steps, as a result of which the HPA responds with exaggerated release and high plasma levels of adrenocorticotrophic hormone (ACTH). The general histopathological manifestation consequent to chronic high ACTH is hyperplasia of the steroid-synthesizing adrenal cells and increased expression of all ACTH-inducible cell factors, reflected in increased steroidogenesis (Waterman *et al.*, 1986) at all possible (i.e., unblocked) steps. Thus, specific to each enzyme defect there is an abnormal adrenal steroid profile, certain steroids being produced in deficiency, and others in excess. Metabolic and developmental disturbances resulting from altered hormonal actions and biochemical and clinical findings allow the specific enzyme defect to be identified.

B. Biochemistry of Steroid Synthesis

The adrenal cortex secretes steroids of three types: mineralocorticoids, glucocorticoids, and sex steroids. Production of the two principal hormonal corticosteroids, aldosterone and cortisol, both essential for normal metabolic functioning and body electrolyte balance, is under separate regulation: aldosterone release is modulated by the renin–angiotensin system, and cortisol release by the hypothalamic–pituitary axis *via* trophic peptide corticotropin (ACTH). Quantitatively the main adrenal product of the androgen series is dehydroepiandrosterone (DHEA), which is secreted and found in the serum free and sulfate-conjugated. The adrenal cortex in some physiologic and pathological states is able to dissociate cortisol and androgen production, but the exact mechanism of endocrine control of this independence is not yet clear.

Steroid 21-hydroxylase is one of five enzymes of steroid synthesis active in the adrenal cortex (not including enzymes of conjugation or interconversion) and required for cortisol biosynthesis. Three of these enzymes support production of steroid intermediates common to corticosteroid and sex steroid pathways, and are expressed in ovary and testis also: (1) cholesterol side-chain cleavage enzyme (or cholesterol desmolase), (2) steroid 17α-hydroxylase/17,20-lyase, and (3) 3β-hydroxy-5-ene steroid dehydrogenase/Δ^5-Δ^4 isomerase. The remaining two enzymes, (4) steroid 21-hydroxylase and (5) steroid 11β-hydroxylase (the latter also catalyzing 18-hydroxylation and 18-/19-dehydrogenation), specifically support formation of corticosteroids.

1. Cholesterol Desmolase (Cytochrome P450scc)

Cleavage of the C-20,C-22 bond on the cholesterol side chain yields isocaproaldehyde (4-methylpentanal), the oxidized C_6 fragment, and a 20-oxosteroid, pregnenolone. Pregnenolone is the common precursor for all free hormonal steroids. This initial step is normally rate-limiting.

2. Steroid 17α- Hydroxylase/17,20-Lyase (Cytochrome P450c17)

The enzyme steroid 17α-hydroxylase/17,20-lyase acts at ring carbon C-17, the site of attachment of the remaining C_2 short side chain (carbons C-20 and C-21) generic to the pregnane (C_{21}) series. This enzyme performs one or both of two sequential conversions: with pregnenolone as substrate, for example, it first makes a hydroxyl group at the 17α position, and then optionally releases 17α-hydroxypregnenolone as product, or, in a further oxidation, cleaves the side-chain bond, leaving a carbonyl group (17-oxo, i.e., keto configuration at C-17) to yield as its final products acetate and the C_{19} steroid DHEA.

Steroid 17α-hydroxylase/17,20-lyase is thus a key enzyme in the balance of hormonal activity, as it controls production both of 17α-hydroxy C_{21} steroids (precursors of cortisol, the principal glucocorticoid) and of C_{19} steroids, i.e., androgenic steroids, which in turn are the precursors of the C_{18} steroids (estrogens). What factors control this enzyme and what are possible differential effects on one or the other activity? With what independence is either activity modulated? Steroids already with the 17α-hydroxyl group appear to be accepted as substrate for conversion to 17-oxo products, and isolated deficiency of the 17,20-lyase function has been identified clinically. A case has been noted of an isolated 17,20-lyase deficiency diagnosed in childhood that changed to a combined 17,20-lyase/17α-hydroxylase deficiency in adulthood (Zachmann *et al.*, 1992).

3. 3β-Hydroxysteroid Dehydrogenase/$\Delta^{5,4}$-Isomerase

Enzyme 3β-hydroxysteroid dehydrogenase/$\Delta^{5,4}$-isomerase (3β-HSD) converts 3β-hydroxy-5-ene steroids to products with the 3-oxo-4-ene configuration seen on all hormonally active steroids except the estrogens. Thus this oxidation/isomerization is crucial. There are known isoforms, and the existence of isoforms is suspected. The dehydrogenase and isomerase components (of at least one form in mammals) are not separable (Ishii-Ohba *et al.*, 1986) but have differing cofactor redox-state dependence *in vitro*.

The enzyme activity is found in adrenal, placenta, testis and ovary, and many nonglandular tissues. The 3β-HSD enzyme competes with steroid 17α-hydroxylase activity for 17-deoxy substrates and competes again with the 17,20-lyase activity for 17α-hydroxy substrates, thus making two early-pathway branchpoints in steroidogenesis. The kinetics of possible isoforms in different tissues thus may result in changed relative amounts of glucocorticoid and sex steroids produced.

Affecting gonadal as well as adrenal function, defects of all three of the early steroidogenic enzymes are forms of CAH in which there is *limited* sex steroid production. In addition to generating symptoms related to glucocorticoid deficit (cholesterol desmolase or 3β-HSD defects) or excess of hypertensinogenic corticoid precursors (17α-hydroxylase defect), they cause pseudohermaphroditism in the male. In contrast, blocks at the 21-hydroxylase or the 11β-hydroxylating steps, which are located more distally, channel corticosteroid precursors into intact androgen pathways, resulting in *virilization* from adrenal steroid overproduction, i.e., pseudohermaphroditism in the female. The 3β-HSD enzyme defect results in high levels of weakly androgenic steroids that are subject to peripheral conversion to more potent forms that masculinize the genetic female; this is the only form of CAH that may result in genital ambiguity in both sexes.

4. Steroid 21-Hydroxylase (Cytochrome P450c21)

Steroid 21-hydroxylase converts steroid intermediate 17α-hydroxyprogesterone (17-OHP) to 11-deoxycortisol (Reichstein's compound S), an immediate precursor of cortisol, and in the 17-deoxysteroid series, converts progesterone to [11-]deoxycorticosterone (DOC) in the pathway to aldosterone.

Corresponding with the histological division of the adrenal cortex into three zones, the zona glomerulosa (ZG), zona fasciculata (ZF), and zona reticularis, the steroid-synthesizing capacities of the cells of each zone are distinct. The primary functional difference is exhibited between the outer, ZG and the zonae fasciculata/reticularis (ZF): (*a*) The expression and/or activity of the aldosterone synthase cytochrome $P450_{aldo}$ (or P450XIB2), which figures in the production of aldosterone under angiotensin II and serum K^+ concentration regulation in the ZG, is not supported in mitochondria of the ZF. (Elements of this contrast differ between the three species most studied: bovine, rat, and human.) (*b*) Cytochrome P450c17 (17α-hydroxylase/17,20-lyase) is not expressed in the ZG (this does not apply for corticosterone-producing species, in which the entire adrenal cortex is absent this activity). (*c*) Tropism and net steroidogenic capacity change between these regions. The low-output ZG is activated by angiotensin II (A II) and serum K^+ concentration and only secondarily modulated by ACTH, whereas ACTH is the primary modulating factor for the high-output cells of the ZF. In the ZF, the largest cortical zone, 21-hydroxylase deficiency limits rates of cortisol synthesis at all levels of ACTH stimulation, resulting in accumulation and oversecretion of 17-OHP, increased conversion of 17-OHP to androstenedione, and elevated production of progesterone from the minor 17-deoxysteroid pathway. A steroid 21-hydroxylase block occurring in the ZG affects synthesis of aldosterone, causing low serum aldosterone. Low aldosterone, possibly in conjunction with natriuretic effects of high serum 17-OHP or other steroids present in abnormal concentration, results in renal salt wasting (inadequate distal tubular/collecting duct reabsorption of sodium from the renal filtrate in exchange for potassium/hydrogen ions). Manifesting as electrolyte imbalance (low [Na^+], very high [K^+], and metabolic acidosis) and hypovolemia, and aggravated by the primary glucocorticoid lack, salt wasting (SW) is a feature of some 75% of cases of 21-hydroxylase deficiency.

5. Steroid 11β-Hydroxylase (Cytochrome $P450_{11\beta}$)

Accepting primarily 21-hydroxylated precursors in the cortisol and 17-deoxysteroid pathways in the ZF, the 11β-hydroxylase (cytochrome $P450_{11\beta}$ or P450XIB1) converts compound S (11-deoxycortisol) to cortisol and DOC to corticosterone (compound B). In the ZG the closely related isoform to P450XIB1, P450XIB2, or aldosterone synthase is expressed; crucial to its distinct glandular nature, this zone uses the B2 enzyme isoform (possibly in conjunction with B1) to successively convert (*a*) DOC to *either* B (i.e., *via* 11β-hydroxylation) *or* to 18-hydroxy-11-deoxycorticosterone (via 18-hydroxylation), then from either of these intermediates to 18-hydroxycorticosterone and (*b*) 18-hydroxycorticosterone to aldosterone. Blocked conversion of DOC in the ZF causes increased secretion and plasma concentration of this 17-deoxysteroid precursor, itself a moderately potent sodium-retaining hormone. Therefore, when steroid 11β-hydroxylase deficiency occurs, in addition to androgen excess and virilization as in 21-hydroxylase deficiency, volume expansion and hypertension result rather than salt wasting.

C. P450 Cytochromes

Of the enzymes just described, in fact, four are hemeproteins of the cytochrome P450 class (Fig. 1). Besides cytochrome P450c21, cytochrome P450c17 is also a microsomal-type P450. Cytochrome P450c17 is the single protein species supporting steroid 17α-hydroxylase and 17,20-lyase activities. Cholesterol side-chain cleavage enzyme, P450scc, governing the initial and rate-limiting step in steroidogenesis, is a mitochondrial cytochrome P450, located on the inner membrane of the organelle. The two isoforms B1 and B2 of cytochrome $P450_{11\beta}$, steroid 11β-hydroxylase and aldosterone synthase, are able to perform steroid 11β- and 18-hydroxylations and also steroid 18- and 19-oxidations, and, like P450scc, are located on the inner mitochondrial membrane.

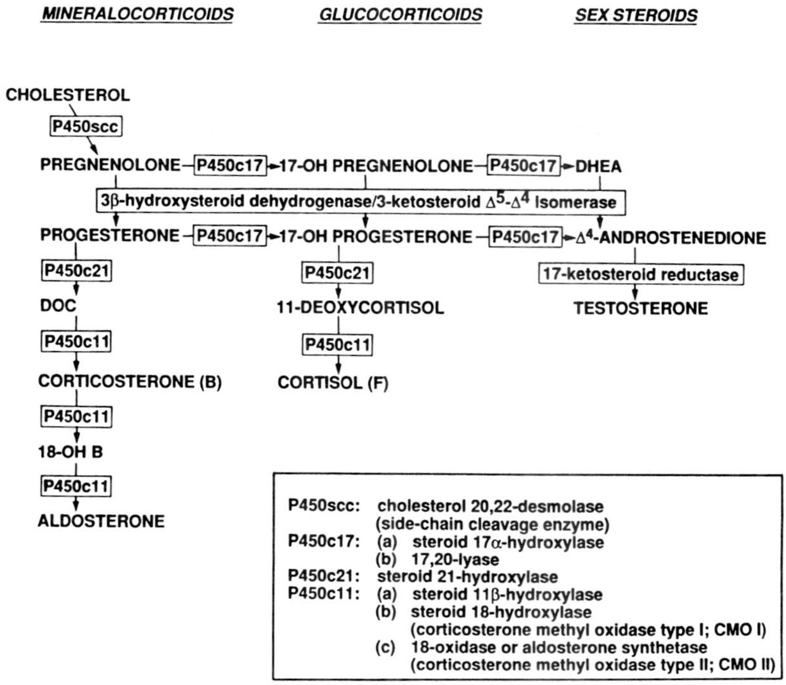

Figure 1 Scheme of adrenal steroidogenesis showing the enzymes responsible for each step as labeled. Abbreviations: 17-OH, 17α-hydroxy; DHEA, dehydroepiandrosterone; DOC, deoxycorticosterone; 18-OH, 18-hydroxy.

Steroid 21-hydroxylase, cytochrome P450c21, has a molecular weight of ~52,000 daltons and in the cell is inserted in the membrane of the endoplasmic reticulum. It accepts the sequential transfer of two electrons from a NADPH-dependent cytochrome P450 reductase, inserts one atom of free oxygen at the C—H bond of C-21 to yield C—OH, and reduces the other to be hydrated to a molecule of water.

Cytochrome P450 purified from bovine adrenocortical microsomes (determined to be one type by single banding on SDS–gel electrophoresis) catalyzed 21-hydroxylation of 17α-hydroxyprogesterone to 11-deoxycortisol (cortisol pathway), and 21-hydroxylation of progesterone to DOC (aldosterone pathway) and current opinion is that a single enzyme protein functions in both pathways in the human adrenal cortex (Kominami *et al.,* 1980). On the other hand, some studies have shown differing enzyme kinetics for conversion of 17α-hydroxysteroids and 17-deoxysteroids (Narasimhulu, 1989); the reason for this has not been clarified.

Expression studies showed no appreciable levels of mRNA for cytochrome P450c21 in human liver, placenta, or testis (although in liver in the mouse (Amor *et al.*, 1985), so extinction of the cytochrome P450c21 gene in all other tissues is difficult to confirm). Presence in kidney of some form of 21-hydroxylase seems likely.

II. Clinical Forms of 21-Hydroxylase Deficiency

Three categories of steroid 21-hydroxylase deficiency are recognized: the classical defect ("virilizing CAH"), which has two phenotypic divisions, *salt-wasting* or *simple virilizing,* depending on the individual capacity to adequately retain sodium, and the attenuated (late-onset or nonclassical defect) (New *et al.*, 1989). In both the salt-wasting and the simple virilizing classical defect, the cardinal feature is ambiguous genitalia in the female at birth, and accelerated androgen-dependent growth and development in either sex from infancy onward. The nonclassical defect has a spectrum of types according to the hyperandrogenic signs manifested, starting with signs as early as 6 months of age, extending through puberty, and including adult patients who lack overt hyperandrogenic symptoms but in whom the biochemical defect can be demonstrated.

A. Classical CAH Due to 21-Hydroxylase Deficiency

Any 21-hydroxylase defect is detected hormonally at all ages by means of an ACTH test (Fig. 2). Increased serum concentration of 17-OHP or increased saliva level of 17-OHP is diagnostic in classical and nonclassical cases. Increased urinary excretion of metabolites of androgens and 17-OHP—17-oxo(keto)steroids, and pregnanetriol—is used more commonly to be measured, but these older tests are still good and are diagnostic.

Filter paper impregnated with capillary blood from the neonate is the sampling format in most classical CAH screening programs: the dried sample may be sent by mail to a central lab (cf. the Guthrie card for hypothyroid screening). Salivary sampling (about 5 ml collected in a sample vial over a 10- to 20-min period) is noninvasive and early morning salivary 17-OHP is elevated in both classical and nonclassical 21-hydroxylase deficiency; the low cost of routine assays for 17-OHP makes this procedure suited to population screening.

In the genetic female with classical 21-hydroxylase deficiency, the developing external genitalia are exposed to excessive concentrations of

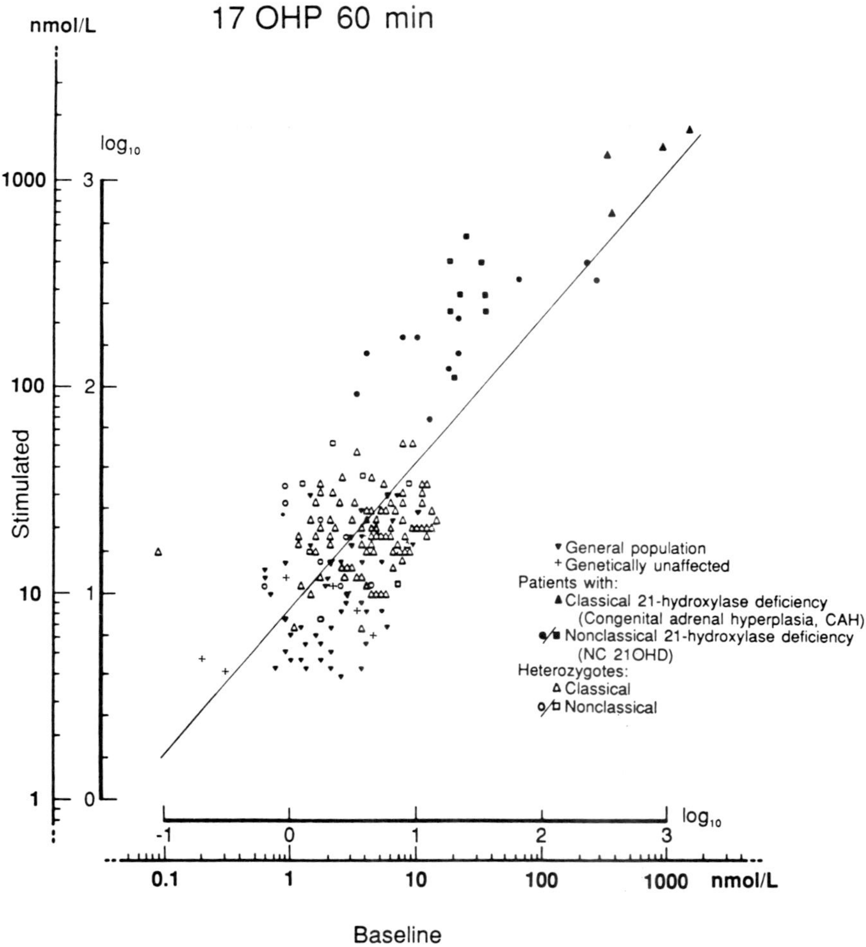

Figure 2 Nomogram for determining 21-hydroxylase deficiency status by graphing baseline versus post 60-min ACTH-stimulated* serum 17α-hydroxyprogesterone concentrations. *, intravenous bolus injection of 0.25 mg Cortrosyn (synthetic ACTH$_{1\text{-}24}$) (New *et al.*, 1983).

adrenal androgens secreted by the hyperplastic fetal adrenal cortex; in the genetic male these serum concentrations are normally furnished by the fetal gonads, and the affected genetic male suffers no abnormality in development of external genitalia. External genitalia in the genetic female range at birth from mildly ambiguous to completely virilized (female pseudohermaphrodite).

Internal genitalia in the female are normal. In males the anti-Müllerian hormone secreted by the fetal testes (Sertoli cells) suppresses the development of the uterus and fallopian tubes from the Müllerian ducts. As genetic females do not have testes, but ovaries, they always have uterus and fallopian tubes despite extreme virilization of the external genitalia. Though circulating adrenal androgens are elevated, Wolffian duct development is not observed in affected females.

Postnatally, continued excess androgen production in boys and girls may result in several symptoms and signs, e.g., (*a*) rapid somatic growth with enhanced muscularity ("infant Hercules") and accelerated epiphyseal maturation (advanced bone age); (*b*) progressive penile or clitoral enlargement; (*c*) early appearance of facial, axillary, and pubic hair; (*d*) deepening voice; (*e*) acne; and (*f*) frontal/temporal balding. In the absence of treatment or in cases of treatment with insufficient hormonal control, there is continuing bone age advancement and early closure of the epiphyses. Thus the patients are tall children but short adults.

The clinical state called salt wasting is the result of the deficiency of aldosterone—the sodium-retaining hormone necessary for maintenance of body fluid volume and electrolyte equilibrium. The salt-wasting form appears to be more frequent than the simple virilizing form of classical 21-hydroxylase deficiency. Hormonal indicators of salt wasting are low serum aldosterone with concomitantly high plasma renin activity (PRA). Urinary pH 1 aldosterone is also low in the salt-wasting patient.

Identification of salt-wasting CAH in males must await an overt episode of adrenal insufficiency because the genitalia are normal and do not flag the possible diagnosis of CAH as in the female with ambiguous genitalia. This may result in delay of diagnosis and potential adrenal crisis requiring immediate therapy. Identification of CAH in the non-salt-wasting male will come with the first signs of androgen excess, later still. The increase in percentage of salt-wasting cases in recent years may be attributed to enhanced ascertainment and increased survival, and to diagnostic and therapeutic advances. Yet a persisting statistic of greater numbers of female cases suggests that identification of CAH, especially in males, is still incomplete.

Past opinion also correlated salt wasting with the extent of virilization in females, but in fact for each degree of virilism, simple virilizing and salt-wasting patients may be found. Thus, because of the importance of ruling out salt wasting in 21-hydroxylase deficiency, even mildly virilized female newborns should be scrutinized for signs of adrenal crisis within the first few weeks of life. In all male infants, the earliest signs of adrenal crisis should be recognized for a baby to receive full endocrine assessment and therapy.

B. Nonclassical 21-Hydroxylase Deficiency

Nonclassical 21-hydroxylase deficiency (NC21-OHD) results from allelic variants of the *CYP21* gene defect milder than those causing virilization of genetic females prenatally (classical 21-OHD); symptoms consequent to androgen elevation occur postnatally. Nonclassical defects are common and show increased frequency in certain ethnic groups.

Clinical expression of even attenuated androgen excess can manifest with specific symptoms of surprisingly rapid onset at any stage postnatally—in infancy, childhood, peripubertally, or in adulthood.

An attenuated, late-onset form of adrenal hyperplasia was first suspected by gynecologists who used glucocorticoids in women with physical signs of hyperandrogenism, including infertility. High levels of non-21-hydroxylated cortisol precursors in the urine, further increasing after ACTH stimulation and suppressing after glucocorticoid administration, confirmed the presence of an adrenal 21-hydroxylase deficiency in a eumenorrheic virilized woman first reported by Decourt *et al.* (1957).

Increasingly precise assessment of steroid 21-hydroxylase defects became possible with the development of a radioimmunoassay for 17-OHP, the major substrate of the 21-hydroxylase enzyme in the adrenal zona fasciculata. Through family studies of classical 21-OHD it became apparent that NC21-OHD was caused by different alleles of the same gene. Thus NC21-OHD is an allelic variant at the 21-hydroxylase locus. Distinct HLA antigen associations for the forms of 21-OHD were determined, some HLA associations for the nonclassical disorder differing according to ethnic composition of the families.

NC21-OHD may present clinically at any age. It can result in premature development of pubic hair in children; to our knowledge, the youngest such patient was noted to have pubic hair at 6 months of age. In a review of 23 cases presenting to The New York Hospital–Cornell Medical Center for evaluation of premature pubarche, 17-OHP responses to ACTH stimulation in the range of NC21-OHD affected were found in 7 children (30%) (Temeck *et al.*, 1987). Other investigators looking at a pediatric high-risk group (children with premature adrenarche) found that only 1 in 15 demonstrated an ACTH-stimulated 17-OHP response in the range of NC21-OHD affected, the remainder falling in the range of heterozygote–homozygous unaffected (Fig. 2).

ACTH testing has led to the diagnosis of NC21-OHD in adolescents suffering with cystic acne. This diagnosis should be sought and low-dose glucocorticoid replacement therapy considered in cases of cystic acne refractory to oral antibiotics and retinoic acid. In one study comparing the responses of 11 female patients with acne and 8 female controls to a 24-h infusion of ACTH, urinary excretion of pregnanetriol was elevated in 6 patients. Yet in another study of young adult females with acne and/or hirsutism, no abnormally high 17-OHP responses to low-dose ACTH stimulation (in this case after a single adrenal-suppressing dose of dexamethasone the night before) suggested NC21-OHD in any of 31 patients. This variance may relate to differences in ethnic composition of the groups, which is not generally a reported feature. Male-pattern baldness in other young women has also been the presenting symptom.

Menarche may be normal or delayed in NC21-OHD. Laparoscopy or abdominal ultrasound examination of adolescent or adult women with NC21-OHD in many cases reveals abnormal enlargement or cystic appearance of the ovaries. Conversely, a subgroup of women with secondary amenorrhea or polycystic ovarian disease (PCO) will be found (when tested with ACTH) to be affected with an adrenal enzyme deficiency (NC21-OHD or nonclassical 3β-HSD deficiency). Serum sex-steroid excess of adrenal origin could disrupt cyclic gonadotropin release or have direct follicular effects in the ovary; either of these conditions could generate formation of ovarian cysts (PCO), which would autonomously sustain androgenemia (Futterweit, 1984).

Retrospective chart analysis of 108 young women presenting to this institution for endocrinologic evaluation of hirsutism and oligomenorrhea revealed NC21-OHD as the etiology in 16 (14%) (Pang *et al.*, 1985). In other studies of hirsute oligomenorrheic women, the prevalence of an adrenal enzyme deficiency ranged from 1.2 to 30%. Again, the variance in frequencies may relate to differences in ethnic composition of the groups studied.

In boys, early beard growth, acne, and growth spurt may be detected. A reliable constellation of physical signs of adrenal (as opposed to testicular) androgen excess in boys is the presence of pubic hair, enlarged phallus, and relatively small testes. In men, mild androgen excess is difficult to evaluate endocrinologically; adrenal sex steroid-induced suppression of the hypothalamic–pituitary–gonadal axis, and direct intraglandular effects in the testis have both been theorized, either possibly resulting in diminished fertility. Oligospermia and subfertility have been reported in men with NC21-OHD and reversal of infertility with glucocorticoid treatment in two men.

A subset of NC-21OHD individuals are overtly asymptomatic when detected (usually as part of a family study), but based on longitudinal follow-up of such patients, it is found that symptoms of hyperandrogenism may wax and wane. The gene defect and the serum androgen elevations in these so-called "cryptic" 21-hydroxylase deficient subjects can be the same as in floridly symptomatic nonclassical patients.

Some children affected with NC21-OHD show advanced bone age and accelerated linear growth velocity, and ultimately may be shorter than the height prediction based on mid-parental height (even when adjusted for linear growth percentiles mapped for the child before any onset of androgen excess symptoms).

III. Genes for Steroid 21-Hydroxylase (Cytochrome P450c21)

The gene for cytochrome P450c21 normally exists in two copies, as an active gene (*CYP21*) and a pseudogene (*CYP21P*) in the haploid genome. These two genes alternate with two related genes, both active: the genes for the two isotypes of the component C4 of serum complement, *C4A* and *C4B*. The pseudogene, *CYP21P*, immediately 3′ of *C4A* formerly was termed the A gene (*CYP21A*), and the active gene, *CYP21*, immediately 3′ of *C4B*, was termed the B gene (*CYP21B*). The arrangement of these genes could have resulted from a duplication event affecting a primordial 21-hydroxylase and serum *C4* gene pair occurring once at some early point in mammalian evolution or occurring independently in different lines of the mammalian radiation. The C4 homologs retain over 99% similarity in the human, and the CYP21 homologs, 98%.

The tandem paired arrangement *C4A-CYP21P–C4B-CYP21* increases the possibility of chromatid strand misalignment during meiosis. Crossing-over will then result in the asymmetrical production of gametes, one type with a single *C4–CYP21* pair and the other with a triplicated *C4–CYP21* pair. Both types have been found. The first mutations analyzed involved deletion of a length of DNA encompassing the entire active or pseudogene gene sequence. Occurring homozygously the haplotype CYP21B-del/C4B-null (also deleted) was associated with severe salt-wasting steroid 21-hydroxylase deficiency, giving evidence that the B (active) sequence was necessary (Fig. 3) for the enzyme activity to be expressed, whereas the converse deletion, CYP21A-del/C4A-null, was observed in individuals who were hormonally normal, giving evidence that the A sequence was not essential. Sequencing of the *CYP21* genes confirmed that the A gene carried deleterious mutations; thus it had ceased to be actively expressed.

Further investigation revealed that the enzyme deficiency could also be caused by nondeletional lesions to the *CYP21* sequence. Molecular characterization of numbers of mutant genes has revealed that sequence aberrations correspond almost without exception to those base differ-

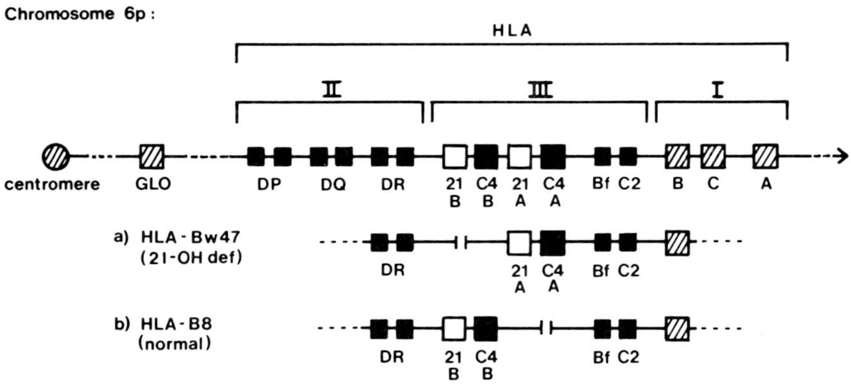

Figure 3 HLA haplotypes with deletions (confirmed by DNA hybrid banding patterns of molecular probes + various digests) from which the paired duplicated configuration of *21* (21-hydroxylase) and *C4* (serum complement component 4) genes was determined and from which it was deduced (*a*) that the *21B* gene is active (this haplotype associated with salt-wasting 21-hydroxylase deficiency) and (*b*) that the *21A* gene is nonessential (this haplotype with endocrinologically normal phenotype). Each of these two haplotypes is serologically C4-half null, the HLA-Bw47 expressing only the C4A isotype and the HLA-B8 expressing only the C4B isotype. There are many more HLA loci than shown here (not to scale). GLO, glyoxalase locus (loosely linked to HLA/21-hydroxylase).

ences already resident on the *CYP21P* pseudogene. There are some 11 mutations in the A pseudogene potentially transferable to the active B gene that would terminate protein synthesis or result in amino acid substitutions. Gene conversions could also involve regulatory segments of the *CYP21* genes, affecting the transcribed levels of mRNA. Other *de novo* point mutations or small-scale inserts/deletions could occur anywhere throughout the structural or regulatory sequences of the gene, again affecting transcription of the gene or processing of mRNA, or resulting in amino acid substitutions altering enzymatic function.

A. Two Homologs: Active Gene *CYP21* and Pseudogene *CYP21P*

The apoprotein of hemeprotein cytochrome P450c21 is encoded by the *CYP21* gene. In the normal haploid genome, *CYP21* is present in two copies, highly correspondent sequences on each sixth chromosome in close proximity and alternating with genes *C4A* and *C4B*. Both of these are active in the human, encoding two isotypes of the fourth component of serum complement. The B homolog, immediately 3′ of *C4B*, encodes the apoprotein. Differences in the nucleotide sequence of the homolog located 3′ of *C4A* preclude its expression. In addition to a number of base

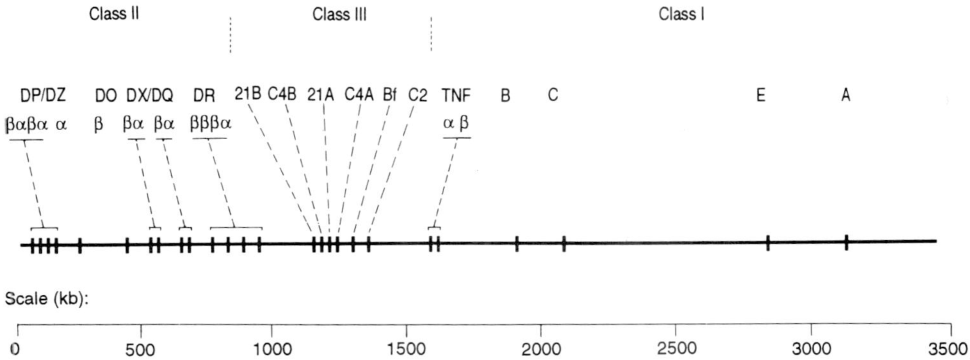

Figure 4 Map of HLA region [to scale—mapped by pulsed-field gel electrophoresis (PFGE)] indicating only some of the major gene loci of each class: class I, antigens A, B, C, and E; class II, major D antigens (DM and DN chains not shown); class III, serum complement component 2 (C2), properdin factor B (Bf), and the paired duplicate arrangement of *21* and *C4* genes. Linkage between HLA-B and 21 is tighter than linkage between HLA-DR and *21*. Mapping by recombinative distance would place the segment C2–Bf–C4A–21A–C4B–21B between the DR and the B loci in the opposite orientation. TNF, tumor necrosis factor(s); kb, kilobase(pair) physical distance.

changes that would cause substitution of the amino acid at the corresponding position of any possible translation product, there is (*a*) an early in-frame stop codon that would limit polypeptide synthesis to a truncated, completely dysfunctional enzyme (assuming a stable translatable mRNA in the cytosol), and (*b*) other base changes that would likely prevent a mature mRNA from being transcribed, processed, or exported from the nucleus. This sequence is categorized and labeled as a pseudogene (*CYP21P*).

1. *CYP21* and *CYP21P* in Alternation with *C4* Genes

Assembled in close linkage on a single chromosomal segment, the major histocompatibility complex (MHC) contains genes encoding cell-surface glycoproteins concerned with antigen presentation to T lymphocytes, recognition of self and nonself, and graft rejection (histocompatibility) (Klein *et al.*, 1990). In certain species such as the human, MHC antigens exhibit an unusual degree of polymorphism. The human MHC, called the HLA (human leukocyte antigen-A, or antigens) system, occupies nearly 4000 kb (4.0 Mb, or about 0.2% of the genome). It is situated in distal p21.3 on human chromosome 6 (Fig. 4).

Antigens called HLA class I (classical antigens HLA-A, B, and C) are expressed at some level in nearly all cell types. Class II antigens (HLA-D/

DR) are more restricted in expression; these are found on the plasma membrane of B lymphocytes, macrophages and dendritic cells, and activated helper T lymphocytes. Class I and II are separated by a gap of about 1000 kb; this is the class III region. At least 36 genes are known to occupy one 680 kb stretch of class III (Trowsdale *et al.*, 1991), within 120 kb of which are located the genes for properdin factor Bf and serum complement component C2, and, alternating with the two isotypes of serum complement component C4, the 21-hydroxylase gene *CYP21* and its highly correspondent pseudogene *CYP21P*. Two recently identified genes in this 120-kb segment are *RD* (Speiser and White, 1989), and another called *XB* (Morel *et al.*, 1989c) or *OSG* (opposite strand gene) (Spence *et al.*, 1989) that actually overlaps the tail end of the active *CYP21* on the complementary strand. A note on nomenclature: In the earlier literature the 21-hydroxylase genes were variously referred to as *21A* and *21B, OH21A* and *OH21B, CA21H-A* and *CA21H-B, P450C21A* and *P450C21B*. Gene labeling for all cytochromes P450 was systematized by the scheme CYP- plus an arabic numeral(s) specifying the P450 gene family. This dictated the names *CYP21A* and *CYP21B* (Nebert *et al.*, 1987). For all of these the letter B denotes the gene 3′ of complement *C4B*, and A the pseudogene, immediately 3′ of *C4A*. The current nomenclature indicates instead the functionality expected at each locus: *CYP21* (active gene) and *CYP21P* (pseudogene) (Nebert *et al.*, 1989).

2. Evolution of the *CYP21/C4* Tandem Array

Human *C4A* and *C4B* are more than 99% identical, and the virtual amino acid sequence deduced for *CYP21P* is 98% identical to that of the P450c21 apoprotein coded by *CYP21*. A gene arrangement similar to the human *C4A–CYP21P–C4B–CYP21* arrangement is known in the mouse, but with differences in the genes inactive or expressed: *Slp–Cyp21–C4–Cyp21P* [Slp, sex-limited protein, originally detected only in adult male mice with certain haplotypes, was thought to be inactive, but may have limited hemolytic function (van den Berg *et al.*, 1992)]; the active *Cyp21* gene and pseudogene positions are reversed in the mouse. Duplication of *C4* is seen in a number of mammals, including a nonhuman primate (Granados *et al.*, 1987), but not others (guinea pig, dog, cat). One view is that the human and mouse arrangements came down from the same tandem array originating in a single duplication event of a primordial **CYP21-*C4* gene unit. This duplication could have occurred sometime around the eutherian radiation in vertebrate evolution (ca. 70 million years ago). An alternative view—that the murine *Slp–C4* and human *C4A–C4B* duplications are independent evolutionary events—is supported with sequence data (Dodds and Law, 1990). Possible further support of this position is provided by occurrence, in the Sardinian population, of a single C4A/single 21-OH gene arrangement. This is putatively an "ancestral" haplotype which, because of nonalignment with the standard double paired arrangement, tends not to enter into recombination and thus has been maintained (Contu *et al.*, 1989) as may be the case for the other noncanonical haplotypes (with *C4* deletion/duplication).

Base changes at matching positions between the human *CYP21* homologs are seen in coding, flanking, and intronic sequences (only 88 in the 5.1-kb length from 1.5 kb upstream to 0.2 kb downstream of the A/B gene), i.e., a divergence of 2%. The cumulative changes expected for two such genes, one free of the selective restraints of having a functional translation product, evolving independently over the time period assumed for a common human-mouse duplication event (position 1 of the two above) would be on the order of 20%. If this position is adopted, it would have to be accepted that some mechanism is causing the *CYP21, Cyp21, C4A/B,* and *Slp/C4* gene pairs to evolve in parallel.

The genetic constitution of the more remotely adjoining HLA class II region, which is still under study, is thought to have some bearing on the events in class III. The class II gene subtypes DP, DN (formerly DZ/DO), DQ, and DR have been further resolved into cognate sequences, some of which are expressed genes and others, pseudogenes (*DPA2* and *DPB2, DQA2* and *DQB2,* and *DRB2*). Full nucleotide sequences of many alleles of the major class I and class II gene are available (Zemmour and Parham, 1992; Marsh and Bodmer, 1992). It has been speculated that these pseudogenes serve as a latent fund of sequence variation for the active gene counterparts (accessed by a mechanism involving nonreciprocal intragenic recombination; see below). Gene interactions in the D/DR region may suggest more about the events that have produced the present organization of 21-hydroxylase and complement *C4* genes, the mechanisms that have conserved correspondence of the *CYP21* gene and pseudogene sequences, and factors now at work in a genetic region of apparent unusually high local activity. The increasing resolving power of recent studies also suggests that such intricate gene interactions may not be specific to this region (probably the best scrutinized of the genome) but may take place throughout many or all chromosomal segments (Brändle *et al.*, 1992). Further data will be necessary to resolve this point.

IV. Mutations and Rearrangements

Two recurrent mechanisms seem to account for the vast majority of *CYP21* gene defects:

1. Unequal crossing-over (breakage and re-joining between chromatid strands misaligned) during meiosis I produces asymmetrical daughter chromatids and gamete pairs, one missing and the other with a redundancy of the unaligned DNA segment in question (Fig. 5). Gene deletions were the first mutations to be identified. The number of *CYP21*-deleted chromosomes to occur as a percentage of all *CYP21* defects may be a function of the population under examination, accounting in different reports for from 5 to 35–40%. Chromosomes with further gene duplications, resulting in triplicated and even quadruplicated *C4–CYP21* pairs, have been identified, putatively complementary division products to deletions created by this mechanism. Misalignment is invited by the tandem *C4A–CYP21P—C4B–CYP21* arrangement and the high degree of conservation persisting in many sections between homologous sequences.

As evidence in support of this mechanism, complementary gametes with triplicate (Werkmeister *et al.*, 1986) and quadruplicate (Collier *et al.*, 1989) *C4–21* pairs have been identified.

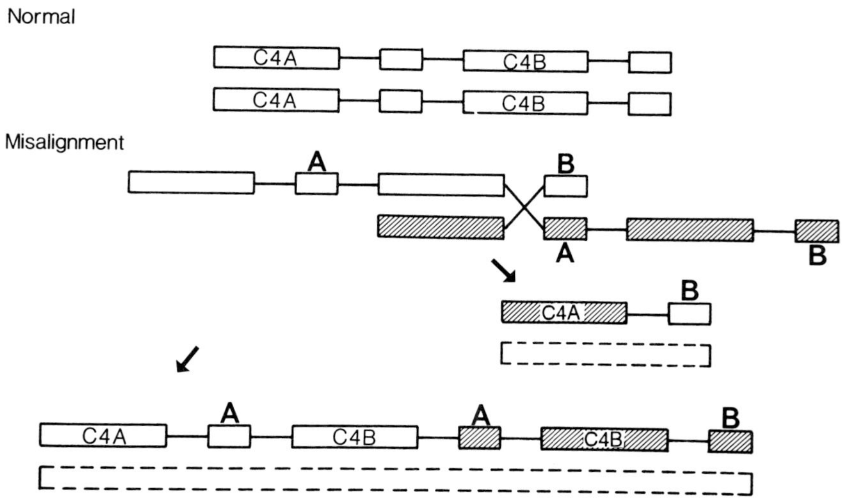

Figure 5 Schematic showing how misalignment of homologous genes (alternating *C4* and 21-hydroxylase in tandem) occurs during meiosis with unequal crossing-over, resulting in complementary gametes, one with a deleted pair (*C4A–21B*) and the other with an extra duplication (*C4A–21A–C4B–21A–C4B–21B*). In subsequent meioses reduction to zero and expansion to four or more pairs are possible. Homogenization of DNA sequences at each locus by gene conversion also occurs, so that the isotypes of *C4* and the number of active *21* and pseudogenes may change.

2. Intragenic recombination events are also known to occur—more complex events whereby short segments of one of two homologs appear to be written over to the other without reciprocal transfer of the replaced sequence, called gene conversion. By this phenomenon, *CYP21* defects are generated by any of the mutations resident on the pseudogene *CYP21P* being transferred to the active *CYP21* gene. Similar transfer events between other related gene pairs have been noted in the literature, but if there is a general mechanism in mammalian cell division and gametogenesis, it is not yet well understood.

Factors that may be promoting exchange of genetic elements in this region have not been clearly identified, and it is not known why gene conversions from *CYP21P* to *CYP21* occur with such frequency. A number of duplicated gene pairs are found nearby in HLA, and for some (e.g., the *DQ* genes), the pseudogene appears possibly to function as a reservoir of diversity for the active sequence.

A. Deletion/Duplication Events

Severe 21-hydroxylase deficiency with phenotypic salt wasting is seen in 15–20% of all occurrences of the rare HLA-A3,Bw47,DR7 haplotype, which also fails to express one C4 isotype. It was thought that a single DNA deletion between HLA-B and HLA-DR resulting in the null *C4* might extend to a structural gene for the adrenal 21-hydroxylase enzyme. A cDNA clone coding for part of the bovine adrenal 21-hydroxylase enzyme, bovine cytochrome P450c21, was used to probe for a homologous sequence(s) in the human genome. DNA from normal subjects and from salt-wasting 21-hydroxylase deficiency patients typed as Bw47 (including one Bw47/Bw47 homozygote) was subjected to several single restriction enzyme digestions, and the band patterns analyzed and compared.

Endonuclease *Taq*I yielded two fragments, 3.2 and 3.7 kb in length, hybridizing with the probe at equal intensity in normal samples; of these the 3.7-kb band was weak or absent in all patients. Consistent with the hypothesized single deletion, this result also suggested that the human P450c21 (21-hydroxylase) gene was present in two copies. Detailed mapping of the region using overlapping cosmid clones established the existence of a coding sequence for P450c21 immediately adjacent (3′) to each of *C4A* and *C4B;* the pseudogene (A gene) carries the 3.2-kb sequence cut by *Taq*I, whereas the 3.7-kb fragment of *Taq*I identifies the active gene (B gene). The four genes of this tandem arrangement are transcribed in the same direction.

1. Deletion of the B Gene

Diminished or absent intensity of the 3.7-kb *Taq*I band signaled absence of the B sequence on one or both chromosomes 6 in all of the individuals hemizygous or homozygous for a DNA segment deletion evidenced by absence of one C4 isotype. This was consistent with the removal of the B gene sequence in the same DNA segment. All of these individuals exhibited the salt-wasting phenotype. This indicated that the B sequence was necessary for adrenal steroid 21-hydroxylation (the sequence carried at the *CYP21* locus was the active coding sequence for a functional steroid 21-hydroxylating enzyme).

2. Deletion of the A Gene

Individuals homozygous for HLA-A1,B8,DR3 with a null C4A isotype normally have no evidence of an adrenal defect. Absence of the 3.2-kb *Taq*I band (identified with the A sequence) in DNA analysis of these subjects suggested the existence in the normal genome of a second copy of the gene (*CYP21A*) not essential for hormonal function. Deletion of *C4A–CYP21A* has been confirmed on the A1,B8,DR3 haplotype, and sequence analysis by Higashi *et al.* (1986) and White *et al.* (1986) of the structural gene gave evidence of the nonexpression of a product; the A gene is now labeled *CYP21P*.

Although *CYP21P* and *CYP21* are about 98% similar in coding sequences, the differences in *CYP21P* are significant: there are two shifts of reading frame, arising from an 8-bp deletion and a single base pair insertion, as well as one point mutation introducing a termination codon, and several point mutations causing nonconservative amino acid substitutions.

The tandem arrangement *C4A–CYP21P–C4B–CYP21* makes misalignment of homologous chromosomal segments possible during meiosis. Unequal crossing-over during gametogenesis could result in one or three *C4–CYP21* sets on a haplotype, producing, for example, the single-set *C4/CYP21* configurations carried on the HLA-A1,B8,DR3 and A3,Bw47,DR7 haplotypes. Presence of a third *C4/CYP21* set on the haplotype HLA-B14,DR1, associated with nonclassical 21-hydroxylase deficiency, first suggested by banding patterns for a third C4 protein on serum electrophoresis, has been confirmed by analysis of DNA from individuals carrying this haplotype. This third *C4/CYP21* set consists of an extra *C4B* gene and an extra *CYP21P* or *CYP21P*-like gene (based on the size of the extra restriction fragments). If the extra *CYP21* gene and the normally present *CYP21P* gene, do not contribute to the development of the nonclassical phenotype, a disease-causing mutation would still be expected on the B gene (as has been found; see below). An interference effect on the expression of (normal or mutant) 21-hydroxylase B genes, a possible involvement of the regulatory sequences of extra *CYP21* gene copies, could be postulated, but no such phenomenon has been noted.

Other 21-hydroxylase deficiency patients negative for HLA-Bw47 have been examined by hybridization analysis using 21-hydroxylase and/or C4 probes. *CYP21* gene deletions are found in approximately one-fourth of chromosomes in 21-hydroxylase CAH, from one-fifth to over one-third patients at different geographical locations; the majority of these deletions include *C4B*. In one family, a chromosomal segment has been noted on which a second *CYP21P* gene has been substituted at the *CYP21* locus. All patients with homozygous deletions of *CYP21* have salt-wasting 21-hydroxylase deficiency. On a chromosome with deleted genes *C4B* and *CYP21* (as on the A3,Bw47,DR7 haplotype), the splicing of *CYP21P*, the A gene, directly to the region 3′ of *CYP21* alters the spacing of endonuclease recognition sites, and may create hybridizing fragments whose lengths may suggest the presence of *CYP21*, i.e., a B gene. This possibility of apparently canonical fragment lengths arising factitiously must be kept in mind in the interpretation of banding patterns in any conventional DNA analysis.

3. New Confirmation of Deletional Mechanism

It is now accepted that classical deficiency alleles of steroid 21-hydroxylase are frequently caused by deletions of *CYP21* (B gene) and also by gene conversions transferring damaging mutations from the *CYP21P* pseudogene to *CYP21*. The placement of recognition sites for certain restriction enzymes can distinguish the normal gene and normal pseudogene in the DNA under analysis [most notably the 3.7-kb (gene) and 3.2-kb (pseudogene) *Taq*I fragments]. Gene conversions involving such sites might be confused with actual deletions of *CYP21*. In order to determine the incidence of this type of gene conversion, a study was constructed examining 15 chromosomes (from 13 families) lacking *Taq*I 3.7-kb fragments (White *et al.*, 1988). A 21-hydroxylase probe failed to hybridize to any fragment obtained after DNA samples were digested with the following other endonuclease enzymes (expected *CYP21* fragment length given in parentheses): *Kpn*I (2.9 kb) in all chromosomes, *Bgl*II/*Eco*RI (2.4 kb) in 14 of 15, *Bgl*II/*Eco*RI (10 kb) in 13 of 15, and *Eco*RI (12 kb) in 13 of 15 chromosomes. When assayed with a C4 probe, 13 of the 15 chromosomes lacked 6.0- or 5.4-kb *Taq*I fragments corresponding to allotypes of *C4B*.

Thus 2 of the 15 chromosomes lacking the 3.7-kb *Taq*I band did not carry actual deletions, whereas 13 of the 15 are missing about 30 kb of DNA, representing one *C4–CYP21* pair, with the *C4A* and a single *CYP21P* or CYP21P-like gene left behind. Further analysis with oligonucleotide probes recognizing mutations on *CYP21P* showed that in all 13 cases the remaining *CYP21* gene did carry the 8-bp deletion characteristic of the *CYP21P* sequence.

4. Variability in Number of Gene Repeats

One study conducted on DNA from members of 32 British families (Collier *et al.*, 1989) sought to uncover the molecular pathology in occurrences of steroid 21-hydroxylase deficiency. The rate of occurrence of deletions of the *CYP21B* gene determined for this population is the highest to date. In 6 of the families there was parental consanguinity and the affected children were HLA homozygous. Affected haplotypes were thus counted as 58 and the results found were as follows:

Mutation Category I: Results of short-range DNA restriction mapping suggested that the *CYP21B* gene was deleted in 25/58 (43%) of disease haplotypes. About 3/4 (76%) of these haplotypes also appeared to have deletions of the *C4B* gene, as shown by diminished relative intensity or absence of the *C4B*-associated bands. Since bands can disappear following gene conversion events, a long-range restriction mapping procedure using a rare cutting endonuclease (*Bss*HII) and pulsed-field gel electrophoresis was followed. This was done to assess *C4–CYP21* gene organization on these haplotypes with greater certainty and to confirm that actual deletions had taken place. *Bss*HII cleaves a segment of about 110 kb that encompasses all of the *C4A–CYP21P–C4B–CYP21* unit between two flanking sites known to be well conserved. Elimination of one of the *C4–CYP21* repeats would reduce this *Bss*HII segment by about 30 kb and produce an 80-kb *Bss*HII band. (Smaller length variations of 6.5 kb due to long and short variants of *C4B* would occur independently.) The appearance of such 80-kb *Bss*HII bands on DNA analysis confirmed *CYP21–C4B* deletions in 19 of the 25 haplotypes first indicated.

Mutation Categories II and III: These categories included haplotypes for which the loss of functional sequence could be attributed to a more restricted deletion event or was the result of an independent mutation. More restricted deletions (category II) accounted for 4 of 58 (6.9%) of mutations, and category III accounted for 29 of 58 (50%). In the category III haplotypes, long-range restriction mapping (*Bss*HII) showed a *C4–CYP21* deletion unrelated to disease in 1, and expansion to three

C4–CYP21 pairs (represented by a 130-kb *Bss*HII band) in 4 (one of these with HLA–B14,DR1, an HLA type with which a third *C4–CYP21* pair is characteristic).

Mutation Category IV: In 6 of the affected haplotypes, both *C4* isotypes were identified and evidence for the existence of two *CYP21* sequences was found (densitometric quantitation, short- and long-range restriction mapping). It was concluded that sequence conversion had occurred in the active *CYP21* gene on these haplotypes because of a *Taq*I (short-range) RFLP indicating the presence only of a *CYP21P* pseudogene, coupled with *Bss*HII (long-range) RFLPs of intact *CYP21–CYP21P* pairs. Presumably these conversions directly caused the expression of an enzyme defect. This mutation type accounted for 6/58 (10%) of the total. Of interest is that 1 of these haplotypes exhibited a *Bss*HII band at 180 kb length, evidently a complement of *four C4–CYP21* pairs.

B. Intragenic Recombination Events

Of equal importance at the DNA level in the development of steroid 21-hydroxylase deficiency is the less well understood mechanism of intragenic recombination, which apparently can involve the reciprocal or nonreciprocal transfer (gene conversion) of *short* genetic segments between two highly related sequences, in this case the transfer of mutations resident on the *CYP21P* pseudogene to the active CYP21 coding sequence, with damaging result (Fig. 6). (Intergenic recombination is subject to negative interference, i.e. the probability of a second recombination event is very low in the proximity of a crossover, and increases with distance. In contrast, intragenic recombinations appear to have a very high positive interference, so that reversing crossovers may be almost immediate.) Enumeration of a number of gene conversion mutations follows.

1. Aberrant Splicing in Processing of mRNA (Intron b)

The splicing mechanism whereby segments of the primary transcript corresponding to introns are excised is central to the nuclear processing that occurs in the production of mature mRNA sequences in the eukaryotic cell. The second intron (intron b) of the *CYP21* gene has the greatest density of noncorrespondent base positions with *CYP21P*. Occurrence of a C → G point mutation (G resident on the pseudogene)–11 bp from the end of the intron for intron b causes a consensus position a further 8 bp

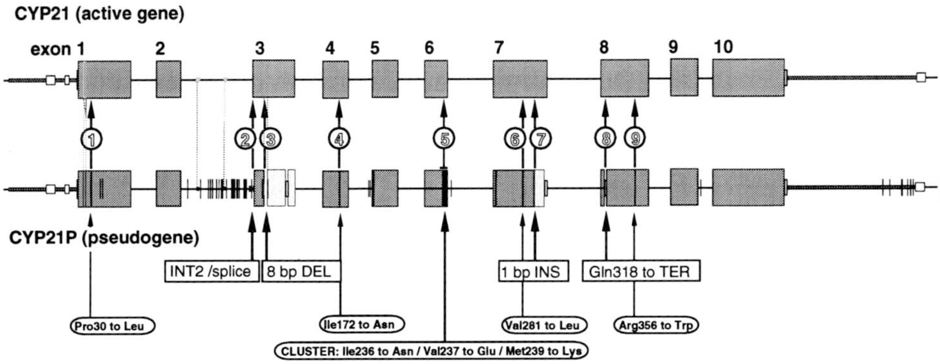

Figure 6 Point mutations and other small-scale DNA mutations occurring in the active *CYP21* gene transferred by gene conversion from the pseudogene *CYP21P*.

upstream to be recognized prematurely as the splice junction acceptor site (Rodrigues *et al.*, 1987; Higashi *et al.*, 1988) (Fig. 7). The 19-bp inclusion brings into frame a spurious 11 amino acid sequence and termination. Two facts are noteworthy: (*a*) the C → G change creates the potential acceptor sequence (C)CCCCAG/ (Mount, 1982) but this does not become the new splice junction; and (*b*) transcript assays have shown that aberrant splicing at a number of AG dinucleotides already resident near the end of this intron occurs at low level in the normal expression of the active gene (Higashi *et al.*, 1988).

2. Amino Acid Substitution $Ile_{172} \rightarrow Asn$ Altering Hydrophobicity of Presumed Membrane-insertion Anchor (Exon 4)

A nondeleted mutant *CYP21B* gene from a patient with classical 21-hydroxylase deficiency was cloned and sequenced (Amor *et al.*, 1988). Instead of the normal B gene triplet ATC encoding isoleucine, codon 172 of this mutant B gene reads AAC, encoding asparagine, a mutation normally present on the A pseudogene (Fig. 8). Transfection of the mutant gene into Y1 cells resulted in normal levels of mRNA, suggesting that transcription of the mutant B gene is unaffected. Thus the consequence of the mutation should be seen at the protein level, the amino acid substitution at position 172 both/either affecting rates of synthesis (altering folding kinetics and completion of the nascent polypeptide chain?), and/or resulting in an encoded P450c21 with a moderate to severe reduction in enzymatic activity. An oligonucleotide probe detected this mutation in 4 of 20 patients.

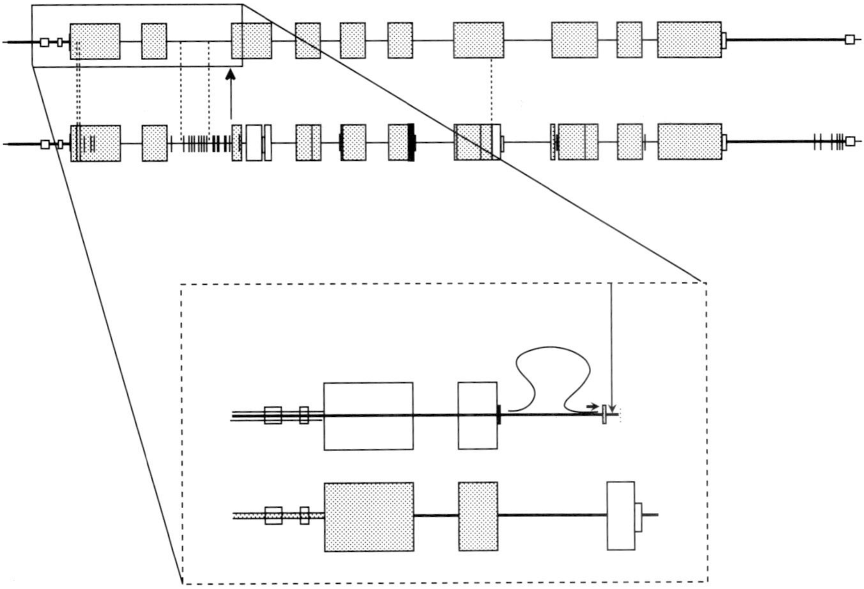

Figure 7 *CYP21* mutations (1). Unusual Splicing (Intron b): (*a*) Point mutation [C → G transversion] near normal end of intron creates a new consensus dinucleotide AG at position -10/11 but this is not the new splice acceptor. (*b*) Instead, apparently activated by this mutation, splicing occurs a further 7 bases upstream (-18/19) at an AG pair (one of several) already resident late in this intron. (*c*). This new, premature acceptor brings the last 19 nucleotide bases of the intron, normally excised, into the processed mRNA transcript; this is functionally equivalent to a (19-bp) insert, which with shift of the reading frame in translation results in an abnormal peptide sequence terminated by the first stop codon encountered in the new reading frame (14 bp into the real exon). (*d*) Transcript assays show aberrant splicing at a number of these premature intron end points to occur at low level in the normal expression of the active (B) gene. (Higashi *et al.*, 1988; Rodrigues *et al.*, 1988).

All patients so far identified with this mutation (including two having a heterozygous deletion of *CYP21B* and in whom studies more precisely correlating 21-hydroxylase phenotype with the Ile_{172} → Asn mutation are possible) retain some ability to make aldosterone. Enzyme function is thus not completely abolished, and the mutation probably correlates with a simple virilizing clinical phenotype. Differences in phenotype between patient 1 (present but decreased ability to synthesize aldosterone) and patients 2 and 3 (who apparently synthesize aldosterone normally) may be due to individual variation in levels of synthesis of the mutant enzyme, or to individual differences in steroid hormone metabolism (end-organ effect, catabolism or excretion). At the level of gene linkage with a pre-

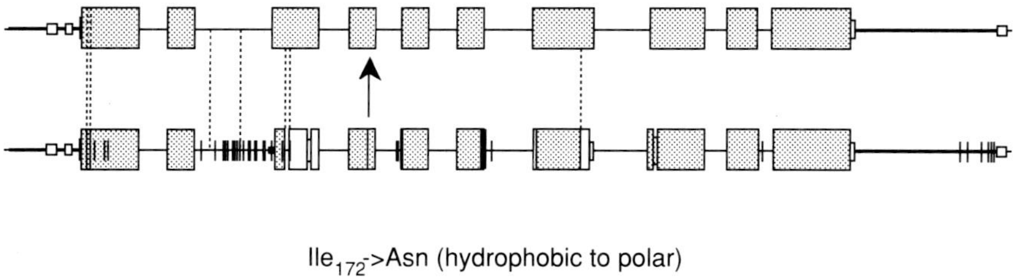

Figure 8 *CYP21* mutations (2). Missense mutation (Exon 4). Point mutation (T → A transversion) in the second position of the triplet (codon 172) produces the nonconservative amino acid substitution isoleucine to asparagine. (Amor *et al.*, 1988; Chiou *et al.*, 1990). Expression studies with clones of this mutation confirm that the enzyme is dysfunctional. Very recent studies of CYP17, the companion microsomal steroidogenic cytochrome P450 enzyme, have established that appropriate anchoring in the endoplasmic reticulum at this N-proximal hydrophobic region is essential for enzymatic function (the first amino acid residues at the N-terminal are less critical as shown by site-directed mutagenesis in this same study (Clark and Waterman (1992).

sumed identical *CYP21* mutation in HLA-identical siblings with 21-hydroxylase deficiency (Stoner *et al.*, 1986), similar variations in aldosterone synthetic capacity have been observed.

3. Cluster of Amino Acid Substitutions (Exon 6)

Toward the end of exon 6, the same T → A transversion occurs three times in the space of 10 bp, each time in the second coding position. This results consistently in replacement of hydrophobic residues (isoleucine-236, valine-237, and methionine-239) with a polar group of each type: (ile →)asparagine [uncharged], (val →)glutamic acid [neg. charged], and (met →)lysine [pos. charged] in a region of the polypeptide already with many charged residues (this has been provisionally identified as the molecular oxygen-binding region of the enzyme) (Fig. 9).

4. Conservative Amino Acid Substitution in Exon 7 (Val281 → Leu) Causing Nonclassic 21-Hydroxylase Deficiency

Association of the nonclassical disorder with HLA-B14,DR1 suggests a recurrent allele traveling on this partial haplotype. This possibility was addressed in a study characterizing the B14,DR1-associated *CYP21* mutation(s) at the molecular level (Speiser *et al.*, 1988).

DNA sequence analysis was performed on a cloned *CYP21B* gene derived from a HLA-B14,DR1 homozygous patient with nonclassical

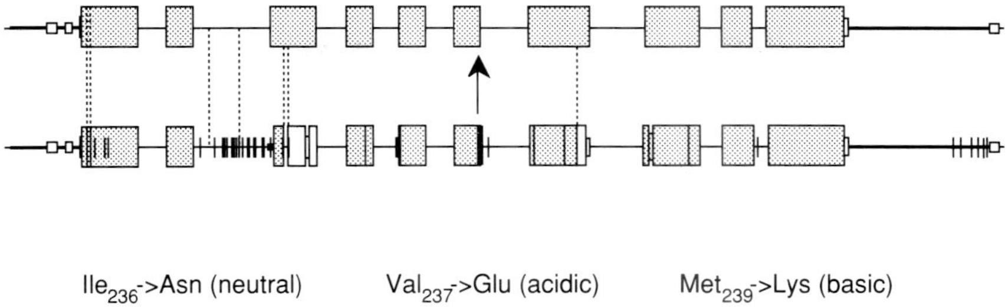

Figure 9 *CYP21* mutations (3). Cluster of 3 missense mutations (exon 6). The same T → A transversion occurs three times in the space of 10 bp, each time in the second coding position. This results consistently in replacement of a hydrophobic group with a polar group within a primary region of the polypeptide already with many charged residues. (Amor *et al.*, 1988; Higashi *et al.*, 1988). *In vitro* expression of this cluster mutation, identified in the genotype of a patient with severe salt-wasting phenotype, determined that the mutant enzyme was devoid of catalytic activity (Tusie-Luna *et al.*, 1990).

21-hydroxylase deficiency. Five deviations from the normal *CYP21* nucleotide sequence were found: (*a*) insertion of 3 nucleotides (CTG) encoding leucine between codons 9 and 10; (*b*) a single base change (GTG → CTG) resulting in a conservative amino acid (aa) substitution, leucine instead of valine (both nonpolar amino acids) at codon 211; (*c*) a cluster of three substitutions between the 8th and 13th bases of intron 6 (CTGTAC → ATGTGT); (*d*) a single base substitution (CTC → CTG), which is silent, leaving intact the leucine residue determined by codon 248; (*e*) a single base change (GTG → TTG) in codon 281 resulting in a second conservative aa substitution, leucine again replacing valine.

The silent base substitution (mutation *d*) was disregarded. The mutation in intron 6 (mutation *c*) also was eliminated from consideration since it begins at the 8th nucleotide of the intron and mutations affecting splicing observed to date have all been within the first 5 nucleotide positions. Comparison with other mammalian species revealed that the leucine repeat between codon 9 and 10 (mutation *a*) is often normally present. Similarly, other amino acid substitutions were seen for valine at codon 211 in the animal P450c21 sequences examined, making it improbable that mutation *b* was significant. On the other hand, valine was uniformly conserved at codon 281, and thus the substitution of leucine caused by mutation *e* was suspected to affect enzymatic function.

A 21-mer oligonucleotide probe was synthesized to correspond to the sequence immediately surrounding the codon 281 GTG → TTG. The probing strategy was to subject DNA samples to *Taq*I digestion to separate the *CYP21* and *CYP21P* sequences as (respectively) the 3.7- and 3.2-kb fragments, and to use labeled oligonucleotide to identify occurrence of the mutation on the active gene by appearance of both bands on autoradiography. DNA samples from nine unrelated nonclassical patients and one normal individual were screened for this mutation. Of the nonclassical patients, three were homozygous and five were heterozygous for HLA-B14,DR1; all were of Ashkenazi Jewish extraction. The 3.7-kb band was visualized in all of these cases, confirming presence of the mutation. The ninth patient was a Yugoslav, in which population nonclassical 21-hydroxylase deficiency occurs with increased frequency but shows no association with HLA-B14,DR1. The DNA sample from this patient showed no evidence of the mutation. DNA from the normal individual also did not reveal any hybridization at the position of the 3.7-kb *Taq*I band. Because the HLA-B14,DR1 haplotype carries an extra 3.2-kb *Taq*I band, a second oligonucleotide probe, specific for the active gene, was used in these genotypes to confirm that the extra 3.2-kb fragment was not cut from an active coding sequence. Lack of hybridization confirmed that the extra sequence was a duplicated *CYP21P* or *CYP21P*-like pseudogene.

Since altered protein structure might not be the only mechanism causing reduced enzymatic activity and emergence of the nonclassical phenotype, transcription of the mutant *CYP21* gene was examined in transfection experiments with the mouse Y1 adrenal cell line. It was determined that there are transcripts of the gene sequence and that protein synthesis results. The product expressed does exhibit some enzyme activity. These experiments cannot assess stability of the transcripts, and the possibility remains that an undetected mutation in a regulatory region could affect the levels of mRNA synthesized in affected adrenal glands. This study nevertheless establishes the functional significance of the $\mathrm{Val}_{281} \rightarrow$ Leu amino acid substitution, and the codon 281 mutation appears to be a consistent molecular genetic marker for the HLA-B14,DR1-associated 21-hydroxylase nonclassical deficiency allele (Fig. 10).

5. Early Termination in Exon 8 (Gln318 → TER)

The single *CYP21B* in the genome of a patient with severe salt wasting and a confirmed heterozygous deletion of the active gene was cloned and sequenced. Several mutations were identified, all normally resident on the A gene (pseudogene), of which the most significant was the base change C → T in codon 318, changing triplet CAG, encoding glutamine, to stop signal TAG (UAG) (a so-called "nonsense" mutation) (Globerman *et al.*, 1988) (Fig. 11). Premature termination of synthesis before the translation of the heme-binding region of the cytochrome P450c21

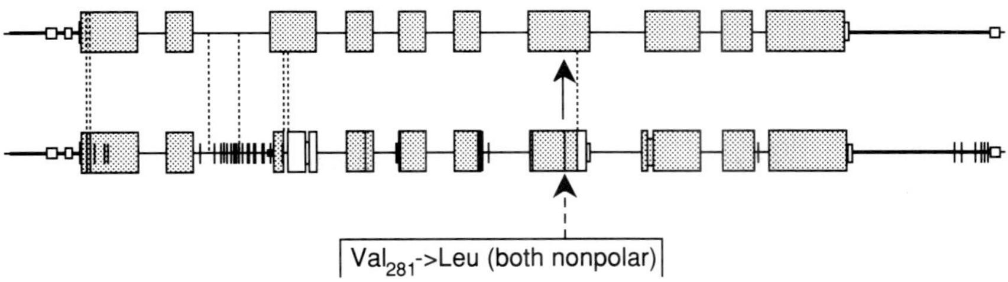

Figure 10 *CYP21* mutations (4). Missense mutation in exon 7. Point mutation (G → T transversion) occurring in the first position changes the specification of codon 281 from valine to leucine. Interspecies sequence comparison shows valine at this location to be strictly conserved. This mutation causes expression of a protein with reduced enzymatic activity and produces the nonclassical 21-hydroxylase deficiency clinical phenotype. It is associated with HLA antigen B14. (Speiser *et al.*, 1988).

protein would result in a completely nonfunctional enzyme. In addition, this type of mutation is liable to affect mRNA stability, resulting in lowered steady-state mRNA levels.

Mutations other than $Gln_{318} \rightarrow$ TER included the nonclassical mutation $Val_{281} \rightarrow$ Leu (see preceding) characterized by Speiser *et al.* (1988).

6. Radical Amino Acid Substitution $Arg_{356} \rightarrow$ Trp late in Exon 8

The mutation $Arg_{353} \rightarrow$ Trp on one haplotype was described by Chiou *et al.* (1990) (Fig. 12). The other haplotype of this patient also contained a point mutation, the same as described in IV,B,2, i.e., $Ile_{172} \rightarrow$ Asn (Amor *et al.*, 1988). Each of these mutations was separately constructed by site-

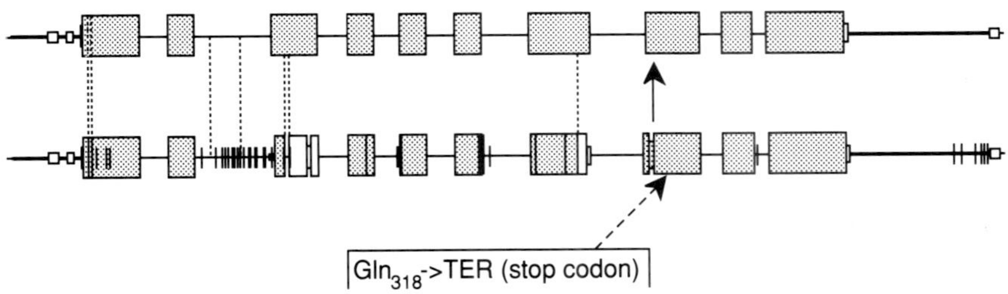

Figure 11 *CYP21* mutations (5). Termination in exon 8 (nonsense mutation). A point mutation (C → T transition) occurring in the first position changes codon 318 to TAG (UAG), specifying termination. End of translation before assembly of the heme-binding region would render any product totally inactive, but also the mRNA transcribed from this mutant gene appears to have reduced stability in the cytoplasm. (Globerman *et al.*, 1988).

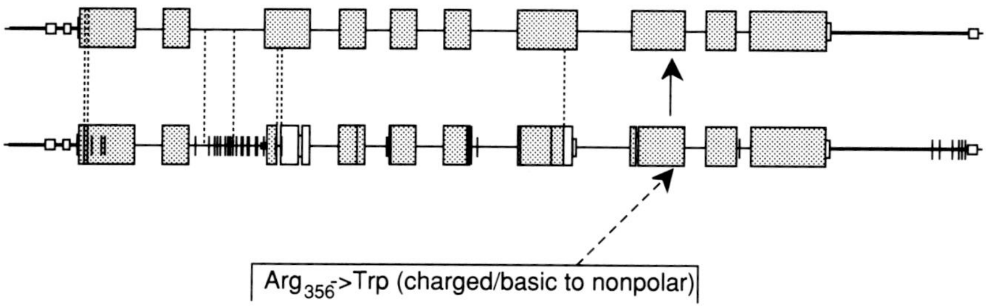

Figure 12 *CYP21* mutations (6). (6) Missense mutation (exon 8). Point mutation (C → T transition) in the first position of codon 356 produces the radical amino acid substitution arginine to tryptophan. (Chiou *et al.*, 1990).

directed mutagenesis of a normal *CYP21* clone. Both failed to express active enzyme activity in COS1 cells. The patients described by Amor *et al.* (1988) showed some aldosterone synthetic capacity and the patient of Chiou *et al.* (1990) was also classified as a simple virilizer (although no hormonal or biochemical data were provided to confirm this diagnosis).

There is some conflict between the clinical syndrome, which implies mild impairment of 21-hydroxylation (since aldosterone is apparently normally synthesized) and the *in vitro* data suggesting no 21-hydroxylase activity. Other clinical/molecular noncorrespondence is documented in a comprehensive study by Speiser *et al.* (1992), in which it is seen that the phenotype is not always as predicted by the genotype (Fig. 13). By using a scoring system that relied on multiple clinical parameters to define the phenotype of 90 patients whose mutations were predictive of one of the three major forms of 21-hydroxylase deficiency—salt-wasting, simple virilizing, and nonclassical—and defining enzymatic compromise on the basis of *in vitro* expression of mutant enzymes, it was found that each group included persons whose phenotype differed significantly from expectation. This study underlines the need for careful, ongoing clinical assessment of all 21-hydroxylase deficiency patients.

7. Other Putative Conversion Events

In another patient, a larger rearrangement involving exons 3–6 includes transfer of the 8-bp deletion in *CYP21P* to the *CYP21* sequence, producing a salt-wasting phenotype (Amor *et al.*, 1988). Transfer of this intragenic deletion to cause salt-wasting 21-hydroxylase deficiency has been noted elsewhere (Harada *et al.*, 1987; Higashi *et al.*, 1988). One of these mutant genes carries two other mutations, each causing an amino acid substitution: serine-269 to threonine, and asparagine-494 to serine

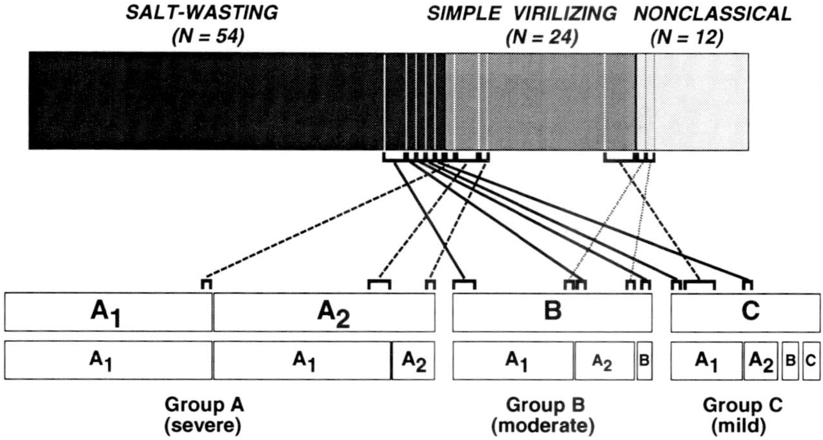

Figure 13 Correspondence and noncorrespondence of clinical phenotype with molecular genotype in 21-hydroxylase deficiency. Overall, type A (severe) mutations result in the SW (salt-wasting) phenotype, type B (moderate) mutations result in the SV (simple virilizing) phenotype, and type C (mild) mutations in the NC (nonclassical) phenotype. Brackets (width indicates number of patients) and oblique lines align instances of each genotype correlating with a clinical type milder or more severe than expected. (From Speiser *et al.*, 1992.) Mutations: A1: deletion; exon 3 DEL; cluster/exon 6; [T ins/F306]; Gln-318 to TER; Arg-356 to Trp. A2: A/C656 to G (intron 2). B: Ile-172 to Asn. C: Pro-30 to Leu; Val-281 to Leu.

(Harada *et al.*, 1987). The $Asn_{494} \rightarrow$ Ser substitution has been noted in two other patients, one with salt wasting, and one a simple virilizer (Rodrigues *et al.*, 1987; Urabe *et al.*, 1990). Comparison with the aa sequence of cytochrome P450c21s of other mammalian species shows that neither Ser_{269} nor Asn_{494} is conserved. Given the frequency with which Asn_{494} occurs, it is almost certainly a polymorphism (P. C. White, personal communication).

C. Recent Assessments of Gene Mutations and Rearrangements

1. Range of Effects of Possible Mutations

Adrenal steroid 21-hydroxylase expression may be null and give rise to classical CAH with the salt-wasting phenotype, as is the case when a deletion encompassing the entire B gene, or a B gene with a particularly damaging lesion to regulatory or coding sequences is present on both haplotypes. However, if at least one haplotype carries a B gene sequence that still allows the expression of a protein product, there may be persistence of some enzymatic activity, with correspondingly reduced hormonal disturbance and clinical expression. Mutations having this effect would obviously be of a range of types including short insertion/deletions within (*a*) regulatory sequences (rate of transcription), (*b*) leader/trailer sequences (reduced nuclear mRNA stability), or (*c*) intronic sequences (aberrant splicing; modification of the mRNA transcript), and point mutations or frame-preserving short insertion/deletions causing (*d*) amino acid substitutions affecting protein structure and function, or (*e*) codon replacements affecting rates of protein translation. Mutations in this range have been identified and they appear to be associated usually with simple virilizing classical or the milder, nonclassical, forms of disease. Certain other mutations may have little functional significance or be silent, and may exist as B gene polymorphisms in the normal population (Fig. 14).

2. HLA Markers in Diagnosis and Predictive Diagnosis

Since association of 21-hydroxylase deficiency with HLA was established in 1977, HLA genotyping has been an invaluable adjunct in the diagnosis and predictive diagnosis of this disorder. Discordance of clinical symptoms in pedigrees had been noted rarely (Rosenbloom and Smith, 1966) previous to this. It was expected that HLA nonidentity would explain this variation. Stoner *et al.* (1986), however, noted clinical discordance especially with regard to salt wasting between HLA-identical siblings in certain families.

Molecular techniques identify the HLA antigen gene sequences and are accordingly more precise than serological procedures, still routinely employed in HLA typing. Even with closely neighboring markers there are inherent limits in association and linkage analysis (Aston *et al.*, 1988). Fine analysis of 21-hydroxylase genes directly as has been carried out in two studies, whose synopses follows, can reveal features of the deficiency alleles at the DNA level clarifying discordance in clinical features of disease in HLA-identical affected subjects.

Clinical discordance-HLA identity (1): In a study of one family with consanguinity (Morel *et al.*, 1989b), DNA analysis by restriction mapping revealed individual differences in three affected family members, the mother and two of three children, each of whom exhibited a clinically distinct form of the disease. These differences occurred despite the extensive identities at nine closely neighboring HLA and complement loci. A P450c21 (21-hydroxylase) cDNA probe hybridized to *Taq*I digests of genomic DNA samples identified the 3.2-kb band (associated with the A pseudogene) in

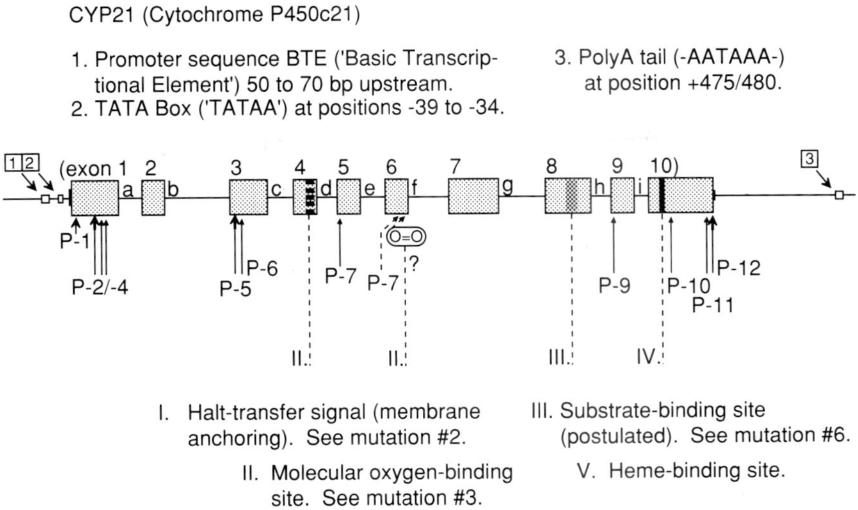

Figure 14 Polymorphisms in the *CYP21* gene. P-1, optional 5th in 4-leucine repeat, Leu$_{9a}$ [---/CTG] . Of two gene clones with published sequence, one has 4 leucine residues and the other has 5, giving two codon numbering conventions (one apart) to the final aa numbered 493/494; P-2, aa substitution, Pro$_{29}$ → Leu P-2: This polymorphism in fact has a deleterious effect (strongly supported by in vitro gene expression studies (Tusie-Luna *et al.*, 1991), and results in a nonclassical (NC) disease phenotype. Notated here as Pro$_{29}$ to Leu, this mutation is referred to in its first report as occurring at codon Pro$_{30}$; P-3 and 4, silent, Leu$_{39}$ [$^{c}/_{T}$TG] and pro$_{45}$ [GC^{c}/A]; P-5, aa alternates, Arg/Lys$_{102}$ [A$^{G}/_{A}$G]; P-6, silent, Leu$_{107}$ [$^{c}/_{T}$TG]; P-7, aa substitution (not confirmed), Asp$_{183}$ → Glu; P-8, silent (not yet observed), Asp$_{234}$[GA^{T}/C]; P-9 to 11, silent, Ser$_{374}$ [TC$^{T}/_{C}$]; Leu$_{433}$ [$^{c}/_{T}$TG]; Pro$_{490}$ [CC$^{G}/_{A}$]; P-12, aa alternates, Ser/Asn$_{493}$ [A$^{G}/_{A}$C].

all family members and the 3.7-kb band (associated with the B functional gene) in all except the older affected child (the index case), who exhibited a severe salt-wasting phenotype. A genomic probe extending outside the sequence recognized by the cDNA probe (necessarily limited to the P450c21 transcript) was then used. Hybridizing of this probe to two *Taq*I fragments immediately downstream of the two fragments mentioned, one of length 2.4 kb following the 3.2-kb (A) fragment and one of 2.5 kb following the 3.7-kb (B) fragment, showed that neither segment had been deleted. Similarly, digestion of samples from all patients with *Bgl*II enzyme and probing (with either probe) showed presence of the larger [10.5-kb (B gene) and 12-kb (A gene)] DNA segments cut by this enzyme. Thus it was concluded that gross deletion had not occurred in any of the family haplotypes, and that the child with salt wasting had received the same mutation (identical by descent) twice, a mutation that would have arisen from a gene conversion event.

Despite the apparent identity of the 21-hydroxylase genotypes of the mother and second affected child based on HLA/complement typing and DNA analysis to this point, use of the enzyme *Eco*RI revealed polymorphism. Interpretation of *Eco*RI banding patterns, however, is difficult, and in this study haplotypic assignment of the bands was not possible. Rearrangements in the pattern for the second child were taken as indication of multiple crossover events in this family. The unusual activity claimed for this genetic region—like the intricate splicing patterns for hemoglobin chain mRNAs and disturbances of these already registered in the literature on the hemoglobinopathies—may be a function of how closely studied is the HLA region itself. It may be that at any site in the genome, lesions sometimes occur that will evade detection by standard analysis, and that in these cases the reason for the clinical expression of disease will be ascertained not only by sequencing of the DNA for the ostensible disease allele(s), but also by examining other elements of the sensitive nuclear and cytoplasmic machinery involved in actualizing the gene message in the cell.

Clinical discordance - HLA identity (2): Marked discordance in clinical features between HLA-identical individuals was identified in three families by Sinnott *et al.* (1989).

- Family 1: boy with simple virilizing 21-hydroxylase defect and two younger sisters showing additional salt-wasting features.
- Family 2: one family member severely affected (virilized female) with very high 17-OHP; asymptomatic female sibling with only slightly elevated 17-OHP levels.
- Family 3: youngest of three male children presenting in neonatal age with severe salt wasting; HLA-identical middle brother asymptomatic with ACTH response in heterozygote range.

DNA from the family members was analyzed. Ratios of hybridization band intensities with CYP21 and C4 probes showed that the two disease haplotypes in family 1 had the gene organization *C4A–21A–C4B–21A–C4B–21B*. One of these is the haplotype carrying HLA-B14,DR1, known to have this triplicate organization and to have a number of small sequence conversions within the *CYP21B* gene affecting function of the 21-hydroxylase enzyme. Gene organization was found to be normal on both affected haplotypes in family 2, i.e., *C4A–21A–C4B–21B*. B gene point mutations not detected with the probes used are expected to underlie disease expression in this family. In family 3, the disease-associated

B47 haplotype of the father appears to show a compound deletion of the *21B* and *C4B* genes, documented in earlier 21-hydroxylase studies involving the extended B47 haplotype. The normal 21-hydroxylase/B47 haplotype of the mother is deduced to have the organization *C4A–21A–C4B/C4A–21B, C4A/C4B* signifying a fusion gene resembling *C4A* at the serotyping level, but resembling *C4B* at the DNA level according to the RFLP patterns obtained. Also of interest in this family is the oldest boy, who shares one HLA haplotype with his siblings. He is biochemically normal and is an HLA-B47 homozygote. However, the mother's B47 haplotype is *not* the one transmitted to the affected child and she also tests normal on ACTH stimulation. Further, DNA sequence analysis of this family has provided data in support of a gene microconversion event having occurred in maternal meiosis in this case (the affected child).

V. Summary

In conclusion, it has now been amply established that the molecular basis of adrenal steroid 21-hydroxylase deficiency involves abnormalities at the locus (loci) of the gene *CYP21,* which codes for the cytochrome P450c21 apoprotein. Characterization of mutations so far suggests that the clinical severity and form of the disease will prove in general to correlate with and be predicted by the extent of the mutation. Deletions, stop codon mutations, frameshifts, and presumably radical amino acid substitutions will result in salt wasting; one nonconservative substitution produces the simple virilizing phenotype, and two conservative substitutions are associated with nonclassical disease. But importantly there will be cases not following this pattern.

Precise determination of the rates of occurrence of specific gene deletions and sequence conversions will require systematic testing of patient DNA. Use of rare-cutting endonuclease and separation of digests by pulsed-field two-dimensional gel electrophoresis permit analysis of the chromosomal region for gene deletions. Identification of the transfer of mutations from *CYP21P* pseudogene, or of the occurrence of any other preexisting sequence, is via construction of exact matching sequences and by use of these synthetic oligonucleotides as probes. Tracing of particular 21-hydroxylase mutations in pedigrees relies on linkage with HLA and use of class I and II probes. Any rare occurrence of recombination events confounding the expected segregation of mutations with the marker alleles can be analyzed with oligonucleotide probes flanking or directed to potential crossover sites.

Improved DNA analysis is especially desirable in light of successes in *prenatal* treatment of classically affected females. In these cases, the effects of androgens on genital development have been lessened by administration in the pregnant mother of steroids able to suppress the fetal adrenal. Analysis of DNA obtained by chorionic villus sampling at 8–10 weeks of gestation, by providing the 21-hydroxylase status of the proband, is of benefit in either confirming affected status by identity with the family index case, or allowing unnecessary steroid treatment of a genetic male or unaffected fetus to be discontinued. Development of panels of oligonucleotide probes able to identify a wide range of CYP21 mutations will be of potential benefit in prenatal evaluation when genetic information on the pedigree is incomplete.

As the number of genotypes analyzed increases, statistics continue to show that total or near-total deletions of the active CYP21 sequence occur with a frequency in the range 10–25% (Morel *et al.*, 1989a; White *et al.*, 1988), or more in some populations (Collier *et al.*, 1989). Nondeletional mutations (presumably arising by reciprocal or nonreciprocal intragenic recombination) currently being identified are found to account for the same percentage of mutations (Morel *et al.*, 1989a). Analysis of nondeletional mutations (by oligonucleotide probe hybridization and direct sequencing of mutant *CYP21* genes) indicates that, of the small number analyzed, almost all mutant *CYP21* genes bear mutations present on *CYP21P* (Owerbach *et al.*, 1992a). The number of point mutations arising from intragenic recombination may thus be very high. On the other hand, *de novo* point mutations occur infrequently (Owerbach *et al.*, 1992b; Helmberg *et al.*, 1992), and further analysis may prove that in instances these also have arisen by intragenic recombination (Helmberg *et al.*, 1992).

Increasingly, the flexibility afforded by novel molecular protocols such as polymerase chain reaction permits analysis of the CYP21 sites directly, without recourse to marker HLA loci (Owerbach *et al.*, 1990; Karaviti *et al.*, 1992; Speiser *et al.*, 1992).

References

Amor, M., Tosi, M., Duponchel, C., Steinmetz, M., and Meo, T. (1985). Liver mRNA probes disclose two cytochrome P-450 genes duplicated in tandem with the complement *C4* loci of the mouse *H-2S* region. *Proc. Natl. Acad. Sci. U.S.A.* **82,** 4453–4457.

Amor, M., Parker, K. L., Globerman, H., New, M. I., and White, P. C. (1988). Mutation in the *CYP21B* gene (Ile-172 → Asn) causes steroid 21-hydroxylase deficiency. *Proc. Natl. Acad. Sci. U.S.A.* **85,** 1600–1604.

Aston, C. E., Sherman, S. L., Morton, N. E., Speiser, P. W., and New, M. I. (1988). Genetic mapping of the 21-hydroxylase locus: Estimation of small recombination frequencies. *Am. J. Hum. Genet.* **43,** 304–310.

Brändle, U., Ono, H., Vincek, V., Klein, D., Golubic, B., and Klein, J. (1992). Trans-species evolution of *Mhc-DRB* haplotype polymorphism in primates: Organization of DRB genes in the chimpanzee. *Immunogenetics* **36,** 39–48.

Carroll, M. C., Campbell, R. D., and Porter, R. R. (1985). Mapping of steroid 21-hydroxylase genes adjacent to complement component C4 genes in HLA, the major histocompatibility complex in man. *Proc. Natl. Acad. Sci. U.S.A.* **82,** 521–525.

Casey, M. L., MacDonald, P. C., and Winkel, C. A. (1985). Deoxycorticosterone biosynthesis in kidney tissue of experimental animals. *Ann. N.Y. Acad. Sci.* **458,** 232–237.

Chiou, S. H., Hu, M. C., and Chung, B.-C. (1990). A missense mutation at Ile -172 → Asn or Arg-356 → Trp causes steroid 21-hydroxylase deficiency. *J. Biol. Chem.* **265,** 3549–3552.

Clark, B., and Waterman, M. R. (1992). Functional expression of bovine 17α- hydroxylase in COS 1 cells is dependent upon the presence of an amino-terminal signal anchor sequence. *J. Biol. Chem.* **267,** 24568–24574.

Collier, S., Sinnott, P. J., Dyer, P. A., Price, D. A., Harris, R., and Strachan, T. (1989). Pulsed-field gel electrophoresis identifies a high degree of variability in the number of tandem 21-hydroxylase and complement C4 gene repeats in 21-hydroxylase deficiency haplotypes. *EMBO J.* **8,** 1393–1402.

Collier, S., Tassabehji, M., and Strachan, T. (1993). *De novo* pathological point mutation at the 21 hydroxylase locus: Implications for gene conversion in the human genome. *Nat. Genet.* **3,** 260–265.

Contu, L., Carcassi, C., and Dausset, J. (1989). The "Sardinian" HLA-A30,B18,DR3,DQw2 haplotype constantly lacks the 21-OHA and C4B genes. Is it an ancestral haplotype without duplication? *Immunogenetics* **30,** 13–17.

Decourt, J., Jayle, M. F., and Baulieu E.-E. (1957). Virilisme cliniquement tardif avec excretion de pregnanetriol et insuffisance de la production de cortisol. *Ann. Endocrinol. (Paris)* **18,** 416–422.

Dodds, A. W., and Law, S. K. A. (1990). The complement component C4 of mammals. *Biochem. J.* **265,** 495–502.

Futterweit, W. (1984). "Polycystic Ovarian Disease," pp. 163–166. Springer-Verlag, New York.

Globerman, H., Amor, M., Parker, K. L., New, M. I., and White, P. C. (1988). Nonsense mutation causing steroid 21-hydroxylase deficiency. *J. Clin. Invest.* **82,** 139–144.

Granados, J., Awdeh, Z. L., Chen, J.-H., Giles, C. M., Balner, H., Yunis, E. J., and Alper, C. A. (1987). There are two *C4* genetic loci and a null allele in the chimpanzee. *Immunogenetics* **26,** 344–350.

Harada, F., Kimura, A., Iwanaga, T., Shimozawa, K., Yata, J., and Sasazuki, T. (1987). Gene conversion-like events cause steroid 21-hydroxylase deficiency in congenital adrenal hyperplasia. *Proc. Natl. Acad. Sci. U.S.A.* **84,** 8091–8094.

Helmberg, A., Tusie-Luna, M.-T., Tabarelli, M., Kofler, R., and White, P. C. (1992). R339H and P453S: CYP21 mutations associated with nonclassical steroid 21-hydroxylase deficiency that are not apparent gene conversions. *Mol. Endocrinol.* **6,** 1318–1322.

Higashi, Y., Yoshioka, H., Yamane, M., Gotoh, O., and Fujii-Kuriyama, Y. (1986). Complete nucleotide sequence of two steroid 21-hydroxylase genes tandemly arranged in human chromosome: A pseudogene and a genuine gene. *Proc. Natl. Acad. Sci. U.S.A.* **85,** 2841–2845.

Higashi, Y., Tanae, A., Inoue, H., and Fujii-Kuriyama, Y. (1988). Aberrant splicing and missense mutations cause steroid 21-hydroxylase deficiency in humans: Possible gene conversion products. *Proc. Natl. Acad. Sci. U.S.A.* **85,** 7486–7490.

Ishii-Ohba, H., Saiki, N., Inano, H., and Tamaoki, B. (1986). Purification and characterization of rat adrenal 3β-hydroxysteroid dehydrogenase with steroid 5-ene-4-ene-isomerase. *J. Steroid Biochem.* **24,** 753–760.

Karaviti, L. P., Mercado, A. B., Mercado, M. B., Speiser, P. W., Buegeleisen, M., Crawford, C., Antonian, L., White, P. C., and New, M. I. (1992). Prenatal diagnosis/treatment in families at risk for infants with steroid 21-hydroxylase deficiency (congenital adrenal hyperplasia). *J. Steroid Biochem. Mol. Biol.* **41,** 445–451.

Klein, J., Bontrop, R. E., Dawkins, R. L., Ehrlich, H. A., Gyllensten, U. B., Heise, E. R., Jones, P. P., Parham, P., Wakeland, E. K., and Watkins, D. I. (1990). Nomenclature for the major histocompatibility complex of different species: A proposal. *Immunogenetics* **31,** 217–219.

Kominami, S., Ochi, H., Kobayashi, Y., and Takemori, S. (1980). Studies on the steroid hydroxylation system in adrenal cortex microsomes. *J. Biol. Chem.* **255,** 3386–3394.

Marsh, S. G. E., and Bodmer, J. G. (1992). HLA class II nucleotide sequences, 1992. *Tissue Antigens* **40,** 229–243.

Matsuo, Y., Iwahashi, K., Kawai, Y., and Ichikawa, Y. (1992). Analysis of the transcripts of cytochrome P-450IID gene subfamily in bovine adrenal gland. *Histochemistry* **97,** 319–322.

Morel, Y., Andre, J., Uring-Lambert, B., Hauptmann, G., Bétuel, H., Tossi, M., Forest, M. G., David, M., Bertrand, J., and Miller, W. L. (1989a). Rearrangements and point mutations of P450c21 gene are distinguished by five restriction endonuclease haplotypes identified by a new probing strategy in 57 families with congenital adrenal hyperplasia. *J. Clin. Invest.* **83,** 527–536.

Morel, Y., David, M., Forest, M. G., Bétuel, H., Hauptmann, G., André, J., Bertrand, J., and Miller, W. L. (1989b). Gene conversions and rearrangements cause discordance between inheritance of forms of 21-hydroxylase deficiency and HLA types. *J. Clin. Endocrinol. Metab.* **68,** 592–599.

Morel, Y., Bristow, J., Gitelman, S. E., and Miller, W. L. (1989c). Transcript encoded on the opposite strand of the human steroid 21-hydroxylase/ complement component C4 gene locus. *Proc. Natl. Acad. Sci. U.S.A.* **86,** 6582–6586.

Mount, S. M. (1982). A catalogue of splice junction sequences. *Nucleic Acids Res.* **10,** 459–472.

Muller-Eberhard, U., Ghizzoni, L., Liem, H. H., New, M., Finlayson, M., and Johnson, E. J. (1985). Evidence that variation among untreated rabbits in hepatic progesterone 21-hydroxylase activity is indicative of enzyme heterogeneity rather than a transient inductive effect. *Ann. N.Y. Acad. Sci.* **458,** 225–231.

Murray, M. (1992). Participation of a cytochrome P450 enzyme from the 2C subfamily in progesterone 21-hydroxylation in sheep liver (short communication). *J. Steroid Biochem. Mol. Biol.* **43,** 591–593.

Narasimhulu, S. (1989). Heterogeneity of the bovine adrenal steroid 21-hydroxylase. *Endocr. Res.* **15,** 67–84.

Nebert, D. W., Adesnik, M., Coon, M. J., Estabrook, R. W., Gonzalez, F. J., Guengerich, F. P., Gunsalus, I. C., Johnson, E. F., Kemper, B., Levin, W., *et al.* (1987). The P450 gene superfamily: Recommended nomenclature. *DNA* **6,** 1–11.

Nebert, D. W., Nelson, D. R., Adesnik, M., Coon, M. J., Estabrook, R. W., Gonzalez, F. J., Guengerich, F. P., Gunsalus, I. C., Johnson, E. F., Kemper, B., *et al.* (1989). The P450 superfamily: Updated listing of all genes and recommended nomenclature for the chromosomal loci. *DNA* **8,** 1–13.

New, M. I., Lorenzen, F., Lerner, A. J., Kohn, B., Oberfield, S. E., Pollack, M. S., Dupont, B., Stoner, E., Levy, D. J., Pang, S., and Levine, L. S. (1983). Genotyping steroid 21-hydroxylase deficiency: Hormonal reference data. *J. Clin. Endocrinol. Metab.* **57,** 320–326.

New, M. I., White, P. C., Pang, S., Dupont, B., and Speiser, P. W. (1989). The adrenal hyperplasias. *In* "The Metabolic Basis of Inherited Disease" (C. V. Scriver, A. L. Beaudet, W. S. Sly, and D. Valle, eds.), 6th ed., Chapter 74, pp. 1881–1917. McGraw-Hill, New York.

Otto, S., Marcus, C., Pidgeon, C., and Jefcoate, C. (1991). A novel adrenocorticotropin-inducible cytochrome P450 from rat adrenal microsomes catalyzes polycyclic aromatic hydrocarbon metabolism. *Endocrinology* **129,** 970–982.

Owerbach, D., Crawford, Y. M., and Draznin, M. B. (1990). Direct analysis of CYP21B genes in 21-hydroxylase deficiency using polymerase chain reaction amplification. *Mol. Endocrinol.* **4,** 125–131.

Owerbach, D., Ballard, A.-L., and Draznin, M. B. (1992a). Salt-wasting congenital adrenal hyperplasia: Detection and characterization of mutations in the steroid 21-hydroxylase gene, CYP21, using the polymerase chain reaction. *J. Clin. Endocrinol. Metab.* **74,** 553–558.

Owerbach, D., Sherman, L., Ballard, A.-L., and Azziz, R. (1992b). Pro-453 to Ser mutation in CYP21 is associated with nonclassic steroid 21-hydroxylase deficiency. *Mol. Endocrinol.* **6,** 1211–1215.

Pang, S., Lerner, A. J., Stoner, E., Levine, L. S., Oberfield, S. E., Engel, I., and New, M. I. (1985). Late-onset adrenal steroid 3β-hydroxysteroid dehydrogenase deficiency. A cause of hirsutism in pubertal and postpubertal women. *J. Clin. Endocrinol. Metab.* **60,** 428–439.

Rodrigues, N. R., Dunham, I., Yu, C. Y., Carroll, M. C., Porter, R. R., and Campbell, R. D. (1987). Molecular characterization of the HLA-linked steroid 21-hydroxylase B gene from an individual with congenital adrenal hyperplasia. *EMBO J.* **6,** 1653–1661.

Rosenbloom, A. L., and Smith, D. W. (1966). Varying expression for salt losing in related patients with congenital adrenal hyperplasia. *Pediatrics* **38,** 215–219.

Sinnott, P. J., Dyer, P. A., Price, D. A., Harris, R., and Strachan, T. (1989). 21-Hydroxylase deficiency families with HLA identical affected and unaffected sibs. *J. Med. Genet.* **26,** 10–17.

Speiser, P. W., and White, P. C. (1989). Structure of the human RD gene: A highly conserved gene in the Class III region of the major histocompatibility complex. *DNA* **8,** 745–751.

Speiser, P. W., New, M. I., and White, P. C. (1988). Molecular genetic analysis of nonclassic steroid 21-hydroxylase deficiency associated with HLA-B14,DR1. *N. Engl. J. Med.* **319,** 19–23.

Speiser, P. W., Agdere, L., Ueshiba, H., White, P. C., and New, M. I. (1991). Aldosterone synthesis in salt-wasting congenital adrenal hyperplasia with complete absence of adrenal 21-hydroxylase. *N. Engl. J. Med.* **324,** 145–149.

Speiser, P. W., Dupont, J., Zhu, D., Serrat, J., Buegeleisen, M., Tusie-Luna, M.-T., Lesser, M., New, M. I., and White, P. C. (1992). Disease expression and molecular genotype in congenital adrenal hyperplasia due to 21-hydroxylase deficiency. *J. Clin. Invest.* **90,** 584–595.

Spence, M. A., Spurr, N. K., and Field, L. L. (1989). Report of the committee on the genetic constitution of chromosomes 7, 8 and 9. *Cytogenet. Cell Genet.* **51,** 149–165.

Stoner, E., DiMartino, J., Kuhnle, U., Levine, L. S., Oberfield, S. E., and New, M. I. (1986). Is salt wasting in congenital adrenal hyperplasia genetic? *Clin. Endocrinol. (Oxford)* **24,** 9–20.

Temeck, J., Pang, S., Nelson, C., and New, M. I. (1987). Genetic defects of steroidogenesis in premature pubarche. *J. Clin. Endocrinol. Metab.* **64,** 609–617.

Trowsdale, J., Ragoussis, J., and Campbell, R. D. (1991). Map of the human MHC. *Immunol. Today* **12,** 443–446.

Tusie-Luna, M.-T., Traktman, P., and White, P. C. (1990). Determination of functional effects of mutations in the steroid 21-hydroxylase gene (CYP21) using recombinant Vaccinia virus. *J. Biol. Chem.* **265,** 20916–20922.

Tusie-Luna, M.-T., Speiser, P. W., Dumić, M., New, M. I., and White, P. C. (1991). A mutation (Pro-30 to Leu) in CYP21 represents a potential nonclassic steroid 21-hydroxylase deficiency allele. *Mol. Endocrinol.* **5,** 685–692.

Urabe K., Kimura, A., Harada, F., Iwanaga, T., and Sasazuki, T. (1990). Gene conversion in steroid 21-hydroxylase genes. *Am. J. Hum. Genet.* **46,** 1178–1186.

van den Berg, C. W., Démant, P., Aerts, P. C., and van Duk, H. (1992). Slp is an essential component of an EDTA-resistant activation pathway of mouse complement. *Proc. Natl. Acad. Sci. U.S.A.* **89,** 10711–10715.

Waterman, M. R., Mason, J. I., Zuber M. X., John, M. E., Rodgers, R. J., and Simpson, E. R. (1986). Control of gene expression of adrenal steroid hydroxylase and related enzymes. *Endocr. Res.* **12,** 393–408.

Werkmeister, J. W., New, M. I., Dupont, B., and White, P. C. (1986). Frequent deletion and duplication of the steroid 21-hydroxylase genes. *Am. J. Hum. Genet.* **39,** 461–469.

White, P. C., Chaplin, D. D., Weis, J. H., Dupont, B., New, M. I., and Seidman, J. G. (1984a). Two steroid 21-hydroxylase genes are located in the murine S region. *Nature (London)* **312,** 465–467.

White, P. C., New, M. I., and Dupont, J. (1984b). Cloning and expression of cDNA encoding a bovine adrenal cytochrome P-450 specific for steroid 21-hydroxylation. *Proc. Natl. Acad. Sci. U.S.A.* **81,** 1986–1990.

White, P. C., New, M. I., and Dupont, B. (1984c). HLA-linked congenital adrenal hyperplasia gene encoding a cytochrome P-450 specific for steroid 21-hydroxylation. *Proc. Natl. Acad. Sci. U.S.A.* **81,** 7505–7509.

White, P. C., Grossberger, D., Onufer, B. J., Chaplin, D. D., New, M. I., Dupont, B., and Strominger, J. L. (1985). Two genes encoding steroid 21-hydroxylase are located near the genes encoding the fourth component of complement in man. *Proc. Natl. Acad. Sci. U.S.A.* **82,** 1089–1093.

White, P. C., New, M. I., and Dupont, B. (1986). Structure of human steroid 21-hydroxylase genes. *Proc. Natl. Acad. Sci. U.S.A.* **83,** 5111–5115.

White, P. C., Vitek, A., Dupont, B., and New, M. I. (1988). Characterization of frequent deletions causing steroid 21-hydroxylase deficiency. *Proc. Natl. Acad. Sci. U.S.A.* **85,** 4436–4440.

Winkel, C. A., Milewich, L., Parker, C. R., Jr., Gant, N. F., Simpson, E. R., and MacDonald, P. C. (1980a). Conversion of plasma progesterone to deoxycorticosterone in men, nonpregnant and pregnant women, and adrenalectomized subjects. Evidence for steroid 21-hydroxylase activity in nonadrenal tissues. *J. Clin. Invest.* **66,** 803–812.

Winkel, C. A., Parker, C. R., Jr., Simpson, E. R., and MacDonald, P. C. (1980b). Production rate of deoxycorticosterone in women during the follicular and luteal phases of the ovarian cycle: The role of extra-adrenal 21-hydroxylation of circulating progesterone in deoxycorticosterone production. *J. Clin. Endocrinol. Metab.* **51,** 1354–1358.

Zachmann, M., Kempken, B., Manella, B., and Navarro, E. (1992). Conversion from pure 17,20-desmolase to combined 17,20-desmolase/17α-hydroxylase deficiency with age. *Acta Endocrinol. (Copenhagen)* **127,** 97–99.

Zemmour, J., and Parham, P. (1992). HLA class I nucleotide sequences, 1992. *Tissue Antigens* **40,** 221–228.

18

The Gene for Anti-Müllerian Hormone

Nathalie Josso, Sandrine Imbeaud, and Jean-Yves Picard

Unité de Recherches sur l'Endocrinologie du Développement
INSERM
École Normale Supérieure
92120 Paris, France

Richard L. Cate

Biogen Inc.
Cambridge, Massachusetts 02142

Molecular Genetics of Sex Determination

I. Introduction

Anti-Müllerian hormone (AMH), also called Müllerian inhibiting substance (MIS) or Müllerian inhibiting factor (MIF), is a glycoprotein dimer, produced by gonadal somatic cells, and responsible for the regression of Müllerian ducts in male fetuses (Jost, 1953). AMH, together with inhibin, activin, and various other factors important for growth and development, belongs to the TGF-β superfamily (Cate *et al.*, 1986).

II. The AMH Gene Product

A. Purification and Processing

Initially purified from incubation media of calf fetal testicular tissue (Picard and Josso, 1984), AMH is now produced by recombinant DNA technology from the culture medium of Chinese hamster ovary cells transfected with the AMH coding region placed under the control of a viral promoter (Wallen *et al.*, 1989) (Fig. 1). More than 98% purity is achieved by immunochromatography on a monoclonal antibody. The hormone is produced as a 140-kDa disulfide-linked full-length dimer but is cleaved by plasmin approximately 110 residues away from the C-terminus to yield a TGF-β-like fragment (Pepinsky *et al.*, 1988). Replacement of the arginine at the processing site by a threonine produces a noncleavable molecule incapable of inducing Müllerian regression (Cate *et al.*, 1990). The mutated AMH is also unable to oppose aromatase induction in the fetal ovary (Fig. 2). Since only a small proportion of AMH released by Sertoli cells or by transfected CHO cells is cleaved, processing probably occurs in the target cell but this has not been experimentally verified.

B. Biological Activity

AMH is known primarily for its eponymous function, namely the inhibition of Müllerian duct derivatives in the male fetus (Jost, 1953). The Müllerian duct regression assay (Picon, 1969), which is based upon the capacity of AMH to inhibit the development of rat fetal Müllerian ducts in organ culture, has been crucial for the localization (Josso, 1974) and characterization (Picard *et al.*, 1978) of the hormone. AMH also opposes the development of other female-specific organs and functions: fetal ovaries exposed to the hormone are stunted and/or masculinized (Vigier *et al.*, 1987; Behringer *et al.*, 1990) and their steroidogenic pathway is

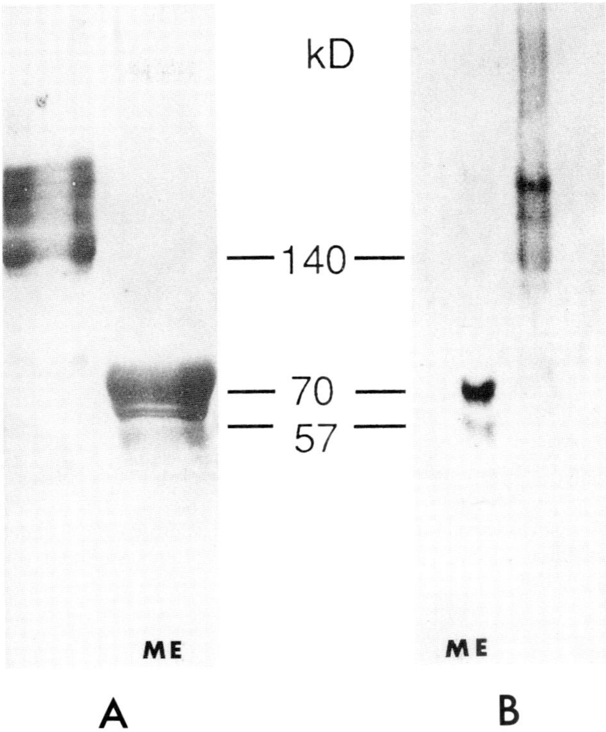

Figure 1 Purified anti-Müllerian hormone, analyzed by polyacrylamide gel electrophoresis on a gradient (4–30%) gel, in the presence of SDS. (A) Bovine AMH (3 μg), (B) Human recombinant AMH (1 μg). In the presence of mercaptoethanol (ME), which cleaves disulfide bonds, AMH migrates as a 70-kDa moiety, the minor 57-kDa band represents the N-terminal fragment. In the absence of mercaptoethanol, 140-kDa dimers and polymers of higher molecular weight are visible.

sex-reversed (Vigier *et al.*, 1989). This has led to the development of a quantitative bioassay for AMH, based on the capacity of AMH to block the cyclic AMP-induced aromatase activity of the fetal ovary in a dose-dependent manner (di Clemente *et al.*, 1992). Both in the Müllerian duct assay (MacLaughlin *et al.*, 1992; Wilson *et al.*, 1993) and in the antiaromatase test (Wilson *et al.*, 1993), the C-terminus is the biologically active moiety. However, the N-terminal proregion enhances the bioactivity of the C-terminus, a novel finding within the TGF-β family (Wilson *et al.*, 1993) (Fig. 2).

Other suggested biological activities of the molecule are controversial. Antiproliferative effects (Fuller *et al.*, 1982; Chin *et al.*, 1991; Parry *et al.*, 1992), oocyte meiosis retardation (Ueno *et al.*, 1988), and inhibition of protein phosphorylation (Hutson *et al.*, 1984) have been obtained only

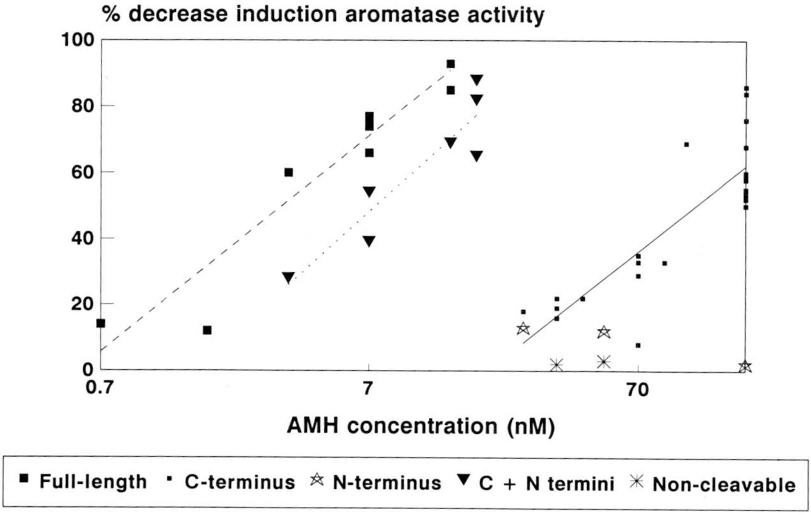

Figure 2 Biological activity of full-length AMH and its proteolytic fragments, measured by the anti-aromatase test described by di Clemente *et al.* (1992). Modified, with permission, from Wilson *et al.*, 1993.

with preparations from a single laboratory, but not with purified native (Rosenwaks *et al.*, 1984) or recombinant (Wallen *et al.*, 1989) AMH from other sources.

III. Structure of the *AMH* Gene

A. Cloning

The *AMH* gene has now been cloned in four species: bovine (Cate *et al.*, 1986; Picard *et al.*, 1986), human (Cate *et al.*, 1986), mouse (Münsterberg and Lovell-Badge, 1991), and rat (Haqq *et al.*, 1992). It is a small gene, spanning 2.75 kb, divided into 5 exons. The last exon is the most interesting, because it exhibits the greatest interspecies homology and is the only part exhibiting homology to the TGF-β family (Cate *et al.*, 1986). It contains an exceptionally high proportion of GC bases, up to 80%.

B. Structure of the *AMH* Promoter

In contrast to the promoters of the *AMH* gene in animal species, which contain canonical or at least almost perfect TATA boxes, the human promoter has only a degenerate (TTAA) site directing the initiation of transcription and exhibits several transcriptional initiation sites (Guerrier *et al.*, 1990). Genes lacking a TATA box, like the so-called "housekeeping" genes are usually ubiquitously expressed, whereas the expression of the *AMH* gene is tightly regulated (see below). A 13-base palindrome, nearly identical to the estrogen response element, which activates the vitellogenin genes of *Xenopus*, is located 1772 bases upstream of the major transcription initiation site and is functional when transfected into HeLa cells (Guerrier *et al.*, 1990), but its physiological significance is not clear at the present time. Several studies have shown that estrogen is required for fetal ovarian differentiation in nonmammalian species (Dorizzi *et al.*, 1991; Elbrecht and Smith, 1992) and opposes AMH effects upon the Müllerian duct (Newbold *et al.*, 1984). One would expect estrogen to exert negative regulation, if any, upon *AMH* gene transcription, but at the time of writing, no experimental data are available to evaluate this possibility.

Regulation of early expression of the *AMH* gene by the product of the *SRY* gene is also a subject of speculation. Two target DNA sequences for the HMG motif contained in the *SRY* gene product have been described (Nasrin *et al.*, 1991; Harley *et al.*, 1992). Homologous sequences are present in the *AMH* promoter (Fig. 3).

C. Gene Mapping

Although involved in male sex differentiation, the *AMH* gene is not located on a sex chromosome, in keeping with the tenet that only the

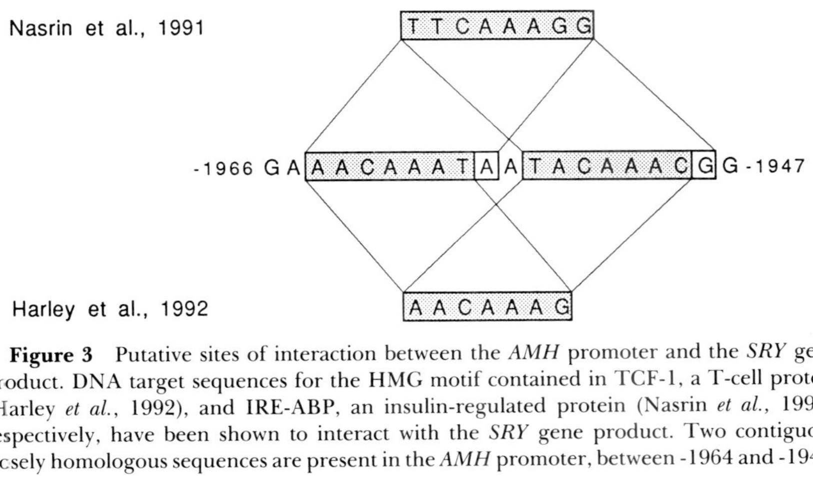

Figure 3 Putative sites of interaction between the *AMH* promoter and the *SRY* gene product. DNA target sequences for the HMG motif contained in TCF-1, a T-cell protein (Harley *et al.*, 1992), and IRE-ABP, an insulin-regulated protein (Nasrin *et al.*, 1991), respectively, have been shown to interact with the *SRY* gene product. Two contiguous clcsely homologous sequences are present in the *AMH* promoter, between -1964 and -1946.

genetic trigger for gonadal sex determination needs to be Y-specific. Preliminary evidence for the autosomal location of the human *AMH* gene was presented by Cate *et al.* (1986), who showed that male and female DNA have a similar hybridization pattern by Southern analysis. Using a panel of human–rodent somatic hybrid cell lines segregating human chromosomes and probes derived from bovine cDNA, Cohen-Haguenauer *et al.* (1987) localized the human gene to chromosome 19. Since the human gene for TGF-β also maps to chromosome 19 (Fujii *et al.*, 1986), and to exclude the possibility of cross-hybridization, *in situ* hybridization was performed. The *AMH* gene maps to the extremity of the short arm of chromosome 19, band 19p13.3, whereas the gene for TGF-β is located on the long arm, subband q13.1 → q13.3. Inhibin subunits, which also belong to the TGF-β family, map to chromosomes 2 and 7 (Barton *et al.*, 1987) and the gene for bone morphogenetic protein 2A is located on human chromosome 20 (Rao *et al.*, 1992).

Recently, data on the location of the *AMH* gene in other species have become available. No association with genes on human chromosome 19 has been detected, indicating complex rearrangements in chromosomal structure during evolution. For mapping in the domestic cattle, DNA probes from bovine AMH and osteonectin were hybridized to bovine–hamster and bovine–mouse hybrid cell lines segregating bovine chromosomes. Bovine osteonectin is homologous to secreted protein cysteine rich (SPARC) in humans, and maps to human chromosome 5. Bovine *AMH* and osteonectin loci were fully concordant to each other and discordant with all other bovine syntenic groups described to date (Rogers *et al.*, 1991).

In the mouse, the *Amh* gene has been mapped to the distal region of chromosome 10, between the phenylhydroxylase and the zinc finger autosomal genes (King *et al.*, 1991). Several polymorphic *Amh*-related sequences were mapped to other chromosomes and could represent pseudogenes.

IV. Expression of the *AMH* Gene

AMH is produced exclusively by gonadal somatic cells in both sexes, but there are major differences: immature Sertoli cells synthesize large amounts, which accumulate in the rough endoplasmic reticulum (Tran and Josso, 1982; Hayashi *et al.*, 1984), whereas granulosa cells synthesize the hormone only after birth, and in low quantities (Vigier *et al.*, 1984).

A. Testicular Expression

In the testis, AMH is produced only by Sertoli cells, in contrast to inhibin and activin, which are synthesized by Sertoli cells and also by interstitial cells (Lee *et al.*, 1989; Roberts *et al.*, 1989; Shaha *et al.*, 1989). In the fetus, AMH is the earliest cognate protein expressed by developing Sertoli cells. Testicular anti-Müllerian activity is found at the time of formation of seminiferous tubules (Picon, 1970; Tran *et al.*, 1977; Vigier *et al.*, 1983) and is sustained up to birth (Fig. 4). AMH is not expressed in the germ cells, nor outside the testis cords, nor in any other fetal tissue. In mice expressing the *W* locus, in which the germ cells fail to reach and colonize the gonadal ridge, AMH expression is normal, indicating that it is not dependent on interaction with germ cells (Münsterberg and Lovell-Badge, 1991).

AMH levels have recently been measured by enzyme-linked immunoassay in human fetal sera (Josso *et al.*, 1993a) (Fig. 5). High levels were found in the serum of all male fetuses studied, including one with a 46,XX karyotype. AMH was not detected in female sera, or in unconcentrated amniotic fluid of either sex. These results are potentially interesting for assessment of testicular function in late fetal life, at a time when testosterone levels are more or less similar in both sexes (Winter *et al.*, 1977).

After birth, Sertoli cells continue to express high levels of AMH during the first years of life, then expression progressively decreases and drops sharply at the beginning of pubertal maturation (Baker *et al.*, 1990; Hudson *et al.*, 1990; Josso *et al.*, 1990, Rey *et al.*, 1993). Nevertheless, low AMH concentrations are found in human adult serum (Hudson *et al.*, 1990) and in bovine rete testis fluid (Vigier *et al.*, 1983), indicating that the gene is not totally repressed in postpubertal Sertoli cells. In the rat testis, Lee *et al.* (1992) detected two discrete AMH transcripts, differing by the extent of polyadenylation, and reported that the smaller one is more abundant postnatally. They suggested that differential polyadenylation may be a means of post-transcriptional regulation of AMH protein expression.

B. Ovarian Expression

Granulosa cells of the ovary and Sertoli cells of the testis share many structural and functional characteristics, including AMH production. However, the chronology is different. No AMH can be detected in fetal ovaries, neither in the rat (Ueno *et al.*, 1989a), where follicles form only

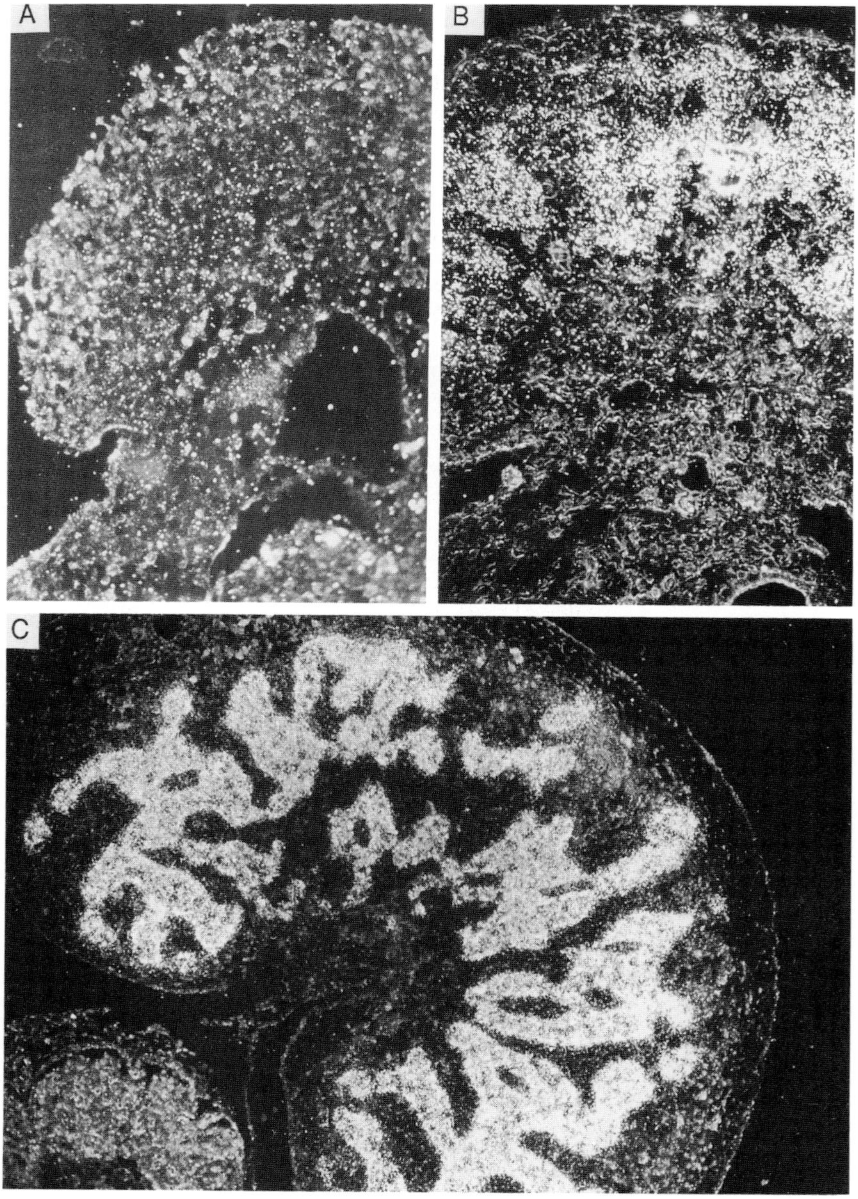

Figure 4 *In situ* hybridization of human fetal testicular tissue, using ^{35}S-labeled sense and antisense *AMH* riboprobes, according to the technique described by Münsterberg and Lovell-Badge (1991). Only the results obtained with the antisense probe are shown, no signal was elicited in any case using the sense probe. (A) 7-week-old fetus (from last menstrual period), with *SRY* DNA sequences detectable by PCR: no *AMH* transcripts are

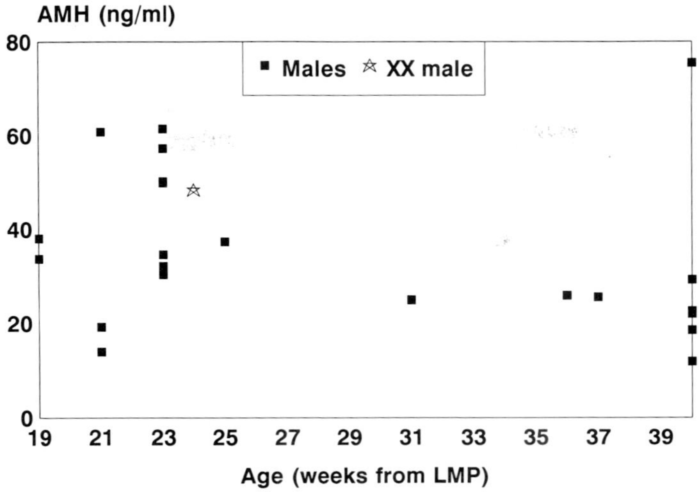

Figure 5 AMH concentration in human male fetuses, including 46,XX male fetus, measured by ELISA, with antibodies against human recombinant AMH, as described by Carré-Eusèbe *et al.* (1992). No AMH was detectable in female fetal serum at any age. With permission from Josso *et al.*, 1993a.

after birth, nor in the ewe, where growing follicles with 2- to 6-cell-deep granulosa cell layers are present already at 120 days gestation (Bézard *et al.*, 1987). In the mouse ovary, AMH transcripts, studied by *in situ* hybridization, are first detectable 6 days after birth (Münsterberg and Lovell-Badge, 1991). As mentioned above, no immunoreactive AMH can be detected in fetal female blood.

AMH expression and immunoreactivity are both dependent on the degree of follicular maturation, rather than on the age of the animal. As the follicle grows, the immunoreactivity of granulosa cells close to the basal lamina progressively fades, whereas that of those lining the antrum and occupying the cumulus oophorus increases (Bézard *et al.*, 1987; Ueno *et al.*, 1989b). AMH production appears to be confined to the stem cell population of granulosa cells, which are incapable of undergoing terminal differentiation under FSH stimulation (Erickson, 1983). Just before ovulation, when the oocyte completes the first meiotic division, AMH immunoreactivity (Ueno *et al.*, 1989b) and transcripts (Münsterberg and Lovell-Badge, 1991) become undetectable in cumulus cells.

---

detectable in this yet undifferentiated testicular tissue. (B) 8 weeks: *AMH* transcripts are detectable in the differentiating seminiferous tubules. (C) 11 weeks: seminiferous tubules are now completely developed and contain abundant *AMH* transcripts. Dark field illumination, ×125, after 4 days exposure. With permission from Josso *et al.*, 1993a.

Whatever the degree of follicular maturation or the age of the animal, the production of AMH by granulosa cells is low compared to that by immature Sertoli cells, as shown by Northern blots of messenger RNA extracted from gonads of both sexes (Picard *et al.*, 1986), and by differences in anti-Müllerian activity (Ueno *et al.*, 1989a). In contrast, when tested at similar concentrations, AMH purified from fetal testis or adult ovarian follicles has similar anti-Müllerian activity (Vigier *et al.*, 1984). Transcription initiation sites are similar in both tissues (Josso *et al.*, 1993b), suggesting that the same promoter drives the gene in the testis and in the ovary.

C. *AMH* Transgenes

Although the mechanisms responsible for the tight chronological regulation of the *AMH* gene are not elucidated at the present time, they play a key role in the normal development of the reproductive tract. In the male, AMH *must* be produced before the end of the Müllerian duct so-called critical period; later, Müllerian ducts lose their responsiveness to the hormone (Picon, 1969). In contrast, females *must not* be exposed to AMH while their gonads and genital ducts are still AMH-responsive. Disruption of the normal pattern of *AMH* gene expression has been achieved by transgenesis. By placing the coding region of the human *AMH* gene under the control of the metallothionein promoter, expressed early in fetal life in most cell types, Behringer *et al.* (1990) obtained a line of transgenic mice characterized by the absence of Müllerian duct derivatives in females. Ovaries were either absent or masculinized.

Others have attempted to prolong the expression of AMH by producing Sertoli cell tumors in mice bearing a transgene linking the SV40 T-antigen gene to the human *AMH* promoter (Peschon *et al.*, 1992). Although an immortal cell line derived from such a tumor displayed features characteristic of normal Sertoli cells, it did not express AMH, suggesting that the promoter does not contain the regulatory sequences responsible for curtailing gene expression at puberty.

V. *AMH* Gene Mutations

A. The Persistent Müllerian Duct Syndrome

The persistent Müllerian duct syndrome (PMDS) is a rare form of inherited male pseudohermaphroditism characterized by the presence of uterus and tubes in otherwise normally virilized males. The condition has been reported in approximately 150 patients to date. Some years ago, it was usually discovered in adult or adolescent patients undergoing surgical repair of cryptorchidism or inguinal hernia. Because early treatment of undescended testes is now advocated, the incidence in children has increased in recent years.

Because the testes are tightly linked to the Fallopian tubes in PMDS, the clinical picture depends on the degree of mobility of the Müllerian derivatives. Usually, the Müllerian derivatives are not fixed in the pelvis and are dragged into the inguinal canal by the descending testis. The inguinal hernia, called "*hernia uteri inguinalis*" in the literature, contains the uterus and tubes. Gentle traction on the uterus draws the controlateral tube and testis in the inguinal canal. Traction is not always necessary, because sometimes both testes are already in the same inguinal canal. This is known as "transverse testicular ectopia," almost never encountered when cases of PMDS are excluded. More rarely, the uterus and tubes are maintained in the pelvis by the round ligament, and the testes are in an "ovarian" position: clinically, the patients are bilaterally cryptorchid and no inguinal hernia is present.

Elucidation of PMDS etiology is feasible only during childhood, when the testis normally expresses large amounts of AMH. A lack or severe

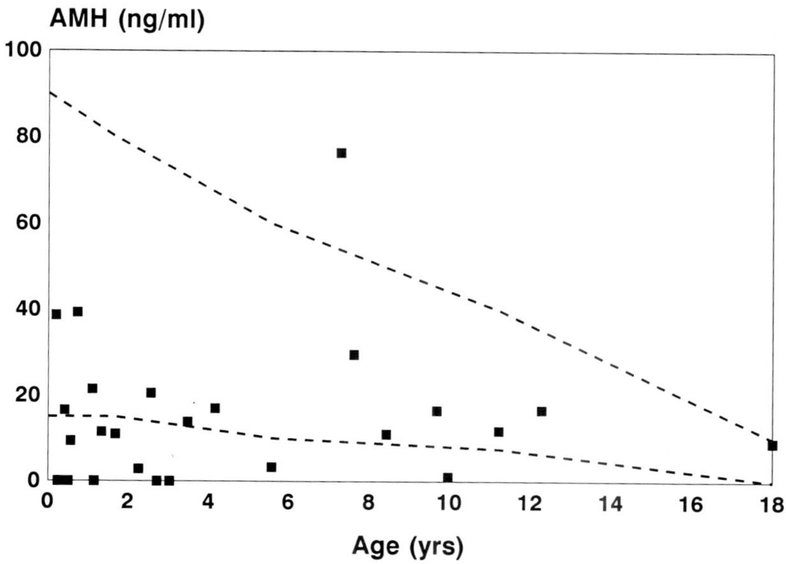

Figure 6 AMH serum concentrations in the persistent Müllerian duct syndrome. The normal range is indicated by the dashed lines.

decrease of testicular AMH production can be assessed by measuring the level of circulating AMH by ELISA (Fig. 6). When levels of circulating AMH are normal, a test of the anti-Müllerian activity of the testicular biopsy is required to distinguish between Müllerian duct insensitivity and *AMH* gene mutation affecting the bioactivity but not the immunoreactivity of the molecule. This test gives valid results in children below the age of 2 years (Donahoe *et al.*, 1977). On the other hand, if serum AMH is undetectable or very low, the likelihood of an *AMH* gene mutation is high.

PMDS is usually transmitted as an autosomal recessive trait, in keeping with the autosomal location of the *AMH* gene. However, two pedigrees with apparently sex-linked PMDS transmission have been reported (Sloan and Walsh, 1976; Naguib *et al.*, 1989). The AMH status of these patients is unknown. X-linked transmission could be explained in the AMH-positive form if the gene for the AMH receptor, which may be involved, were located on the X chromosome. Against this hypothesis is the fact that, in dogs, AMH-positive PMDS is transmitted as an autosomal recessive trait (Meyers-Wallen *et al.*, 1989) and one would expect X-linkage to be conserved between species. The problem will be resolved when X-linked cases of PMDS become available for testing and the AMH receptor is cloned.

B. Genetic Analysis of *AMH* Gene Mutations

At the time of writing (January 1993), our laboratory has analyzed DNA from 23 PMDS patients with low or undetectable serum AMH concentrations, and from their families. Three siblings were homozygous for a stop mutation in the fifth exon; the truncated protein was detectable in the culture medium of transfected CHO cells (Knebelmann *et al.*, 1991). One compound heterozygote had a microdeletion in the maternal allele and a stop mutation in the paternal one (Carré-Eusèbe *et al.*, 1992). In the other patients, various missense mutations were distributed over the whole length of the gene (Fig. 7), particularly exon 1 (Loeff *et al.*, 1993). This is in keeping with the recent demonstration that the N-terminal fragment of AMH plays an important role in preserving the bioactivity of the molecule (Wilson *et al.*, 1993). However, caution must be exercised before clinical abnormalities are ascribed to particular nucleotide changes, because the gene is highly polymorphic and many base substitutions have no deleterious effects on protein activity. Proteins affected by missense mutations were analyzed by hydrophobic cluster analysis (Lemesle-Varloot *et al.*, 1990) in an effort to predict modifications of secondary structures likely to affect protein stability, but definitive conclusions must

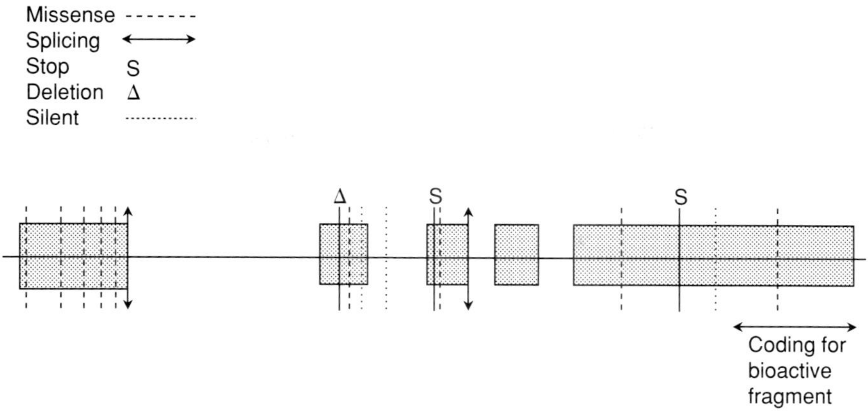

Figure 7 Nature and localization of polymorphism and mutations of the *AMH* gene, observed in patients affected by the persistent Müllerian duct syndrome and in their families.

await study of the mutated gene product in a suitable expression system. In the families of patients homozygous for an *AMH* gene mutation, heterozygous males were normal. One mother found to be homozygous for the mutation affecting her PMDS son was also clinically unaffected. PMDS patients undergo normal pubertal maturation, but information regarding fertility is lacking. In contrast, PMDS-affected dogs with unilateral or bilateral scrotal testes are fertile, despite relatively low sperm counts (Meyers-Wallen *et al.,* 1989).

In conclusion, neither experimental data nor clinical investigation support the hypothesis that AMH is involved in physiological functions other than Müllerian duct regression. This is surprising, given the relationship of AMH to the TGF-β superfamily and the production of AMH by adult Sertoli and granulosa cells. However, fact must prevail over speculation and, until proven otherwise, the *AMH* gene product must be content with its eponymous function: repression of the female component of the reproductive tract.

References

Baker, M. L., Metcalfe, S. A., and Hutson, J. M. (1990). Serum levels of Müllerian inhibiting substance in boys from birth to 18 years, as determined by enzyme immunoassay. *J. Clin. Endocrinol. Metab.* **70,** 11–15.

Barton, D. E., Yang-Feng, T. L., Mason, A. J., Seeburg, P. H., and Francke, U. (1987). INHA, INHBA, and INHBB, the loci for the three subunits of inhibin, mapped in mouse and man. *Cytogenet. Cell. Genet.* **46,** 578.

Behringer, R. R., Cate, R. L., Froelick, G. J., Palmiter, R. D., and Brinster, R. L. (1990). Abnormal sexual development in transgenic mice chronically expressing Müllerian inhibiting substance. *Nature (London)* **345,** 167–170.

Bézard, J., Vigier, B., Tran, D., Mauléon, P., and Josso, N. (1987). Immunocytochemical study of anti-Müllerian hormone in sheep ovarian follicles during fetal and postnatal development. *J. Reprod. Fertil.* **80,** 509–516.

Carré-Eusèbe, D., Imbeaud, S., Harbison, M., New, M. I., Josso, N., and Picard, J. Y. (1992). Variants of the anti-Müllerian hormone gene in a compound heterozygote with the persistent Müllerian duct syndrome and his family. *Hum. Genet.* **90,** 389–394.

Cate, R. L., Mattaliano, R. J., Hession, C., Tizard, R., Farber, N. M., Cheung, A., Ninfa, E. G., Frey, A. Z., Gash, D. J., Chow, E. P., Fisher, R. A., Bertonis, J. M., Torres, G., Wallner, B. P., Ramachandran, K. L., Ragin, R. C., Manganaro, T. F., MacLaughlin, D. T., and Donahoe, P. K. (1986). Isolation of the bovine and human genes for Müllerian inhibiting substance and expression of the human gene in animal cells. *Cell (Cambridge, Mass.)* **45,** 685–698.

Cate, R. L., Donahoe, P. K., and MacLaughlin, D. T. (1990). Müllerian-inhibiting substance. *In* "Handbook of Experimental Pharmacology" (M. B. Sporn and A. B. Roberts, eds.), Vol. 95, Part II, pp. 179–210. Springer-Verlag, Berlin.

Chin, T. W., Parry, R. L., and Donahoe, P. K. (1991). Human Müllerian Inhibiting Substance inhibits tumor growth in vitro and in vivo. *Cancer Res.* **51,** 2101–2106.

Cohen-Haguenauer, O., Picard, J. Y., Mattei, M. G., Serero, S., Nguyen, V. C., de Tand, M. F., Guerrier, D., Hors-Cayla, M. C., Josso, N., and Férzal, J. (1987). Mapping of the gene for anti-Müllerian hormone to the short arm of human chromosome 19. *Cytogenet. Cell. Genet.* **44,** 2–6.

di Clemente, N., Ghaffari, S., Pepinsky, R. B., Pieau, C., Josso, N., Cate, R. L., and Vigier, B. (1992). A quantitative and interspecific test for biological activity of anti-Müllerian hormone: The fetal ovary aromatase assay. *Development (Cambridge, UK)* **114,** 721–727.

Donahoe, P. K., Ito, Y., Morikawa, Y., and Hendren, W. H., III (1977). Müllerian inhibiting substance in human testes after birth. *J. Pediatr. Surg.* **12,** 323–330.

Dorizzi, M., Mignot, T. M., Guichard, A., Desvages, G., and Pieau, C. (1991). Involvement of oestrogens in sexual differentiation of gonads as a function of temperature in turtles. *Differentiation (Berlin)* **47,** 9–17.

Elbrecht, A., and Smith, R. G. (1992). Aromatase enzyme activity and sex determination in chickens. *Science* **255,** 467–470.

Erickson, G. F. (1983). Primary cultures of ovarian cells in serum-free medium as models of hormone-dependent differentiation. *Mol. Cell. Endocrinol.* **29,** 21–49.

Fujii, D. M., Brissenden, J. E., Derynck, R., and Francke, U. (1986). Transforming growth factor-β gene maps to human chromosome 19 long arm and to mouse chromosome 7. *Somatic Cell. Mol. Genet.* **12,** 281–288.

Fuller, A. F., Guy, S., Budzik, G. P., and Donahoe, P. K. (1982). Müllerian inhibiting substance inhibits colony growth of a human ovarian carcinoma cell line. *J. Clin. Endocrinol. Metab.* **54,** 1051–1055.

Guerrier, D., Boussin, L., Mader, S., Josso, N., Kahn, A., and Picard, J. Y. (1990). Expression of the gene for anti-Müllerian hormone. *J. Reprod. Fertil.* **88,** 695–706.

Haqq, C., Lee, M. M., Tizard, R., Wysk, M., DeMarinis, J., Donahoe, P. K., and Cate, R. L. (1992). Isolation of the rat gene for Müllerian inhibiting substance. *Genomics* **12,** 665–669.

Harley, V. R., Jackson, D. I., Hextall, P. J., Hawkins, J. R., Berkovitz, G. D., Sockanathan, S., Lovell-Badge, R., and Goodfellow, P. N. (1992). DNA binding activity of recombinant SRY from normal males and XY females. *Science* **255,** 453–456.

Hayashi, H., Shima, H., Hayashi, K., Trelstad, R. L., and Donahoe, P. K. (1984). Immunocytochemical localization of Müllerian inhibiting substance in the rough endoplasmic reticulum and Golgi apparatus in Sertoli cells of the neonatal calf testis using a monoclonal antibody. *J. Histochem. Cytochem.* **32,** 649–654.

Hudson, P. L., Dougas, I., Donahoe, P. K., Cate, R. L., Epstein, J., Pepinsky, R. B., and MacLaughlin, D. T. (1990). An immunoassay to detect human Müllerian inhibiting substance in males and females during normal development. *J. Clin. Endocrinol. Metab.* **70,** 16–22.

Hutson, J. M., Fallat, M. E., Kamagata, S., Donahoe, P. K., and Budzik, G. P. (1984). Phosphorylation events during Müllerian duct regression. *Science* **233,** 586–589.

Josso, N. (1974). Müllerian inhibiting activity of human fetal testicular cells deprived of germ cells by in vitro irradiation. *Pediatr. Res.* **8,** 755–758.

Josso, N., Legeai, L., Forest, M. G., Chaussain, J. L., and Brauner, R. (1990). An enzyme-linked immunoassay for anti-Müllerian hormone: A new tool for the evaluation of testicular function in infants and children. *J. Clin. Endocrinol. Metab.* **70,** 23–27.

Josso, N., Lamarre, I., Picard, J. Y., Berta, P., Davies, N., Morichon, N., Peschanski, M., and Jeny, R. (1993a). Anti-Müllerian hormone in early development. *Early Hum. Dev.* **33,** 91–99.

Josso, N., Cate, R. L., Picard, J. Y., Vigier, B., di Clemente, N., Wilson, C., Imbeaud, S., Pepinsky, R. B., Guerrier, D., Boussin, L., Legeai, L., and Carré-Eusèbe, D. (1993b). Anti-Müllerian hormone, the Jost factor. *Recent Prog. Horm. Res.* **48,** 1–59.

Jost, A. (1953). Problems of fetal endocrinology: The gonadal and hypophyseal hormones. *Recent Prog. Horm. Res.* **8,** 379–418.

King, T. R., Lee, B. K., Behringer, R. R., and Eicher, E. M. (1991). Mapping anti-Müllerian hormone (Amh) and related sequences in the mouse: Identification of a new region of homology between MMU10 and HSA19p. *Genomics* **11,** 273–283.

Knebelmann, B., Boussin, L., Guerrier, D., Legeai, L., Kahn, A., Josso, N., and Picard, J. Y. (1991). Anti-Müllerian hormone Bruxelles: A nonsense mutation associated with the persistent Müllerian duct syndrome. *Proc. Natl. Acad. Sci. U.S.A.* **88,** 3767–3771.

Lee, M. M., Cate, R. L., Donahoe, P. K., and Waneck, G. L. (1992). Developmentally regulated polyadenylation of two discrete messenger ribonucleic acids for Müllerian inhibiting substance. *Endocrinology (Baltimore)* **130,** 847–853.

Lee, W., Mason, A. J., Schwall, R., Szonyi, E., and Mather, J. P. (1989). Secretion of activin by interstitial cells in the testis. *Science* **243,** 396–397.

Lemesle-Varloot, L., Henrissat, B., Gaboriaud, C., Bissery, V., Morgat, A., and Mornon, J. P. (1990). Hydrophobic cluster analysis—procedures to derive structural and functional information from 2-D-representation of protein sequences. *Biochimie* **72,** 555–574.

Loeff, D. S., Imbeaud, S., Reyes, H. M., Meller, J. L. and Rosenthal, I. M. (1993). Surgical and genetic aspects of persistent Müllerian duct syndrome. *J. Pediatr. Surg.* (in press).

MacLaughlin, D. T., Hudson, P., Graciano, A. L., Kenneally, M. K., Ragin, R. C., Manganaro, T. F., and Donahoe, P. K. (1992). Müllerian duct regression and antiproliferative activities of Müllerian inhibiting substance reside in its carboxy-terminal domain. *Endocrinology (Baltimore)* **313,** 291–296.

Meyers-Wallen, V. N., Donahoe, P. K., Ueno, S., Manganaro, T. F., and Patterson, D. F. (1989). Müllerian inhibiting substance is present in testes of dogs with persistent Müllerian duct syndrome. *Biol. Reprod.* **41,** 881–888.

Münsterberg, A., and Lovell-Badge, R. (1991). Expression of the mouse anti-Müllerian hormone gene suggests a role in both male and female sex differentiation. *Development (Cambridge, UK)* **113,** 613–624.

Naguib, K. K., Teebi, A. S., Al-Awadi, S. A., El-Khalifa, M. Y., and Mahfouz, E. S. (1989). Familial uterine hernia syndrome: report of an Arab family with four affected males. *Am. J. Hum. Genet.* **33,** 180–181.

Nasrin, N., Buggs, C., Kong, X. F., Carnazza, J., Goebl, M., and Alexander-Bridges, M. (1991). DNA-binding properties of the product of the testis-determining gene and a related protein. *Nature (London)* **354,** 317–320.

Newbold, R. R., Suzuki, Y., and MacLachlan, J. A. (1984). Müllerian duct maintenance in heterotypic organ culture after in vivo exposure to diethylstilbestrol. *Endocrinology (Baltimore)* **115,** 1863–1868.

Parry, R. L., Chin, T., Epstein, J., Hudson, P. L., Powell, D. M., and Donahoe, P. K. (1992). Recombinant human Müllerian inhibiting substance inhibits human ocular melanoma cell lines in vitro and in vivo. *Cancer Res.* **52,** 1182–1186.

Pepinsky, R. B., Sinclair, L. K., Chow, E. P., Mattaliano, R. J., Manganaro, T. F., Donahoe, P. K., and Cate, R. L. (1988). Proteolytic processing of Müllerian inhibiting substance produces a transforming growth factor-β-like fragment. *J. Biol. Chem.* **263,** 18961–18965.

Peschon, J. J., Behringer, R. R., Cate, R. L., Harwood, K. A., Idzerda, R. L., Brinster, R. L., and Palmiter, R. D. (1992). Directed expression of an oncogene to Sertoli cells in transgenic mice using Müllerian inhibiting substance regulatory sequences. *Mol. Endocrinol.* **6,** 1403–1411.

Picard, J. Y., and Josso, N. (1984). Purification of testicular anti-Müllerian hormone allowing direct visualization of the pure glycoprotein and determination of yield and purification factor. *Mol. Cell. Endocrinol.* **34,** 23–29.

Picard, J. Y., Tran, D., and Josso, N. (1978). Biosynthesis of labelled anti-Müllerian hormone by fetal testes: Evidence for the glycoprotein nature of the hormone and for its disulfide-bonded structure. *Mol. Cell. Endocrinol.* **12,** 17–30.

Picard, J. Y., Benarous, R., Guerrier, D., Josso, N., and Kahn, A. (1986). Cloning and expression of cDNA for anti-Müllerian hormone. *Proc. Natl. Acad. Sci. U.S.A.* **83,** 5464–5468.

Picon, R. (1969). Action du testicule foetal sur le développement in vitro des canaux de Müller chez le rat. *Arch. Anat. Microsc. Morphol. Exp.* **58,** 1–19.

Picon, R. (1970). Modifications, chez le rat, au cours du développement du testicule, de son action inhibitrice sur les canaux de Müller in vitro. *C. R. Hebd. Acad. Sci. Seances Sér. D* **271,** 2370–2372.

Rao, V. V. N. G., Löffler, C., Wozney, J. M., and Hansmann, I. (1992). The gene for bone morphogenetic protein-2a (BMP2a) is localized to human chromosome-20p12 by radioactive and nonradioactive in situ hybridization. *Hum. Genet.* **90,** 299–302.

Rey, R., Lordereau-Richard, I., Carel, J. C., Barbet, P., Cate, R. L., Roger, M., Chaussain, J. L., and Josso, N. (1993). Anti-Müllerian hormone and testosterone serum levels are inversely related during normal and precocious pubertal development. *J. Clin. Endocrinol. Metab.* (in press).

Roberts, V., Meunier, H., Sawchenko, P. E., and Vale, W. (1989). Differential production and regulation of inhibin subunits in rat testicular cell types. *Endocrinology (Baltimore)* **125,** 2350–2359.

Rogers, D. S., Gallagher, D. S., and Womack, J. E. (1991). Somatic cell mapping of the genes for anti-Müllerian hormone and osteonectin in cattle: Identification of a new bovine syntenic group. *Genomics* **9,** 298–300.

Rosenwaks, Z., Liu, H. C., Picard, J. Y., and Josso, N. (1984). Anti-Müllerian hormone is not cytotoxic to human endometrial cancer in tissue culture. *J. Clin. Endocrinol. Metab.* **59,** 166–169.

Shaha, C., Morris, P. L., Chen, C. L. C., Vale, W., and Bardin, C. W. (1989). Immunostainable inhibin subunits are in multiple types of testicular cells. *Endocrinology (Baltimore)* **125,** 1941–1950.

Sloan, W. R., and Walsh, P. C. (1976). Familial persistent müllerian duct syndrome. *J. Urol.* **115,** 459–461.

Tran, D., and Josso, N. (1982). Localization of anti-Müllerian hormone in the rough endoplasmic reticulum of the developing bovine Sertoli cell using immunocytochemistry with a monoclonal antibody. *Endocrinology (Baltimore)* **111,** 1562–567.

Tran, D., Meusy-Dessole, N., and Josso, N. (1977). Anti-Müllerian hormone is a functional marker of foetal Sertoli cells. *Nature (London)* **269,** 411–412.

Ueno, S., Manganaro, T. F., and Donahoe, P. K. (1988). Human recombinant Müllerian inhibiting substance inhibition of rat oocyte meiosis is reversed by epidermal growth factor in vitro. *Endocrinology (Baltimore)* **123,** 1652–1659.

Ueno, S., Takahashi, M., Manganaro, T. F., Ragin, R. C., and Donahoe, P. K. (1989a). Cellular localization of Müllerian inhibiting substance in the developing rat ovary. *Endocrinology (Baltimore)* **124,** 1000–1006.

Ueno, S., Kuroda, T., MacLaughlin, D. T., Ragin, R. c., Manganaro, T. F., and Donahoe, P. K. (1989b). Müllerian inhibiting substance in the adult rat ovary during various stages of the estrous cycle. *Endocrinology (Baltimore)* **125,** 1060–1066.

Vigier, B., Tran, D., Du Mesnil du Buisson, F., Heyman, Y., and Josso, N. (1983). Use of monoclonal antibody techniques to study the ontogeny of bovine anti-Müllerian hormone. *J. Reprod. Fertil.* **69,** 207–214.

Vigier, B., Picard, J. Y., Tran, D., Legeai, L., and Josso, N. (1984). Production of anti-Müllerian hormone: Another homology between Sertoli and granulosa cells. *Endocrinology (Baltimore)* **114,** 1315–1320.

Vigier, B., Watrin, F., Magre, S., Tran, D., and Josso, N. (1987). Purified bovine AMH induces a characteristic freemartin effect in fetal rat prospective ovaries exposed to it in vitro. *Development (Cambridge, UK)* **100,** 43–55.

Vigier, B., Forest, M. G., Eychenne, B., Bézard, J., Garrigou, O., Robel, P., and Josso, N. (1989). Anti-Müllerian hormone produces endocrine sex-reversal of fetal ovaries. *Proc. Natl. Acad. Sci. U.S.A.* **86,** 3684–3688.

Wallen, J., Cate, R. L., Kiefer, D. M., Riemen, M. W., Martinez, D., Hoffman, R. M., Donahoe, P. K., von Hoff, D. D., Pepinsky, R. B., and Oliff, A. (1989). Minimal antiproliferative effect of recombinant Müllerian inhibiting substance on gynecological tumor cell lines and tumor explants. *Cancer Res.* **49,** 2005–2011.

Wilson, C., di Clemente, N., Ehrenfels, C., Pepinsky, R. B., Josso, N., Vigier, B., and Cate, R. L. (1993). Müllerian inhibiting substance requires its N-terminal domain for maintenance of biological activity, a novel finding within the TGF-β superfamily. *Mol. Endocrinol.* **7,** 247–257.

Winter, J. S. D., Faiman, C., and Reyes, F. I. (1977). Sex steroid production by the human fetus: Its role in morphogenesis and control by gonadotropins. *In* "Morphogenesis and Malformation of the Genital System" (D. B. Bergsma and R. J. Blandau, eds.), pp. 41–58. Alan R. Liss, New York.

19

Müllerian-Inhibiting Substance: Critical Roles in Sexual Differentiation

Michael L. Gustafson and Patricia K. Donahoe
Pediatric Surgical Research Laboratories and the Department of Surgery
Massachusetts General Hospital and Harvard Medical School
Boston, Massachusetts 02114

Molecular Genetics of Sex Determination

I. Introduction

In 1947, Alfred Jost described the existence of a "Müllerian inhibitor," which he hypothesized was responsible for regression of the Müllerian ducts in the male mammalian embryo. This hormone, first called Müllerian-inhibiting substance (MIS) (Josso, 1972), was later shown to be a high-molecular-weight protein secreted by the Sertoli cell (Blanchard and Josso, 1974). Subsequent efforts to investigate the various functions of MIS during fetal and postnatal sexual development relied upon crude, and then homogeneous, MIS preparations purified from bovine testes (Budzik *et al.*, 1980, 1983; Picard and Josso, 1984). However, successful cloning of the full-length bovine and human MIS genes in 1986 by Cate *et al.* allowed production of highly purified and biologically active recombinant human protein for such studies (Pepinsky *et al.*, 1988; MacLaughlin *et al.*, 1991; Ragin *et al.*, 1992). This material has also been used to study the proteolytic processing steps necessary to activate MIS. Cloning of the mouse (Munsterberg and Lovell-Badge, 1991) and rat (Haqq *et al.*, 1992) MIS genes and sequencing of their upstream regions has likewise facilitated investigation of the developmental and sexually dimorphic regulation of MIS expression, which appears to be an essential downstream step in the *SRY*-induced pathway of male sexual differentiation (Haqq *et al.*, 1993). Most recently, cloning of a candidate MIS receptor gene (He *et al.*, 1993) is allowing analysis of the intracellular mechanisms by which MIS controls cellular differentiation and proliferation. In this review, we emphasize the molecular approaches and tools that have been used to advance our understanding of MIS from a "Müllerian inhibitor" as described insightfully by Jost (1947) to a hormone now believed to be critical for several facets of sexual differentiation.

II. Biochemistry of MIS

A. Protein Structure

MIS was first isolated and characterized from fetal (Picard and Josso, 1984) and postnatal (Budzik *et al.*, 1983) bovine testes, with more recent analyses performed on human MIS produced recombinantly (rhMIS) (Cate *et al.*, 1986; Pepinsky *et al.*, 1988; MacLaughlin *et al.*, 1991). MIS is a 140,000 MW glycoprotein composed of two identical 70-kDa disulfide-linked subunits. Each subunit is initially synthesized as a 553 (rat) to 575 (bovine) amino acid precursor, with a short leader sequence containing a putative signal peptide (von Heijne, 1986) and a prosequence (Fig. 1). There is marked homology between the protein sequences of bovine, rat, and human MIS, particularly of 11 to 12 cysteine residues and their carboxy (C)-terminal domains; human MIS shares 108 of 112 C-terminal amino acids with bovine MIS and 104 of 112 amino acids with rat MIS.

The dimeric structure of MIS and the presence of seven conserved cysteine residues in its C-terminal domain place MIS within a large family of peptide growth factors having diverse roles in cell regulation and differentiation. Members of this family include the transforming growth factor-βs (Derynck *et al.*, 1985), inhibins (Mason *et al.*, 1985), activins (Ling *et al.*, 1986; Vale *et al.*, 1986), Vg-1 (Weeks and Melton, 1987), decapentaplegia complex (Padgett *et al.*, 1987), and bone morphogenesis factors (Wozney *et al.*, 1988). A number of these growth factors exist as latent dimeric precursors, possessing 20–40% C-terminal amino acid homology, that undergo proteolytic cleavage to release biologically active disulfide-linked C-terminal fragments (Cate *et al.*, 1990) (Fig. 1). The structural conservation of these proteins over an extended evolutionary period attests to their functional importance in cell growth and regulation. As described below, however, "bioactive" C-terminal rhMIS has been isolated only recently and purified for biological characterization (MacLaughlin *et al.*, 1992a).

rhMIS is prepared in our laboratory from media of Chinese hamster ovary (CHO) cells transfected with the human MIS gene driven by an SV40 early promoter (Cate *et al.*, 1986; MacLaughlin *et al.*, 1991); purification is by immunoaffinity chromatography with a murine monoclonal anti-rhMIS antibody column (MacLaughlin *et al.*, 1991) and low pH elution. Resulting preparations are 90–95% purified, undergo time-dependent proteolytic cleavage at 37°C, and are biologically active in a urogenital ridge bioassay. A preelution high-salt wash step further increases the rhMIS purity and increases its antiproliferative effects in anti-tumor assays (Chin *et al.*, 1991; Ragin *et al.*, 1992).

In reduced SDS–polyacrylamide gels, rhMIS appears primarily as a doublet at 70 kDa, with two minor bands at 57 and 12.5 kDa (Ragin *et al.*, 1992). These two bands correspond to amino (N)- and C-terminal MIS fragments and can be best appreciated after plasmin cleavage of holo-rhMIS (Pepinsky *et al.*, 1988). Amino acid (aa) sequence analysis has demonstrated that this cleavage occurs at arginine 427–serine 428 (of a total 535 aa for rhMIS) at a monobasic consensus cleavage site (Benoit *et al.*, 1987), rather than at a dibasic or tetrabasic site as in other members of the TGF-β family (Fig. 1). Since MIS N- or C-terminal specific antibodies fail to separate these fragments by immunoaffinity techniques, it is believed that they remain in noncovalent association

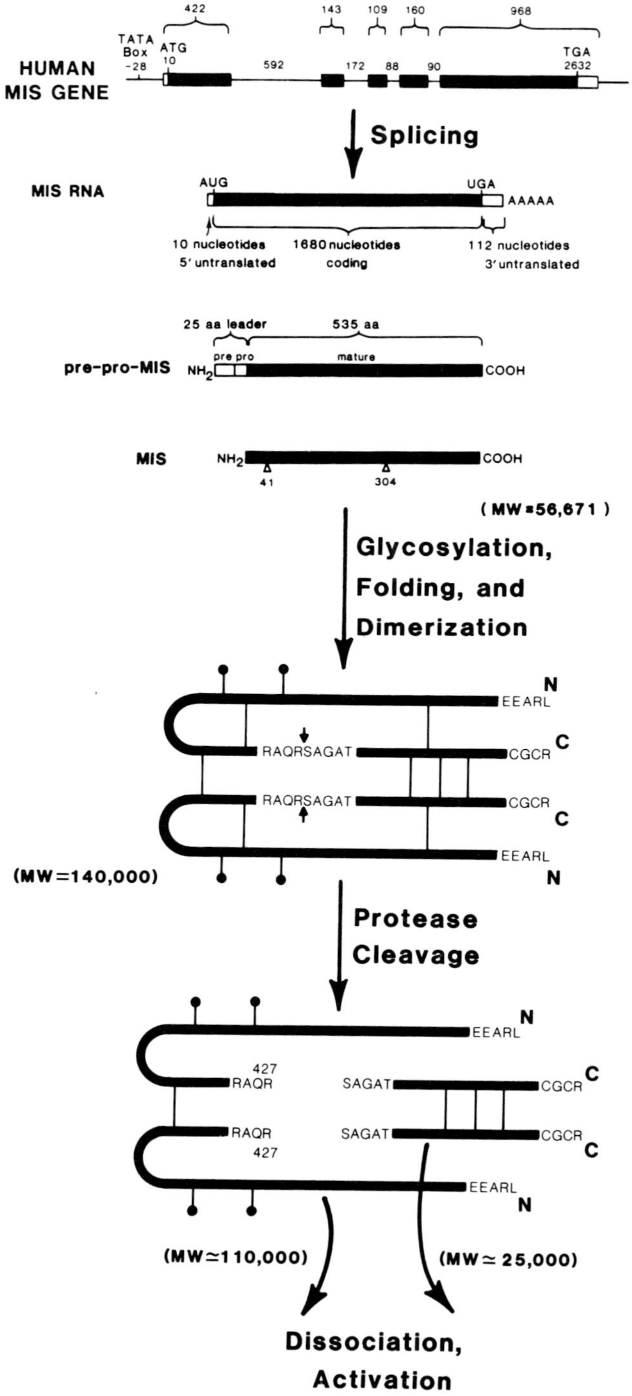

HUMAN MIS GENE
TATA Box -28
ATG 10
422
592
143
172
109
88
160
90
968
TGA 2632
Splicing
MIS RNA
AUG
UGA
AAAAA
10 nucleotides 5′ untranslated
1680 nucleotides coding
112 nucleotides 3′ untranslated
25 aa leader
535 aa
pre pro
mature
pre-pro-MIS
NH2
COOH
MIS
NH2
COOH
41
304
(MW=56,671)
Glycosylation, Folding, and Dimerization
N
EEARL
RAQRSAGAT
CGCR
C
RAQRSAGAT
CGCR
C
EEARL
N
(MW=140,000)
Protease Cleavage
EEARL
N
427
RAQR
SAGAT
CGCR
C
RAQR
SAGAT
CGCR
C
427
EEARL
N
(MW≃110,000)
(MW≃ 25,000)
Dissociation, Activation

after cleavage (Cate *et al.*, 1990; MacLaughlin *et al.*, 1992a). Mutation of the above Arg-427 site to a threonine results in noncleavable MIS. The demonstration that this form is biologically inactive has provided important experimental evidence that proteolytic processing of this hormone is requisite *in vivo* (Cate *et al.*, 1990). Additional experiments have shown that prolonged exposure of rhMIS to plasmin results in further cleavage of the N-terminal piece to 34- and 22-kDa fragments (Ragin *et al.*, 1992), with the secondary cleavage site likely at arginine 229–serine 230. The biological significance of this cleavage is still unclear, but preparations containing such secondary fragments remain biologically active. The carbohydrate content of rhMIS approaches 18% by weight, with two potential N-linked glycosylation sites within the N-terminus; removal of such carbohydrate with endoglycosidase F does not alter bioactivity (Cate *et al.*, 1990). Immunohistochemical studies suggest that MIS is produced in the rough endoplasmic reticulum and then transported to the Golgi apparatus for glycosylation, prior to secretion from Golgi-derived vacuoles (Hayashi *et al.*, 1984).

Highly purified rhMIS C-terminus is now being produced for functional analysis. This has been accomplished by timed plasmin cleavage of holo-rhMIS followed by size exclusion chromatography in the presence of 1 *M* acetic acid (MacLaughlin *et al.*, 1992a). This process separates the earlier eluting N-terminus from the smaller C-terminal fragment (Fig. 2). Contrary to earlier evidence that both MIS fragments must be present for full biological function (Pepinsky *et al.*, 1988; Cate *et al.*, 1990), C-terminal rhMIS, without contaminating N-terminal fragments, has recently been shown to have full bioactivity in Müllerian duct regression and antiproliferative tumor assays (MacLaughlin *et al.*, 1992a). Amino-terminal MIS, meanwhile, is inactive in such studies at equimolar concentrations and does not enhance the observed C-terminal effects when the two fragments are added together. Therefore, as for other TGF-β-like molecules, the bioactivities of MIS appear to reside primarily in its C-terminal domain. However, Wilson *et al.* (1993) have recently reported enhanced bioactivity of C-terminal MIS in an ovarian aromatase bioassay after addition of a tenfold molar excess of MIS N-terminus, a novel finding within the TGF-β family. Efforts are

---

Figure 1 Human MIS transcription, translation, processing, and activation: The human MIS gene is composed of five exons. Translation of MIS mRNA produces a pre-prohormone from which a 25 amino acid leader sequence is removed. After glycosylation, folding, and dimerization, MIS is then cleaved at a monobasic consensus site (Arg 427–Ser 428) to produce amino- and carboxy-terminal fragments, which remain noncovalently associated until MIS activation occurs.

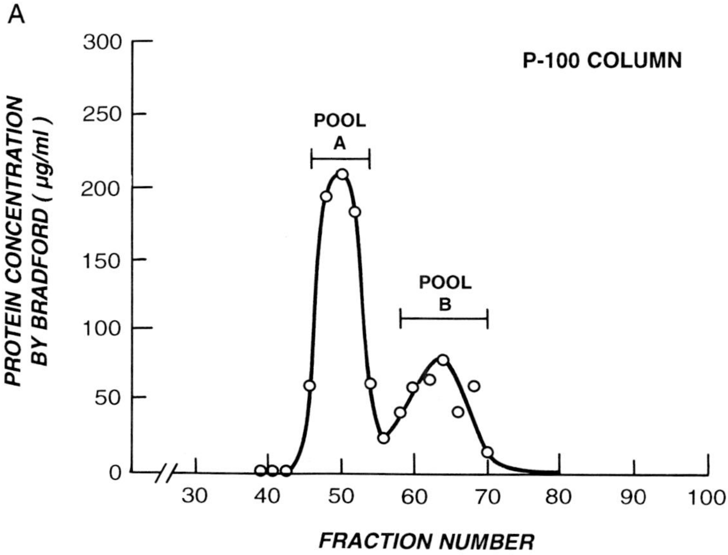

Figure 2 Purification of carboxy-terminal rhMIS: (A) A representative P-100 column chromatogram of a plasmin-cleaved rhMIS preparation. The protein content of each fraction is given on the ordinate (μg/ml) and the fraction number on the abscissa. The positions of the pools for amino-terminal (pool A) and carboxy-terminal (pool B) fragments are indicated by brackets. (B) Polyacrylamide gel electrophoresis of rhMIS under reducing conditions. Lanes 1 and 2 represent rhMIS before (IAP) and after (Plasmin) plasmin treatment; plasmin cleavage produces characteristic 55-, 34-, and 12.5–14-kDa bands. Lanes 3 and 4 represent pools A and B after P-100 fractionation of plasmin-cleaved rhMIS; lane 3 contains the N-terminal 55- and 34-kDa fragments, whereas lane 4 contains only the C-terminal 12.5- to 14-kDa bands. Adapted from MacLaughlin *et al.* (1992a).

under way to identify and isolate the endogenous proteases that activate MIS *in vivo* and that may confer an additional degree of tissue specificity to MIS in the physiologic setting (Losty, MacLaughlin, and Donahoe, unpublished data).

B. Mechanism of Action

Although the precise intracellular mechanisms of MIS activity have not been fully elucidated, evidence has been accumulating that this fetal regressor acts by altering membrane phosphorylation. Early work in this area focused on the antagonism observed between epidermal growth factor (EGF) and MIS in several biological systems. EGF, alone among

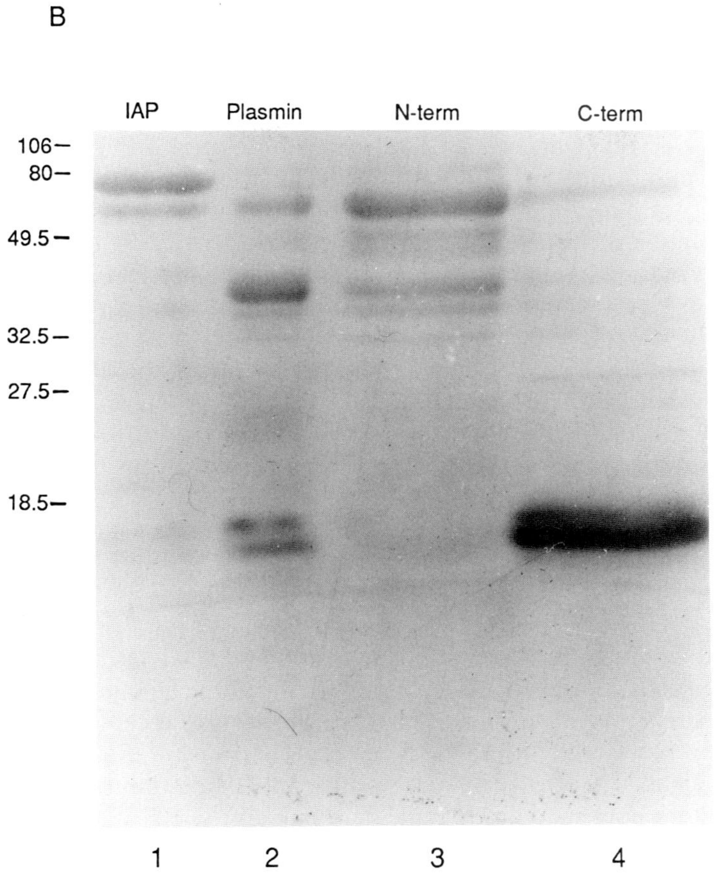

Figure 2 (*Continued*)

such proteins as platelet derived growth factor, insulin, nerve growth factor, and fibroblast growth factor, was initially shown to inhibit MIS-induced Müllerian duct regression in the organ culture bioassay (Hutson *et al.*, 1984; Budzik *et al.*, 1985). We have since shown that bovine MIS and holo-rhMIS can both inhibit the *in vitro* and *in vivo* growth of an EGF receptor-rich human vulvar epidermoid carcinoma cell line, A431 (Coughlin *et al.*, 1987; Chin *et al.*, 1991). Similar nonhomogeneous MIS preparations from both sources can inhibit EGF receptor tyrosine autophosphorylation of A431 membranes (Coughlin *et al.*, 1987; Cigarroa *et al.*, 1989) and fetal rat lung membranes (Catlin *et al.*,

1991). This effect, which is not due to competition for EGF binding sites, could not be reproduced with highly purified MIS preparations from either source, however. Further antagonism between MIS and EGF *in vitro* has been observed on germinal vesicle breakdown of cultured rat oocytes (Ueno *et al.*, 1988) and proliferative and progesterone production of human granulosa–luteal cells (Kim *et al.*, 1992). Furthermore, recent experiments have confirmed the original observations of MIS inhibition of EGF receptor tyrosine kinase activity, by use of highly purified C-terminal rhMIS (Maggard, Donahoe, and MacLaughlin, unpublished data). These findings suggest that membrane preparations of A431 cells lack the enzymes necessary to activate MIS. Processed, activated, and purified rhMIS C-terminus appears to be capable of blocking autophosphorylation of the EGF receptor. The interactions between the MIS receptor (see below) and the EGF receptor should provide a paradigm for understanding the downstream molecular events that mediate MIS activity.

C. Receptor Characterization

Early studies with MIS-responsive cell lines and fetal rat lung indicate that MIS exerts its effects on cell growth and differentiation via a specific cell surface receptor. Efforts to characterize this receptor, however, have been hampered by the low level of its expression in sensitive cells and the small size of the MIS target, the fetal Müllerian duct. Both of these factors have made it difficult to isolate and purify native receptor by traditional biochemical techniques. Specific and competitive binding to the cell lines A431 and HTB-111 (a human endometrial carcinoma) have been reported with labeled holo-rhMIS and an MIS anti-idiotypic antibody (generated against a murine anti-bovine MIS monoclonal antibody) (MacLaughlin *et al.*, 1992b). However, classical receptor studies have been limited by inability to radiolabel holo-rhMIS or C-terminal rhMIS without marked reduction of biological activity and binding (M. L. Gustafson, *et al.*, unpublished data). Based on sequence alignment with TGF-β and its recently reported three dimensional structure (Daopin *et al.*, 1992), it is hypothesized that an essential tyrosine residue in the "heel of the hand" receptor recognition site for MIS is altered by iodination. Therefore, other methods of producing and labeling MIS for receptor studies are being pursued in our laboratory.

Since COS cell expression cloning has been unsuccessful in isolating the MIS receptor from several cDNA libraries (constructed from A431, fetal lung, and fetal urogenital ridge RNA), we recently chose an alternative strategy for receptor identification. Probes from the newly cloned mouse activin and human TGF-β type II receptors (Matthews and Vale, 1991; Lin *et al.*, 1992) and the homologous *daf-1* product of *Caenorhabditis elegans* (Georgi *et al.*, 1990) were used to amplify urogenital ridge cDNA in the polymerase chain reaction (PCR). Each of these three proteins, which together constitute an emerging family of transmembrane serine/threonine kinase receptors, has a small extracellular ligand-binding domain, a single hydrophobic transmembrane domain, and a highly homologous intracellular kinase domain (Massague, 1992). Since MIS belongs to the TGF-β superfamily, we reasoned that the MIS receptor might be a similar serine/threonine kinase.

Therefore, the polymerase chain reaction and degenerate oligonucleotides designed from the homologous kinase domains of the above receptors were used to isolate four novel transmembrane serine/threonine kinases from a fetal rat urogenital ridge library (He *et al.*, 1993). Each of these proteins has greatest similarity with one another and with other members of this serine/threonine kinase family within the intracellular domain; this region has 12 subdomains of highly conserved residues, including a putative ATP binding site, and several characteristic serine/threonine kinase motifs. One of these four clones is an excellent candidate for the MIS receptor, based on specific localization by *in situ* hybridization to mesenchyme of the 14.5- to 15-day male fetal rat Müllerian duct and to oocytes of preantral and antral follicles of pubertal and adult ovaries (Fig. 3). Although both of these sites are known targets for MIS action, demonstration of specific MIS binding to this receptor clone has been hampered by decreased activity of labeled MIS preparations. It will be interesting to evaluate how EGF, via its tyrosine kinase receptor, and MIS, via its candidate serine/threonine kinase receptor, exert opposite effects on many biological processes via changes in membrane phosphorylation. Information on where and how their respective downstream signal transduction pathways intersect will bring new understanding to events regulating cellular proliferation and differentiation.

III. Ontogeny of MIS

A. Testis

The sexually dimorphic patterns of MIS production by the developing testis and ovary represent one of the critical developmental "switches" during early sexual development. Immature Sertoli cells begin pro-

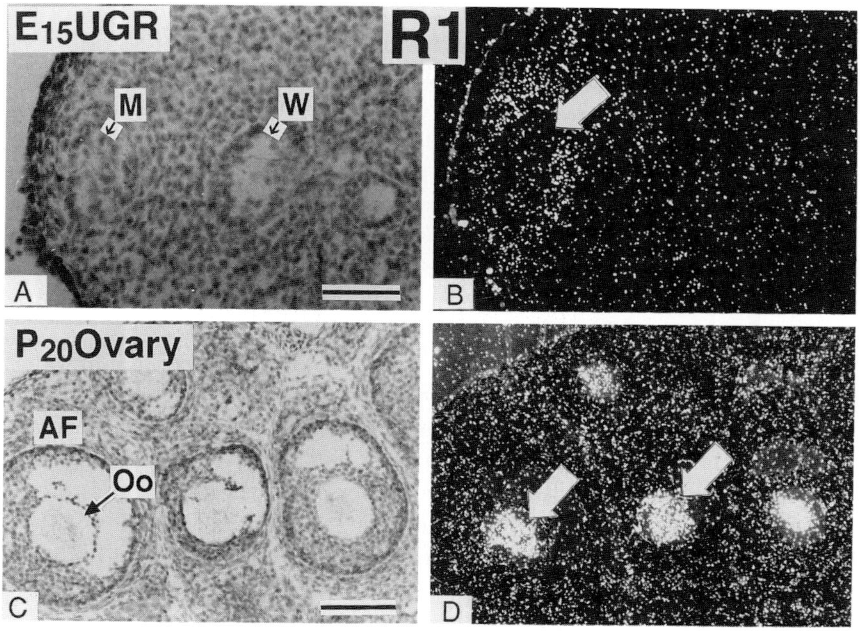

Figure 3 *In situ* hybridization of the candidate MIS receptor: A novel serine/threonine kinase transmembrane receptor (designated R1) with homology to the activin and TGF-β type II receptors localizes to two known sites of MIS action. (A,B) Light and dark field views of R1 hybridization signal in the 15-day (E15) fetal male urogenital ridge (UGR); R1 signal is observed over mesenchyme surrounding the Müllerian duct (M) (heavy arrow), but not the adjacent Wolffian duct (W). (C,D) R1 signal is also intense over the oocytes (Oo) (heavy arrow) of preantral and antral follicles (AF) of the postnatal day 20 (P20) ovary, with less intense signal seen in adjacent granulosa cells. (Bar: 100 μm) Adapted from He *et al.* (1993).

ducing MIS on gestational day 13 in the rat, prior to seminiferous tubule formation (Josso *et al.*, 1985; Hirobe *et al.*, 1992). As shown in Fig. 4, diffuse MIS signal is found over the gonad on the morning of day 13 (Fig. 4A); it is not until the afternoon of day 13, however, that primitive testicular cords are first observed and become associated with patchy MIS signal (Fig. 4B). By the following afternoon, intense MIS mRNA expression is evident within Sertoli cells of more mature testicular cords, whereas germ cells and interstitial cells have no MIS signal (Fig. 4C). MIS secretion then continues at relatively high levels until the fifth day after birth, after which levels decrease to a low baseline by puberty. These observations in the rat have been documented by bioassay of testicular fragments (Donahoe *et al.*, 1976; Ueno *et al.*, 1989a),

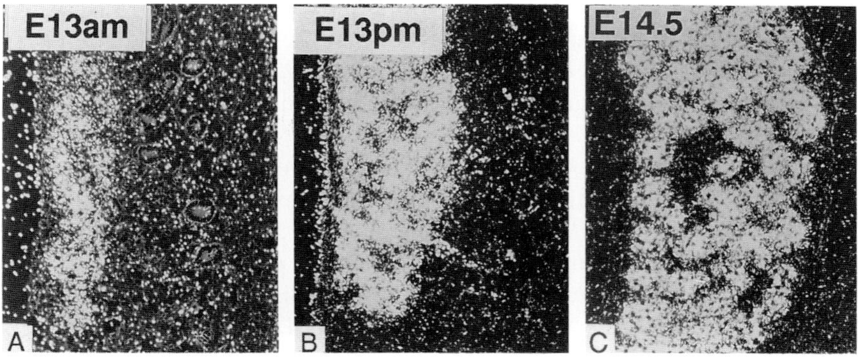

Figure 4 MIS *in situ* hybridization of the developing testis: (A) On the morning of fetal day 13 (E13am), when germ cells are scattered throughout the gonad and somatic cells are not yet organized, MIS signal is seen diffusely over the testis. (B) Later that day (E13pm), MIS expression increases in a patchy distribution as some somatic cells begin to form early testis cords. (C) By fetal day 14.5 (E14.5), testis cord formation is observed throughout the gonad, with intense MIS signal localized to Sertoli cells of the testis cords. Adapted from Hirobe *et al.* (1992).

immunohistochemistry (Tran *et al.*, 1987; Ueno *et al.*, 1989a; Kuroda *et al.*, 1990), Northern analysis (Kuroda *et al.*, 1990; Lee *et al.*, 1992), and most recently by *in situ* hybridization (Hirobe *et al.*, 1992). An identical pattern of MIS expression beginning on gestational day 12.5 has also been reported in the mouse (Munsterberg and Lovell-Badge, 1991). In the human male, MIS secretion is believed to begin by at least day 51 of gestation (Taguchi *et al.*, 1984). MIS then remains detectable until age 2 years by bioassay (Donahoe *et al.*, 1977a), age 6 years by immunocytochemistry (Tran *et al.*, 1987), and throughout adulthood at low levels by serum ELISA (Hudson *et al.*, 1990).

B. Ovary

It has only been within the last 10 years that the ontogeny and effects of MIS in the fetal and adult ovary have been appreciated. MIS was first detected in the chick ovary by bioassay (Hutson *et al.*, 1981), and then in bovine follicular fluid by RIA (Necklaws *et al.*, 1986). Subsequently, the site of MIS production was localized by immunocytochemistry to granulosa cells of the bovine ovary (Takahashi *et al.*, 1986a). Unlike the fetal rat testis, however, the developing fetal ovary has no detectable MIS expression by bioassay, immunohistochemistry, or Northern and *in situ* analyses (Ueno *et al.*, 1989a; Hirobe *et al.*, 1992). MIS expression is first detected in the rat ovary on postnatal day 7 by immunohistochemistry, day 4 by bioassay, and day 3 by mRNA studies; MIS mRNA transcripts then peak slightly during the peripubertal period and remain elevated throughout adulthood.

Earlier immunohistochemical and more recent *in situ* analyses have provided excellent cellular localization of MIS expression during follicular maturation (Ueno *et al.*, 1989b; Hirobe *et al.*, 1992). As shown in Fig. 5, MIS immunostaining is present in larger primary, preantral, and small antral follicles, but is not detectable within primordial, larger antral, or atretic follicles, or the corpus luteum. The earliest MIS signal is observed over the single layer of cuboidal granulosa cells of the larger primary follicles, with greatest message subsequently seen in granulosa cells closest to the oocyte of preantral follicles and in cumulus cells and periluminal granulosa cells of antral follicles; very little message, meanwhile, localizes to the cells lining the basement membrane of antral follicles. Such a pattern of MIS expression by granulosa cells suggests that a correlation exists with their mitotic activity (Hirobe *et al.*, 1992). In addition, immunohistochemistry has demonstrated that the MIS signal observed in granulosa cells of large preovulatory follicles on the morning (1000 AM) and afternoon (1600 PM) of proestrus decreases rapidly by 1 h after the LH surge (2300 PM) and the resumption of oocyte meiosis (Ueno *et al.*, 1989b). As described below, these combined findings suggest a potential role for MIS in the regulation of follicular development and oocyte maturation.

IV. Regulation of MIS Expression

A. Background

Genomic and/or cDNA clones have been isolated for MIS of the bovine and human (Cate *et al.*, 1986), mouse (Munsterberg and Lovell-Badge, 1991), and rat (Haqq *et al.*, 1992). The human and mouse MIS genes have been subsequently mapped to the short arm of chromosomes 19 and 10, respectively (Cohen-Hagenauer *et al.*, 1987; King *et al.*, 1991). There is striking homology of the MIS gene and its five exons across species, with 65–80% nucleotide identity between the coding regions and marked conservation of the intron–exon structure. In addition, there is 75% nucleotide conservation within the proximal 250 bp of the upstream 5′ regions, which decreases to 20% further upstream. Intensive investigation is under way to identify the *cis*-acting elements and

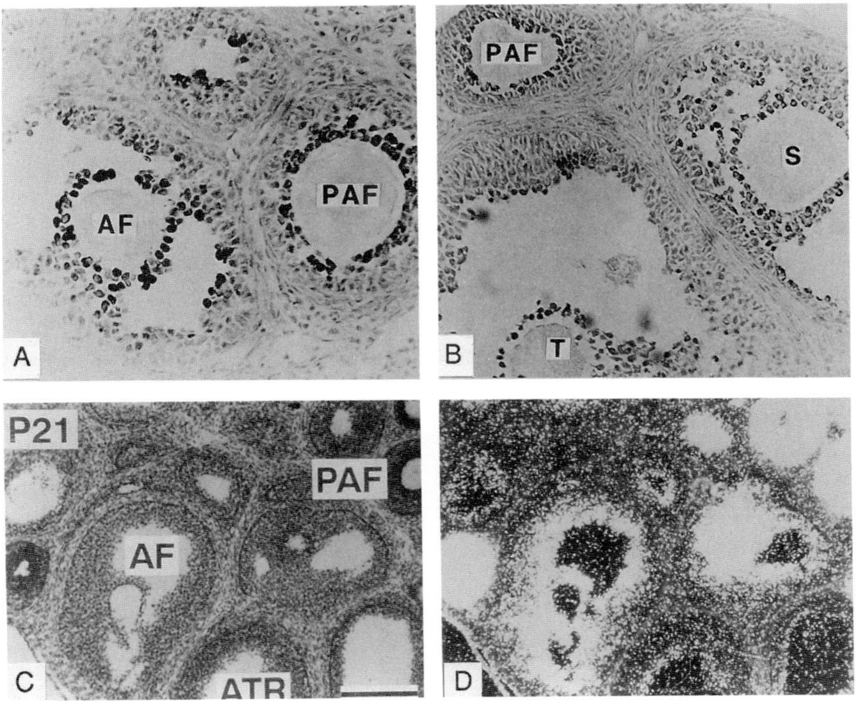

Figure 5 MIS in the peripubertal rat ovary: (A,B) Immunohistochemistry of the 27-day (A) and 21-day (B) postnatal rat ovary shows more intense staining of the inner granulosa cells and cumulus cells of preantral (PAF) and smaller antral (AF) follicles than that of larger antral follicles (S and T) (200×). (C,D) Light and dark field views of *in situ* hybridization of the 21-day postnatal ovary demonstrates MIS mRNA in the granulosa cells of preantral and antral follicles, but not of atretic follicles (ATR). Adapted from Ueno *et al.* (1989a) and Hirobe *et al.* (1992).

trans-acting factors responsible for the sexually dimorphic regulation of MIS secretion during development.

B. Sex-Determining Gene (*SRY*)

Since MIS is one of the first proteins produced by the developing testis prior to morphological differentiation, it seemed logical to search for a link between the putative sex-determining gene, *SRY*, and MIS in the molecular pathways leading to male gonadal differentiation. The tissue specificity and complementary ontogeny of *SRY* and MIS expression suggest that MIS transcription may be induced by *SRY*, and that MIS may then act to support continued testicular and sexual differentiation. In the mouse, *Sry* is expressed in the urogenital ridge between fetal days 10.5 and 12.5 (Koopman *et al.*, 1990) (corresponding to fetal days 11.5–13.5 in the rat), immediately prior to the onset of testicular differentiation and the expression of MIS mRNA and protein (Munsterberg and Lovell-Badge, 1991). In the rat, *SRY* expression has not been fully characterized, but sex cord formation occurs between fetal day 13, when *in situ* studies show that MIS mRNA is first expressed (Hirobe *et al.*, 1992), and fetal day 14.5, when discrete seminiferous tubule formation is first observed (Fig. 4).

The *SRY* gene encodes a protein containing an HMG (high mobility group)-box motif thought to be the site of DNA binding. To test our hypothesis that *SRY* is a transcriptional activator for MIS expression during early gonadogenesis, the 85-residue *SRY* HMG-box (bp 579–835 as identified by Sinclair *et al.*, 1990) was expressed in *Escherichia coli* as a thrombin-cleavable fusion protein with glutathione *S*-transferase. After purification by glutathione agarose binding, thrombin cleavage, and FPLC, the fidelity and folding of this protein (designated SRY-p) was verified by gel electrophoresis, N-terminal sequencing, amino acid analysis, and circular dichroism. This protein was then used for DNase protection assays of a human MIS promoter fragment from −139 to −14 bp (Haqq *et al.*, 1993). SRY-p protected nucleotides −75 to −45 from DNase digestion, with a dissociation constant of 8 n*M* (Fig. 6). We have now designated this sequence SRY*e*, as the putative SRY-binding element (Fig. 7) (Haqq *et al.*, 1993).

DNase protection analysis was subsequently performed with SRY-p and a fragment of the rat P450 aromatase promoter from −329 to −181 bp (Fig. 6). A *cis*-acting element in this region, homologous to the human MIS promoter SRY*e* sequence and also designated SRY*e* (Fig. 7), interacts with SRY-p at a dissociation constant of less than 4 n*M*. No SRY-p binding activity could be detected with control promoter fragments of the SV40 early region and the α-gonadotropin gene. We hypothesize that SRY acts differently on the MIS and P450 aromatase genes, since MIS expression is upregulated and aromatase activity downregulated during normal male embryogenesis (Fig. 8). Intensive studies are under way to uncover the nuances of DNA binding that might account for this differential regulation despite similar SRY binding sites.

Nuclear extracts from rat urogenital ridges dissected at successive stages of embryogenesis display SRY*e* binding activity by gel shift analysis (Fig. 9) until day 16 of gestation (Haqq *et al.*, 1993), which correlates well with the initiation of MIS expression. As shown, the strong binding

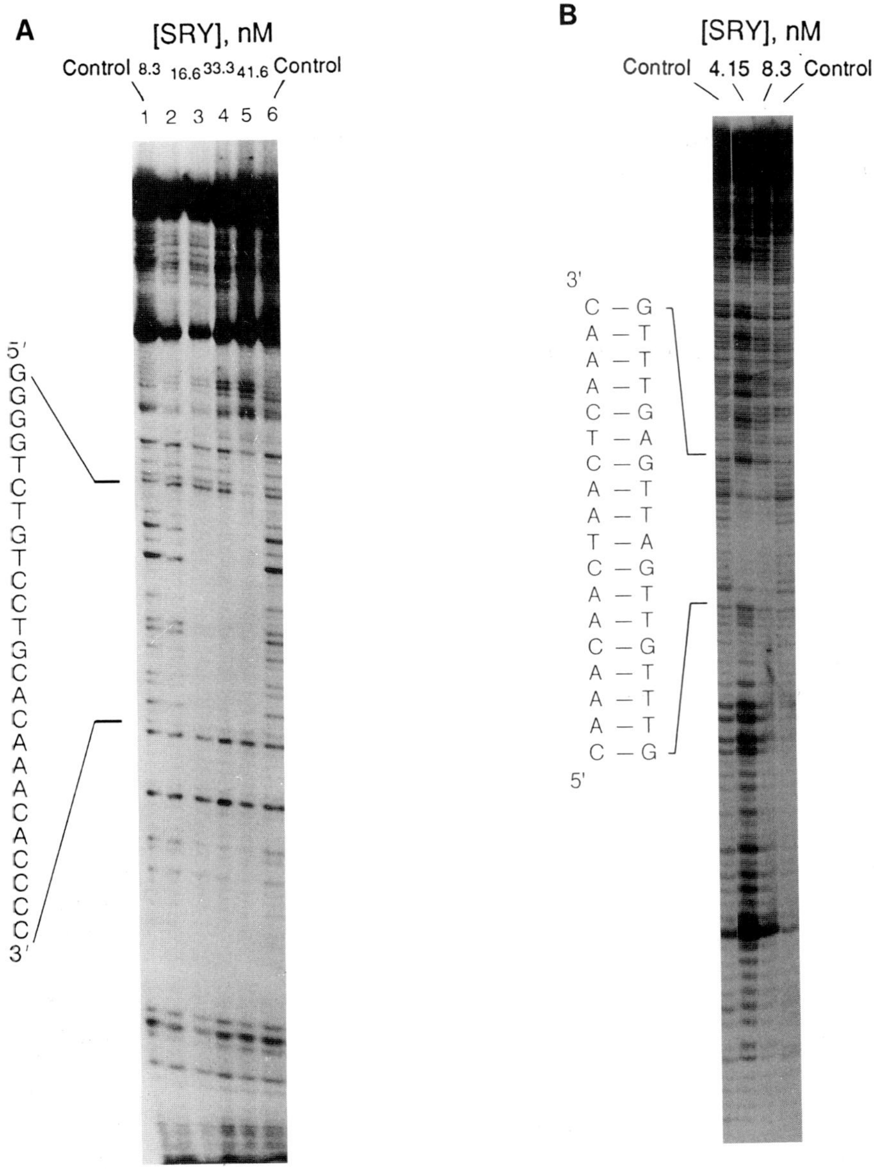

Figure 6 DNase footprint of the human MIS and rat aromatase promoters with SRY-p: (A) The SRY HMG-box protein (SRY-p) protects a portion of the human MIS promoter fragment (−139 to −4) from DNase digestion, with a physiologic affinity of 8 n*M*. The region of DNase protection and its DNA sequence is indicated. (B) SRY-p also protects a region of the rat aromatase promoter from DNase digestion, although at a somewhat lower affinity (<4 n*M*). The region of DNase protection and its DNA sequence is likewise indicated. Adapted from Haqq *et al.* (1993).

A. MIS promoters

```
                                                                              M3
Human:   AGGCC___CA_GCTCTATCaCTGGGgAgGGAGATAGGCTGCCA____GGGACAGAAAGGG |CTcTTTG| AGA_AG
Bovine:  AGGCCACACACACTGTATCGCTGctcA_GGAGATAGGCaGGCAcgTtGGaACAGAAgGGG |CT__TTG| AG__gG
Rat:     _CGGC_______CTCTgcaGTTGGG__________GGCTGGC_tcTgaGGACAGAAA_Gc |CT_TTTG| AGAcAG

Human:   __GCCAC___tCTGCCtGGagTG____GgGgCGCCgGGCACTGTCCCCCAAGGTCgCGgC___AGAGG_AGATA
Bovine:  __aCCAtaggCCTcCCaGG________ctcaCagCAGGCACcagCCttCAAGGTCAtGTCccagGAGG_AGATA
Rat:     ccGCCtCccaCCTGatgGGcaTGcaaaGcGcaGCCAGGCACTGT_CCCCAAGGTCACcTC___AGgGGttGATA

                   SRYe
Human:   |GgGGtCTGTCCTG___CACAAACACCCC| acCTTCCACTcggctCA |ctTAAg| GCAGG_CAGCCCAGCC  (-7)
Bovine:  |G_GGaCcGcCCTGcacCACAAACAgCtC| tGCTcCCtCT________ |TATAAa| GtAGGGCAGCCCAGCC  (-7)
Rat:     |GaGG__TGTCC___ctCcCAAgCACtCt| gGC_TCgAaT____gCA  |TATAA_| GCAGGaCcaCCCAGaC  (-7)
```

B. P-450 Aromatase promoters

```
                                SRYe
Rat:  CTTTACCTGCT |CTTG_GTTT_GTTAGTTGAGTTTGC| TTTAAATAAGGAGGATTGCCTCAGC_____
Chi:  GTTTAATTT__ |CTTGTGTTTTGGTTAGC_AATTTGC| AAGGAGTAACCCTCAGTACTGCCAGCCAAGC

Rat:  AAATGCTGCTGATGAAATCACAATTGAGTATGCACGTCACTCTACCCACTCAAGGGCAAG
Chi:  AATGCTGGAGATGAAGGTCTGACAGCCAAAGCATCCAAAAGCATGCTTCCATATCAAGGTC

Rat:  ATGATAAGGTTCTATCAGACCAACCGCTGAACAGGACCTGAGTCTCCCAAGGTCATCCTT
Chi:  AGAGAAGCTTTGTTATTTCCAGCCAAAGAGTGTCTGCTGCACATCTCTTTCACCTTGTG

Rat:  GTTTTGACTTGTAACCACAAATTTGTCTTGCCTTGTCAC |TATAAA| ACATCTGTCCATTCC   (-68)
Chi:  ATGTAACCTGTCCTGGCAAGCAGTCCAGGTCACTCGCAT |TATAAA| ACACAGCTACCTCAG   (-6 )
```

Figure 7 Proximal promoters of the MIS and P450 aromatase genes: (A) Proximal region of the human, bovine, and rat MIS gene promoters. The two SRY-p binding sites, SRY*e* (high affinity) and M3 (low affinity), and the TATA-box are boxed and in bold type. (B) Proximal region of the rat and chicken P450 aromatase gene promoters, with the SRY*e* and TATA-box sequences boxed and in bold type. Adapted from Haqq *et al.* (1993).

activity observed on fetal day 12 could be competed by rabbit polyclonal antibody raised to SRY-p, but not by a nonspecific rabbit antibody. In addition, nuclear extracts from embryonic male or female limbs and from 14-day female fetal urogenital ridges display no SRY*e* binding. By further mobility shift analysis, we have identified the minimal region of SRY*e* sufficient for SRY-p binding to be a 15-bp sequence from −63 to −49 (Haqq *et al.*, 1993). This region is capable of forming a single specific complex with SRY-p as a 15-bp oligonucleotide (5′GGGGT**G**TTTGTGCAG3′). A sequence closely related to this 15-bp region is present further upstream in the MIS gene from −145 to −131 bp (designated M3; 5′GGGCT**C**TTTG**A**GAAG3′). This region, however, does not exhibit DNase protection under conditions in which SRY*e* is fully occupied by SRY-p. Mutational analysis also demonstrated that C to G and A to T transversions (bold type) in the M3 sequence enhanced SRY-p binding two- and fivefold, respectively. Conversely, T-to-A transversion of the middle T in the thymidine triplet decreases SRY-p binding by at least 100-fold.

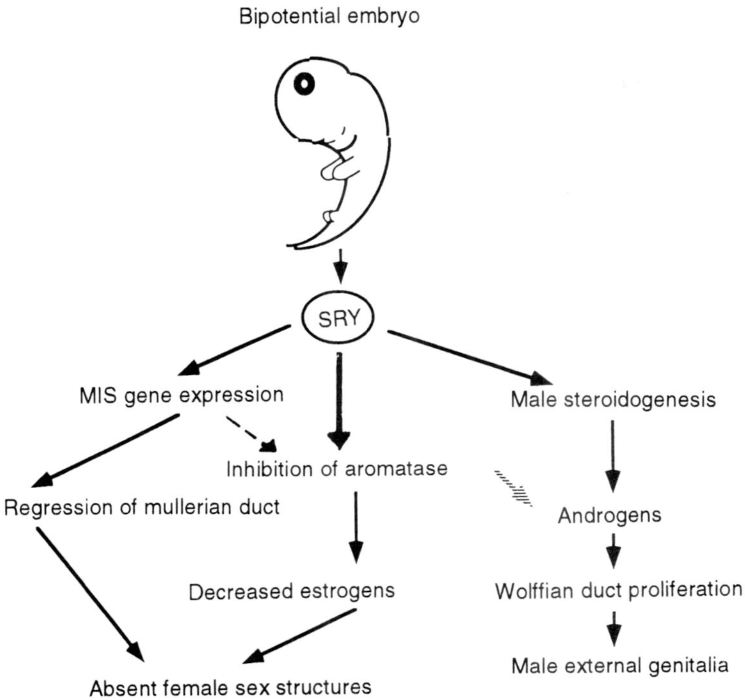

Figure 8 Schematic diagram of pathways leading to sexual dimorphism: Potential interactions between SRY, MIS, aromatase, and androgens are indicated. Regression of the Müllerian ducts and maintenance of the Wolffian ducts are critical steps for normal male sexual differentiation. Note the proposed indirect role of aromatase inhibition in this male pattern of development. Adapted from Haqq *et al.* (1993).

C. Additional Upstream Regulatory Elements

Cloning and sequencing of the human MIS promoter from −3.6 kb to the transcription start site by Guerrier *et al.* (1990) has greatly facilitated the search for upstream regulatory elements. When we compared the 5′ upstream regions of the human, bovine, and newly cloned rat MIS (Haqq *et al.*, 1992) genes, we found no canonical transcription factor binding sites except for a single SP-1 site (CCGCCC) in the human and bovine genes. Its significance is unknown since SP-1 sites are usually multiple and occur with AP-1 sites, which are not found within this region. A sequence with one mismatch for the cAMP response element is present at −262 bp in the human but not the bovine or rat promoters, whereas the rat gene has a sequence with one mismatch for the consensus binding site for homeobox proteins (such as fushi tarazu) in the *Drosophila* ultrabithorax promoter (T-Hox; see Fig. 10). Finally, a 13-bp

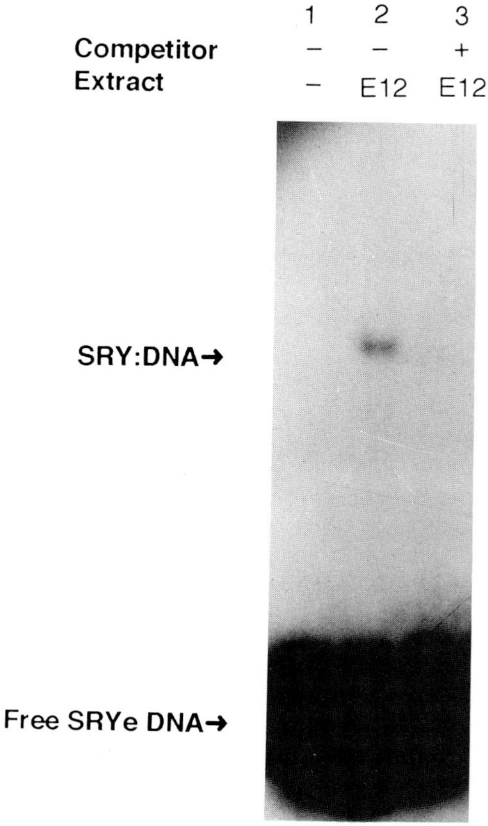

Figure 9 Ontogeny of SRY*e* binding activity: Nuclear extracts from pooled embryonic day 12 (E12) rat urogenital ridges demonstrate SRY*e* binding activity (lane 2), which can be competed by anti-SRY-p polyclonal antibody (lane 3). Adapted from Haqq *et al.* (1993).

palindrome that differs from the estrogen response element (ERE) of the vitellogenin gene by only one nucleotide is present at −1772 bp of the human gene (Guerrier *et al.*, 1990). When these authors inserted 2–4 copies of this sequence 5′ of a reporter gene in the presence of estrogen receptors, they observed estrogen inducibility, although similar results could not be obtained with a 182-bp fragment of the human gene encompassing this ERE.

Alignment of the human, bovine, and rat MIS promoters reveals several additional sequences that are highly conserved, but do not correspond to known regulatory elements (Fig. 10). These sequences have

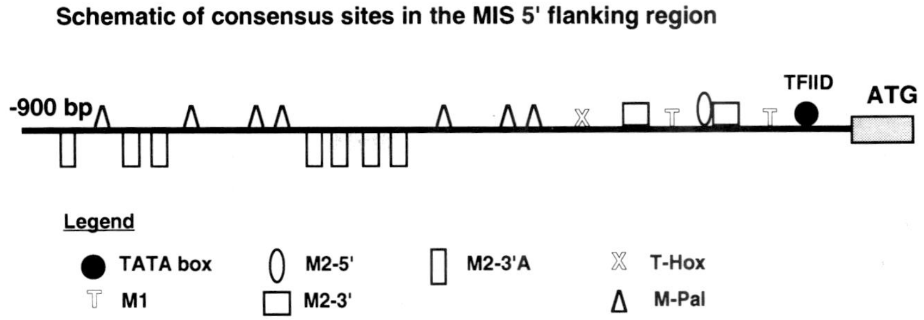

Figure 10 Schematic diagram of consensus sites in the MIS upstream region: Several potential regulatory elements in the 5′ upstream region of the MIS gene between the TATA-box and −900 bp are represented in this figure, including the M1 and M2 sequences.

been designated M1, M2, etc., from proximal to distal. As a first step, large fragments of the MIS promoter or synthetic duplex oligonucleotides corresponding to these sequences were incubated with nuclear extracts of various fetal and adult tissues prior to gel shift analysis; sequences displaying the expected tissue and developmental specificity for MIS expression were then examined further. For example, the nonamer 5′GGAGATAGG3′ (M1), present twice in the upstream region and homologous to the repeated androgen binding protein enhancer sequence 5′GGAGAA3′, binds to several nongonadal nuclear extracts (Haqq *et al.*). Although it is therefore probably not involved with the regulation of tissue-specific expression, M1 may still be important for basal or age-specific MIS secretion.

The next most proximal sequence (M2) is found at −102 to −78 bp in the rat (Fig. 10). The 5′ end of M2 is a unique sequence with dyad symmetry (5′GCCAGGCACTGTC3′), whereas the 3′ end (5′CCCAAGGTCA3′) is repeated once at −200 bp with one mismatch and seven times between −600 and −900 bp with a related motif (M2-3′A). Nuclear extracts from fetal and postnatal testes and adult ovaries consistently produce an M2 mobility shift that correlates extremely well with *in vivo* MIS expression. An M2 protein, which binds to M2 in gel shift assays and protects the M2 sequence during DNase digestion, has recently been purified from newborn bovine testes (Haqq *et al.*) and is currently being cloned to determine its functional significance in MIS regulation. An additional palindromic sequence (M-Pal) of unknown significance is repeated several times in the MIS promoter (Fig. 10), but sex-, tissue-, or ontogeny-specific binding of nuclear extracts to this element has not been observed.

D. Translational Control

The 5′ untranslated region of MIS mRNA contains only 10 (bovine and human) or 8 (mouse and rat) nucleotides, and therefore is unlikely to affect the rate of protein synthesis. The 3′ untranslated region is likewise relatively short, extending from 62 nucleotides in the rat to 112 in the human (Fig. 1). An alternative method of MIS mRNA translational regulation, however, may exist via differential polyadenylation of MIS transcripts. By a series of oligonucleotide-directed RNase H mapping experiments, Lee *et al.* (1992) demonstrated that two distinct MIS mRNA species, differing only at their 3′ ends in the extent of polyadenylation, are present in a developmentally regulated pattern in the rat testis. A 2.0-kb transcript is abundant in the 14-day fetal testis, decreases markedly by the time of birth, and is barely detectable postnatally; the smaller 1.8-kb transcript, meanwhile, becomes detectable by day 18 of gestation, peaks in abundance immediately before birth, decreases rapidly within the first week of life, and then remains detectable throughout adulthood (Fig. 11). The pattern of expression of this second transcript closely resembles that of MIS secretion itself. Although the biological significance of these mRNA differences is yet to be determined, differential polyadenylation could affect the stability or translatability of such messages (Sachs, 1990).

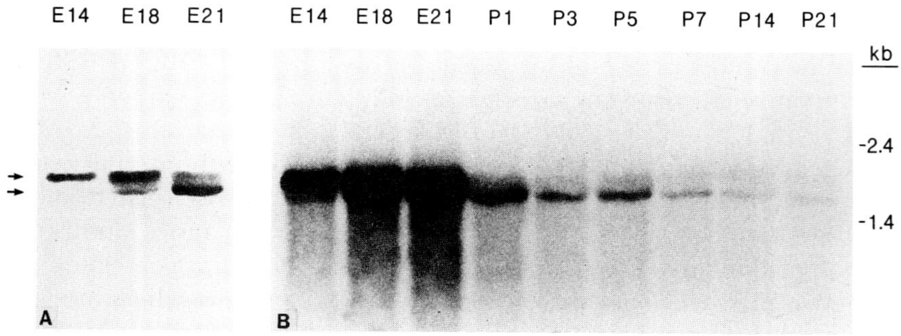

Figure 11 MIS mRNA ontogeny in the rat testis: Ten micrograms of total RNA from embryonic day 14 (E14) to postnatal day 21 (P21) rat testes is present in each lane of this Northern blot. (A) Two discrete MIS mRNA species (arrows) appear to be developmentally regulated prior to birth; a 2.0-kb species is abundant in embryonic testes but barely detectable after birth, whereas the 1.8-kb species becomes detectable at E18 and abundant at E21. (B) Longer exposure of this same blot reveals that 1.8-kb MIS mRNA decreases during the first postnatal week to a low baseline subsequently seen through adulthood (data not shown). Adapted from Lee *et al.* (1992).

E. Gonadotropins and Steroids

Earliest studies on the regulation of MIS secretion by gonadotropins were conducted in the late 1970s. Testes of newborn rats treated *in utero* with LHRH antiserum were initially observed to have increased MIS bioactivity compared to buffer-treated controls (Bercu *et al.*, 1978). In a follow-up study, treatment of these newborn animals with FSH for 5 days reversed the increased MIS expression, whereas hCG injection produced no change (Bercu *et al.*, 1979). Taken together, these results suggested that the hypothalamic–pituitary–gonadal axis plays a role in regulating MIS expression *in vivo*. FSH inhibition of MIS secretion by fetal and neonatal testes was subsequently confirmed by immunohistochemical (Fig. 12) and Northern analyses and was shown to be transcriptionally mediated (Kuroda *et al.*, 1990, 1991).

The lack of a primary or immortalized Sertoli or granulosa cell line that continues to produce MIS in culture has limited further study of LH/FSH effects (Vigier *et al.*, 1985; LaQuaglia *et al.*, 1986). LaQuaglia *et al.* (1986) demonstrated that FSH increases cAMP in the media of Sertoli cells, whereas MIS mRNA and protein production both fall rapidly after the cells are placed in culture. On the other hand, Voutilainen and Miller (1987), using a 27-mer probe from the fifth exon of the human MIS gene and Northern/dot blot analyses, found that FSH and hCG had no effect on MIS mRNA levels in cultured human fetal testicular cells, whereas dibutyryl cAMP induced a fivefold increase. Moreover, cultured adult human granulosa cells collected after hormonal stimulation for *in vitro* fertilization had a ninefold increase in MIS mRNA in response to dibutyryl cAMP, a fivefold increase after hCG exposure, and no change with FSH treatment.

The role of steroid hormones in the regulation of MIS remains somewhat ambiguous. Kuroda *et al.* (1991) evaluated MIS expression in the 21-day fetal rat testis with holo-, N-, and C-terminal MIS-specific antibodies after *in utero* fetal injection of LH, FSH, testosterone, estradiol, or diethylstilbesterol. Treatment with FSH resulted in decreased MIS immunostaining in the testis with all three antibodies, consistent with decreased MIS expression. On the other hand, hCG or testosterone exposure did not change holo-MIS staining, but N- and C-terminal immunoreactivity was significantly increased. Consequently, we hypothesize that testosterone and hCG/LH (most likely acting via increased testosterone secretion) regulate MIS post-translationally by enhancing its proteolytic cleavage. Estradiol had no effect on fetal MIS expression in this study, although it is known to block the regressive effects of MIS on the fetal male Müllerian duct, without altering testicular MIS secretion (Hutson *et al.*, 1982; Newbold *et al.*, 1984). The exact mechanism of this

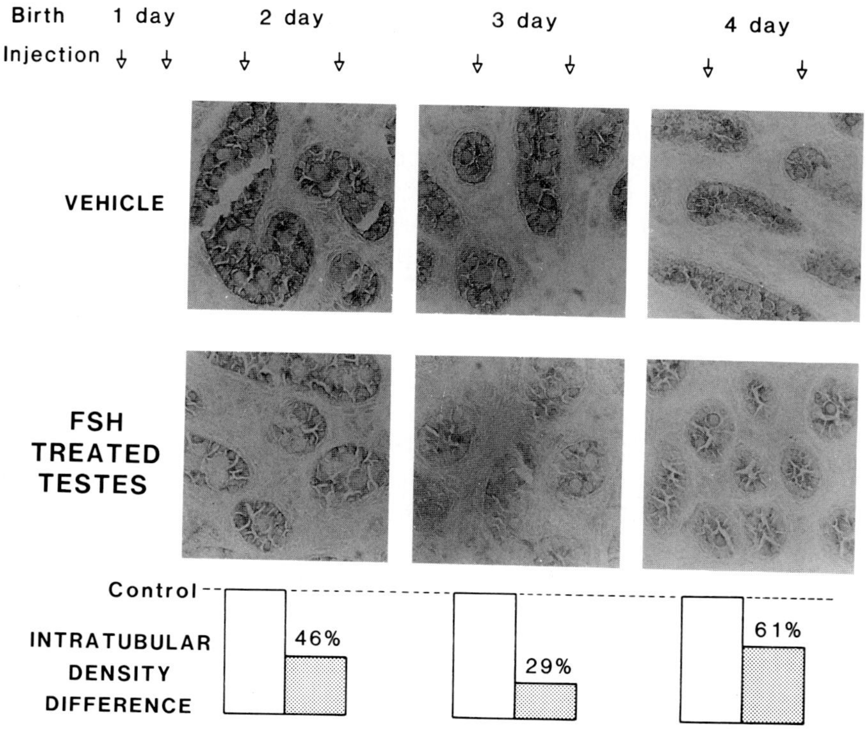

Figure 12 FSH inhibition of MIS expression: Testes of newborn rat pups treated with FSH or vehicle-control injections twice a day for 2, 3, or 4 days were harvested for immunohistochemistry. FSH-treated testes had decreased MIS immunostaining compared to controls. Adapted from Kuroda *et al.* (1990).

MIS–estrogen antagonism is still unclear, but inhibition at the receptor level is suspected (MacLaughlin *et al.*, 1983).

V. Biological Roles for MIS

A. Müllerian Duct Regression

During early embryonic development, the Müllerian ducts develop as paired longitudinal invaginations of coelomic epithelium along the anterolateral aspect of the retroperitoneal urogenital ridge, in close proximity to the Wolffian ducts. In the absence of MIS in the female, these structures ultimately give rise to the fallopian tubes, uterus, and upper one-third of the vagina; in the male, however, MIS produced by

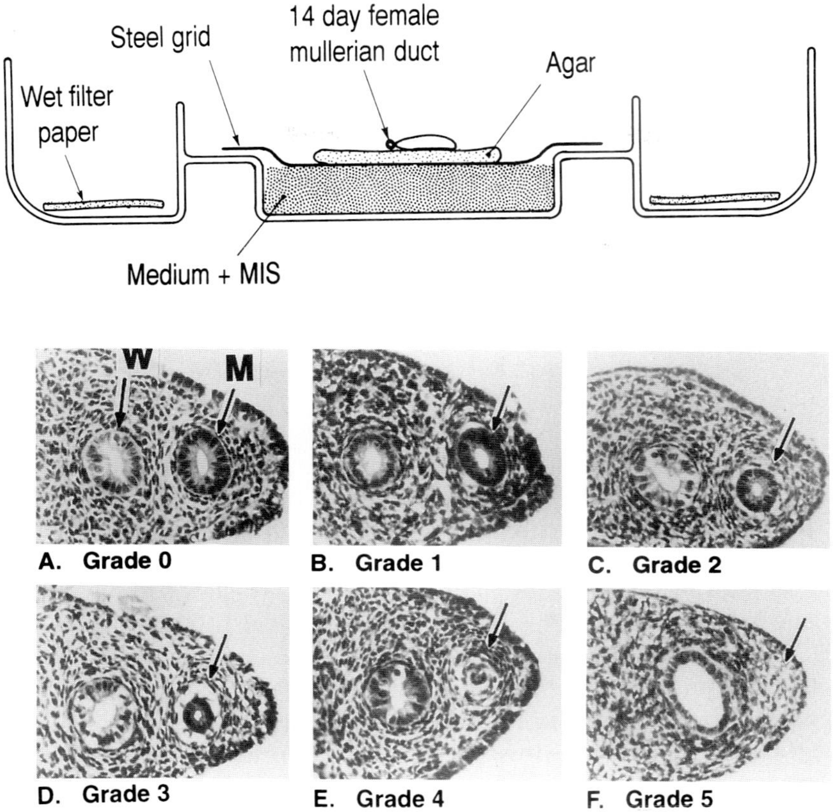

Figure 13 Bioassay for MIS activity: The schematic diagram (top) depicts the standard organ culture assay for MIS activity, which utilizes 14.5-day female fetal rat urogenital ridges. (A) Section of a control ridge in the absence of MIS, demonstrating normal architecture of the Müllerian (M) and Wolffian (W) ducts. (B–D) Intermediate grade 1–3 regression of the Müllerian duct is seen after 72 h incubation with rhMIS; mesenchymal condensation around the duct, basement membrane breakdown, and loss of epithelial cells are characteristic findings. (E–F) Further grade 4–5 regression of the cultured Müllerian duct; some tubular epithelial cells have migrated into the surrounding mesenchyme.

the organizing gonad causes Müllerian duct regression. This *in vivo* function of MIS was exploited by Picon in 1969 to establish an *in vitro* assay for MIS biological activity. As later modified by Donahoe *et al.* (1977b), this assay involves culturing a female 14.5-day fetal rat urogenital ridge at an air–media interface with tissue fragments (such as fetal rat testis) or over MIS-containing media for 72 h (Fig. 13). After sectioning and staining of the cephalic urogenital ridge, regression of its Müllerian duct is rated from 0 (no regression) to 5 (complete regression). The organ culture bioassay requires 1.5–2.0 μg rhMIS/ml of media to cause complete regression. Still regarded as the MIS "gold standard," this system has been used in our laboratory to evaluate MIS activity in rat, mouse, bovine, human, chicken, marsupial, and canine gonads, to continually monitor the quality of rhMIS produced, and to screen for biologically active preparations that are presumably capable of binding to receptor and activating intracellular transduction pathways.

A developmental window of Müllerian duct responsiveness to MIS, before or after which MIS exposure will not induce full regression, appears to exist in all species. For example, if the 14.5-day fetal rat urogenital ridge is exposed to MIS for less than 12 h, no Müllerian duct regression is observed *in vitro* 72 h later; on the other hand, if MIS exposure occurs for 24 h or more, regression is complete and irreversible (Donahoe *et al.*, 1977b). Likewise, ridges from female fetuses older than 15 days exhibit decreased or no response to MIS in culture. Thus, it appears that the presence or absence of MIS exposure between 14 and 15 days of fetal development determines the fate of the male and female Müllerian duct (Picon, 1969; Donahoe *et al.*, 1977b; Josso *et al.*, 1977). In a study by Taguchi *et al.* (1984) regression of the Müllerian duct in the human fetus began by at least day 51 postovulation and was completed by day 64, with regression progressive and irreversible during this narrow interval. Josso *et al.* (1977) also found that urogenital ridges from human female fetuses greater than 30 mm in crown–rump length were no longer responsive to MIS *in vitro*.

Light (Fig. 13) and electron microscopic studies of Müllerian duct regression reveal that an early event in this process is condensation of periductal mesenchymal cells into a whorl around the epithelium (Price *et al.*, 1979). This is associated with breakdown of the periductal basement membrane and extension of epithelial cell processes through the resulting gaps (Trelstad *et al.*, 1982). Some of the epithelial cells subsequently die an apoptotic death (Price *et al.*, 1979), with autophagy of adjacent cells, whereas others migrate into the surrounding mesenchyme (Trelstad *et al.*, 1982; Hayashi *et al.*, 1982). On a biochemical level, increased hyaluronidase activity and a decrease in hyaluronic acid, laminin, heparan sulfate, fibronectin, and type IV collagen within the basement membrane and mesenchymal matrix have all been documented during duct involution (Hayashi *et al.*, 1982; Donahoe *et al.*, 1984). A recent study by Tsuji *et al.* (1992), in which mesenchymal cells surrounding the 15.5-day fetal rat Müllerian duct were cultured separately from the epithelial components and then tested for their sensitivity to MIS, supports the earlier histochemical evidence that Müllerian duct regression is a mesenchymally mediated event. In this study, addition of MIS resulted in significant inhibition of tritiated thymidine incorporation of mesothelial, but not epithelial, cells. Coincidentally, it is during this same interval that mRNA for the putative MIS receptor localizes to mesenchyme surrounding the fetal rat Müllerian duct (Fig. 3B).

B. Gonadal Differentiation

A potential role for MIS in gonadal differentiation was first suggested by the phenomenon of "freemartinism," which refers to a constellation of abnormalities in bovine genetic females that have placental anastomoses to their male twins. In addition to Müllerian duct agenesis, these XX females have ovaries with markedly reduced volume, germ cell depletion, and primitive seminiferous tubules (Jost *et al.*, 1972). External genitalia develop normally in these animals, however, suggesting that the observed changes are secondary to MIS rather than testosterone. By treatment of prospective ovaries from 14-day fetal rats with purified bovine MIS for 3–10 days, Vigier *et al.* (1987) could reproduce these characteristic features *in vitro* in a time- and dose-dependent manner, including formation of seminiferous-like cords with polarized epithelial cells resembling fetal Sertoli cells. Their results suggest that MIS in this model inhibits the normal oogonial proliferation that occurs between 14 and 17 days gestation. Fetal ovaries at 20 days gestation, meanwhile, are not affected by MIS and form normal ovarian follicles.

It was subsequently demonstrated that the above morphological changes are associated with inhibition of aromatase biosynthesis and the production of testosterone instead of estradiol within the MIS-treated ovaries (Vigier *et al.*, 1989). This observation led to the development of an alternative quantitative bioassay for MIS (DiClemente *et al.*, 1992). In the fetal ovary aromatase assay, 16-day fetal rat ovaries cocultured with MIS for 3 days are assayed for reduction of cAMP-induced aromatase activity, which correlates inversely with the log of the MIS concentration. Stimulation of basal aromatase activity with cAMP is required to detect MIS inhibitory effects in this system, with a half-maximal response dose of 0.42 μg rhMIS/ml. However, cAMP-stimulated aromatase activity in the fetal rat testis is not suppressed by MIS.

Female transgenic mice chronically expressing MIS at supraphysiologic levels (Behringer *et al.*, 1990) nearly all had a blind vagina and altered ovarian development. Uterus and oviducts were absent. Animals with the highest levels of MIS expression had ovaries with reduced numbers of germ cells by postnatal day 2, disappearance of germ cells and failure of follicle formation by day 9, and formation of seminiferous-like tubules by day 16, followed by disappearance of the gonad entirely by adulthood. Five of 21 transgenic males with the highest MIS levels had feminized external genitalia, mammary gland development, small germ-cell-free undescended testes, and underdeveloped epididymides; seminal vesicles were absent. The other males had normal testes and external genitalia and were fertile. One working hypothesis derived from these studies is that the testis-determining factor initiates Sertoli cell differentiation and MIS secretion (see Section IV,B), with MIS acting in an autocrine fashion to regulate further Sertoli cell function. The Sertoli cell subsequently effects Leydig cell differentiation and testosterone production, germ cell proliferation, and seminiferous tubule formation. Because each of the above experimental models involves exposing early gonadal somatic cells to MIS, it is further hypothesized that granulosa cells can be modified by MIS to become "Sertoli-like" and form seminiferous tubule-like structures (Behringer *et al.*, 1990).

C. Germ Cell Meiosis

The fate of primordial germ cells in human embryos follows two distinct paths distinguished by the timing of germ cell entry into meiosis. In the female embryo, germ cells undergo multiple mitotic divisions and differentiation between the third and seventh months to form primary oocytes, which are then arrested in prophase of the first meiotic division by birth. At puberty, when MIS is expressed in developing follicles, only selected oocytes resume meiosis with the start of each cycle. In sharp contrast, male germ cells undergo mitosis shortly after birth (while MIS levels are still high) to form prespermatogonia, which then remain quiescent until shortly before puberty. A burst of mitotic activity at this time, when MIS is decreasing to low adult levels, generates primary spermatocytes that undergo two meiotic divisions to form four haploid spermatids.

The sexually dimorphic pattern of MIS expression is currently being studied for its role in gonadal regulation. Table I illustrates several of the possible relationships that exist between mitotic and meiotic activities of gonadal cells and MIS production in the rat model (see Sections III,A and III,B). In male and female rats, MIS secretion is greatest when germ cell meiotic activity is inhibited, with subsequent reduction

Table I
Correlation of MIS with Germ and Somatic Cell Mitotic/Meiotic Activity

	Male (+SRY)		Female (−SRY)	
	Sertoli cells	Germ cells	Granulosa cells	Germ cells
E_{10-11}[a]		Mitosis (Gonocytes/prespermatogonia)		Mitosis (oogonia)
$E_{13.5-14}$	(++) MIS			
E_{16}-P_2[b]	(++) MIS; cell division	No meiotic division; cessation of mitosis		First meiotic division (oocytes); cessation of mitosis
P_{3-8}	(+) MIS; cessation of cell division	Mitosis (spermatogonia)	(+) MIS; cell division in preantral/antral follicles	Meiotic arrest (dictyotene)
P_{8-15}	Low MIS	(Spermatocytes)		
P_{15}-Adult	Low MIS	Two meiotic divisions (spermatids)	(+) MIS in preantral/antral follicles; (−) MIS in large preovulatory follicles	Resumption of meiosis with ovulation

[a] E, Embryonic day.
[b] P, Postnatal day.

in MIS expression accompanied by initiation or resumption of meiosis. Periods of greatest somatic cell mitosis, on the other hand, correspond to times of peak MIS production, in rapidly growing fetal testes cords and early preovulatory follicles.

Several additional lines of evidence support a role for MIS in the regulation of germ cell maturation. By using spontaneous germinal vesicle breakdown as the endpoint, Takahashi *et al.* (1986b) found that purified bovine MIS inhibits resumption of meiosis in denuded or cumulus-enclosed rat oocytes *in vitro*. This effect of MIS on the oocyte is dose-dependent, reversible, and cAMP independent, and is not significantly altered by FSH, LH, progesterone, estradiol, or testosterone. rhMIS preparations have similar effects in the presence of NP-40, a stabilizing detergent that presumably acts to decrease MIS aggregation. Contrary to our experience, Tsafriri *et al.* (1988) have attributed the inhibitory activity seen in this assay to NP-40 rather than MIS. However, at the concentration used in these experiments NP-40 alone has no effects on oocyte meiosis or on Müllerian duct regression; in addition, MIS activity on the oocyte can be blocked by addition of either anti-MIS antibody or EGF (Ueno *et al.*, 1988). Similarly, Taketo *et al.* (1991) have reported an interesting correlation between germ cell development in B6.Y^{DOM} mouse ovotestes, which are thought to result from delayed testicular differentiation, and MIS expression. MIS-positive testis cords have central germ cells arrested at the prespermatogonia stage, whereas MIS-negative tubules contain polar germ cells in meiotic prophase. These correlations, although not conclusive, encourage further studies to develop rhMIS as a novel contraceptive agent and suggest that MIS should be tested as a therapeutic against germ cell tumors.

D. Fetal Lung Maturation

Fetal lung development is sexually dimorphic in many species, with males lagging behind females of identical gestational age and newborn respiratory distress syndrome more common in premature males (Torday *et al.*, 1985). Although testosterone is known to inhibit synthesis of phosphatidylcholine, a major component of pulmonary surfactant, by immature pulmonary fibroblasts, we suspected that MIS could also play a role in fetal lung maturation. Therefore, the effect of MIS on pulmonary surfactant turnover was examined *in vitro* (Catlin *et al.*, 1988). Fragments of 17.5-day female fetal rat lungs incubated with fetal testes, bovine MIS, or rhMIS had reduced disaturated phosphatidylcholine accumulation compared to control fragments cultured with fetal ovaries or buffer alone. These results were later recapitulated *in utero* when nanomolar concentrations of rhMIS were injected into the dorsal subcutaneous tissue of 19-day rat fetuses; pulmonary disaturated phosphatidylcholine was found to be decreased when measured 48 to 72 h later (Catlin *et al.*, 1990). MIS has also been shown to inhibit EGF receptor tyrosine autophosphorylation of fetal rat lung plasma membranes (Catlin *et al.*, 1991).

Visualization of MIS binding and trafficking in fetal lung tissue has recently been accomplished with laser confocal microscopy (Catlin *et al.*, 1992). Fragments of 21-day male fetal rat lungs were incubated for 1 to 24 h with or without rhMIS. After extensive washing, fixation, freezing, and sectioning of lung and control tissues, a primary rabbit anti-rhMIS and secondary FITC-conjugated donkey anti-rabbit antibody were used for MIS localization. Fetal lung incubated with rhMIS and stained with these antibodies revealed punctate membrane fluorescence at 1 h, cytosolic and nuclear signal by 6 h (Fig. 14), and predominantly nuclear localization by 24 h. These findings are consistent with specific adsorptive endocytosis of membrane-bound MIS by fetal lung membranes, but how and why nuclear trafficking of this large protein occurs, remain a mystery.

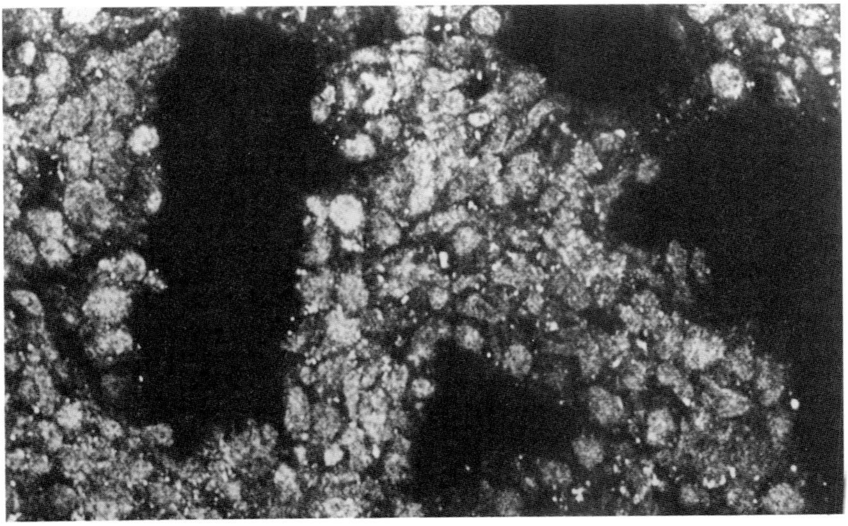

Figure 14 Localization of MIS binding to fetal rat lung tissue: Male 21-day fetal rat lungs were incubated with rhMIS for 6 h, anti-rhMIS primary antibody, and FITC-conjugated secondary antibody prior to visualization with confocal microscopy. By 6 h, MIS is trafficked intracellularly to the cytoplasm and nucleus. By 24 h, MIS signal is primarily intranuclear (data not shown). Adapted from Catlin *et al.* (1992), with permission of Wiley-Liss.

E. Testicular Descent

We first noted that human undescended testes had diminished MIS bioactivity when compared to age-matched descended gonads (Donahoe *et al.*, 1977a). Coupled with the observation that patients with persistent Müllerian duct syndrome (PMDS) have an increased incidence of undescended testes, we reasoned that MIS might play a role in testicular descent. Subsequently Yamanaka *et al.* (1991) demonstrated that serum MIS levels were lower in patients with undescended testes than they were in normal age-matched controls. It remains possible, however, that such differences could be related to overall impaired hormonal function of the abnormally located gonad. Josso *et al.* (1983) likewise argue that gonadal descent is incomplete in PMDS patients secondary to mechanical restraint by the retained Müllerian structures. Normal gonadal migration in the male fetus is thought to occur as a two-step process (Hutson and Donahoe, 1986). It is hypothesized that MIS regulates the first stage of transabdominal movement, which is completed by the 12th week of human development and most likely occurs due to rapid expansion of the gubernaculum. The second stage of inguinal and transscrotal descent occurs between the 7th fetal month and birth and is thought to be a neuronally mediated, testosterone-driven event. Once the gubernaculum can be evaluated for MIS receptor expression, or the MIS or MIS receptor genes can be experimentally disrupted, definitive information about the role of MIS in testicular descent should be forthcoming.

VI. Antiproliferative Effects of MIS

The central role of MIS as a negative embryonic growth regulator led us to investigate its use as an antiproliferative agent and tumor inhibitor (Donahoe *et al.*, 1979). Earliest studies focused on tumors of the female reproductive tract, including the uterus and fallopian tubes, as well as on ovarian epithelial carcinomas. The latter group was included because these tumors arise from the surface epithelium of the ovary, which is a derivative of the same coelomic epithelium that invaginates to form the Müllerian ducts. These tumors account for 95% of ovarian malignancies and primarily occur in postmenopausal patients after the cessation of MIS expression. Growth of one such ovarian serous cystadenocarcinoma cell line (HOC-21) was inhibited by partially purified bovine MIS in monolayer microcytotoxicity tests (Donahoe *et al.*, 1979) and clonogenic soft agar assays (Fuller *et al.*, 1982). Moreover, after subcutaneous injection in nude mice, growth of HOC-21 and human endometrial carcinoma (HEC-1) cells was inhibited by preincubation with bovine MIS (Donahoe *et al.*, 1981; Fuller *et al.*, 1984). These results were confirmed with several primary ovarian, endometrial, and tubal carcinomas (Fuller *et al.*, 1985).

Although early crude testicular extracts were characterized by consistent antitumor activity, increasingly pure MIS preparations demonstrated only inconsistent inhibitory effects. For example, Wallen *et al.* (1989) reported that immunoaffinity-purified rhMIS had antiproliferative activity against only 3 of 13 primary human ovarian cancer cell lines in a colony inhibition assay, with no inhibition of a number of established human gynecologic tumor cell lines. During this same period in our laboratory, less homogeneous rhMIS purified by ion exchange and carbohydrate affinity chromatography had inhibitory effects against an endometrial carcinoma cell line *in vivo,* whereas more homogeneous immunoaffinity-purified rhMIS had no such activity.

The addition of a preelution high-salt wash step to the immunoaffinity purification procedure in 1990 (see Section II,A) resulted in separation of a copurifying triplet of low-molecular-weight (14K–20K) CHO-derived proteins from the rhMIS (Ragin *et al.*, 1992). These proteins are present in media from wild-type CHO cells but not in fetal calf serum, are not recognized on Western analysis by anti-rhMIS antibodies, increase the proteolytic cleavage of MIS preps at 37°C, and have growth-stimulatory effects that mask the inhibitory activity of purified rhMIS. As shown in Fig. 15, recent preparations of holo-rhMIS devoid of the salt-eluted proteins demonstrate inhibitory effects against several genital tract tumors (Chin *et al.*, 1991) and ocular melanomas (Parry *et al.*, 1992) in colony-inhibition and subrenal capsule assays. For such *in vivo* studies, holo-rhMIS is delivered via intraperitoneal miniosmotic Alzet pumps, with the relatively constant serum MIS levels achieved 24 h after implantation maintained for 8 days. Serum concentrations of rhMIS that are approximately 100-fold lower than those used in assays *in vitro* have been effective *in vivo.*

The recent finding of MIS-responsive ocular melanomas suggests that other neural crest tumors should be evaluated and that the spectrum of MIS-responsive cancers may be much wider than initially anticipated. The effects of MIS on cell cycle kinetics of one of these ocular melanoma cell lines was also recently reported (Parry *et al.*, 1992). As shown in Table II, buffer-treated OM431 cells density-arrested in G1 (84%) progress through the cell cycle after passage to 30% confluency (56% G1), whereas similarly passaged but rhMIS-treated cells are delayed in G1 at 24 h (86%). Most recently, we have shown that cleaved, separated,

A

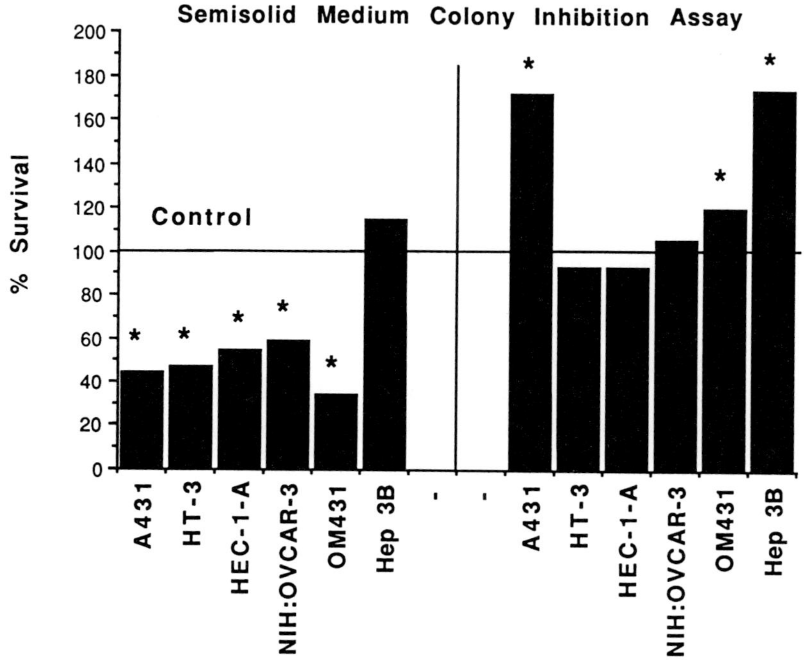

Figure 15 *In vitro* antiproliferative effects of rhMIS: (A) A semisolid medium (double-layer) colony inhibition assay demonstrates antiproliferative effects of rhMIS (left) against several Müllerian-derived human carcinoma cell lines (e.g., A431, HT-3, HEC-1-A, and NIH:OVCAR-3), as well as an ocular melanoma cell line (OM431). The "triplet" of proteins separated from MIS by a preelution salt-wash step has growth-stimulatory or no effects on these same lines (right). A hepatocarcinoma cell line (Hep 3B) is not effected by MIS. (* $P < 0.05$ when compared with controls.) (B) Dose-dependent inhibition of A431 colony formation by increasing concentration of crude rhMIS preparations (left) and immunoaffinity-purified rhMIS (right). (* $P < 0.05$ when compared with controls.) Adapted from Chin *et al.* (1991).

and highly purified rhMIS C-terminus possesses antitumor activity against the A431 cell line (MacLaughlin *et al.*, 1992a) and has similar cell cycle effects on A431 cells (our unpublished data). Thus, the previously inconsistent effects seen with MIS were most probably due to the proportion of cleaved and activated C-terminal fragment in each preparation tested.

The antiproliferative activity of C-terminal MIS is currently being examined in animal models, by use of delivery systems as described above. One potential limitation of such preparations *in vivo*, however, is rapid degradation and greatly reduced half-life, since the noncovalently asso-

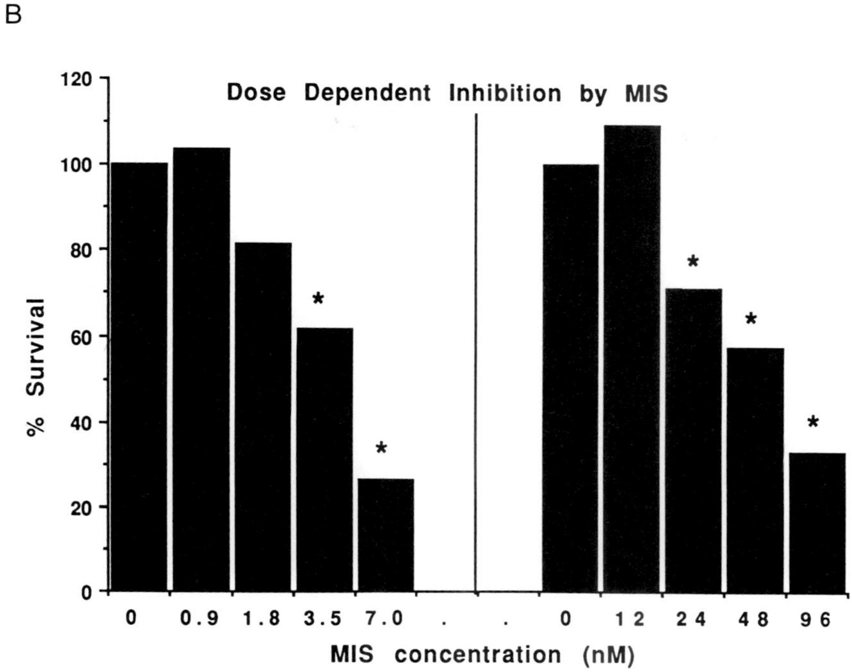

Figure 15 (*Continued*)

Table II
The Effect of MIS on OM431 Cell Cycle Kinetics during Monolayer Growth[a]

OM431[b]	G1 (%)	S (%)	G2/M (%)
Density arrest[c]	84	8	8
Buffer control[d]	56	42	2
Holo-rhMIS[e]	86	10	4

[a] Adapted from Parry *et al.* (1992).
[b] A human ocular melanoma cell line.
[c] Percentage of cells in G1, S, or G2/M phases of the cell cycle after density arrest, as determined by FACS analysis.
d After passage to 30% confluency and treatment with buffer alone, cells progressed in the cell cycle (56.2% G1).
e After passage to 30% confluency and treatment with rhMIS, cells were delayed in G1 phase of the cell cycle at 24 h (86% G1).

ciated N-terminal fragment may be essential for stability in serum. An alternative method of "delivering" MIS to susceptible tumors would be retroviral transfection. Studies are under way to evaluate the effects of MIS gene transfection on the growth characteristics of such tumor cells. For example, an ocular melanoma cell line permanently transfected with an MIS cDNA (designated OM431.MIS) forms significantly fewer pulmonary metastases after tail-vein injection of SCID mice than does the wild-type nontransfected line (Boveri *et al.*, 1993) (Fig. 16). Likewise, the MIS-transfected CHO cell line forms much smaller colonies in

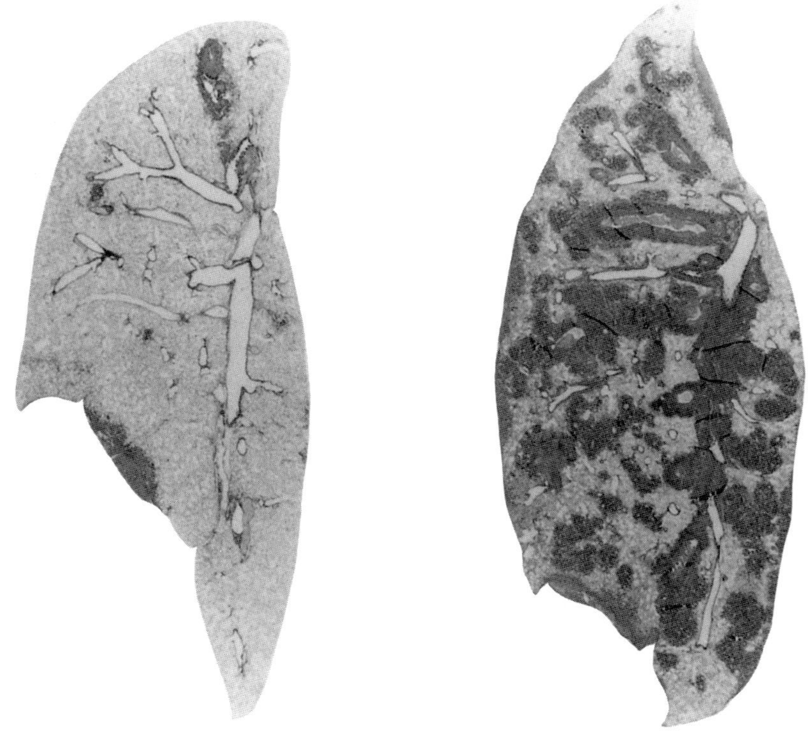

Figure 16 Transfection of the MIS gene inhibits metastatic pulmonary growth: A human ocular melanoma cell line (OM431) was transfected with the cDNA for human MIS (designated OM431.MIS) and then injected into the tail veins of immunodeficient SCID mice. Pulmonary metastases were greatly inhibited by MIS transfection, as seen in representative coronal pulmonary sections from these animals. OM431.MIS cells occupied significantly less relative area of the lung (14.14 ± 2.94%) compared to wild-type OM431 cells (35.83 ± 3.06%) ($P < 0.0001$). Adapted from Boveri *et al.* (1993).

assays *in vitro* than does the wild-type CHO line or a CHO line transfected with a cDNA for noncleavable mutant MIS; smaller tumor nodules are also formed by MIS-transfected CHO cells in subrenal capsule and pulmonary metastasis assays (Boveri *et al.*, 1993). Recent progress in cloning a candidate MIS receptor may make it possible to preselect receptor-positive tumors for MIS therapy.

VII. Studies on Human MIS

Generation of murine monoclonal and rabbit polyclonal antibodies against rhMIS has facilitated the development of a highly sensitive (0.5–1.0 ng/ml) MIS enzyme-linked immunosorbent assay (ELISA) in our laboratory (Hudson *et al.*, 1990). Two other groups established similar assays. Baker *et al.* (1990) in Australia used monoclonal antibody against bovine MIS and could detect 6.25 ng/ml, and Josso *et al.* (1990) in France used polyclonal anti-bovine MIS and could detect 2 ng/ml. We have used this assay to examine serum MIS levels in the normal population and in a variety of clinical settings. Serum MIS values for males and females at various ages throughout life are shown in Fig. 17 (Hudson *et al.*, 1990). MIS is present at high levels in normal males for several years postnatally (10–70, ng/ml), before falling to a low baseline level that persists from puberty throughout adulthood (1–5 ng/ml). In contrast, serum MIS is undetectable in newborn females. MIS then becomes detectable at similarly low serum concentrations (1–5 ng/ml) between puberty and the onset of menopause, when it again falls below measurable levels.

Serum MIS levels have proven useful in the diagnostic evaluation and clinical management of patients with a variety of intersex and gonadal anomalies (Josso *et al.*, 1991; Gustafson *et al.*, 1993). For example, we have reported the use of serum MIS as a sensitive and specific testicular marker in boys with suspected anorchia or male pseudohermaphroditism. In such patients, serum MIS is often more specific than serum testosterone is for the detection of testicular tissue (Gustafson *et al.*, 1993). This is true before and after the neonatal androgen surge, when male and female testosterone levels are indistinguishable. Serum MIS may therefore obviate the need for diagnostic hCG stimulation in many cases. MIS is also produced by ovotestes, dysgenetic gonads, and intra-abdominal testes in patients with testicular feminization; in this subgroup, postoperative MIS levels can be determined to confirm complete removal of potentially premalignant gonadal tissue.

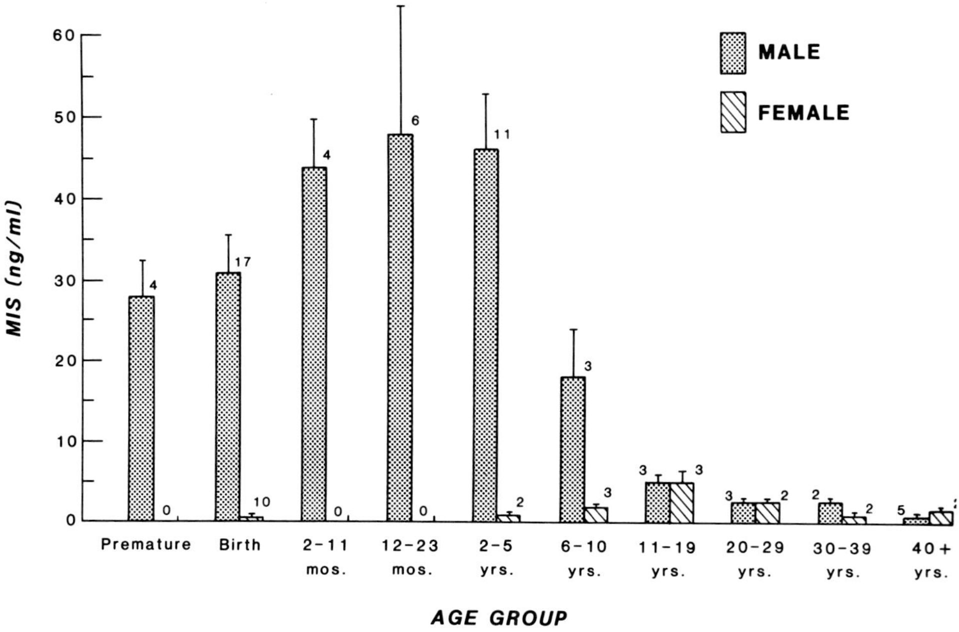

Figure 17 The ontogeny of serum MIS levels in human males and females: Serum MIS levels as determined by ELISA are represented on the ordinate (ng/ml), with age groups on the abscissa. Numerals above the columns indicate the number of samples per group; brackets denote standard error of the mean. Adapted from Hudson *et al.* (1990).

Serum MIS has also been reported to be a useful tumor marker for sex cord tumors of the ovary in juveniles, adolescents, and adults (Gustafson *et al.*, 1992, 1993). For example, MIS levels correlated extremely well with overall tumor burden and disease status in the patient with a rare "sex cord tumor with annular tubules" represented in Fig. 18. In this woman, serum MIS was approximately 1000 times normal at the time of presentation (22,900 pmol/liter = 3200 ng/ml) and fell after each of her first three resections. MIS later increased with each recurrence and fell after each tumor debulking procedure. In this and in subsequent patients with sex cord (granulosa or Sertoli cell) tumors of the ovary, serum MIS has been a useful indicator of persistent or recurrent disease and a sensitive marker to monitor therapeutic responses.

The MIS ELISA has been helpful in evaluating the etiology of the rare PMDS. Patients with this syndrome are phenotypically normal 46,XY males who typically present with uterus inguinale (retained uterus and fallopian tubes in an inguinal hernia sac) and unilateral or bilateral cryptorchidism. Since these males may have either normal or undetectable serum MIS (Gustafson *et al.*, 1993), the disorder is believed to be heterogeneous, with some patients having a defect in MIS secretion and others having end-organ unresponsiveness. Testicular biopsies in five PMDS patients were positive for MIS mRNA by Northern analysis, but only two of the five specimens were positive for bioactive MIS (Guerrier *et al.*, 1989). The remaining three patients were siblings who were later identified as having an MIS "nonsense" mutation (Knebelmann *et al.*, 1991). This produced a premature stop codon in the fifth exon of the MIS gene, resulting in an inactive protein lacking 178 of the C-terminal amino acids. An additional PMDS patient has been identified as having compound heterozygous MIS gene mutations, consisting of a 14-bp deletion in the second exon of his maternal allele and a stop mutation in the third exon of his paternal allele (Carre-Eusebe *et al.*, 1992). The other MIS-positive patients most probably represent MIS receptor defects. A PMDS dog model, in which autosomal recessive inheritance and normal production of bioactive MIS are observed, is similarly thought to involve an MIS receptor abnormality (Meyers-Wallen *et al.*, 1989). Cloning of the human and dog MIS receptor genes will allow further genetic analysis of this syndrome.

VIII. Summary

Understanding of Müllerian Inhibiting Substance has progressed substantially over the last 20 years from that of a pure "Müllerian inhibitor" to that of a carefully regulated cell-growth and differentiation factor. Production of rhMIS for biochemical analysis has revealed similarities to the TGF-β family of growth modulators, including bioactivity of its C-terminal domain. Study of the activin and TGF-β receptors, meanwhile, is providing new insight into possible mechanisms of MIS action. The sexually dimorphic expression of MIS has been characterized and correlated with postulated *in vivo* functions in the fetus and adult. Such biological roles for MIS include primary gonadal differentiation and involution of the female reproductive ducts during normal male sexual development, regulation of germ cell meiosis, and modulation of fetal lung maturation and surfactant production *in utero*. Recent purification protocols have restored the antitumor activity of rhMIS, which suggests that clinical trials of this hormone or MIS gene therapy should be rigorously pursued for certain human malignancies. Serum MIS determination has also become clinically applicable in the management of a wide variety of intersex and gonadal anomalies. Finally, the control of MIS secretion during sexual differentiation, postnatal development, puberty,

A

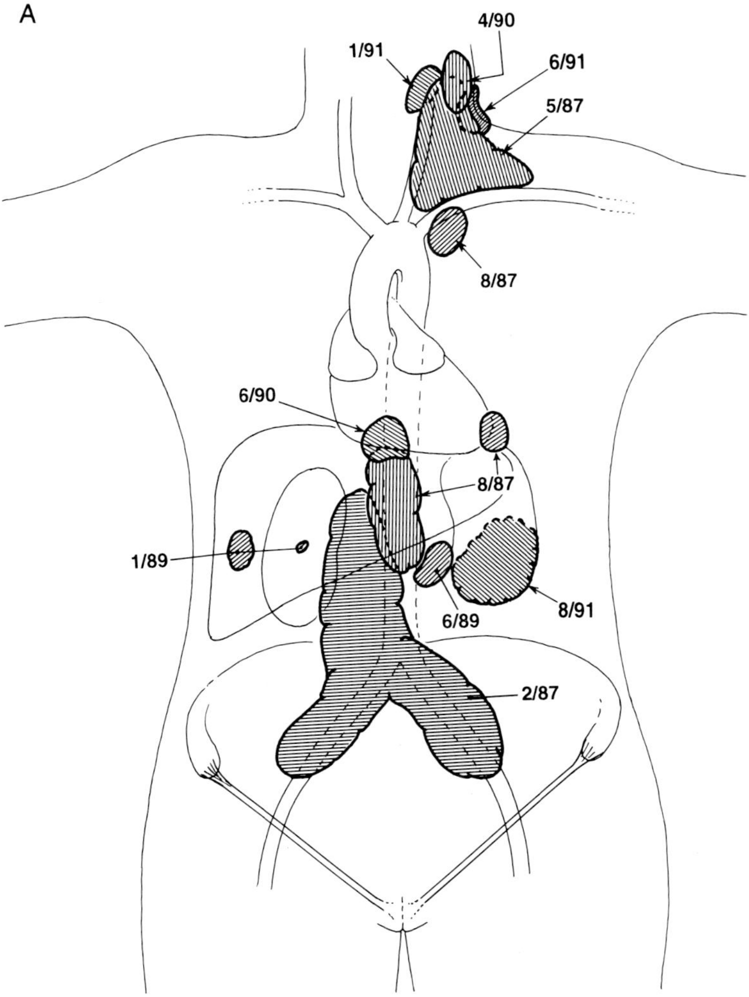

Figure 18 Serum MIS as a novel tumor marker for ovarian sex cord tumors: (A) Eleven episodes of tumor recurrence, with corresponding dates of surgical resection, are schematically shown in a woman with widely metastatic ovarian sex cord tumor with annular tubules. (B) This patient's initial serum MIS level was 22,900 pmol/liter (3200 ng/ml), which is approximately 1000 times normal (7–35 pmol/liter; 1–5 ng/ml). Serum MIS levels fell after tumor debulking in this patient and then increased with each subsequent recurrence. Serum MIS has been a sensitive marker for both persistent and recurrent disease in such patients. Adapted from Gustafson *et al.* (1992).

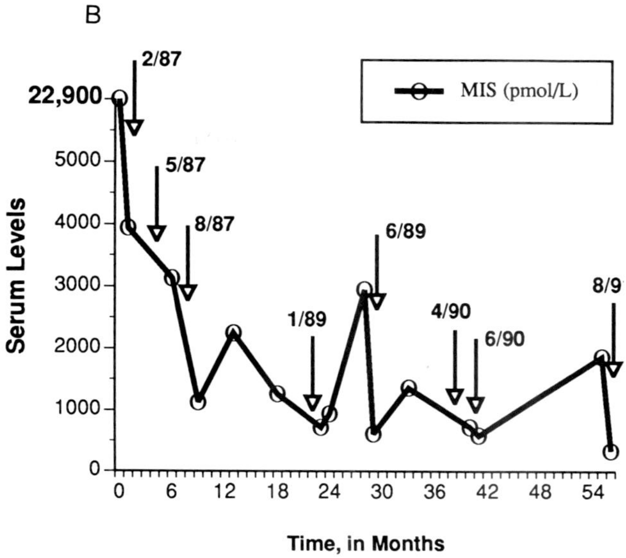

Figure 18 (*Continued*)

and adulthood remains an area of intensive investigation; *SRY*, gonadotropins, steroids, other yet unidentified factors, and site-specific proteolytic enzymes probably all play a role in this exquisite regulation. The role of MIS in the downstream pathway of *SRY*, leading to male gonadal differentiation, provides some insight into the "black box" between genetic and gonadal sex, and brings another level of understanding to the biological mystery of sex differentiation in mammals.

References

Baker, M. L., Metcalfe, S. A., and Hutson, J. M. (1990). Serum levels of Müllerian Inhibiting Substance in boys from birth to 18 years, as determined by enzyme immunoassay. *J. Clin. Endocrinol. Metab.* **70,** 11–15.

Behringer, R. R., Cate, R. L., Froelick, G. J., Palmiter, R. D., and Brinster, R. L. (1990). Abnormal sexual development in transgenic mice chronically expressing Müllerian Inhibiting Substance. *Nature* (*London*) **345,** 167–170.

Benoit, R., Ling, N., and Esch, F. (1987). A new prosomatostatin-derived peptide reveals a pattern for prohormone cleavage at monobasic sites. *Science* **238,** 1126–1129.

Bercu, B. B., Morikawa, Y., Jackson, I., and Donahoe, P. K. (1978). Increased secretion of Müllerian Inhibiting Substance after immunological blockade of endogenous luteinizing hormone releasing hormone in the rat. *Pediatr. Res.* **12,** 139–142.

Bercu, B. B., Morikawa, Y., Jackson, I. M., and Donahoe, P. K. (1979). Inhibition of Müllerian Inhibiting Substance secretion by FSH. *Pediatr. Res.* **13,** 246–249.

Blanchard, M., and Josso, N. (1974). Source of anti-Müllerian hormone synthesized by the fetal testis: Müllerian inhibiting activity of fetal bovine Sertoli cells in tissue culture. *Pediatr. Res.* **8,** 968–971.

Boveri, J. F., Parry, R. L., Ruffin, W. K., Gustafson, M. L., Lee, K. W., He, W. W., and Donahoe, P. K. (1993). Transfection of the Müllerian Inhibiting Substance gene inhibits local and metastatic tumor growth. *Int. J. Oncol.* **2,** 135–143.

Budzik, G. P., Swann, D. A., Hayashi, A., and Donahoe, P. K. (1980). Enhanced purification of Müllerian Inhibiting Substance by lectin affinity chromatography. *Cell (Cambridge, Mass.)* **21,** 909–915.

Budzik, G. P., Powell, S. M., Kamagata, S., and Donahoe, P. K. (1983). Müllerian Inhibiting Substance fractionation by dye affinity chromatography. *Cell (Cambridge, Mass.)* **34,** 307–314.

Budzik, G. P., Donahoe, P. K., and Hutson, J. M. (1985). A possible purification of Müllerian Inhibiting Substance and a model for its mechanism of action. *In* "Developmental Mechanisms: Normal and Abnormal" (J. W. Lash, ed.), pp. 207–223. Alan R. Liss, New York.

Carre-Eusebe, D., Imbeaud, S., Harbison, M., New, M. I., Josso, N., and Picard, J. Y. (1992). Variants of the anti-Müllerian hormone gene in a compound heterozygote with the persistent Müllerian duct syndrome and his family. *Hum. Genet.* **90,** 389–394.

Cate, R. L., Mattaliano, R. J., Hession, C., Tizard, R., Farber, N. M., Cheung, A., Ninfa, E. G., Frey, A. Z., Gash, D. J., Chow, E. P., Fisher, R. A., Bertonis, J. M., Torres, G., Wallner, B. P., Ramachandran, K. L., Ragin, R. C., Manganaro, T. F., MacLaughlin, D. T., and Donahoe, P. K. (1986). Isolation of the bovine and human genes for Müllerian Inhibiting Substance and expression of the human gene in animal cells. *Cell (Cambridge, Mass.)* **45,** 685–698.

Cate, R. L., Donahoe, P. K., and MacLaughlin, D. T. (1990). Müllerian Inhibiting Substance. *In* "Handbook of Experimental Pharmacology" (M. B. Sporn, and A. B. Roberts, eds.), Vol. 95(2), pp. 179–210. Springer-Verlag, Heidelberg.

Catlin, E. A., Manganaro, T. F., and Donahoe, P. K. (1988). Müllerian Inhibiting Substance depresses accumulation *in vitro* of disaturated phosphatidylcholine in fetal rat lung. *Am. J. Obstet. Gynecol.* **159,** 1299–1303.

Catlin, E. A., Powell, S. M., Manganaro, T. F., Hudson, P. L., Ragin, R. C., Epstein, J., and Donahoe, P. K. (1990). Sex specific fetal lung development and Müllerian Inhibiting Substance. *Am. Rev. Respir. Dis.* **141,** 466–470.

Catlin, E. A., Uitvlugt, N. D., Donahoe, P. K., Powell, D. M., Hayashi, M., and MacLaughlin, D. T. (1991). Müllerian Inhibiting Substance blocks epidermal growth factor receptor phosphorylation in fetal rat lung membranes. *Metab., Clin. Exp.* **40,** 1178–1184.

Catlin, E. A., Ezzell, R., Donahoe, P. K., Manganaro, T., Ebb, R. G., and MacLaughlin, D. T. (1992). Müllerian Inhibiting Substance binding and uptake. *Dev. Dyn.* **193,** 295–299.

Chin, T., Parry, R. L., and Donahoe, P. K. (1991). Human Müllerian Inhibiting Substance inhibits tumor growth *in vitro* and *in vivo*. *Cancer Res.* **51,** 2101–2106.

Cigarroa, F. G., Coughlin, J., Donahoe, P., White, M., Uitvlugt, N., and MacLaughlin, D. T. (1989). Recombinant human Müllerian Inhibiting Substance inhibits epidermal growth factor receptor tyrosine kinase. *Growth Factors* **1,** 179–191.

Cohen-Hagenauer, O., Picard, J. Y., Mattei, M. G., Serero, S., Nguyen, V. C., de Tand, M. F., Guerrier, D., Hors-Cayla, M. C., Josso, N., and Frezal, J. (1987). Mapping of the gene for anti-Müllerian hormone to the short arm of the human chromosome 19. *Cytogenet. Cell. Genet.* **44,** 2–6.

Coughlin, J. P., Donahoe, P. K., Budzik, G. P., and MacLaughlin, D. T. (1987). Müllerian

Inhibiting Substance blocks autophosphorylation of the EGF receptor by inhibiting tyrosine kinase. *Mol. Cell. Endocrinol.* **49,** 75–86.

Daopin, S., Piez, K. A., Ogawa, Y., and Davies, D. R. (1992). Crystal structure of transforming growth factor-beta-2: An unusual fold for the superfamily. *Science* **257,** 369–373.

Derynck, R., Jarrett, J. A., Chen, E. Y., Eaton, D. H., Bell, J. R., Assoian, R. K., Roberts, A. B., Sporn, M. B., and Goeddel, D. V. (1985). Human transforming growth factor-B complementary DNA sequence and expression in normal and transformed cells. *Nature (London)* **316,** 701–705.

DiClemente, N., Ghaffari, S., Pepinsky, R. B., Pieau, C., Josso, N., Cate, R. L., and Vigier, B. (1992). A quantitative and interspecific test for biological activity of anti-Müllerian hormone: The fetal ovary aromatase assay. *Development (Cambridge, UK)* **114,** 721–727.

Donahoe, P. K., Ito, Y., Marfatia, S., and Hendren, W. H. (1976). The production of Müllerian Inhibiting Substance by the fetal, neonatal, and adult rat. *Biol. Reprod.* **15,** 329–334.

Donahoe, P. K., Ito, Y., Morikawa, Y., and Hendren, W. H. (1977a). Müllerian Inhibiting Substance in human testes after birth. *J. Pediatr. Surg.* **12,** 323–330.

Donahoe, P. K., Ito, Y., and Hendren, W. H. (1977b). A graded organ culture assay for the detection of Müllerian Inhibiting Substance. *J. Surg. Res.* **23,** 141–148.

Donahoe, P. K., Swann, D. A., Hayashi, A., and Sullivan, M. D. (1979). Müllerian duct regression in the embryo is correlated with cytotoxic activity against a human ovarian cancer. *Science* **205,** 913–915.

Donahoe, P. K., Fuller, A. F., Jr., Scully, R. E., Guy, S. R., and Budzik, G. P. (1981). Müllerian Inhibiting Substance inhibits growth of a human ovarian cancer in nude mice. *Ann. Surg.* **194,** 472–480.

Donahoe, P. K., Budzik, G. P., Trelstad, R. L., Schwartz, B. R., Fallat, M. E., and Hutson, J. M. (1984). Molecular dissection of Müllerian duct regression. *In* "The Role of Extracellular Matrix in Development" (R. L. Trelstad, ed.), pp. 573–595. Alan R. Liss, New York.

Fuller, A. F., Jr., Guy, S. R., Budzik, G. P., and Donahoe, P. K. (1982). Müllerian Inhibiting Substance inhibits colony growth of a human ovarian cancer cell line. *J. Clin. Endocrinol. Metab.* **54,** 1051–1055.

Fuller, A. F., Jr., Budzik, G. P., Krane, I. M., and Donahoe, P. K. (1984). Müllerian Inhibiting Substance inhibition of a human endometrial carcinoma cell line xenografted in nude mice. *Gynecol. Oncol.* **17,** 124–132.

Fuller, A. F., Jr., Krane, I. M., Budzik, G. P., and Donahoe, P. K. (1985). Müllerian Inhibiting Substance reduction of colony growth of human gynecologic cancers in a stem cell assay. *Gynecol. Oncol.* **22,** 135–148.

Georgi, L. L., Albert, P. S., and Riddle, D. L. (1990). daf-1, a *C. elegans* gene controlling dauer larva development, encodes a novel receptor protein kinase. *Cell (Cambridge, Mass.)* **61,** 635–645.

Guerrier, D., Tran, D., VanderWinden, J. M., Hideux, S., Van Outryve, L., Legeai, L., Bouchard, M., Van Vliet, G., De Laet, M. H., Picard, J. Y., Kahn, A., and Josso, N. (1989). The persistent Müllerian duct syndrome: A molecular approach. *J. Clin. Endocrinol. Metab.* **68,** 46–52.

Guerrier, D., Boussin, L., Mader, S., Josso, N., Kahn, A., and Picard, J. Y. (1990). Expression of the gene for anti-Müllerian hormone. *J. Reprod. Fertil.* **88,** 695–706.

Gustafson, M. L., Lee, M. M., Scully, R. E., Moncure, A. C., Hirakawa, T., Goodman, A., Muntz, H. G., Donahoe, P. K., MacLaughlin, D. T., and Fuller, A. F. (1992). Müllerian

Inhibiting Substance: A novel tumor marker for ovarian sex cord tumor with annular tubules. *N. Engl. J. Med.* **326,** 466–471.

Gustafson, M. L., Lee, M. M., Asmundson, L., MacLaughlin, D. T., and Donahoe, P. K. (1993). Müllerian Inhibiting Substance in the diagnosis and management of intersex and gonadal anomalies. *J. Pediatr. Surg.* **28,** 439–444.

Haqq, C. M., Lee, M. M., Tizard, R., Wysk, M., DeMarinis, J., Donahoe, P. K., and Cate, R. L. (1992). Isolation of the rat gene for Müllerian Inhibiting Substance. *Genomics* **12,** 665–669.

Haqq, C. M., King, C.-Y., Donahoe, P. K., and Weiss, M. A. (1993). SRY recognizes conserved DNA sites in sex-specific promoters. *Proc. Natl. Acad. Sci. U.S.A.* **90,** 1097–1101.

Hayashi, A., Donahoe, P. K., Budzik, G. P., and Trelstad, R. L. (1982). Periductal and matrix glycosaminoglycans in rat Müllerian duct development and regression. *Dev. Biol.* **92,** 16–26.

Hayashi, M., Shima, H., Hayashi, K., Trelstad, R. L., and Donahoe, P. K. (1984). Immunocytochemical localization of Müllerian Inhibiting Substance in the rough endoplasmic reticulum and Golgi apparatus in Sertoli cells of the neonatal calf testis using a monoclonal antibody. *J. Histochem. Cytochem.* **32,** 649–654.

He, W. W., Gustafson, M. L., Hirobe, S., and Donahoe, P. K. (1993). The developmental expression of four novel serine/threonine kinase receptors homologous to the Activin/TGF-β II receptor family. *Dev. Dyn.,* **196,** 133–142.

Hirobe, S., He, W. W., Lee, M. M., and Donahoe, P. K. (1992). Müllerian Inhibiting Substance messenger ribonucleic acid expression in granulosa and Sertoli cells coincides with their mitotic activity. *Endocrinology* (*Baltimore*) **131,** 854–862.

Hudson, P. L., Dougas, I., Donahoe, P. K., Cate, R. L., Epstein, J., Pepinsky, R. B., and MacLaughlin, D. T. (1990). An immunoassay to detect human Müllerian Inhibiting Substance in males and females during normal development. *J. Clin. Endocrinol. Metab.* **70,** 16–22.

Hutson, J. M., and Donahoe, P. K. (1986). The hormonal control of testicular descent. *Endocr. Rev.* **7,** 270–283.

Hutson, J. M., Ikawa, H., and Donahoe, P. K. (1981). The ontogeny of Müllerian Inhibiting Substance in the gonads of the chicken. *J. Pediatr. Surg.* **16,** 822–827.

Hutson, J. M., Ikawa, H., and Donahoe, P. K. (1982). Estrogen inhibition of Müllerian Inhibiting Substance in the chick embryo. *J. Pediatr. Surg.* **17,** 953–959.

Hutson, J. M., Fallat, M. E., Kamagata, S., Donahoe, P. K., and Budzik, G. P. (1984). Phosphorylation events during Müllerian duct regression. *Science* **223,** 586–588.

Josso, N. (1972). Permeability of membranes to the Müllerian-Inhibiting Substance synthesized by the human fetal testis *in vitro:* A clue to its biochemical nature. *J. Clin. Endocrinol. Metab.* **34,** 265–270.

Josso, N., Picard, J. Y., and Tran, D. (1977). The anti-Müllerian hormone. *Recent Prog. Horm. Res.* **33,** 117–167.

Josso, N., Fekete, C., Cachin, O., Nezelof, C., and Rappaport, R. (1983). Persistence of Müllerian ducts in male pseudohermaphroditism, and its relationship to cryptorchidism. *Clin. Endocrinol.* (*Oxford*) **19,** 247–258.

Josso, N., Vigier, B., Tran, D., and Picard, J. Y. (1985). Initiation of production of anti-Müllerian hormone by the fetal gonad. *Arch. Anat. Microsc. Morphol. Exp.* **74,** 96–100.

Josso, N., Legeai, L., Forest, M. G., Chaussain, J. L., and Brauner, R. (1990). An enzyme linked immunoassay for anti-Müllerian hormone: A new tool for the evaluation of testicular function in infants and children. *J. Clin. Endocrinol. Metab.* **70,** 23–27.

Josso, N., Boussin, L., Knebelmann, B., Nihoul-Fekete, C., and Picard, J. Y. (1991). Anti-Müllerian hormone and intersex states. *Trends Endocrinol. Metab.* **2,** 227–233.

Jost, A. (1947). Recherches sur la différenciation sexuelle de l'embryon de lapin. *Arch. Anat. Microsc. Morphol. Exp.* **36,** 271–315.

Jost, A., Vigier, B., and Prepin, J. (1972). Freemartins in cattle: The first steps of sexual organogenesis. *J. Reprod. Fertil.* **29,** 349–379.

Kim, J. H., Seibel, M. M., MacLaughlin, D. T., Donahoe, P. K., Ransil, B. J., Hametz, P. A., and Richards, C. J. (1992). The inhibitory effects of Müllerian-Inhibiting Substance on epidermal growth factor induced proliferation and progesterone production of human granulosa-luteal cells. *J. Clin. Endocrinol. Metab.* **75,** 911–917.

King, T. R., Lee, B. K., Behringer, R. R., and Eicher, E. M. (1991). Mapping anti-Müllerian hormone (Amh) and related sequences in the mouse: Identification of a new region of homology between MMU10 and HSA19p. *Genomics* **11,** 273–283.

Knebelmann, B., Boussin, L., Guerrier, D., Legeai, L., Kahn, A., Josso, N., and Picard, J. Y. (1991). Anti-Müllerian hormone Bruxelles: A nonsense mutation associated with the persistent Müllerian duct syndrome. *Proc. Natl. Acad. Sci. U.S.A.* **88,** 3767–3771.

Koopman, P., Munsterberg, A., Capel, B., Vivian, N., and Lovell-Badge, R. (1990). Expression of a candidate sex-determining gene during mouse testis differentiation. *Nature (London)* **348,** 450–452.

Kuroda, T., Lee, M. M., Haqq, C. M., Powell, D. M., Manganaro, T. F., and Donahoe, P. K. (1990). Müllerian Inhibiting Substance ontogeny and its modulation by FSH in the rat testes. *Endocrinology (Baltimore)* **127,** 1825–1832.

Kuroda, T., Lee, M. M., Ragin, R. C., and Donahoe, P. K. (1991). MIS production and cleavage is modulated by gonadotropins and steroids. *Endocrinology (Baltimore)* **129,** 2985–2993.

LaQuaglia, M., Shima, H., Hudson, P., Takahashi, M., and Donahoe, P. K. (1986). Sertoli cell production of Müllerian Inhibiting Substance *in vitro*. *J. Urol.* **136,** 219–224.

Lee, M. M., Cate, R. L., Donahoe, P. K., and Waneck, G. L. (1992). Developmentally regulated polyadenylation of two discrete messenger ribonucleic acids for Müllerian Inhibiting Substance. *Endocrinology (Baltimore)* **130,** 847–853.

Lin, H. Y., Wang, X.-F., Ng-Eaton, E., Weinberg, R. A., and Lodish, H. F. (1992). Expression cloning of the TGF-β type II receptor, a functional transmembrane serine/threonine kinase. *Cell (Cambridge, Mass.)* **88,** 1–20.

Ling, N., Ying, S. Y., Ueno, N., Shimasaki, S., Esch, F., Hotta, M., and Guillemin, R. (1986). Pituitary FSH is released by a heterodimer of the β-subunits from the two forms of inhibin. *Nature (London)* **321,** 779–782.

MacLaughlin, D. T., Hutson, J. M., and Donahoe, P. K. (1983). Specific estradiol binding in embryonic Müllerian ducts: A potential modulator of regression in the male and female chick. *Endocrinology (Baltimore)* **113,** 141–145.

MacLaughlin, D. T., Epstein, J., and Donahoe, P. K. (1991). Bioassay, purification, cloning and expression of Müllerian Inhibiting Substance. *In* "Methods in Enzymology" (D. Barnes *et al.*, eds.), Vol. 198, pp. 358–369. Academic Press, San Diego.

MacLaughlin, D. T., Hudson, P. L., Graciano, A. L., Kenneally, M. K., Ragin, R. C., Manganaro, T. F., and Donahoe, P. K. (1992a). Müllerian duct regression and antiproliferative bioactivities of Müllerian Inhibiting Substance reside in its carboxy-terminal domain. *Endocrinology (Baltimore)* **131,** 291–296.

MacLaughlin, D. T., Levin, R. K., Catlin, E. A., Taylor, L. A., Preffer, F. I., and Donahoe, P. K. (1992b). Identification of Müllerian Inhibiting Substance specific binding in human cell lines. *Horm. Metab. Res.* **24,** 570–575.

Mason, A., Hayflick, J., Ling, N., Esch, F., Ueno, N., Ying, S. Y., Guillemin, R., Niale, H., and Seeburg, P. (1985). Complementary DNA sequence of ovarian follicular fluid inhibin show precursor structure and homology with transforming growth factor-β. *Nature (London)* **318,** 659–663.

Massague, J. (1992). Receptors for the TGF-β family. *Cell (Cambridge, Mass.)* **69,** 1067–1070.

Matthews, L. S., and Vale, W. W. (1991). Expression cloning of an activin receptor, a predicated transmembrane serine kinase. *Cell (Cambridge, Mass.)* **65,** 973–982.

Meyers-Wallen, V. N., Donahoe, P. K., Ueno, S., Manganaro, T. F., and Patterson, D. F. (1989). Müllerian Inhibiting Substance is present in testes of dogs with persistent Müllerian duct syndrome. *Biol. Reprod.* **41,** 881–888.

Munsterberg, A., and Lovell-Badge, R. (1991). Expression of the mouse anti-Müllerian hormone gene suggests a role in both male and female sexual differentiation. *Development (Cambridge, UK)* **13,** 613–624.

Necklaws, E. C., LaQuaglia, M. P., MacLaughlin, D. T., Hudson, P., Mudgett-Hunter, M., and Donahoe, P. K. (1986). Detection of Müllerian Inhibiting Substance in biological samples by a solid phase sandwich radioimmunoassay. *Endocrinology (Baltimore)* **118,** 791–796.

Newbold, R. R., Suzuki, Y., and McLachlan, J. A. (1984). Müllerian duct maintenance in heterotypic organ culture after *in vivo* exposure to diethylstilbestrol. *Endocrinology (Baltimore)* **115,** 1863–1868.

Padgett, R. W., St. Johnston, R. D., and Gelbart, W. M. (1987). A transcript from a *Drosophila* pattern gene predicts a protein homologous to the transforming growth factor-β family. *Nature (London)* **325,** 81–84.

Parry, R. L., Chin, T. W., Epstein, J., Powell, D. M., Hudson, P. L., and Donahoe, P. K. (1992). Recombinant human Müllerian Inhibiting Substance inhibits human ocular melanoma cell lines *in vitro* and *in vivo*. *Cancer Res.* **52,** 1182–1186.

Pepinsky, R. B., Sinclair, L. K., Chow, E. P., Mattaliano, R. J., Manganaro, T. F., Donahoe, P. K., and Cate, R. L. (1988). Proteolytic processing of Müllerian Inhibiting Substance produces a transforming growth factor-β-like fragment. *J. Biol. Chem.* **263,** 18961–18964.

Picard, J. Y., and Josso, N. (1984). Purification of testicular anti-Müllerian hormone allowing direct visualization of the pure glycoprotein and determination of yield and purification factor. *Mol. Endocrinol.* **34,** 23–29.

Picon, R. (1969). Action du testicule foetal sur le développement in vitro des canaux de Muller chez le rat. *Arch. Anat. Microsc. Morphol. Exp.* **58,** 1–19.

Price, J. M., Donahoe, P. K., and Ito, Y. (1979). Involution of the female Müllerian duct of the fetal rat in the organ-culture assay for the detection of Müllerian Inhibiting Substance. *Am. J. Anat.* **156,** 265–283.

Ragin, R. C., Donahoe, P. K., Kenneally, M. K., Ahmad, M., and MacLaughlin, D. T. (1992). Human Müllerian Inhibiting Substance: Enhanced purification imparts biochemical stability and restores antiproliferative effects. *Protein Express. Purif.* **3,** 236–345.

Sachs, A. (1990). The role of poly(A) in the translation and stability of mRNA. *Curr. Opin. Cell Biol.* **2,** 1092–1098.

Sinclair, A. H., Berta, P., Palmer, M. S., Ross Hawins, J., Griffiths, B. L., Smith, M. J., Foster, J. W., Frischauf, A. M., Lovell-Badge, R., and Goodfellow, P. N. (1990). A gene from the human sex-determining region encodes a protein with homology to a conserved DNA-binding motif. *Nature (London)* **346,** 240–244.

Taguchi, O., Cunha, G. R., Lawrence, W. D., and Robboy, S. J. (1984). Timing and irreversibility of Müllerian duct inhibition in the embryonic reproductive tract of the human male. *Dev. Biol.* **106,** 394–398.

Takahashi, M., Hayashi, M., Manganaro, T. F., and Donahoe, P. K. (1986a). The ontogeny of Müllerian Inhibiting Substance in granulosa cells of the bovine ovarian follicle. *Biol. Reprod.* **35,** 447–453.

Takahashi, M., Koide, S. S., and Donahoe, P. K. (1986b). Müllerian Inhibiting Substance as oocyte meiosis inhibitor. *Mol. Cell. Endocrinol.* **47,** 225–234.

Taketo, T., Saeed, J., Nishioka, Y., and Donahoe, P. K. (1991). Delay of testicular differentiation in the B6.Y^{DOM} ovotestis demonstrated by immunocytochemical staining for Müllerian Inhibiting Substance. *Dev. Biol.* **146,** 386–395.

Torday, J. S., Nielsen, H. C., Fenel, M. D., and Avery, M. E. (1985). Sex differences in fetal lung maturation. *Am. Rev. Respir. Dis.* **123,** 205–208.

Tran, D., Picard, J. Y., Campargue, J., and Josso, N. (1987). Immunocytochemical detection of anti-Müllerian hormone in Sertoli cells of various mammalian species including human. *J. Histochem. Cytochem.* **35,** 733–743.

Trelstad, R. L., Hayashi, A., Hayashi, K., and Donahoe, P. K. (1982). The epithelial-mesenchymal interface of the male rat Müllerian duct: Loss of basement membrane integrity and ductal regression. *Dev. Biol.* **92,** 27–40.

Tsafriri, A., Picard, J. Y., and Josso, N. (1988). Immunopurified anti-Müllerian hormone does not inhibit spontaneous resumption of meiosis *in vitro* of rat oocytes. *Biol. Reprod.* **38,** 481–486.

Tsuji, M., Shima, H., Yonemura, C. Y., Brody, J., Donahoe, P. K., and Cunha, G. R. (1992). Effect of human recombinant Müllerian Inhibiting Substance on isolated epithelial and mesenchymal cells during Müllerian duct regression in the rat. *Endocrinology (Baltimore)* **131,** 1481–1488.

Ueno, S., Manganaro, T. F., and Donahoe, P. K. (1988). Human recombinant Müllerian Inhibiting Substance inhibition of rat oocyte meiosis is reversed by epidermal growth factor *in vitro. Endocrinology (Baltimore)* **123,** 1652–1659.

Ueno, S., Takahashi, M., Manganaro, T. F., Ragin, R. C., and Donahoe, P. K. (1989a). Cellular localization of Müllerian Inhibiting Substance in the developing rat ovary. *Endocrinology (Baltimore)* **124,** 1000–1006.

Ueno, S., Kuroda, T., MacLaughlin, D. T., Ragin, R. C., Manganaro, T. F., and Donahoe, P. K. (1989b). Müllerian Inhibiting Substance in the adult rat ovary during various stages of the estrous cycle. *Endocrinology (Baltimore)* **125,** 1060–1066.

Vale, W., Rivier, J., Vaughan, J., McClintock, R., Corrigan, A., Woo, W., Karr, D., and Spiess, J. (1986). Purification and characterization of an FSH releasing protein from porcine ovarian follicular fluid. *Nature (London)* **321,** 776–779.

Vigier, B., Picard, J. Y., Campargue, J., Forest, M. G., Heyman, Y., and Josso, N. (1985). Secretion of anti-Müllerian hormone by immature bovine Sertoli cells in primary culture, studied by a competition-type radioimmunoassay: Lack of modulation by either FSH or testosterone. *Mol. Cell. Endocrinol.* **43,** 141–150.

Vigier, B., Watrin, F., Magre, S., Tran, D., and Josso, N. (1987). Purified bovine AMH induces a characteristic freemartin effect in fetal rat prospective ovaries exposed to it *in vitro. Development (Cambridge, UK)* **100,** 43–55.

Vigier, B., Forest, M. G., Eychenne, B., Bézard, J., Garrigou, O., Robel, P., and Josso, N (1989). Anti-Müllerian hormone produces endocrine sex reversal of fetal ovaries. *Proc Natl. Acad. Sci. U.S.A.* **86,** 3684–3688.

von Heijne, G. (1986). A new method for predicting signal sequence cleavage sites. *Nucleic Acids Res.* **14,** 4683–4690.

Voutilainen, R., and Miller, L. M. (1987). Human Müllerian inhibitory factor messenger ribonucleic acid is hormonally regulated in the fetal testis and in adult granulosa cells. *Mol. Endocrinol.* **1,** 604–608.

Wallen, J. W., Cate, R. L., Kiefer, D. M., Riemen, M. W., Martinez, D., Hoffman, R. M., Donahoe, P. K., Von Hoff, D. D., Pepinsky, B., and Oliff, A. (1989). Minimal antiproliferative effect of recombinant Müllerian Inhibiting Substance on gynecological tumor cell lines and tumor explants. *Cancer Res.* **49,** 2005–2011.

Weeks, D. L., and Melton, D. A. (1987). A maternal mRNA localized to the vegetal hemisphere in *Xenopus* eggs codes for a growth factor related to TGF-β. *Cell (Cambridge, Mass.)* **51,** 861–867.

Wilson, C. A., di Clemente, N., Ehrenfels, C., Pepinsky, R. B., Josso, N., Vigier, B., and Cate, R. L. (1993). Müllerian inhibiting substance requires its N-terminal domain for maintenance of biological activity, a novel finding within the transforming growth factor-β superfamily. *Mol. Endocrinol.* **7,** 247–257.

Wozney, J. M., Rosen, V., Celeste, A. J., Mitsock, L. M., Whitters, M. J., Kriz, R. W., Hewick, R. M., and Wang, E. A. (1988). Novel regulators of bone formation: Molecular clones and activities. *Science* **242,** 1528–1534.

Yamanaka, J., Baker, M., Metcalfe, S., and Hutson, J. M. (1991). Serum levels of Müllerian Inhibiting Substance in boys with cryptorchidism. *J. Pediatr. Surg.* **26,** 621–623.

Index

▮

■